Persistence of Forage Legumes

Persistence of Forage Legumes

Proceedings of a trilateral workshop held in Honolulu, Hawaii, 18-22 July 1988. The trilateral workshop was held under the auspices and support of the Australian/United States of America Agreement on Scientific and Technical Cooperation and the New Zealand/United States of America Science and Technology Agreement.

Editors

G. C. Marten
A. G. Matches
R. F Barnes
R. W. Brougham
R. J. Clements
G. W. Sheath

American Society of Agronomy, Inc.
Crop Science Society of America, Inc.
Soil Science Society of America, Inc.

Madison, Wisconsin, USA

1989

Second Printing 1989

American Society of Agronomy, Inc.
Crop Science Society of America, Inc.
Soil Science Society of America, Inc.

Library of Congress Cataloging-in-Publications Data

Persistence of forage legumes.

1. Legumes--Congresses. 2. Forage plants--Congresses. I. Marten, G. C. II. American Society of Agronomy.
SB203.P28 1989 633.3 89-146
ISBN 0-89118-098-2

Printed in the United States of America

Table of Contents

ACKNOWLEDGMENT.. ix

APPRECIATION EXTENDED................................... ix

EXECUTIVE SUMMARY....................................... xi

RECOMMENDATIONS FOR COLLABORATIVE RESEARCH PROJECTS OR EXCHANGES AMONG SCIENTISTS FROM AUSTRALIA, NEW ZEALAND, AND THE UNITED STATES OF AMERICA....................... xiii

PREFACE... xvi

LIST OF PARTICIPANTS.................................... xviii

LIST OF CONTRIBUTORS NOT IN ATTENDANCE.................. xxiii

OVERVIEW OF PROBLEMS WITH LEGUMES

Sown Pastures and Legume Persistence: An Australian Overview
D. Gramshaw, J.W. Read, W.J. Collins, and E.D. Carter...................................... 1

Overview of Legume Persistence in New Zealand
G.W. Sheath and R.J.M. Hay.......................... 23

A Survey of Legume Production and Persistence in the United States
A.G. Matches.. 37

Legume Persistence Problems in Hawaii: An Overview
P.P. Rotar.. 45

General Discussion...................................... 67

DEVELOPMENT AND GROWTH CHARACTERISTICS OF LEGUMES

The Adaptation, Regeneration, and Persistence of Annual Legumes in Temperate Pasture
K.F.M. Reed, M.J. Mathison, and E.J. Crawford..... 69

Development and Growth Characteristics of Temperate Perennial Legumes
M.B. Forde, M.J.M. Hay, and J.L. Brock............ 91

Selection for Root Type in Red Clover
R.R. Smith.. 111

Tropical Forage Legume Development, Diversity, and Methodology for Determining Persistence
A.E. Kretschmer, Jr. 117

Demography of Pasture Legumes
R.M. Jones and E.D. Carter........................ 139

Rooting Characteristics of Legumes
A.G. Matches.. 159

General Discussion.................................... 173

MAJOR EDAPHIC AND CLIMATIC STRESSES

Climatic and Edaphic Constraints to the Persistence of Legumes in Pastures
Z. Hochman and K.R. Helyar........................ 177

Environmental Selection of Legumes
D. Scott, J.H. Hoglund, J.R. Crush, and J.M. Keoghan.. 205

Major Edaphic and Climatic Stresses in the United States
D.R. Buxton... 217

Rhizobial Ecology in Tropical Pasture Systems
P. Woomer and B.B. Bohlool........................ 233

General Discussion.................................... 247

CULTURAL PRACTICES AND PLANT COMPETITION

Cultural Practices Influencing Legume Establishment and Persistence in Australia
D. Gramshaw and M.A. Gilbert...................... 249

Aspects that Limit the Survival of Legume Seedlings
W.L. Lowther, J.H. Hoglund, and M.J. Macfarlane.... 265

Legume Establishment and Harvest Management in the U.S.A.
C.C. Sheaffer....................................... 277

Growth and Competition as Factors in the Persistence of Legumes in Pastures
M.J. Fisher and P.K. Thornton..................... 293

Competition from Associated Species on White and Red Clover in Grazed Swards
R.J.M. Hay and W.F. Hunt.......................... 311

Effect of Competition on Legume Persistence
C.C. Sheaffer....................................... 327

General Discussion.................................... 335

PLANT-ANIMAL INTERFACE

The Plant-Animal Interface and Legume Persistence--
An Australian Perspective
M.L. Curll and R.M. Jones.......................... 339

Plant-Animal Factors Influencing Legume Persistence
G.W. Sheath and J. Hodgson......................... 361

Legume Persistence Under Grazing in Stressful
Environments of the United States
C.S. Hoveland...................................... 375

A Case Study of White Clover/Ryegrass Introductions
into Kukuyu Grass on a Commercial Cattle Ranch in Hawaii
B.J. Smith... 387

General Discussion................................. 395

MAJOR PESTS AND DISEASES

Diseases of Pasture Legumes in Australia
J.A.G. Irwin....................................... 399

Arthropod Pests and the Persistence of Pasture Legumes in
Australia
P.G. Allen... 419

Initiatives in Pest and Disease Control in New Zealand
Towards Improving Legume Production and Persistence
R.N. Watson, R.A. Skipp, and B.I.P. Barratt........ 441

Diseases and Forage Stand Persistence in the
United States
K.T. Leath... 465

Insects that Reduce Persistence and Productivity of
Forage Legumes in the USA
R.C. Berberet and A.K. Dowdy....................... 481

General Discussion................................. 501

GENETICS AND BREEDING FOR PERSISTENCE

Developing Persistent Pasture Legume Cultivars for
Australia
R.J. Clements...................................... 505

Breeding for Legume Persistence in New Zealand
J.R. Caradus and W.M. Williams..................... 523

Breeding and Genetics of Legume Persistence
R.R. Smith and A.E. Kretschmer, Jr. 541

General Discussion................................. 553

AREAS OF COLLABORATIVE WORK

Collaborative Research Among Scientists in Australia, New Zealand, and the United States of America
R.F Barnes.. 555

SUMMARY OF THE TRILATERAL WORKSHOP ON PERSISTENCE OF FORAGE LEGUMES
G.C. Marten.. 569

ACKNOWLEDGMENT

The following institutions provided funds and/or helped organize the trilateral workshop:

Australian Department of Industry, Technology, and Commerce, Canberra, ACT, Australia (Australia/USA Agreement on Scientific and Technical Cooperation)
CSIRO Division of Tropical Crops and Pastures, Brisbane, Queensland, Australia
Department of Scientific and Industrial Research, Wellington, New Zealand
New Zealand Ministry of Agriculture and Fisheries, Wellington, New Zealand
New Zealand/United States of America Science and Technology Council
Hellaby Trust, New Zealand
Grassland Memorial Trust, New Zealand
C. Alma Baker Trust, New Zealand
Trimble Trust, New Zealand
National Science Foundation, Washington, DC, USA
U.S. Department of Agriculture, Agricultural Research Service, Washington, DC, USA
Department of Agronomy, Horticulture, and Entomology, Texas Tech University, Lubbock, TX, USA
Department of Agronomy and Soil Science, University of Hawaii, Honolulu, HI, USA
American Society of Agronomy, Inc., Madison, WI, USA
Crop Science Society of America, Inc., Madison, WI, USA
Soil Science Society of America, Inc., Madison, WI, USA

APPRECIATION EXTENDED

The participants extended their sincere appreciation to the Department of Agronomy and Soil Science at the University of Hawaii (especially to Samir El-Swaify, Peter Rotar, and Burton Smith) for the excellence of their hosting arrangements and the spectacular tour of the diverse grazing lands on the "Big Island" of Hawaii that concluded the workshop.

EXECUTIVE SUMMARY

Forage legumes are unique among crop plants. They contribute essential soil N, by fixing it from the air, for their own growth and associated or succeeding grasses and other crops. They also supply high-quality herbage for grazing or conserved feeding or ruminant livestock that helps ensure an economical and healthful supply of dairy products, meat, and animal fiber nationally and internationally. Further, forage legumes provide a more continuous supply of feed throughout the growing year than is possible with grasses alone. In addition, forage legumes aid in prevention of soil erosion by provision of year-round ground cover, and lessen contamination of groundwater, especially by nitrates. Finally, forage legumes enhance the beauty of the environment and promote healthy populations of numerous wild birds and mammals.

Support of research in the persistence of forage legumes is not only prudent, but essential for truly sustainable agriculture. If we increase persistence of forage legumes, we will be in a position to more effectively manage our agricultural production systems and reduce the risks associated with the ever-changing economic and environmental stresses. Support for increased collaborative interdisciplinary research among and within our countries is essential to ensure improvement of legume persistence that will aid our economic and social well-being.

Whereas no commodity group or organization is likely to promote fiscal support of this biological treasure, mostly because its economic and social value is usually hidden within systems that lead to marketed animal products, we will handicap future generations if we do not engender much greater research activity in forage legume persistence immediately.

Top priority research objectives for legume persistence identified by Australian, New Zealand, and USA scientists during the Trilateral Workshop included the following:

1. Forage legume germplasm development (including evaluation of genetic stocks and breeding of improved cultivars).
2. Disease, insect, and other pest biology, resistance, and control in forage legume species.
3. Legume plant population dynamics (demography).
4. Soil nutrient stress in legumes such as acidification, Al toxicity, and P deficiency.

5. Basic adaptive mechanisms by forage legumes to stressful environments.
6. Dynamics of the plant-animal interface during grazing of forage legumes by ruminants.

Information exchange must accompany these research endeavors; financial support will also be required to foster such exchange. We must look beyond the immediate grain futures market and recognize the potentially dominant role of the forage legume. Enhancement of the persistence of forage legumes will contribute substantially to international stability of agriculture.

RECOMMENDATIONS FOR COLLABORATIVE RESEARCH PROJECTS OR EXCHANGES AMONG SCIENTISTS FROM AUSTRALIA, NEW ZEALAND, AND THE UNITED STATES OF AMERICA

The participants in the Workshop were surveyed to determine their priorities for future collaborative research in forage legume persistence. The six general areas outlined in the Executive Summary received the most support.

Participants also supported the development of a report on forage legume cultivars, similar to the "Forage Grass Varieties of the United States," which includes cultivar name, where developed, agronomic characteristics, and genetic base. Dr. Peter Rotar of the University of Hawaii at Manoa may be able to compile such a report if financial support and personnel assistance can be obtained. The Australian Register of Herbage Plant Cultivars could serve as a model.

Specific researchable problems were identified by participants. Whereas most of the researchable problems and information needs fell within the six general problem areas, a few others were identified. Areas listed by scientists as having potential for international collaboration are outlined below.

Country	Specific Researchable Problems and/or Exchanges
Problem Area 1 -	Forage Legume Germplasm Development
Australia	Trilateral collection of white clover (*Trifolium repens* L.) germplasm in the Mediterranean region in autumn, 1989 and subsequent evaluation for persistence.
New Zealand	Exchange of white clover germplasm; joint collection of white clover germplasm from Yugoslavia; development of screening options; coordinated evaluation of germplasm across environmental gradients.
USA	Coordinated collection of forage legume germplasm; evaluation of red clover (*T. pratense* L.) and alfalfa (*Medicago sativa* L.) germplasm for foliar and other diseases; evaluation of palatability variation among genotypes within legume species that have low palatability.

Problem Area 2 - <u>Legume Plant Population Dynamics (Demography)</u>

Australia and New Zealand	Short-term personnel exchanges and continuing information exchange.
USA	Coordinated determination of pest effects on stand longevity; persistence of legumes in conserved crop-pasture systems.

Problem Area 3 - <u>Soil Nutrient Stress in Legumes</u>

Australia and USA	Development of legume models and expert systems on soil acidification and lime requirements; short-term personnel exchanges.
New Zealand	Acidification and Al toxicity tolerance; establishment of birdsfoot trefoil (*Lotus corniculatus* L.) in acid, low-fertility soils; legume inoculation in difficult soils; adaptation of perennial lupine and annual legumes in acid, low-fertility soils.

Problem Area 4 - <u>Basic Adaptative Mechanisms by Forage Legumes to Stressful Environments</u>

Australia	Information exchange and short-term personnel exchanges to study root mechanisms of legumes generally and to study stress adaptive mechanisms and the genetics of adaptations specifically in tropical legumes.
New Zealand	Root dynamics of white and red clovers under grazing.
USA	Influence of root characteristics on water stress of legumes; effects of drought on growth, morphology, disease resistance, and physiology of red clover, alfalfa, and other legumes; influence of plant morphology on persistance; influence of enzyme kinetics on legume plant tolerance to environmental stress (especially heat and drought).

Problem Area 5 - <u>Disease, Insect, and Other Pest Biology, Resistance, and Control in Legumes</u>

Australia	Collaboration at all levels in legume insect and disease studies; specific research with USA on siratro (*Macroptilium atropurpureum*) rust strain differentiation.

New Zealand	Impact of Sitona on alfalfa nodulation; quantitative assessment of virus effects on white clover and its genetic improvement for virus resistance.
USA	Legume root insect-pathogen interactions, especially related to Sitona, root organisms and nodulation; influence of Verticillium wilt of alfalfa on persistence; biocontrol of forage legume diseases by use of bacterial antagonists.
Problem Area 6-	Dynamics of the Plant-Animal Interface During Grazing
Australia	Information exchange generally; joint experimentation with tropical legumes and coordinated total effort (workshop needed to determine priorities).
New Zealand	Plant-animal impact on selective grazing behavior and plant competition.
USA	Assessment of intake variation among and within grazed legume species.

Other Problems

Australia	Modelling relative to plant competition in tropical legume pasture systems; personnel exchanges on regional/national legume surveys, including use of satellite imagery for vegetation mapping.
New Zealand	Impact of grass endophytes on legume persistence.
USA	Seed coating and vegetative propagation systems for forage legumes.

Those wishing to initiate or become involved in collaborative efforts in any of the above exchanges or research problems are urged to contact participants of this workshop who appear to have similar areas of interest. Also, the Proceedings editors may serve as initial contacts. They will attempt to identify appropriate scientists in the three countries who might be interested in collaborative research or exchanges.

PREFACE

Legumes are unique among forages in that they generally have two major advantages compared to grasses: (i) only legumes can fix significant amounts of atmospheric N, thereby obviating the need for fossil-fuel-energy consuming synthetic N fertilizers; and (ii) legumes have greater nutrient intake potential by ruminants, thereby allowing more efficient animal production. More than 60% of the feed that ruminant livestock, horses, swine, and poultry consume in the USA is derived from forages. Forages supply more than 80% of the nutrients consumed in the nation's leading agricultural industry, beef cattle production. The principal forage legume in the USA is alfalfa (*Medicago sativa* L.) (13 million ha), but substantial areas have also been sown to red clover (*Trifolium pratense* L.) (6 million ha), white clover (*T. repens* L.) (5 million ha), and other miscellaneous legumes. In New Zealand, 9 million ha of white clover-based pastures provide the main grazing resource for livestock industries. There are 23 million ha of improved pastures in southern Australia, principally based on subterranean clover (*T. subterraneum* spp. *subterraneum*) and white clover, and 4.5 million ha of sown pastures in northern Australia, of which 1 million ha includes tropical legumes such as stylo (*Stylosanthes* spp.) and siratro (*Macroptilium atropurpureum*).

If problems in legume establishment and persistence could be overcome, then the hundreds of millions of hectares of grazing and forage cropland in the three countries could be better used to increase efficiency of animal production, as well as conserve soil, water, and wildlife resources. Legume forages, when rotated with other crops such as corn (*Zea mays* L.), wheat (*Triticum aestivum* L.), and sorghum (*Sorghum bicolor* (L.) Moench), also serve as economical sources of N for grain crop production; therefore, enhancement of legume persistence could improve the efficiency of grain production too.

In 1984, a trilateral workshop on Forage Legumes for Energy-Efficient Animal Production was held in New Zealand. Scientists at that workshop identified the solution to poor persistence of forage legumes as a key priority for further research. Thus, this workshop in Hawaii was organized to provide an opportunity to discuss the problem in depth and to enable direct examination of both temperate and tropical forages within a single geographic location.

This trilateral workshop, with the goal of defining and narrowing the gaps in knowledge in legume persistence, included 33 papers presented and discussed by 33 scientists from Australia, New Zealand, and the USA. The workshop consisted of invited papers and extensive discussions in eight areas: (i) overview of problems with legumes, (ii) development and growth characteristics of legumes, (iii) major edaphic and climatic stresses, (iv) cultural practices and plant competition, (v) plant-animal interface, (vi) major pests and diseases, (vii) genetics and breeding for persistence, and (viii) areas of collaborative work.

The workshop objectives were: (i) to document problems of poor forage legume persistence in each country and their economic consequences, (ii) to review what is known of constraints to forage legume persistence in each country, (iii) to exchange information on concepts, methods, approaches, and recent advances regarding forage legume persistence, (iv) to compile, interpret, and document pertinent data on persistence of forage legumes, (v) to develop a consensus on important gaps in biological information needed to allow modelling of forage legume persistence, and (vi) to enable key scientists of Australia, New Zealand, and the USA conducting research on forage legume persistence to meet and to promote exchange of information and future research collaboration.

This publication should prove helpful as further research is planned. Grateful appreciation is extended to the agencies, institutions, and professional scientific and educational societies that provided funds or assisted in the organization of the workshop and publication of the Proceedings. The editorial committee also expresses its appreciation to the authors and workshop participants from Australia, New Zealand, and the USA.

Editorial Committee

G. C. Marten
U.S. Department of Agriculture, Agricultural Research Service at the Univ. of Minnesota, St. Paul, MN, USA

A. G. Matches
Department of Agronomy, Horticulture, and Entomology, Texas Tech Univ., Lubbock, TX, USA

R. F Barnes
American Society of Agronomy, Crop Science Society of America, and Soil Science Society of America, Madison, WI, USA

R. W. Brougham
DSIR, Grasslands Division, Palmerston North, New Zealand

R. J. Clements
CSIRO, Division of Tropical Crops and Pastures, St. Lucia, Queensland, Australia

G. W. Sheath
Ministry of Agriculture and Fisheries, Whatawhata Research Centre, Hamilton, New Zealand

LIST OF PARTICIPANTS

Peter G. Allen
Entomologist
South Australian Dep. of Agriculture
Northfield Research Laboratories
GPO Box 1671
Adelaide, SA 5001, Australia

Robert F Barnes
Agronomist
ASA-CSSA-SSSA
677 S. Segoe Rd.
Madison, WI 53711, USA

Richard C. Berberet
Entomologist
Dep. of Entomology
Oklahoma State University
Stillwater, OK 74078-0464, USA

B. Ben Bohlool
Microbiologist
Dep. of Agronomy and Soil Science
University of Hawaii
1000 Holomua Ave.
Paia, Maui, HI 96779, USA

Raymond W. Brougham
Agronomist/Ecologist
P. O. Box 8094
Palmerston North, New Zealand

Dwayne R. Buxton
Plant Physiologist
USDA-ARS, Dep. of Agronomy
Iowa State University
Ames, IA 50011, USA

John R. Caradus
Pasture Plant Breeder
Grasslands Division
DSIR
Palmerston North, New Zealand

Robert J. Clements
Pasture Agronomist/Plant Breeder
CSIRO Div. of Tropical Crops & Pastures
306 Carmody Rd.
St. Lucia, Queensland 4067, Australia

Michael L. Curll
Pasture Agronomist
NSW Dep. of Agriculture
Agric. Research and Advisory Stn.
Glen Innes, NSW 2370, Australia

Samir El-Swaify
Soil Scientist
Dep. of Agronomy and Soil Science
Sherman Laboratory
University of Hawaii
1910 East-West Road
Honolulu, HI 96822, USA

Myles J. Fisher
Ecophysiologist
CIAT (Centro Int. de Agric. Tropical)
Apartado Aereo 6713
Cali, Colombia, South America

Margot B. Forde
Botanist
Grasslands Division
DSIR
Palmerston North, New Zealand

David Gramshaw
Pasture Agronomist/Ecologist
Queensland Dep. of Primary Industries
GPO Box 46
Brisbane, Queensland 4001, Australia

Michael J. M. Hay
Plant Nutritionist
Grasslands Division
DSIR
Palmerston North, New Zealand

R. John M. Hay
Pasture Agronomist
Grasslands Division
DSIR
Palmerston North, New Zealand

John Hodgson
Pasture Agronomist
Agronomy Dep.
Massey University
Palmerston North, New Zealand

Zvi Hochman
Agronomist
NSW Dep. of Agriculture
North Coast Agric. Institute
Wollongbar, NSW 2480, Australia

Carl S. Hoveland
Agronomist
Agronomy Dep.
University of Georgia
Athens, GA 30602, USA

John A. G. Irwin
Plant Pathologist
Botany Dep.
University of Queensland
St. Lucia, Queensland 4067, Australia

Richard M. Jones
Pasture Agronomist
CSIRO Div. of Tropical Crops and Pastures
306 Carmody Rd.
St. Lucia, Queensland 4067, Australia

Albert E. Kretschmer, Jr.
Agronomist
University of Florida
IFAS Agric. Res. and Education Center
P. O. Box 248
Fort Pierce, FL 34954, USA

Kenneth T. Leath
Plant Pathologist
USDA-ARS and Professor,
Plant Pathology Dep.
Pennsylvania State University
University Park, PA 16802, USA

William L. Lowther
Rhizobiologist/Agronomist
Ministry of Agriculture & Fisheries
Invermay Agriculture Centre
Mosgiel, New Zealand

Gordon C. Marten
Agronomist
USDA-ARS
411 Borlaug Hall
University of Minnesota
1991 Upper Buford Circle
St. Paul, MN 55108, USA

Arthur G. (Jerry) Matches
Thornton Professor
Dep. of Agronomy, Horticulture, and Entomology
Texas Tech University
P. O. Box 4169
Lubbock, TX 79409, USA

Kevin F. M. Reed
Pasture Agronomist
Victorian Dep. of Agriculture and Rural Affairs
Pastoral Research Institute
P. O. Box 180
Hamilton, Victoria 3300, Australia

Peter B. Rotar
Agronomist
Dep. of Agronomy and Soil Science
University of Hawaii at Manoa
1910 East-West Road
Honolulu, HI 96822, USA

David Scott
Pasture Agronomist
Grasslands Division
DSIR
Lincoln, New Zealand

Craig C. Sheaffer
Agronomist
Dep. of Agronomy and Plant Genetics
University of Minnesota
St. Paul, MN 55108, USA

Gavin W. Sheath
Pasture Agronomist
Ministry of Agriculture and Fisheries
Whatawhata Research Centre
Hamilton, New Zealand

Burton J. (Burt) Smith
Extension Specialist in Pastures and Livestock Management
University of Hawaii
P. O. Box 237
Kamuela, HI 96743, USA

Richard R. Smith
Geneticist, USDA-ARS
Agronomy Dep.
U.S. Dairy Forage Research Ctr.
University of Wisconsin
1925 Linden Drive West
Madison, WI 53706, USA

Richard N. Watson
Entomologist/Nematologist
Ministry of Agriculture and Fisheries
Ruakura Agriculture Centre
Hamilton, New Zealand

Paul Woomer
Microbiologist
Dep. of Agronomy and Soil Science
University of Hawaii
1000 Holomua Avenue
Paia, Maui, HI 96779, USA

Guests at the Conference

Mr. Andres Alvarez
Graduate Student
Dep. of Agronomy and Soil Science
University of Hawaii
Honolulu, HI, USA

Dr. Nguyen Hue
Soil Scientist
Dep. of Agronomy and Soil Science
University of Hawaii
Honolulu, HI, USA

Dr. George Love
Range Ecologist
USDA-SCS
Honolulu, HI, USA

Ms. Karen Oglesby
Graduate Student
Dep. of Agronomy and Soil Science
University of Hawaii
Honolulu, HI, USA

Dr. Russell Yost
Soil Scientist
Dep. of Agronomy and Soil Science
University of Hawaii
Honolulu, HI, USA

LIST OF CONTRIBUTORS NOT IN ATTENDANCE

B. I. P. Barratt
Ministry of Agriculture and Fisheries
Invermay Research Centre
Mosgiel, Otago
New Zealand

J. L. Brock
Grasslands Division
DSIR
Palmerston North
New Zealand

E. D. Carter
Waite Agricultural Research Inst.
Univ. of Adelaide
Glen Osmond 5064, S.A.
Australia

W. J. Collins
Dep. of Agriculture
South Perth 6151, W.A.
Australia

E. J. Crawford
Dep. of Agric.
Box 1671
Adelaide, S.A. 5001
Australia

J. R. Crush
Grasslands Division
DSIR
Palmerton North
New Zealand

A. K. Dowdy
Dep. of Entomology
Oklahoma State University
Stillwater, OK 74078-0464
USA

M. A. Gilbert
Dep. of Primary Industries
Mareeba 4880, Queensland
Australia

K. R. Helyar
Agricultural Research Centre
Wollongbar, N.S.W. 2480
Australia

J. H. Hoglund
Grasslands Division
DSIR
Lincoln
New Zealand

W. F. Hunt
Grasslands Division
DSIR
Palmerston North
New Zealand

J. M. Keoghan
Ministry of Agriculture & Fisheries
Invermay Research Centre
Mosgiel, Otago
New Zealand

M. J. Mathison
Dep. of Agriculture
Box 361
Mt. Barker, S.A. 5251
Australia

M. J. Macfarlane
Ministry of Agriculture & Fisheries
Whatawhata Research Centre
Hamilton, New Zealand

J. W. Read
Dep. of Agriculture
Sydney 2000, N.S.W.
Australia

R. A. Skipp
Plant Diseases Division
DSIR
Palmerston North
New Zealand

P. K. Thornton
Division of Animal Resource Mgt.
The Edinburgh School of Agriculture
West Mains Road
Edinburgh, EH9 3JG
United Kingdom

W. M. Williams
Grasslands Division
DSIR
Palmerston North
New Zealand

SOWN PASTURES AND LEGUME PERSISTENCE: AN AUSTRALIAN OVERVIEW

D. Gramshaw, J.W. Read, W.J. Collins and E.D. Carter

SUMMARY

Recent trends in sown pasture development, the changing pasture production environment, and the current persistence status of important temperate and tropical legumes in Australia are broadly reviewed. Since 1970, a sustained decline in farmer's terms of trade and rapid changes in the relative profitability of livestock and cropping enterprises have directly or indirectly influenced the persistence of many pasture legumes. Reduced P fertilizer use, over-grazing, expansion of crop areas, and the introduction of new cropping technologies are implicated in the decline of legumes. Land degradation problems have also emerged and include soil acidification, salinization, waterlogging, and compaction. Occurrence of new diseases and insect pests, or increased prevalence and severity of those previously known, has substantially affected the persistence of subterranean clover (Trifolium subterraneum L.), annual medics (Medicago spp.), lucerne (Medicago sativa L.) and stylosanthes (Stylosanthes spp.) in recent years; these are key species used in Australia over a wide adaptational range. Major factors limiting persistence of each important legume are briefly documented.

PERSPECTIVE

Sown Pastures in Australia

Historically, the most important feature of Australian agriculture has been the evolution and expansion of sown pasture technologies (Donald, 1965). Commencing in the 1920s, the sown pasture revolution accelerated after 1945 and peaked nationally in the early 1970s at 28m (million) ha (ABS). The area subsequently declined to a low of 24m ha in 1981 but has since steadily returned to near 28m ha in 1987. About 85% of the present development on an area basis is in southern Australia (23.5m ha) with most of the balance (4.5m ha) in Queensland (Table 1). These statistics substantially underestimate the total area of land on which sown pastures are grown because they exclude ley pastures that are under crop. They also exclude the extensive areas where introduced species have become naturalized.

Menz (1984) projected a potential area under sown pasture each year for Australia approaching 30m ha, claiming limited remaining scope for expansion in southern Australia and little immediate opportunity for new development in northern Australia. However,

Table 1. Area (million ha) of sown pastures in Australia in 1986-87 and trends[1] over the previous 3 yr.

	Legume pastures	Grass pastures	Total pastures
Western Australia	7.00	-	7.00
South Australia	3.47	0.11	3.58
Victoria	4.32 Δ	1.49 ∇	5.81 Δ
New South Wales	4.96 Δ	0.98 ∇	5.94 Δ
Queensland	1.31 Δ	3.03 Δ	4.34 Δ
Tasmania	0.90	-	0.90
Northern ,Territory	<0.10	<0.10	<0.10
AUSTRALIA	22.1	5.6	27.7

1 Δ upward; ∇ downward

Walker & Weston (1989) estimate a realistic potential for a further 17m ha of sown pasture development in Queensland. A further 6m ha of development in the Northern Territory and northern Western Australia is also feasible (E.J. Weston, 1988, personal communication). The national potential in the longer-term may therefore be near 50m ha. The contrasting remaining potentials for southern and northern Australia reflect the shorter history of development in northern Australia.

Although sown pastures in Australia occupy only 7% of the total pastoral and agricultural land, they have a large impact on livestock and crop production because they are concentrated within the most productive humid and sub-humid regions (Figs. 1 and 2). Menz (1984) estimated that 60% of sown pastures are in the wheat-sheep zone and 38% are in the high rainfall, near-coastal zone. He also estimated that 60% of the sown pasture area in Australia was grazed by sheep and most of the remainder by beef cattle.

Legume Pastures

Pasture legumes are especially important in Australia where most of the animal production and grain cropping occurs in ley-farming or, by developed-world standards, relatively low-input grazing systems. There is strong reliance on legume-dominant pastures to contribute N for enhanced animal and crop production.

To provide an overview of the roles and persistence of pasture legumes in Australia it is convenient to use, as did Helyar (1985), the bioclimatic classification of Nix (1982) (Fig. 2). For simplicity, some of the original bioclimatic zones have been combined to give a total of 12 zones. The more important legumes variously adapted within each zone are listed in Tables 2 and 3. These include recently released legumes for which there is limited experience of persistence under farming conditions. The listings exclude legumes of minor value, those of restricted application and forage crop legumes; the latter are used mainly as annual crops and persistence beyond the first year is seldom important.

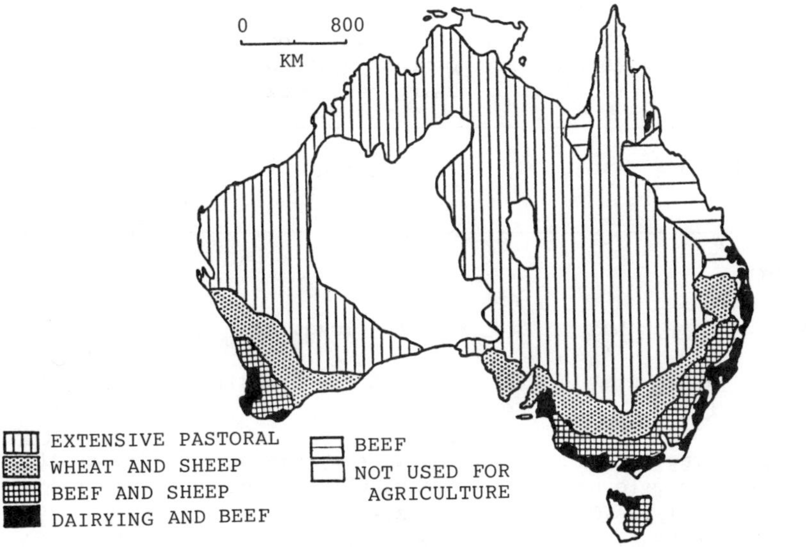

Fig. 1. Farming and pasture zones (adapted from Davidson, 1985).

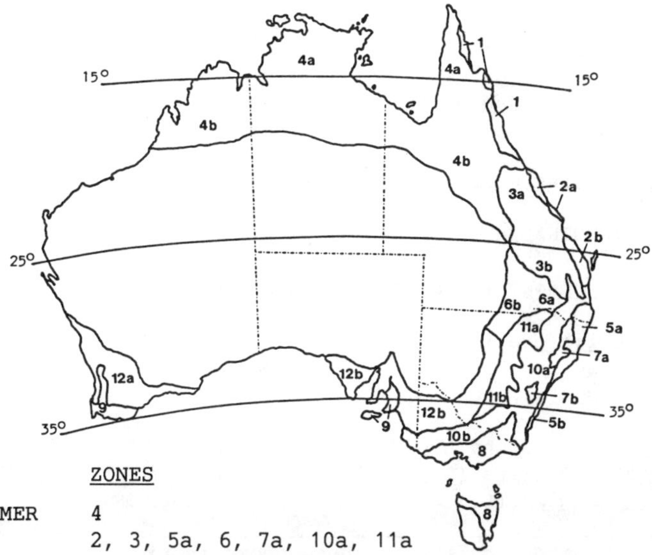

RAINFALL SEASONALITY	ZONES
STRONGLY SUMMER	4
SUMMER	2, 3, 5a, 6, 7a, 10a, 11a
NON-SEASONAL	1, 5b
WINTER	7b, 8, 10b, 11b, 12(north-eastern parts)
STRONGLY WINTER	9, 12(southern parts)

Fig. 2. Bioclimatic zones of Australia adapted from Nix (1982) and Laut et al. (1980). The numbered zones embrace the areas suitable for sown pasture legumes as specified in Tables 2 and 3. Moisture declines from humid on the coastal fringes to semi-arid at the inland limit for sown legumes. Temperatures grade from megatherm in the north to mesotherm in the south.

Table 2. Temperate legumes used or naturalized in pastures and their adaptational range within Australia. Only dominant species(D), other important species with specialized adaptation, and promising new species(N) are listed.

Scientific name	Common name	Zone(s) of adaptation[1]
PERENNIAL		
Trifolium repens L.	White clover(D)	2, 5, 7, 8, 9, 10
Medicago sativa L.	Lucerne (alfalfa)(D)	3b, 5, 6a, 7, 8, 9, 10, 11, 12b
Trifolium fragiferum L.	Strawberry clover	8, 9, 10b
Trifolium semipilosum Fres.	Kenya white clover	1, 2b, 5a
Lotus pedunculatus Cab.	Greater lotus(N)	2b, 5a
Trifolium pratense L.	Red clover	5b, 7, 8
ANNUAL, NEUTRAL/ACID SOILS		
Trifolium subterraneum L. (sown and naturalized)	Subterranean clover(D)	7, 8, 9, 10, 11b, 12
Trifolium resupinatum L.	Persian clover	
Trifolium balansae Boiss.	Balansa clover(N)	
Ornithopus spp.	Serradellas	9, 12
Medicago polymorpha L.	Burr medic(N)[2]	11, 12
Medicago murex Willd.	Murex medic(N)	10, 11, 12
ANNUAL, NEUTRAL/ALKALINE SOILS		
Medicago truncatula Gaertn. var. *truncatula*	Barrel medic(D)	3b, 6, 9, 11a, 12
Medicago littoralis Rhode	Strand medic	12a
Medicago rugosa Desr.	Gamma medic	
Medicago scutellata (L.) Mill.	Snail medic	3b, 6, 11a, 12
Medicago polymorpha L. *Medicago laciniata* (L.) Mill. *Medicago minima* (L.) Bartal.	) Naturalized burr,) cut-leaf and) woolly burr medic	3b, 6, 11, 12

1 Zone numbers relate to those shown in Fig. 2.
2 Also adapted to alkaline soils.

Seven sown legumes dominate Australia's legume-based pastures, namely subterranean clover, barrel medic, white clover, lucerne (or alfalfa), Caribbean stylo, shrubby stylo, and siratro. The other listed species are nevertheless collectively important, although individually are confined to specialized production niches.

Table 3. Tropical legumes used or naturalized in pastures and their adaptational range within Australia. Only dominant species(D), other important species with specialized adaptation, and promising new species(N) are listed.

Scientific name	Common name	Zone(s) of adaptation[1]
PERENNIAL TWINING		
Macroptilium atropurpureum (DC.) Urb	Siratro(D)	2, 3b, 5a
Centrosema pubescens Benth.	Centro	1
Pueraria phaseoloides (Roxb.) Benth.	Peuro	1
Desmodium intortum (Mill.) Urb.	Greenleaf desmodium	1, 2b, 5a
Desmodium uncinatum (Jacq.) DC.	Silverleaf desmodium	1, 5a
Neonotonia wightii (Wight & Arn.) Lackey	Glycine	1, 2b, 5a
PERENNIAL NON-TWINING		
Stylosanthes scabra Vog.	Shrubby stylo(D)	2, 3, 4
Stylosanthes guianensis (Aubl.) Swartz var guianensis	Common stylo	1, 2a, 4a
Stylosanthes guianensis (Aubl.) Swartz var intermedia	Fine stem stylo	3
Lotononis bainesii Baker	Lotononis	2b, 5a
Aeschynomene falcata (Poir) DC. Prodr.	Bargoo jointvetch	2b, 5a
Vigna parkeri Bak.	Creeping vigna(N)	1, 2b
Cassia rotundifolia Pers.	Round-leafed cassia(N)	2, 3, 4
Leucaena leucocephala (Lam.) de Wit	Leucaena	2, 4a, 5a
ANNUAL		
Stylosanthes hamata (L.) Taub.	Caribbean stylo(D)	3a, 4
Stylosanthes humilis H.B.K.	Townsville stylo	3a, 4b
Aeschynomene americana L.	American jointvetch(N)	1, 2

1 Zone numbers relate to those shown in Fig. 2.

Subterranean clovers (neutral to acid soils) and barrel and other annual medics (neutral to alkaline soils) are vital agents for N accretion, especially in the wheat-sheep zone; these temperate, self-regenerating annuals enhance animal and crop production in a range of crop and pasture sequences (Puckridge & French, 1983). White clover supports beef, sheep and dairy production in high rainfall or irrigated environments over a wide geographical range from tableland areas in north Queensland to south west Western Australia. Lucerne is grown widely on neutral to alkaline soils as a valuable hay and/or grazing plant at latitudes >23°S. The two stylo species and siratro contribute mainly to beef production in northern Australia. Notably, 11 tropical legume genera are used in northern Australia, but only four temperate genera in southern Australia (Tables 2 and 3). However, a wider range of cultivars and species within the temperate genera, particularly for subterranean clover and annual medics, provides genetic variation for adaptation within differing production environments.

Extensive legume pastures in Australia are grazed year-round at relatively constant stocking rates (Wheeler & Freer, 1986) because seasonality of pasture production and economic constraints preclude more sophisticated management (Christian, 1987). Intensive animal production is restricted to high rainfall or irrigated pastures (0.8m ha irrigated) used either for dairying or animal fattening. Hay production, excluding that from irrigated lucerne, is primarily opportunistic to help combat short-term feed deficits, rather than to systematically maintain continuity of feed (Wheeler & Freer, 1986). Less than 10% of all sown pastures are cut for hay (ABS, 1986), and these usually only once a year. Silage accounts for less than 3% of conserved fodder. Most soils on which legume-based pastures are grown require applied P to realize the growth potential of the legume. Sulfur, Ca, K and a range of trace elements are important on some soils. Further details are provided by Hochman & Helyar and by Gramshaw & Gilbert (these Proceedings).

Statistics on the area of sown pastures containing legumes (Table 1) indicate that: 21m ha or 88% of the sown pasture area in southern Australia is legume-based; 1.3m ha or 30% of the sown pasture area in Queensland is legume-based; the 3 yr (1985-87) indicative trends in the area of legume pastures are upward in Victoria, New South Wales, and Queensland, but essentially static in South Australia and Western Australia.

These data provide no insight into the status of legumes in pastures across Australia. However, a number of reports (e.g., Carter et al., 1982; Gillespie, 1983; Dear & Loveland, 1985) refer to deteriorating legume content and productivity in annual legume pastures in southern Australia. One of us (W.J. Collins) estimates the current 7.0m ha of sown pastures in Western Australia comprises 6.2m ha sown at some stage to subterranean clover (some now without legume) and 0.7m ha of annual medic, mainly burr medic. The 70% of pastures in Queensland without legume may include some pastures that were initially legume-based, but mostly reflect the substantial areas of pasture (e.g., brigalow and gidgee) in which only grasses have been sown and for which there are few legumes with proven commercial persistence.

A CHANGING PASTURE ENVIRONMENT

Economic, biological, and technological developments since 1970 within livestock and cropping enterprises have had substantial impact on the productivity and persistence of legume pastures. These are briefly reviewed as a necessary background for considering the current and possible future status of the important legume species.

Economic Trends

Prior to 1970 the expansion of sown pastures nationally was assessed by Menz (1984) to be largely independent of fluctuations in economic variables. However, this changed dramatically in 1974 when a slump in beef prices, mediocre wool prices and more profitable cropping options led to either a slowing in pasture development or a depression in the area under sown pastures over all Australian states (Fig. 3). This coincided with peak livestock numbers (Peel, 1986) generated by previously favorable prices for livestock and livestock products; the resultant increased grazing pressures most likely decreased pasture productivity and stability.

In 1976, the Australian Government revised its policy on superphosphate pricing leading to a steep rise in cost. A dramatic reduction ensued in both the area of pasture fertilized and the rate of fertilizer applied (Easter et al., 1982; Fig. 4). Continued escalation in fertilizer cost since 1976 has been associated with a sustained lower rate of fertilizer application, not only on pastures, but also on the steadily expanding area sown to wheat (Fig. 4). Remaining government fertilizer subsidies were removed from July 1988 further increasing the cost of superphosphate.

A sustained decline in farmer's terms of trade has paralleled the volatility in changing pasture and cropping practices since the early 1970s (Williams, 1986). This has enforced adoption of lower-cost management practices, often at the expense of maintaining longer-term productivity. The marked influence of farm costs and

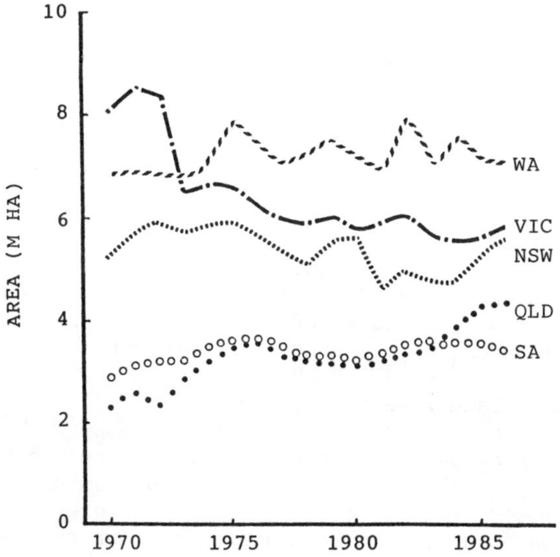

Fig. 3. Trends in the area of sown pastures in Australian states 1970-1987 (ABS Statistics). Areas in Tasmania and the Northern Territory comprise less than one million ha and are not shown.

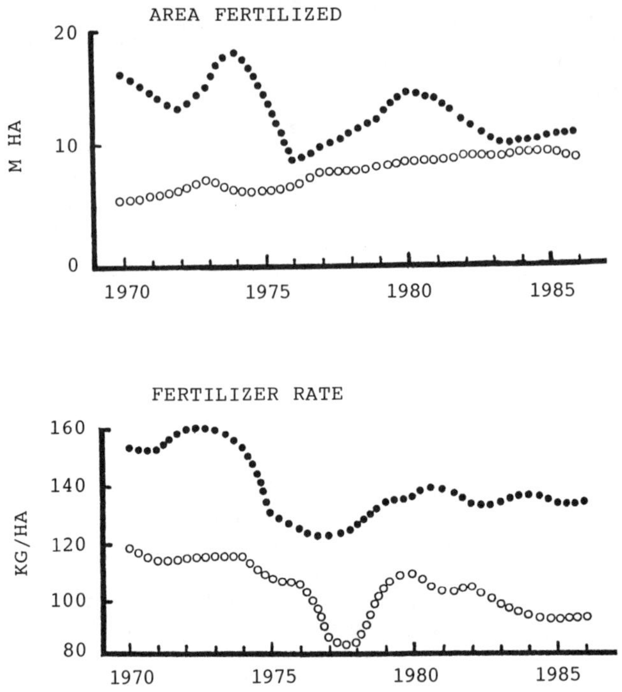

Fig. 4. Use of fertilizers other than solely nitrogenous fertilizers on native and sown pastures (• • •) and wheat (∘ ∘ ∘) in Australia 1970-1987 (ABS Statistics).

product prices on sown pasture development is highlighted by a recent analysis of the south eastern New South Wales region (Vere & Muir, 1986). In this study, the farm cost inflation of the 1970s was identified as a particularly powerful disincentive to pasture improvement activity.

Recently prices for beef and wool have become buoyant and grain prices have been depressed. The latest statistics showing an increasing area of legume pastures in some states (Table 1) no doubt reflect these changing economic circumstances.

Insect Pests and Diseases

Prior to 1970, few major diseases had been discerned on pasture legumes in Australia. In the early 1970s, anthracnose (*Colletotrichum gloeosporoides*) of *Stylosanthes* species became devastating as a pathogen in northern Australia, anthracnose/ crown rot (*Colletotrichum trifolii*) and Phytophthora root rot (*Phytophthora megasperma* f. sp. *medicaginis*) were identified as debilitating pathogens of lucerne in the subtropics, and clover scorch (*Kabatiella caulivora*) become widespread and damaging across

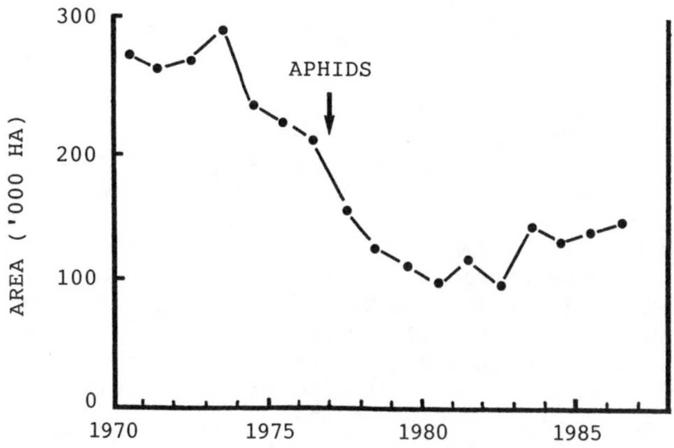

Fig. 5. Area of lucerne cut for hay in Australia (1970-1987).

southern Australia (Irwin, these Proceedings). Root rots causing mortality of Trifolium species also gained recognition and importance during the 1970s, although the causal organisms appear to differ geographically and include Phytophthora, Pythium, Fusarium, Rhizoctonia, and Aphanomyces spp. (Johnstone & Barbetti, 1987; Irwin, these Proceedings).

The advent of spotted alfalfa (Therioaphis trifolii f. maculata) and blue-green (Acyrthosiphon kondoi) aphids in rapid succession from 1977 had initial catastrophic affects throughout Australia on the ubiquitous but susceptible Hunter River lucerne, as well as affecting annual medics, subterranean clover and other temperate legumes (Allen, these Proceedings). Trends in the area of lucerne cut for hay reflect an accelerated decline post-1977 with little recovery until 1984 (Fig. 5); many damaged dryland lucerne stands were absorbed within the simultaneously expanding crop area. The leucaena psyllid (Heteropsylla cubana) appeared in north Queensland in 1986.

Land Degradation

Soil acidification, salinization, waterlogging, and compaction are becoming increasing problems in pasture and pasture-crop systems.

Soil acidification is of particular concern in southern Australia (Hochman & Helyar, these Proceedings). Overall, some 10m ha of crop-subterranean clover farming systems appear to be at risk (Coventry, 1985). Cregan (1980) estimated that 7m ha of light-textured soil could eventually have decreased soil pH to 5.5, or less, with some 4.5m ha of this area within Western Australia. Acidification leads to a range of legume-symbiont nutrition problems, including an increased requirement for P fertilizer (Batten & Osbourne, 1983), and also increases the susceptibility of legumes to root rot diseases (Johnstone & Barbetti, 1987). Tropical legumes, excepting glycines and desmodiums, are generally better adapted to acid soils than are temperate legumes.

Secondary soil salinization, caused by the removal of deep-rooted tree species and their replacement with shallower-rooted introduced species, is increasing within sown pasture and cropping lands (Poole, 1983; Reeves, 1987). Surveys in Western Australia in 1979 indicated that nearly 2% of cleared land was affected (Poole, 1983). Salinity is also a problem in irrigated pastures (Grieve et al., 1986). Lowered water usage by pastures and crops in recharge areas has led to waterlogging in many regions (Reeves, 1987).

Soil structural decline and compaction is emerging as an important form of land degradation associated with conventional cultivation methods and trampling by stock (Poole, 1983; Reeves, 1987). The extent of the problem is poorly documented nationally, although it is estimated in Western Australia that soil compaction is costing $50m per annum in lost agricultural production (G.A. Robertson, cited by Reeves [1987]).

New Production Technologies

Since 1970 there have been major changes in crop production techniques involving more cost-effective mechanization and reduced or minimum tillage (Donald, 1982; Reeves, 1987). Crop stubble retention, more frequent cropping or longer cropping sequences, pre- and post-sowing herbicides, greater N fertilizer use on crops and the incorporation of grain legume crops (e.g., lupins, chick-peas, and mung beans) in cereal rotations all feature in varying combinations within the new technologies. These innovations are strongly influencing the persistence of legumes in pasture-crop systems (Carter et al., 1982; Carter, 1987; Reeves, 1987).

Wider adoption of tropically adapted Bos indicus cattle in northern Australia is beginning to impose increased grazing pressure on native grazing lands (Burrows et al., 1986), and possibly also on associated sown legume-based pastures. Increasing use of a range of feed supplements and rumen modifiers to improve utilization of low quality feed may exacerbate the problems of overgrazing.

STATUS OF IMPORTANT LEGUMES

This section gives an overview of persistence of important legumes used in Australian pastures. Comments are largely confined to situations where the species are generally adapted and are in commercial use, and exclude the problems encountered when the species are grown in marginal environments. Persistence and adequate yield are both necessary if a legume is to viably enhance production systems. The definition of persistence is therefore taken to be "productive persistence."

Temperate Perennials

White clover, grown on an estimated 6m ha, has recently come under scrutiny due to its fluctuating presence and productivity in pastures throughout its region of adaptation (Curll, 1987). In the summer rainfall environments of south eastern Queensland and northern New South Wales, white clover behaves largely as an annual, due to poor survival of stolons and taproots over summer, and is favored by heavy summer grazing which improves seedling regeneration (Jones & Carter, these Proceedings). Pythium middletonii, the organism implicated in the rotting of stolons and roots, appears to be less debilitating on naturalized white clover (Irwin & Jones, 1977). Additional factors thought to be important for persistence in southern Australian states include erratic rainfall sequences, rugose leaf curls, a range of viruses, clover rot (Sclerotinia trifoliorum), red-legged earth mite, blue oat mite (Penthaleus major) and reduced rates of P fertilizer.

Lucerne contributes widely in Australia as a hay crop, mostly under irrigation, and as a grazed pasture legume either in pure stands or in association with a range of temperate or tropical grasses. Productive persistence depends on the survival of originally established plants, since self-regenerating seedlings seldom survive, and persistence contrasts strongly between subtropical (short-term) and temperate environments (longer-term) (Leach, 1978). Major factors limiting lucerne persistence include: inadequate grazing or cutting management and associated competition from either volunteer or sown species; drought and waterlogging; crown rots (Colletotrichum trifolii and Acrocalymma medicaginis), root rots and cankers (Phytophthora megasperma var. medicaginis and Rhizoctonia solani), and wilts (Fusarium spp. and Corynebacterium insidiosum); and lucerne aphids, red-legged earth mite (Halotydeus destructor), lucerne flea (Sminthrus viridis) and sitona weevil (Sitona discoideus). Most of these factors decrease persistence over a period of time but catastrophic stand failures may occur from disease or insect attack at the seedling stage, after drought-breaking rains and when waterlogging coincides with high soil temperatures (Cameron, 1973; Irwin, 1977; Leach, 1978; Johnstone & Barbetti, 1987; Allen, 1987).

The crisis caused by lucerne aphids in 1977 was rapidly averted by the importation into Australia of aphid resistant cultivars from North America (Allen, these Proceedings). However, the local breeding of multiple disease and aphid resistant cultivars, particularly cv. Trifecta (Queensland) and cv. Aurora (New South Wales), provided lucernes with better persistence and production than most imported cultivars (Clements, these Proceedings). Improved lucerne persistence is sought in current breeding programs; tolerance to red-legged earth mite, lucerne flea, and sitona weevil have been detected and selection is occurring for improved waterlogging and salinity tolerance (Clements, these Proceedings).

Four other temperate perennials are used in specific environments. Strawberry clover has a role on waterlogging-prone or moderately saline soils, especially on rendzina soils in south eastern South Australia and south western Victoria or within surface-irrigated pastures; it persists well except under dry conditions in summer when careful grazing is necessary. Red clover is used in high-rainfall or irrigated pastures on the south eastern mainland or in Tasmania and controlled grazing is needed to maintain it in pastures. Kenya white clover has been a persistent legume in the humid subtropics (Jones & Jones, 1982), its attraction being a greater tolerance of moisture stress and adaptation to more acid soils than white clover; it is sometimes difficult to establish and its vigor in pastures is variable (Cook et al., 1985). Grasslands Maku lotus successfully contributes a legume component in pastures on wet sites in south eastern

Queensland and northern New South Wales; it has persisted in appropriate niches for more than 8 yr, even though it has low seed production and relies on perennation by stolons and rhizomes (B.G. Cook, 1988, personal communication).

Temperate Annuals

Persistence of self-regenerating annual legumes within distinctly winter rainfall (Mediterranean-type) climates in parts of southern Australia is determined by seed production capacity, controlled strongly through maturity date/length of growing season relationships; by seed conservation through hardseededness; and by soil adaptation (Reed et al., these Proceedings). However, the importance of maturity date becomes equivocal where rainfall patterns are less winter seasonal and winter rainfall is less reliable, as in the non-Mediterranean climates of inland New South Wales (Cornish, 1985; Wolfe, 1985; Young, 1987) and in southern Queensland. Diseases, insect pests, grazing management, fertilizer practice and the type of production system all influence persistence through direct or indirect effects on seed production, seed survival, and seedling regeneration. Subterranean clover has an apparent persistence advantage over other annuals through its unique burr burial ability (Collins et al., 1976), although annual medics and serradellas generally have higher levels of hardseededness than most Trifolium spp. (Bolland, 1985, 1986). Various authors (Carter et al., 1982; Gillespie, 1983; Dear & Loveland, 1985; Carter, 1987) have emphasized recent deterioration of annual legume pastures in southern Australia. Their evidence suggests that soil seed reserves in commercial pastures, especially leys, are often inadequate for effective regeneration.

Deterioration of subterranean clover in Western Australia (Gillespie, 1983) has been linked to changes in winter rainfall characteristics; to increased cropping frequency or longer cropping phases; to changing crop and pasture management practices (details given earlier); to poor grazing management; and to diseases and insect pests, particularly clover scorch, root rots, blue-green aphids and red-legged earth mites. Many of the adaptational attributes required for persistent subterranean clover are being addressed in current breeding and selection work, including the improvement of seed production capacity and the level of hardseededness (Collins & Stern, 1987). Subterranean clover cultivars such as Daliak, Dalkeith, Trikkala, Esperance, Junee, Green Range, and Karridale provide improved combinations of disease and aphid resistances and seed attributes, but there remains considerable potential for further improvement. Cultivars resistant to red-legged earth mite or adapted to neutral to alkaline soils, although sought, are not yet available.

Factors similar to those limiting the regeneration of subterranean clover are also implicated with medic pastures (Carter et al., 1982; Carter, 1987), except phoma black stem (Phoma medicaginis) is the major disease and additional major insect pests are spotted aphid, lucerne flea, and sitona weevil. Medic regeneration is more disadvantaged by heavy grazing because of greater pod accessibility, particularly with the larger-podded snail and gamma medics. Less than 2% of ingested medic seed is voided in sheep feces (Carter, 1980).

The medic species (Table 2), and their cultivars, provide genotypes variously adapted to degrees of aridity and mainly to neutral to alkaline soils differing widely in surface texture: for example; barrel medic on a range of soil textures, snail medic on self-mulching clays and strand medic on sand plain or mallee soils (Reed et al., these Proceedings). Combined tolerance or resistance to blue-green and spotted aphids is now available in new cultivars of most of the important traditional species; barrel medic (cv. Paraggio, Sephi, and Parabinga), snail medic (cv. Sava and Kelson) and gamma medic (cv. Sapo and Paraponto). Only snail and gamma medics have tolerances to lucerne flea, red-legged earth mite and sitona weevil. Burr medics (M. polymorpha var. brevispina cv. Circle Valley and Serena) have provided more acid tolerant legumes adapted to soils where subterranean clover and barrel medic have failed to persist; currently there is an estimated 0.7m ha sown to burr medic in Western Australia with a projection of 2.5m ha within 10 yr. M. murex cv. Zodiac, which was released in 1988, will introduce a unique medic that is tolerant of acid, sandy-surfaced soils down to pH 5.0 (Reed et al., these Proceedings). It is hard-seeded and tolerant of blue-green aphid and of red-legged earth mite at the seedling stage.

Perceptions on the important attributes required for the persistence of medics, and perhaps other annual species, may be imperfect. Young (1987) noted that the survival mechanisms of the successfully naturalized cutleaf, burr and woolly burr medics in western New South Wales are enigmatic. Flowering varies among them from early to mid-season, drought tolerance of seedlings ranges from better to worse than barrel medics and the species differ markedly in hardseededness!

Persian clover is increasingly used in Victoria and South Australia. Some deficiencies of cv. Marral have been overcome in a new cultivar, Kyambro, which has higher hardseededness, some tolerance of leaf and stem rust (Uromyces sp.) and earlier maturity for drier environments (A.D. Craig, 1988, personal communication).

Balansa clover cv. Paradana, a recently released alternative to subterranean clover where rainfall is more than 450 mm, has persisted experimentally in South Australia for from 5-8 yr under heavy grazing and cropping. A high tolerance of clover scorch and root rots, prolific seeding, high hardseededness and tolerance of aphids and waterlogging are its key attributes.

Rossiter and Ozanne (1970) reported that rose clover (T. hirtum), cupped clover (T. cherleri), and serradella (mainly O. compressus) were sown in Western Australia on nearly 0.3m ha. Few pastures containing rose or cupped clover now exist. Poor growth rates and seed production under heavy grazing are implicated (Taylor & Rossiter, 1974). There has been a recent resurgence in interest in the serradellas on light-textured soils due to their high hardseededness, tolerance of acidity, relative freedom from diseases and insect pests (except red-legged earth mite) and the availability of new, earlier maturing cultivars. However, much remains to be learned about their role and management for persistence within differing production systems and environments.

Apart from the requirement for disease and pest resistances, this brief consideration of annual temperate legumes highlights the changing requirement for improved persistence attributes during the last decade. High seed production and greater hardseededness have become more important within pasture-crop systems. Acid tolerant and hardseeded medics, balansa clover, and possibly serradellas seem capable of complementing subterranean clover in many areas traditionally sown only to this species.

Tropical Twining and Scrambling Perennials

Legumes in this group in commercial use (Table 3) are all sensitive in varying degrees to heavy, continuous grazing and have moderate to high moisture and nutrient requirements (Kretschmer, these Proceedings). Careful management is required to maximize persistence of these legumes.

Centro, puero, glycine, and the desmodiums have few major disease and insect problems, except when glycine and desmodium roots are damaged by amnemus weevil (Amnemus spp.) before summer dry spells (Irwin, Allen, these Proceedings). Commercial contribution of these species is constrained mainly by specific soil or climatic adaptation and, where they are adapted, sometimes by a reluctance of graziers to adopt the necessary grazing regimes. A survey of dryland dairy farms on the Atherton Tablelands in north Queensland (Spackman, 1978) showed that only 38% of pastures had a vigorous legume component and 45% had insufficient or no legume for regeneration; inadequate grazing management and fertilizer use were implicated. A similar situation is claimed for south eastern Queensland (Lowe & Hamilton, 1986). Twining tropical legumes have performed poorly in commercial dairy situations (Murtagh et al., 1980) and alternative legumes tolerant of heavy grazing may be preferable (Lowe & Hamilton, 1986).

Siratro, the only twining legume with potential in both the medium rainfall subtropics and wetter environments, has been one of the major disappointments of tropical pasture development. Released in 1962, its commercial adoption is well below expectation and, where sown, its persistence has often been poor. Failure to maintain light stocking or to use strategic spelling, and inadequate maintenance P and possibly other nutrients, are contributing factors (Brown, 1983; Jones & Bunch, 1988a, b). Rust (Uromyces appendiculatus) has been noted to reduce siratro yield by 25% (Jones et al., 1982), but there is no evidence that epidemics have led to serious plant mortality. Some dry summer seasons in the late 1960s and during the 1970s, combined with reduced beef and increased fertilizer prices later in this period, possibly contributed to commercial failure leading to the now decreased grazier confidence in the siratro technology.

Tropical Non-Twining Perennials

This group, comprising herbaceous, shrub, and tree legumes (Table 3), generally has lower nutrient demands and better grazing tolerances than the twining species. Shrubby stylo and round-leafed cassia have potential over large areas of northern Australia (Walker & McKeague, 1986; Cook, 1988). The remaining species have narrow, though contrasting adaptation.

Stylosanthes has provided an array of genotypes able to persist on highly infertile soils (but also responsive to enhanced P status), to resist heavy grazing and to survive water stress (Gardener, 1984). Susceptibility of the genus to anthracnose

(Colletotrichum gloeosporoides) has substantially limited the usefulness of many of the genotypes. An extensive pathogenic specialization of the fungus (Irwin, these Proceedings) also potentially threatens tolerance presently found in three perennials (Seca shrubby, Cook common, and Oxley fine stem) and a sole annual (Verano caribbean stylo).

Persistence of Seca stylo is assured by strong perennation and supportive seedling regenerative mechanisms over a wide environmental range (McKeon & Mott, 1984; Gardener, 1984). Deficiencies are its slow establishment and unreliable seedling persistence in inland, low-input surface sowings (Gramshaw & McKeon, 1986, unpublished data) and a failure, in common with other stylos, to improve animal production on infertile soils (Winks, 1984; C.P. Miller, 1988, personal communication). Low rates of P fertilizer or P supplements offer potentially cost-effective methods to achieve productive persistence and extend the application of stylos (Winks et al., 1977; Hendricksen et al., 1986).

Cook stylo can contribute in high rainfall areas under moderate grazing pressure in an essentially pioneering role. Oxley fine-stem stylo is well adapted on well-drained, sandy surfaced soils in subtropical regions when stocking pressure is maintained to control grass competition. Lotononis, jointvetch, and creeping vigna all persist well under heavy grazing on their preferred soils in the subtropics. The first two species are persistent on light-textured, infertile soils, their use being mainly restricted by seed supply (Wilson et al., 1982; Cameron, 1985); however, collapse of established lotononis stands has occurred in the coastal lowlands of south east Queensland. Creeping vigna, only recently released for use on moist soils of low to moderate fertility, has survival mechanisms that should ensure persistence (Jones & Clements, 1987).

Round-leafed cassia promises to have a widespread role for beef production in medium rainfall areas of northern Australia (Cook, 1988). Perennation, prolific seeding, rapid germination, and vigorous seedling growth provide an impressive combination of survival attributes.

Leucaena is the most persistent tropical legume in commercial use and the only one well adapted to clay soils. Exacting establishment requirements for seedling survival and fears of mimosine toxicity at high animal intake levels have slowed its adoption (Wildin, 1983; Lesleighter & Shelton, 1986), although the discovery of rumen bacteria that eliminates the toxicity problem (Jones & Megarrity, 1983) has renewed grazier interest. The leucaena psyllid has created some uncertainty for the future of leucaena; ungrazed plants in coastal areas have died after prolonged psyllid infestation, although no serious mortality has been reported in grazed plantings in inland localities. Leucaena leucocephela germplasm shows little resistance to the psyllid although some tolerance occurs in other Leucaena species (Bray & Woodroffe, 1988).

Tropical Annuals

Verano caribbean stylo is the only annual (sometimes biennial) tropical legume with a history of stable contribution to beef production on infertile soils in the seasonally dry tropics of northern Australia. Its success is due to tolerance to anthracnose, high seed yields, rapid germination rates, good seedling vigor, facultative perennation and strong competitive ability with grasses (Torssell et al., 1976; Gardener, 1984). These attributes permit some latitude in grazing management and ensure adequate regeneration and persistence under more fertile and productive conditions compared with other herbaceous stylos. Cool nights limit growth and seed set, constraining Verano to latitudes lower than 23°S in Queensland.

Townsville stylo, a strict annual, has many of the attributes of Verano, except being a poorer competitor with grasses under lenient grazing (Gillard & Fisher, 1978). Unfortunately, its persistence and contribution as a sown and widely naturalized species is now seriously diminished by anthracnose (Staples et al., 1986).

Glenn American jointvetch, adapted to >1000 mm rainfall and tolerant of waterlogging, has proven highly productive and demonstrates ability to rapidly colonize pastures following strip-seeding (Bishop et al., 1985). Early experience suggests that regeneration is not a problem, especially if adequate nutrients are supplied, establishment is encouraged by heavy grazing to control grass growth prior to the opening summer rains, and moderate grazing occurs during reproduction.

Adequacy of Tropical Legumes

While there exists a reasonable range of tropical legume options for high and medium rainfall coastal and near coastal regions in northern Australia, the range for inland areas is quite limited (Gramshaw & Walker, 1988). New legumes that persist on clay soils in the semi-arid zone and in ley-farming systems in the medium rainfall subtropics are major needs (Staples et al., 1986). The restricted range of commercial species available for some other environments should cause concern in view of previous experiences with unforeseen diseases.

CONCLUSION

A naive conclusion from this overview would be that many of the key pasture legumes used in Australia have been grossly deficient in environmentally and agronomically adaptive traits. Moderating this, however, is the obvious increased complexity now prevalent within Australian pasture and pasture-crop production systems, a complexity generated by rapidly changing production technologies and management practices that are strongly conditioned by economic forces. Exacerbating factors are the land degradation problems now evident in southern Australia where pasture development has had a relatively longer history, and the occurrence or increasing importance of major diseases and pests in recent years with both temperate and tropical legume species.

The rapid adjustments in management inputs and land use experienced in the last decade are likely to continue under stringent economic circumstances and in association with fluctuating product prices. Other diseases and insect pests almost certainly will increase in importance, or will emerge unexpectedly, which will necessitate improved or alternative legumes. Important factors in the success of legumes in the future will be efficient P nutrition and the development of legumes that can cope with periods of heavy grazing and innovative cropping practices. Alternatively, economically acceptable grazing and cropping practices that favor the persistence and contribution of legumes need to be identified and demonstrated on a commercial scale. Legumes are needed that are persistent in the inland subtropics of northern Australia within ley or permanent pastures.

REFERENCES

ABS. Various dates. Australian Bureau of Statistics, Canberra.

Allen, P.G. 1987. Insect pests of pasture in perspective. p. 211-225. In J.L. Wheeler et al. (ed.) Temperate pastures. Aust. Wool Corp. Tech. Publ. AWC, CSIRO, Melbourne.

Batten, G.D., and G.J. Osbourne. 1983. Pastures in rotation with crops in relation to phosphorus in row bearing situations. p. 22-46. In Phosphorus for pastures. Report of a Specialist Workshop, Aust. Wool Corp., Sydney.

Bishop, H.G., D.H. Ludke, and M.T. Rutherford. 1985. Glenn jointvetch: A new pasture legume for Queensland coastal areas. Queensl. Agric. J. 11: 241-245.

Bolland, M.D.A. 1985. Serradella (Ornithopus sp.): maturity range and hard seed studies of some strains of five species. Aust. J. Exp. Agric. 25: 580-587.

Bolland, M.D.A. 1986. A laboratory assessment of seed softening patterns for hard seeds of Trifolium subterranean subspp. subterraneum and brachycalycinum, and of annual medics. J. Aust. Inst. Agric. Sci. 52: 91-94.

Bray, R.A., and T.D. Woodroffe. 1988. Resistance of some Leucaena species to the leucaena psyllid. Trop. Grassl. 22: 11-16.

Brown, R.F. 1983. Siratro: The main points of debate. p. 59-63. In R.F. Brown (ed.) Siratro in south east Queensland. Queensl. Dep. Prim. Indust. Conf. and Workshop Series QC 83002.

Burrows, W.H., J.G. McIvor, and M.H. Andrew. 1986. Management of Australian savannas. p. 1-10. In G.J. Murtagh and R.M. Jones (ed.) Proc. 3rd Aust. Conf. Trop. Past., Occas. Publ. 3., Trop. Grassl. Soc. Aust. Watson Ferguson, Brisbane.

Cameron, D.G. 1973. Lucerne in wet soils - the effect of stage of regrowth, cultivar, air temperature and root temperature. Aust. J. Agric. Res. 24: 851-861.

Cameron, D.G. 1985. Tropical and subtropical pasture legumes. 6. Lotononis (Lotononis bainesii): a very useful but enigmatic legume. Queensl. Agric. J. 111: 69-72.

Carter, E.D. 1980. The survival of medic seeds following ingestion of intact pods by sheep. p. 178. In Proc. Aust. Agron. Conf., Lawes, Queensl.

Carter, E.D. 1987. Establishment and natural regeneration of annual pastures. p. 35-51. In J.L. Wheeler et al. (ed.) Temperate pastures. Aust. Wool Corp. Tech. Publ. AWC, CSIRO, Melbourne.

Carter, E.D., E.C. Wolfe, and C.M. Francis. 1982. Problems of maintaining pastures in the cereal-livestock areas of southern Australia. p. 68-82. In Proc. Aust. Agron. Conf., Wagga Wagga, N.S.W.

Christian, K.R. 1987. Matching pasture production and animal requirements. p. 463-476. In J.L. Wheeler et al. (ed.) Temperate pastures. Aust. Wool Corp. Tech. Publ. AWC, CSIRO, Melbourne.

Collins, W.J., C.M. Francis, and B.J. Quinlivan. 1976. The interrelation of burr burial, seed yield and dormancy in strains of subterranean clover. Aust. J. Agric. Res. 27: 787-797.

Collins, W.J., and W.R. Stern. 1987. The national subterranean clover improvement program - progress and directions. p. 276-278. In J.L. Wheeler et al. (ed.) Temperate pastures. Aust. Wool Corp. Tech. Publ. AWC, CSIRO, Melbourne.

Cook, B.G. 1988. Persistent new legumes for intensive grazing. 2. Wynn round-leafed cassia. Queensl. Agric. J. 114: 119-121.

Cook, B.G., J.C. Mulder, and B. Powell. 1985. A survey to assess the influence of environment and management on frequency, vigour and chemical composition of *Trifolium semipilosum* cv. Safari in south eastern Queensland. Trop. Grassl. 19: 49-58.

Cornish, P.S. 1985. Adaptation of annual *Medicago* to a non-Mediterranean climate. II. Relationships between maturity grading and forage production, seed pod production and pasture regeneration. p. 17-22. In Z. Hochman (ed.) The ecology and agronomy of annual medics. N.S.W. Dep. Agric. Tech. Bull. 32.

Coventry, D. 1985. Integrating pastures with crops for sustained productivity. p. 40-44. In W.J. McDonald (ed.) Production strategies for sustained profitability. Proc. 1st Annu. Conf. Grassl. Soc., N.S.W.

Cregan, P.D. 1980. Soil acidity and associated problems - Guidelines for farmer recommendations. N.S.W. Dep. Agric. Agbull. 7.

Curll, M.L. 1987. National white clover improvement. Proc. Specialist Workshop, Armidale. Aust. Wool Corp., N.S.W. Dep. Agric., Sydney.

Dear, B.S., and B. Loveland. 1985. A survey of seed reserves of subterranean clover pastures on the Southern Tablelands of New South Wales. p. 214. In Proc. 3rd Aust. Agron. Conf., Hobart, Tasmania.

Donald, C.M. 1965. The progress of Australian agriculture and the role of pastures in environmental change. Aust. J. Sci. 27: 187-198.

Donald, C.M. 1982. Innovation in Australian agriculture. p. 55-82. In D.B. Williams (ed.) Agriculture in the Australian economy. Sydney Univ. Press, Sydney.

Easter, C.D., C.J. Robinson, and B.G. Moir. 1982. Government assistance for the consumption of nitrogenous and phosphatic fertilizers, B.A.E. Occas. Paper No. 71, AGPS, Canberra.

Gardener, C.J. 1984. The dynamics of *Stylosanthes* pastures. p. 333-357. In H.M. Stace and L.A. Edye (ed.) The biology and agronomy of *Stylosanthes*. Academic Press, Australia.

Gillard, P., and M.J. Fisher. 1978. The ecology of Townsville stylo-based pastures in northern Australia. p. 340-352. In J.R. Wilson (ed.) Plant relations in pastures. CSIRO, Melbourne.

Gillespie, D.J. 1983. Pasture deterioration - causes and cures. J. Agric. West. Aust. 1: 3-8.

Gramshaw, D., and B. Walker. 1988. Sown pasture development in Queensland. Queensl. Agric. J. 114: 93-101.

Grieve, A.M., E. Dunford, D. Marston, R.E. Martin, and P. Slavich. 1986. Effects of waterlogging and soil salinity on irrigated agriculture in the Murray Valley: a review. Aust. J. Exp. Agric. 26: 761-777.

Helyar, K.R. 1985. The distribution and use of forage legumes in Australia. p. 2-19. In R.F. Barnes et al. (ed.) Forage legumes for energy-efficient animal production. Proc. of a Trilateral Workshop, Palmerston North, New Zealand. USDA-ARS.

Hendricksen, R.E., R.W. McLean, and R.W. Dicker. 1986. The role of supplements and diet selection in beef production and management of tropical pastures in Australia. p. 56-67. In G.J. Murtagh and R.M. Jones (ed.) Proc. 3rd Aust. Conf. Trop. Past., Occas. Publ. 3., Trop. Grassl. Soc. Aust. Watson Ferguson, Brisbane.

Irwin, J.A.G. 1977. Factors contributing to poor lucerne persistence in southern Queensland. Aust. J. Exp. Agric. Anim. Husb. 17: 998-1003.

Irwin, J.A.G., and R.M. Jones. 1977. The role of fungi and nematodes as factors associated with death of white clover (*Trifolium repens*) stolons over summer in south eastern Queensland. Aust. J. Exp. Agric. Anim. Husb. 17: 789-794.

Johnstone, G.R., and M.J. Barbetti. 1987. Impact of fungal and virus diseases on pasture. p. 235-248. In J.L. Wheeler et al. (ed.) Temperate pastures. Aust. Wool Corp. Tech. Publ. AWC, CSIRO.

Jones, R.J., T.J. Hall, and R. Reid. 1982. Rust on *Macroptilium atropurpureum*. CSIRO Div. Trop. Crops and Past., Annu. Rep. 1981-82, p. 113.

Jones, R.J., and R.M. Jones. 1982. Observations on the persistence and potential for beef production of pastures based on *Trifolium semipilosum* and *Leucaena leucocephala* in subtropical coastal Queensland. Trop. Grassl. 16: 24-29.

Jones, R.J., and R.G. Megarrity. 1983. Comparative toxicity responses of goats fed on *Leucaena leucocephala* in Australia and Hawaii. Aust. J. Agric. Res. 34: 781-790.

Jones, R.M., and G.A. Bunch. 1988a. The effect of stocking rate on the population dynamics of Siratro in Siratro (*Macroptilium atropurpureum*) - setaria (*Setaria sphacelata*) pastures in south east Queensland. I. Survival of plants and stolons. Aust. J. Agric. Res. 39: 209-219.

Jones, R.M., and G.A. Bunch. 1988b. The effect of stocking rate on the population dynamics of Siratro in Siratro (*Macroptilium atropurpureum*) - setaria (*Setaria sphacelata*) pastures in south east Queensland. II. Seed set, soil seed reserves, seedling recruitment and seedling survival. Aust. J. Agric. Res. 39: 221-234.

Jones, R.M., and R.J. Clements. 1987. Persistence and productivity of Centrosema virginianum and Vigna parkeri cv. Shaw under grazing on the coastal lowlands of south east Queensland. Trop. Grassl. 21: 55-63.

Laut, O., D. Firth, and T.A. Paine. 1980. Provisional environmental regions of Australia. Vol. 1. The regions. CSIRO, Melbourne.

Leach, G.J. 1978. The ecology of lucerne pastures. p. 290-308. In John R. Wilson (ed.) Plant relations in pastures. CSIRO, Melbourne.

Lesleighter, L.C., and H.M. Shelton. 1986. Adoption of the shrub legume Leucaena leucocephala in central and south east Queensland. Trop. Grassl. 20: 97-106.

Lowe, K.F., and B.A. Hamilton. 1986. Dairy pastures in the Australian tropics and subtropics. p. 68-79. In G.J. Murtagh and R.M. Jones (ed.) Proc. 3rd Aust. Conf. Trop. Past., Occas. Publ. 3., Trop. Grassl. Soc. Aust. Watson Ferguson, Brisbane.

McKeon, G.M., and J.J. Mott. 1984. Seed biology of Stylosanthes. p. 311-332. In H.M. Stace and L.A. Edye (ed.) The biology and agronomy of Stylosanthes. Academic Press, Australia.

Menz, K. 1984. Australia's pasture resource base: History and current issues. Paper presented to 28th Annu. Conf. Aust. Agric. Econ. Soc., Sydney, N.S.W.

Murtagh, G.J., A.G. Kaiser, D.O. Huett, and R.M. Hughes. 1980. Summer-grazing components of a pasture system in a subtropical environment. 1. Pasture growth, carrying capacity and milk production. J. Agric. Sci. 94: 645-663.

Nix, H.A. 1982. Environmental determinants of biogeography and evolution in Terra Australis. p. 47-66. In W.R. Barker and P.J.M. Greenslade (ed.) Evolution of the flora and fauna of arid Australia. Peacock Press, Adelaide.

Peel, L.J. 1986. History of the Australian pastoral industries. p. 41-57. In G. Alexander and O.B. Williams (ed.) The pastoral industries of Australia. Sydney Univ. Press, Sydney.

Poole, M.L. 1983. Long-term sustainable production systems in agriculture. Paper presented to 53rd ANZAAS Congress, Perth, W.A.

Puckridge, D.W., and R.J. French. 1983. The annual legume pasture in cereal-ley farming systems of southern Australia - A review. Agric. Econ. Environ. 9: 229-67.

Reeves, T.G. 1987. Pastures in cropping systems. p. 501-515. In J.L. Wheeler (ed.) Temperate pastures. Aust. Wool Corp. Tech. Publ. AWC, CSIRO.

Rossiter, R.C., and P.G. Ozanne. 1970. South-western temperate forests, woodlands and heaths. p. 199-218. In R. Milton Moore (ed.) Australian grasslands. ANU Press, Canberra.

Spackman, G.B. 1978. A survey of dairy pastures on the Atherton Tablelands. Queensl. Dep. Prim. Indust. Agric. Br. Rep. P-4-78.

Staples, I.B., R. Reid, and G.P.M. Wilson. 1986. Plant introduction for specific needs in northern Australia. p. 29-37. In G.J. Murtagh and R.M. Jones (ed.) Proc. 3rd Aust. Conf. Trop. Past., Occas. Publ. 3., Trop. Grassl. Soc. Aust. Watson Ferguson, Brisbane.

Taylor, G.B., and R.C. Rossiter. 1974. Persistence of several annual legumes in mixtures under continuous grazing in the south west of Western Australia. Aust. J. Exp. Agric. Anim. Husb. 14: 632-639.

Torssell, B.W.R., J.R. Ive, and R.B. Cunningham. 1976. Competition and population dynamics in legume-grass swards with Stylosanthes hamata (L.) Taub. (sens. lat.) and Stylosanthes humilis (H.B.K.) Aust. J. Agric. Res. 27: 71-83.

Vere, D.T., and A.M. Muir. 1986. Pasture improvement adoption in south-eastern New South Wales. Rev. Mkt. Agric. Econ. 54: 19-31.

Walker, B., and P.J. McKeague. 1986. Development and evaluation of Stylosanthes pastures for seasonally dry areas of Queensland, Australia. p. 1286-1288. In Proc. XVth Int. Grassl. Congr., Kyoto.

Walker, B., and E.J. Weston. 1989. Pasture development in Queensland. Trop. Grassl. 23: in press.

Wheeler, J.L., and M. Freer. 1986. Pasture and forage: The feed base for pastoral industries. p. 165-182. In G. Alexander and O.B. Williams (ed.) The pastoral industries of Australia. Sydney Univ. Press, Sydney.

Wildin, J.H. 1983. Adoption of Leucaena for cattle grazing in Australia. p. 801-803. In Proc. XIV Int. Grassl. Congr., Kentucky. Westview Press, Colorado.

Williams, D.B. 1986. Agriculture in the Australian economy. p. 1-19. In L.W. Martinelli (ed.) Science for agriculture : The way ahead. AIAS, Melbourne.

Wilson, G.P.M., R.M. Jones, and B.G. Cook. 1982. Persistence of jointvetch (Aeschynomene falcata) in experimental sowings in the Australian subtropics. Trop. Grassl. 16: 155-156.

Winks, L. 1984. Cattle growth in the dry tropics of Australia. Aust. Meat Res. Comm. Rev. No. 45: 1-43.

Winks, L., F.C. Lamberth, and P.K. O'Rourke. 1977. The effect of a phosphorus supplement on the performance of steers grazing Townsville stylo-based pasture in north Queensland. Aust. J. Exp. Agric. Anim. Husb. 19: 357-366.

Wolfe, E.C. 1985. Subterranean clover and annual medics - boundaries and common ground. p. 23-28. In Z. Hochman (ed.) The ecology and agronomy of annual medics. N.S.W. Dep. Agric. Tech. Bull. 32.

Young, R.R. 1987. Legumes in western New South Wales. p. 29-39. In B.H. Downing (ed.) Proc. pasture evaluation workshop, Cobar. N.S.W. Dep. Agric. Misc. Bull. 30.

DISCUSSION

Matches: How is alfalfa being used in Australia?

Gramshaw: Alfalfa is used for hay production under irrigation and for grazing and opportunistic hay production under dryland conditions. Before the aphids in 1977, the grazed area was much greater than the irrigated hay area but, since then, large areas of the dryland alfalfa have been cropped. Currently, there is obvious renewed interest in dryland sowings.

Allen: Most alfalfa stands on acid, low fertility sands in southern Australia not suitable for cropping were lost due to aphids. The area is being resown to aphid-resistant cultivars.

Rotar: What are the relative persistencies of lucerne in northern and southern Australia?

Gramshaw: A half-life of 1 to 2 yr in northern Australia and about 2 to 6 yr in southern Australia is common.

Clements: In southern Australia, some lucerne pastures in ideal environments have persisted up to 40 yr.

Hoveland: In view of the erratic nature of white clover's contribution in Australia, is the acreage of this legume decreasing?

Curll: The area of white clover in Australia is currently about 6 million ha and is remaining at about that level. White clover persistence is affected by grazing management, with excessively heavy and light stocking both being disadvantageous.

Kretschmer: Does P deficiency actually cause tropical legume death in pastures or does it just reduce production?

Gramshaw: Maintenance applications of P are generally essential to keep legume plants in pastures in Australia, except for species like stylo that have low P requirements. With declining P availability, lowered legume productivity usually precedes the loss of plants.

Hochman: In a long-term experiment on the north coast of New South Wales, legume content in a continuously grazed Kenya clover/kikuyu pasture is correlated with annual P application rates.

Barnes: What conflicts, if any, are there in Australia associated with environmental groups, particularly in relation to land degradation?

Gramshaw: Environmental groups are becoming more activist on visually obvious issues, such as unique flora and fauna and severe soil erosion. Issues like soil acidification that cause slow and less perceptible environmental degradation have not, to my knowledge, evoked much interest.

OVERVIEW OF LEGUME PERSISTENCE IN NEW ZEALAND

G.W. Sheath and R.J.M. Hay

SUMMARY

An overview of legume persistence in New Zealand is presented. Five agro-climatic zones are identified; defining the context of persistence is stressed; and the factors leading to poor persistence of commonly used legumes are discussed. Major constraints to persistence are: white clover (*Trifolium repens L.*) - moisture stress; red clover *(T. pratense L.)* - intensive grazing; subterranean clover *(T. subterraneum L.)* - variable rainfall; lotus *(Lotus pedunculatus Cav.)* - competition; lucerne *(Medicago sativa L.)* - pests and diseases. In seeking legumes better adapted to stress, the need to consider lesser known/used legumes that are residents of natural grasslands is emphasized.

INTRODUCTION

In providing an overview of legume persistence in New Zealand, we intend to:

1. Describe the legumes that are present and/or sought within the various agro-climatic zones.
2. Define if and when persistence is perceived as a problem.
3. Briefly outline the strengths and weaknesses that determine persistence of the main legumes used in livestock farming.
4. Highlight the fact that many of the traditionally used legumes are unsuitable for extreme conditions and that there is merit in searching for better legumes within a wider range of genera.

AGRO-CLIMATIC ZONES

As illustrated in Fig.1, there are numerous agro-climatic zones that can be defined. This diversity is a function of variable climate and land resource. Most important are variations in summer rainfall, winter temperature, soil type, land contour and altitude. The pastoral systems associated with many of these zones were described by Sheath and Harris (1985).

Subtropical

This zone is distinguished by the occurrence of C_4 grasses such as kikuyu (*Pennisetum clandestinum Hochst. ex Chiov.*) and paspalum (*Paspalum dilatatum Poir.*) in permanent grazed pastures. White clover (*Trifolium repens L.*) is most commonly sought as the base legume, but *Lotus pedunculatus* has been sown on poorly drained podzol soils (Levy, 1970); red clover (*T. pratense L.*) has been encouraged as a forage source in mixed pastures for finishing animals (Ussher, 1986); and annual legumes have been tried within forage farming systems (Taylor *et al.*, 1979).

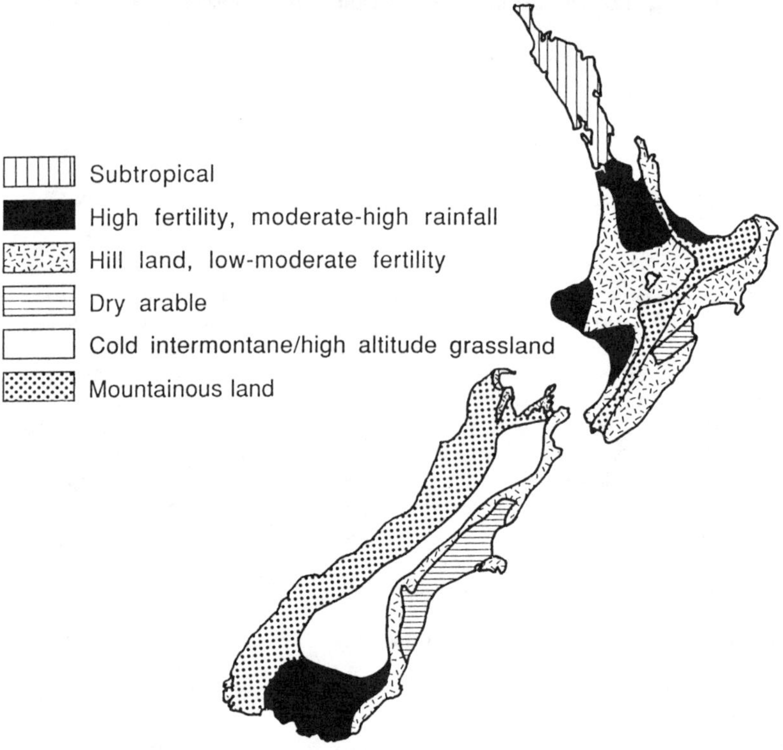

Figure 1. Agro-climatic zones of New Zealand.

High Soil Fertility/Moderate-High Rainfall

In the North Island, factory supply dairy farming dominates flat contoured land, while sheep (*Ovis aries*) and beef (*Bos taurus*) finishing systems occupy rolling contoured land of <15-20° slope. Within both systems, the use of white clover dominates (Lancashire, 1985). New Zealand wild white clover would have formed much of the historical base of these populations, but in the last 40 yr, "Grasslands Huia" would have been sown. More recently the use of "Grasslands Pitau" has been advocated (Goold & Douglas, 1976).

In the South Island, sheep and beef finishing systems occupy this zone. Again white clover dominates, although an expansion of mixed grass-red clover pastures has been encouraged for finishing livestock (Hay & Ryan, 1983).

Dry Arable

Mixed livestock and cropping enterprises include a pasture component within rotating land use patterns (Levy, 1970). These are short-term pastures (<5-10 years) and have white clover as their sown base. Subterranean clover (*T. subterraneum*) is a carryover from earlier land use and is often a common volunteer legume. Within this sub-hygrous zone (500-800 mm annual rainfall) lucerne (*Medicago sativa L.*) is used as a monoculture forage for *in-situ* grazing and/or conservation (Douglas, 1986).

Low-Moderate Soil Fertility/Hill Land

Land topography of this zone is highly variable in both slope and aspect and has a marked impact on soil fertility, soil moisture and micro-climate. While summer drought is commonly experienced in low rainfall, eastern regions (750-1000 mm annually), localized drought can also occur within wetter, western regions (1000-2000 mm annually) as a result of slope and aspect factors (Bircham & Gillingham, 1986). Within these contrasting hill environments, similar legume species are found, but their proportion varies (Rumball & Esler, 1968). White clover is generally sought as the dominant legume, but where conditions are unsuitable, opportunistic annual legumes become abundant, viz: *Lotus angustissimus, L. sauveolens T. subterraneum, T. glomeratum, T. dubium* and *T. striatum.*

Cold Inter-Montane and High Altitude Land

These zones are characterized by a 4-5 month winter period where little to no forage growth occurs. The dry inter-montanne basins (400-700 mm annual rainfall) of the South Island contain volunteer annuals (*T. arvense, T. striatum*) that are not consciously sought or cultivated for pastoral use. On suitable soils, lucerne is used for grazing and forage conservation and more recently, *L. corniculatus* is proving to be a successful cultivated legume.

Mountain lands are farmed to an altitude of 1000 m asl; above this altitude low temperature reduces legume survival (Musgrave, 1977; Floate *et al.*, 1985). Improvement of natural grasslands has involved oversowing of white clover, *T. hybridum* and red clover where fertilizer has been applied (Scott *et al.*, 1985); and *L. pedunculatus* where acid, low P status soils are involved (Scott & Mills, 1981).

PERSISTENCE - A PROBLEM OR NOT?

In New Zealand, pasture legumes are almost completely utilized as an N_2-fix source within grass dominated pastures (Lancashire, 1985). During the early stages of land and pasture development, legumes can dominate and fix large quantities of N (e.g., 680 kg N $ha^{-1}yr^{-1}$; Sears *et al.*, 1965). However, where composition is in long- term equilibrium with soil nutrient status, legume production may range from 150-5000 kg DM $ha^{-1}yr^{-1}$ and associated fixed N from 34-342 kg $ha^{-1}yr^{-1}$ (Hoglund *et al.*, 1979). This variation reflects the impact of climate, soil, competition and grazing management on both legume abundance and growth.

The content of legume within a forage source is dependent on the number of growth units (e.g., plants, stolon segments) and their size. Unit size should not be considered a direct indicator of persistence, even though it may influence longevity by determining relative competitiveness. On this basis, we will confine our discussion of persistence to those factors that determine the presence/absence of growth units.

For perennial legumes that exist as discrete plants (e.g., lucerne, red clover), poor persistence can be clearly defined as a reduction in plant density. This is most obvious where legumes are grown as a monoculture forage. Defining persistence for non-discrete plants that vegetatively propagate by stolons or rhizomes (e.g., white clover, *L. pedunculatus*) is more difficult. Frequency of stolon and rhizome units is the net balance of formation and loss (Sheath, 1980a; Chapman, 1983; Turkington & Burdon, 1983). Perceived persistence is therefore determined by positive and negative vectors which means it has a bi-directional mode. For these situations, it is not the death of a stolon unit that is viewed as persistence, but the relative rate between death and replacement.

White clover exemplifies these processes in that there is regular (1-2 yr) turnover of stolon units and associated root (Chapman, 1983; Turkington & Burdon, 1983; Hay *et al.*, 1988). Although white clover is botanically classified as a perennial, there may well be merit in accepting the suggestion of Hollowell (1966), that agronomically, white clover should be considered as a vegetative annual. To a certain extent, *L. pedunculatus* operates in a similar manner (Sheath, 1980a). Certainly, maintenance of white clover populations within undisturbed pasture (moist environment; continuous grazing) is achieved by vegetative production of offspring (Turkington *et al.*, 1979; Chapman, 1983), but where local or general disturbances occur seed and seedlings can contribute to population persistence (Table 1). High levels of buried legume seed exist in both intensively stocked lowland and unploughable hill pastures (Suckling & Charlton, 1978). In fact where regular disturbance is experienced (e.g., drought) adapted genotypes develop a strong annual reseeding habit (Blaser & Killinger, 1950; Jones, 1980).

There can be no argument that discrete plants of annual species are not persistent; by their very nature they die within the year. With annuals, persistence is more appropriately considered in terms of population regeneration and therefore reseeding, germination and seedling survival are important factors (Sheath & Macfarlane, 1988b).

The term or concept of persistence is loosely used. Obviously it needs to be closely defined in the context of the legume species being considered.

Table 1. Density of white clover stolon growing points ($/m^2$) before and after summer droughts; and white clover seedling density in winter ($/m^2$) (taken from Macfarlane et al., 1988).

	Stolon density				Seedling density	
	Dec 1981	March 1983	Dec 1982	May 1983	June 1982	June 1983
WEF[1]	3243	1310	1560	318	61	242
HC[2]	2625	1300	1780	505	23	173
Huia[3]	1535	1113	1380	428	16	125
Lousiana S1[4]	798	428	615	480	12	95
Haifa[5]	430	329	415	313	21	62

[1] profuse seeding, small leaved.
[2] small leaved.
[3] medium leaved.
[4] profuse seeding, medium leaved.
[5] large leaved.

SPECIFIC LEGUMES

White Clover

In the establishment and maintenance of white clover based pastures, correction of nutrient deficiencies (e.g., P, S, Mo) is necessary (During, 1984). It has commonly been argued that such corrections are necessary to establish a legume base that drives a nitrogen cycle and allows the development of a stable, dominant grass component (Levy, 1970). This has primarily been based on the view that white clover is at a competitive disadvantage with grasses because of a poorer exploratory root structure and distribution (Caradus, 1980). To date no genotypes of white clover have been identified as being better adapted to low nutrient status (Caradus, 1986), although there is genetic variation in Al tolerance (Caradus *et al.*, 1987).

It can be expected that cessation of nutrient input will ultimately lead to a decline in white clover abundance. However, relative rates of decline are not clear. Recent evidence suggests that ryegrass content deteriorates more rapidly than white clover when P inputs are reduced (Table 2). It is possible that reduced vigour of ryegrass allows greater opportunity for white clover spread and that this compensates for any competitive disadvantage that white clover may face in nutrient uptake. This would support Turkington and Harper's (1979) thesis that white clover abundance is strongly influenced by soil N and its effect on the aggressiveness of companion grasses.

Table 2. Effect of withholding superphosphate fertilizer on the abundance (% point frequency) of pasture species in hill country (Sheath & Gillingham, 1988, unpublished data)

	Lolium perenne		*Agrostis repens*		*Trifolium repens*		*T. subterraneum*	
	M[1]	W[2]	M	W	M	W	M	W
1984	54	50	34	30	41	34**	13	12
1985	54	53	28	28	36	33	10	9
1986	56	48*	28	31	36	34	11	11
1987	57	47**	31	33	39	35	11	13

*,** Significant at the 0.05 and 0.01 probability levels, respectively.

[1]History of superphosphate maintenance dressings (100,200,300,500,1000 kg $ha^{-1}yr^{-1}$) since 1979.

[2]Previous dressings ceased in 1983.

Whereas grass species abundance is more sensitive to soil nitrogen status, white clover is equally sensitive to soil moisture stress (Ledgard *et al.*, 1987). Current New Zealand cultivars of white clover fail to persist in semi arid environments and low moisture-retentive soils (Greenwood & Sheath, 1982; Williams *et al.*, 1977; Chapman *et al.*, 1985). There is debate as to the form of more suitable genotypes for drought conditions; types with large nodal roots being advocated by Woodfield and Caradus (1987) and dense, prolific seeding types by Macfarlane and Sheath (1984). If either large roots (e.g., Haifa) or free-seeding (e.g., Louisiana SI) is associated with large leaved types, then they are unsuitable where intensive grazing occurs within New Zealand's farming systems (Chapman *et al.*, 1986).

Clear evidence exists that under intensive sheep grazing, small-medium leaf types are most persistent (Williams & Caradus, 1979; Macfarlane & Sheath, 1984). It has been well recognized that strongly stoloniferous types are better able to persist and colonise bare ground (Turkington *et al.*, 1979; Turkington & Burdon, 1983). Given the morphological plasticity of white clover genotypes, there is no clear New Zealand evidence that supports the contention of Frame and Newbould (1986) that white clover should be rotationally grazed to ensure persistence. In fact, where the inherent dominance of spring grass growth is restricted, white clover abundance is enhanced by set stocking and/or frequent grazings (Brougham *et al.*, 1978; Hay & Baxter, 1984). It has been recently argued that the effect of grazing management on white clover abundance is mediated through the size and vigor of stolon fragments following the annual break-up of plants in spring (Hay *et al.*, 1988).

Pests and diseases of white clover will be covered in more detail in this proceedings (see Watson *et al.*), but in brief, the main problems are associated with root pruning by *Costelytra zealandica*; root infections by nematodes *Heterodera trifolii*, *Meloidogyne hapla*, and *Ditylenchus dipsaci*. Not only is white clover damaged by *C. zealandica*, but its presence encourages greater damage to the total pasture (Kain *et al.*, 1979). *H. trifolii* and *M. hapla* infection may impact more on vigour than persistence in that they depress N_2 fixation and the production of both legume and grass

(Steele *et al.*, 1985). More clearly, *D. dipsaci* can lead to poor survival of susceptible cultivars (Cooper & Williams, 1983).

Stresses of plant nutrition, competition and grazing management mainly influence persistence by restricting stolon formation. Soil moisture stress and insect damage are the only factors that clearly impact on their survival. In areas that regularly experience severe summer drought, white clover is probably an inappropriate legume species.

Red Clover

While red clover is the second most commonly sown legume in New Zealand, its content is generally low in permanent mixed pastures after 2-3 yr (Lancashire, 1985). One of the most important uses of red clover is as a pioneer legume in oversowing development of moist tussock hill and high country. It does not persist well in sub-hygrous environments (500-800 mm; Greenwood & Sheath, 1982) and is intolerant of close or frequent sheep grazing (Brougham, 1960). In association with prairie grass (*Bromus catharticus*) and bull-beef grazing, reduced red clover persistence was most marked where hard summer grazing was employed (Cosgrove & Brougham, 1985). Greater resistance to crown rot (*Sclerotinia trifoliorum*) and *D. dipsaci* was developed in the late flowering tetraploid "Grasslands Pawera" (Anderson, 1973), but it too, can rapidly succumb to these secondary agents when additive factors (e.g., intensive grazing) debilitate the plant. More persistent early flowering lines of red clover have been developed (e.g., G21, G22, and lines of Swiss origin) which give higher winter growth rates.

Subterranean Clover

While low quantities of subterranean clover are sown (<100 t annually), the residual carryover from earlier sowings (Saxby, 1956) are most evident in dry environments where persistence of white clover is marginal to non-existent. Resident populations are of the Mt Barker and Tallarook types (i.e., mid-late season flowering; Suckling *et al.*, 1983; Macfarlane & Sheath, 1984) and they are generally smaller and denser than equivalent Mt Barker or Tallarook cultivars. Populations of early season cultivars are not persistent (Sheath & Richardson, 1983).

Recent cultivar evaluations have indicated that open erect cultivars (e.g., Woogenellup, Clare) that elevate runners into the grazing zone, reseed and regenerate poorly (Sheath & Macfarlane, 1988a). Similarly, cultivars that readily germinate following early summer rains have poor seedling survival and population persistence (Sheath & Macfarlane, 1988b). Unlike conditions in Mediterranean environments, subterranean clover commonly faces summer rains that promote "false strikes."

In mixed pastures, subterranean clover should be considered as an opportunistic plant. Its content is variable between years depending on the severity of drought and the recovering vigour of surviving perennial plants (Sheath & Boom, 1985). Therefore, in addition to grazing constraints on reseeding and summer "false strikes," competition against seedlings impacts on short term population levels. However, repeated establishment of volunteer subterranean clover populations indicate that buried hardseed ensures long term persistence of suitable genotypes (Suckling & Charlton, 1978; Suckling *et al.*, 1983; Macfarlane & Sheath, 1984).

Lotus pedunculatus

Tolerance of low soil pH, Al toxicity, low soil phosphorus and water-logged soils makes *L. pedunculatus* a competitive and persistent plant when one or a combination of these constraints dominate growth (Sheath, 1981). Under these conditions *L. pedunculatus* can be intensively grazed and remain persistent (Scott & Mills, 1981). Where soil conditions are modified and the competitive ability of companion species improved, then the persistence of *L. pedunculatus* is very sensitive to grazing management. Due to very slow regrowth potential, it is vulnerable to competition if grazing is frequent and/or severe (Sheath, 1980b).

In contrast to white clover, the root feeding insect, *C. zealandica*, is not a problem (Kain *et al.*, 1979). *Lotus pedunculatus* produces a feed inhibiting chemical which not only is self protecting, but also reduces the impact of the pest on companion species.

Lucerne

Lucerne is the major forage legume grown as a monoculture in New Zealand and has been actively promoted for dryland areas during this century. The area grown in lucerne peaked at 220 000 ha during the mid 1970s, but has since declined to approximately 100 000 ha (Douglas, 1986). Poor plant persistence has been a major contributor to this decline. Pest problems associated with *D. dipsaci* and *Graphognathus leucoloma* were further exacerbated by the arrival of *Sitona discoideus*, *Acyrthosiphon kondoi*, *A. pisum*, and *Therioaphis trifolii* during the 1970 and 1980s (Kain & Troughton, 1982; Goldson *et al.*, 1985). Bacterial wilt (*Corynebacterium insidiosum*) also causes major damage to stands, and while verticillium wilt, crown rot and virus diseases are recognized as being widespread, they are not considered to be serious diseases (Douglas, 1986). Multiple resistance breeding programs are occurring and have assisted in improving persistence of lucerne stands (Dunbier & Easton, 1982).

Cutting or grazing lucerne when 10% flowering occurs, is the recommended management to ensure satisfactory persistence, production and quality (O'Connor, 1970). Frequent and/or continuous grazing reduces plant survival and spelled regrowth periods of at least 5-6 weeks are desirable (Janson, 1982). Recommended spelling periods are difficult to achieve during early-mid spring where lactating sheep are involved in lucerne dominated feed systems.

Lucerne is grown in semi-arid to sub-humid environments (300-800 mm annually) and on free draining/low moisture retentive soils (e.g., volcanic ash, stony soils). Correction of soil nutrient deficiencies (P, S, K, Mo, B) is necessary where they exist (During, 1984). New Zealand soils are naturally devoid of *Rhizobium meliloti*, therefore inoculation is required (Greenwood,

1964). Nodulation failures have led to stand declines, but the advent of quality assessment of new inoculation/pelletting methods and strains have reduced the likelihood of such failures (Hale, 1983).

Alternative Legumes

Previous sections have identified the role, strengths, and weaknesses of legumes commonly used in New Zealand agriculture. Because of physical or economic reasons, many of the constraints to persistence (e.g., drought, nutrient deficiencies) may not be overcome with traditional legumes and alternatives may need to be sought. The use of inappropriate species, and the slowness in recognising the role of alternative legumes may well reflect the conservatism of researchers and extension. But in addition, the particular drive in New Zealand to maximize production per ha has ensured the continued need for legumes to be compatible with efficient and hence, intensive grazing systems.

Currently, there is considerable economic pressure on New Zealand farmers to reduce input costs and this must ultimately lead to increased stress on pasture plants. Identification of alternative legumes tolerant of current and future stresses is an urgent priority. To date, work has very much remained within known or used species common to improved Western block agriculture e.g., *T. repens*, *T. pratense*, *M. sativa*. While one cannot be critical of this work, emphasis must be placed in assessing the role of lesser used species (e.g., *Lotus corniculatus*, *Coronilla varia*, *Medicago polymorpha*, *L. tenuis*, *T. balansae*, *T. fragiferum*, *T. ambiguum*, *T. michelianum*, and *T. dubium*) and lesser known species (e.g., *Trigonella ruthenica*, *Hedysarum coronarium*, *T. amabile*, *Ornithopus sativus*, *Lupinus polyphyluss*). There must be numerous genera of legumes that exist as residents in natural and semi-improved grasslands that warrant investigation where persistence of traditionally used legumes is poor. It is expected that this workshop will assist in identifying potential plant material and the means by which it can be utilized.

CONCLUSION

There are few pastoral areas in New Zealand that do not contain one or several legume species. If one fails to persist, then there is sufficient plasticity within the range of available legumes for a more adapted species/genotype to fill the vacant niche (Scott *et al.*, 1985; Chapman & Macfarlane, 1985; Hoglund & White, 1985). Provided there is genetic diversity, it could therefore be argued, that persistence is not a problem in achieving legume based pastures. This unfortunately is not the case if the importance of persistence is gauged on the level of forage production being sought. Where natural resources and input expenditure are to be efficiently utilised in systems dependant on the N_2-fixing ability of legumes, then the most cost effective legume should be farmed to its fullest extent. The important constraints to persistence and achieving this, are: white clover - moisture stress, red clover - intensive grazing, subterranean clover - variable rainfall, *Lotus pedunculatus* -competition, Lucerne - pests and diseases. If these constraints remain unmanageable, then the identification and use of alternative legumes should receive attention.

REFERENCES

Anderson, L.B. 1973. Relative performance of late-flowering tetraploid red clover "Grasslands 4706", five diploid red clovers and white clover. N.Z. J. Exp. Agric. 1: 233-237.

Bircham, J.S., and A.G. Gillingham. 1986. A soil water balance model for sloping land. N.Z. J. Agric. Res. 29: 315-323.

Blaser, R.E., and G.W. Killinger. 1950. Life history studies of Louisiana white clover (*Trifolium repens*). Agron. J. 42: 215-220.

Brougham, R.W. 1960. Effects of frequent hard grazing at different times of the year on the productivity and species yield of a grass-clover pasture. N.Z. J. Agric. Res. 3: 125-136.

Brougham, R.W., P.R. Ball, and W.M. Williams. 1978. The ecology and management of white clover based pastures. p.309-324. In J.R. Wilson (ed.) Plant relations in pasture. CSIRO. Publ., Melbourne, Australia.

Caradus, J.R. 1980. Distinguishing between grass and legume species for efficiency of phosphorus use. N.Z. J. Agric. Res. 23: 75-81.

Caradus, J.R. 1986. Variations in partitioning and percentage nitrogen and phosphorus content of leaf, stolon, root of white clover genotypes. N.Z. J. Agric. Res. 29: 367-379.

Caradus, J.R., A.D. Mackay, and M.W. Pritchard. 1987. Towards improving the aluminium tolerance of white clover. Proc. N.Z. Grassl. Assoc. 48: 163-169.

Caradus, J.R., and R.W. Snaydon. 1986. Response to phosphorus of white clover. 1. Field studies. N.Z. J. Agric. Res. 29: 155-162.

Chapman, D.F. 1983. Growth and demography of *Trifolium repens* stolons in grazed hill swards. J. Appl. Ecol. 20: 590-608.

Chapman, D.F., and M.J. Macfarlane. 1985. Pasture growth limitations in hill country and choice of species. In R.E. Burgess and J.L. Brock (ed.) Using herbage cultivars. N.Z. Grassl. Assoc. 3.

Chapman, D.F., G.W. Sheath, M.J. Macfarlane, and P.J. Rumball. 1986. Performance of subterranean and white clover varieties in dry hill country. Proc. N.Z. Grassl. Assoc. 47: 53-62.

Cooper, B.M., and W.M. Williams. 1983. White clover evaluations in Northland. N.Z. J. Exp. Agric. 11: 209-214.

Cosgrove, G.P., and R.W. Brougham. 1985. Grazing management influences on seasonality and performance of ryegrass and red clover in a mixture. Proc. N.Z. Grassl. Assoc. 46: 71-76.

Douglas, J.A. 1986. The production and utilisation of lucerne in New Zealand. Grass Forage Sci. 41: 81-128.

Dunbier, M.W., and H.S. Easton. 1982. Cultivar development. p.117-120. In F.B. Wynn-Williams (ed.) Lucerne for the 80's. pp.117-120. Spec. Publ. 1. Agron. Soc. N.Z.

During, C. 1984. Fertilisers and soils in New Zealand farming. Government Printer, Wellington.

Floate, M.J.S., P.D. McIntosh, and W.H. Risk. 1985. Effects of fertiliser and environment on lotus production on high country acid soils in Otago. Proc. N.Z. Grassl. Assoc. 46: 111-118.

Frame, J., and P. Newbould. 1986. Agronomy of white clover. Adv. Agron. 40: 1-88.

Goldson, S.L., C.B. Dyson, and J.R. Proffitt. 1985. The effect of *Sitona discoideus* on lucerne yields in New Zealand. Bull. Entomol. Res. 75: 429-442.

Goold, G.J., and J.A. Douglas. 1976. Comparison of "Grasslands Huia" and "Grasslands Pitau" white clover. N.Z. J. Exp. Agric. 4: 135-141.

Greenwood, P.B., and G.W. Sheath. 1982. Suitability of some pasture species within sub-humid areas of Otago. 2. Legumes. N.Z. J. Exp. Agric. 10: 371-376.

Greenwood, R.M. 1964. Populations of rhizobia in New Zealand soils. Proc. N.Z. Grassl. Assoc. 26: 95-101.

Hale, C.N. 1983. Lucerne cultivars and their rhizobium requirements. N.Z. J. Exp. Agric. 11: 161-163.

Hay, M.J.M., J.L. Brock, and V.J. Thomas. 1988. Seasonal and sheep grazing effects on branching structure and dry weight of white clover in mixed swards. Proc. N.Z. Grassl. Assoc. 49: 197-201.

Hay, R.J.M., and G.S. Baxter. 1984. Spring management of pasture to increase summer white clover growth. Proc. Lincoln College Farmers Conf. 34: 137-137.

Hay, R.J.M., and D.L. Ryan. 1983. An evaluation of Pawera red clover with perennial grasses in a summer-dry environment. Proc. N.Z. Grassl. Assoc. 44: 91-97.

Hoglund, J.H., J.R. Crush, J.L. Brock, R. Ball, and R.A. Carran. 1979 Nitrogen fixation in pasture. XII General discussion. N.Z. J. Exp. Agric. 7: 45-51.

Hoglund, J.H., and J.G.H. White. 1985. Environmental and agronomic constraints in dryland pasture and choice of species. p.39-44 In R.E. Burgess, and J.L. Brock (ed.) Using herbage cultivars. N.Z. Grassl. Assoc. 3.

Hollowell, E.A. 1966. White clover *Trifolium repens L.*, annual or perennial? Proc. Int. Grassl. Congr. 10.

Janson, C.G. 1982. Lucerne grazing management research. p.85-90. In R.B. Wynn-Williams (ed.) Lucerne for the 80s. Spec. Publ. 1. Agron. Soc. N.Z.

Jones, R.M. 1980. Survival of seedlings and primary taproots of white clover in subtropical pastures in south-east Queensland. Trop. Grassl. 14: 19-21.

Kain, W.M., R. East, and J.A. Douglas. 1979. *Costelytra zealandica* - pasture species relationships on the pumice soils of central North Island of New Zealand. Proc. 2nd Aust. Conf. Grassl. Invertebrate Ecol. 88-91.

Kain, W.M., and T.E.T. Trought. 1982. Insect pests of lucerne in New Zealand. p.49-59. In R.B. Wynn-Williams (ed.) Lucerne for the 80's. Spec. Publ. 1. Agron. Soc. N.Z.

Lancashire, J.A. 1985. Distribution and use of forage legumes in New Zealand. p.20-34. In R. Barnes *et al.* (ed.) Forage legumes for energy-efficient animal production. USDA, Washington, D.C.

Ledgard, S.F., G.J. Brier, and R.A. Litter. 1987. Legume production and nitrogen fixation in hill pasture communities. N.Z. J. Agric. Res. 30: 413-424.

Levy, E.B. 1970. Grasslands of New Zealand. Government Printer, Wellington.

Macfarlane, M.J., and G.W. Sheath. 1984. Clover - What types for dry hill country? Proc. N.Z. Grassl. Assoc. 45: 140-150.

Macfarlane, M.J., G.W. Sheath, and A.W. McGowan. 1989. Evaluation of clovers in dry hill country. 3. White clover at Whatawhata. N.Z. J. Agric. Res. (in press).

Musgrave, D.J. 1977. An evaluation of various legumes at high altitude. Proc. N.Z. Grassl. Assoc. 38: 126-132.

O'Connor, K.F. 1970. Influence of grazing management on herbage and animal production from lucerne. Proc. N.Z. Grassl. Assoc. 32: 108-116.

Rumball, P.J., and A.E. Esler. 1968. Pasture pattern on grazed slopes. N.Z. J. Agric. Res. 11: 575-588.

Saxby, S.H. 1956. This history of subterranean clover in New Zealand. N.Z. J. Agric. 92: 518-527.

Scott, D., J.M. Koeghan, G.G. Cossens, L.A. Maunsell, M.J.S. Floate, B.J. Wills, and G. Douglas. 1985. Limitation to pasture production and choice of species. p.9-15. In R.E. Burgess, and J.L. Brock (ed.) Using herbage cultivars. N.Z. Grassl. Assoc. 3.

Scott, R.S., and E.G. Mills. 1981. Establishment and management of "Grasslands Maku" lotus in acid, low-fertility tussock grasslands. Proc. N.Z. Grassl. Assoc. 42: 131-141.

Sears, P.D., V.C. Goodall, and R.H. Jackman. 1965. Pasture growth and soil fertility. VIII The influence of grasses, white clover, fertiliser on production of an impoverished soil. N.Z. J. Agric. Res. 8: 270-288.

Sheath, G.W. 1980a. Effects of season and defoliation on the growth habit of *Lotus pedunculatus Cav.* "Grasslands Maku". N.Z. J. Agric. Res. 23: 191-200.

Sheath, G.W. 1980b. Production and regrowth characteristics of *Lotus pedunculatus Cav.* "Grasslands Maku". N.Z. J. Agric. Res. 23: 201-210.

Sheath, G.W. 1981. *Lotus pedunculatus* - an agricultural plant? Proc. N.Z. Grassl. Assoc. 42: 160-170.

Sheath, G.W., and R.C. Boom. 1985. Effects of November-April grazing pressure on hill country pastures. 2. Pasture species composition. N.Z. J. Exp. Agric. 13: 329-340.

Sheath, G.W., and A.J. Harris. 1985. Environmental and management limitations of legume based forage systems in New Zealand. p.110-115. In R.F. Barnes *et al.*, (ed.) Forage legumes for energy-efficient animal production. USDA, Washington D.C.

Sheath, G.W., and M.J. Macfarlane. 1989a. Evaluation of clovers in hill country. 1. Regeneration and production of subterranean clover at Whatawhata. N.Z. J. Agric. Res. (in press).

Sheath, G.W., and M.J. Macfarlane. 1989b. Evaluation of clovers in hill country. 2. Components of subterranean clover regeneration at Whatawhata. N.Z. J. Agric. Res. (in press).

Sheath, G.W., and S. Richardson. 1983. Morphology, flowering and persistence of subterranean clover cultivars grown in North Island hill country : a preliminary note. N.Z. J. Exp. Agric. 11: 205-208.
Steele, K.W., R.N. Watson, P.M. Bonish, R.A. Littler, and G.W. Yeates. 1985. Effect of invertebrates on nitrogen fixation in temperate pastures. Proc. Int. Grassl. Congr. 15: 450-451.
Suckling, F.E.T., and J.F.L. Charlton 1978. A review of the significance of buried legume seeds with particular reference to New Zealand agriculture. N.Z. J. Exp. Agric. 6: 211-215.
Suckling, F.E.T., M.B. Forde, and W.M. Williams. 1983. Naturalised subterranean clover in New Zealand. N.Z. J. Agric. Res. 26: 35-44.
Taylor, A.O., K.A. Hughes, and B.J. Hunt. 1979. Annual cool season legumes for forage. N.Z. J. Exp. Agric. 7: 149-152.
Turkington, R., and J.J. Burdon 1983. The biology of Canadian weeds. 57. *Trifolium repens L.* Can. J. Plant Sci. 63: 243-266.
Turkington, R., M.A. Cahn, A. Vardy, and H.L. Harper. 1979. The growth, distribution and neighbour relationships of *Trifolium repens* in a permanent pasture. III. Establishment and growth in natural and perturbed sites. J. Ecol. 67: 231-243.
Turkington, R., and J.L. Harper. 1979. The growth, distribution and neighbour relationships of *Trifolium repens* in a permanent pasture. I. Ordination, pattern and contact. J. Ecol. 67: 201-208.
Williams, W.M., L.B. Anderson, and B.M. Cooper. 1977. Evaluation of clovers on sandy coastal soil. Proc. N.Z. Grassl. Assoc. 39: 130-138.
Williams, W.M., and J.R. Caradus. 1979. Performance of white clover lines on New Zealand hill country. Proc. N.Z. Grassl. Assoc. 40: 162-169.
Woodfield, D.R., and J.R. Caradus. 1987. Adaptation of white clover to moisture stress. Proc. N.Z. Grassl. Assoc. 48: 143-150.
Ussher, G.R. 1986. Case study information reflecting farmer use of Pawera red clover in northern Northland. Proc. N.Z. Grassl. Assoc. 47: 179-182.

DISCUSSION

Curll: There seems to be little emphasis on the value of legume forage for animal production in New Zealand. In Australia, legumes are important not only for N supply but also for increasing the quality of the animal's diet.

Sheath: Our grass components of pasture are generally of high quality and so we do not have a general reliance on legumes to improve diet quality. In animal finishing systems, there may be specific attempts to increase clover content in summer to improve growth rates of young stock.

Reed: What are the "showy" subclover cultivars which do not persist?

Sheath: 'Larisa', 'Clare', 'Woogenellup', 'Seaton Park' are cultivars whose runners are readily grazed by sheep and this is independent of maturity date.

Hoveland: No reference was made of virus diseases. This is one of the major reasons for stand losses of white clover in the USA.

Sheath: Virus infection does occur, but there is no definite information that suggests they are a primary agent in lack of persistence. It may well be that there are insufficient additive stresses that combine to cause plant death.

Kretschmer: Is it better to have a shorter persistent legume than a high yielding non-persistent legume?

Sheath: It depends on the cost of replacement. In long-term pastures, it is better to have more persistent legumes even if there is a cost in production potential.

Rotar: We seem to be putting legumes in less than optimum environments. With greater stresses being put on the legume there is greater need for site specificity in breeding.

Sheaffer: Are we as scientists aware of the goals of producers?

Sheath: Producers are interested in net returns. Maximum productivity and legume persistence may not always be their primary goal. Costs of production influence producers' decisions on whether to include legumes in their farming enterprise and to use improved management.

A SURVEY OF LEGUME PRODUCTION AND PERSISTENCE IN THE UNITED STATES

Arthur G. Matches

SUMMARY

In 1988, 26 forage specialists from 20 states representing different regions in the USA were surveyed. For their areas, 43% of the respondents indicated that legumes were extensively grown in pastures and 57% said that legumes were not extensively grown. Over the next 5 to 10 yrs, 68% of the respondents expect legume use to increase, 32% expect use to remain the same, and none expect a decline in legume use. Problems identified with legumes were (i) lack of persistence, especially for red clover (*Trifolium pratense* L.) and white clover (*T. repens* L.), (ii) difficulty in successful establishment especially with birdsfoot trefoil (*Lotus corniculatus* L.) and strawberry clover (*T. fragiferum* L.), (iii) lack of competitiveness with grasses, (iv) stand losses from heaving, (v) inadequate heat, cold and/or drought tolerance, (vi) lack of adequate disease, insect and nematode resistance, (vii) poor tolerance to soils with low pH and high aluminum content, (viii) allelopathic effects of some grasses on legume establishment and growth, (ix) not able to withstand continuous grazing, (x) low percentages of hard seed, and (xi) low N_2 fixation. Availability of more persistent legumes would benefit grassland agriculture in the USA.

INTRODUCTION

In 1985, Knight published a review on the distribution and use of forage legumes in the USA. Also, a comprehensive review by Burns (1985) about legume-based forage systems in the southern USA appeared in the same publication. These reviews are recommended for those not acquainted with forage-legume production and use in the USA.

This review is compiled from the results of a survey that I conducted in 1988 of forage specialists in different regions and within regions in the USA. Comments of the participants serve as examples of concerns that scientists have about problems associated with legume production and persistence in their areas. This survey will identify problems of forage legumes that should be addressed in workshops and in research.

NORTHEASTERN USA

Major pasture legumes in the northeastern USA are white clover which usually volunteers, red clover, birdsfoot trefoil, and alfalfa (*Medicago sativa* L.) which is grown mainly for hay (G. A. Jung, 1988, personal communication). Jung believes that about 90% of the pastures could be expected to contain legumes, especially white clover. Legume establishment is considered a problem only with birdsfoot trefoil which lacks good seedling vigor as compared to red clover. Variability over time in white clover stands and production are major concerns in the Northeast, and a combination of management, fertility (particularly P), and environmental stresses (high summer temperature, lack of water) seem to interact to cause this problem. Another problem, according to Jung, is the poor persistence of alfalfa stands when grown with orchardgrass (*Dactylis glomerata* L.) and fertilized with livestock manure. Some dairy farmers have lost alfalfa stands after only 1 to 2 yr of fertilizing with manure. Because of joint agency (USDA-ARS, SCS, Extension) educational programs on intensive grazing systems for the Northeast, legume use in pastures is expected to increase over the next 5 to 10 yr.

SOUTHEASTERN USA

In Virginia, H. E. White (1988, personal communication) lists the white clovers (ladino and white Dutch) and red clover as the legumes most extensively used for pasture. Tall fescue (*Festuca arundinacea* Schreb.), one of their main pasture grasses, is competitive and is a primary contributor to poor legume persistence. White clover is so shallow rooted that dry periods may cause serious loss of stands. Red clover is a productive legume once established, but diseases of the root and crown usually take it out after 2 yr. White would like to have red clovers that are more resistant to diseases. In the future, he expects legume use to stay about the same as at present, but alfalfa is beginning to be used more widely as a grazing legume in Virginia.

Legumes are grown throughout North Carolina, but only about 25% of the pastures contain appreciable amounts of legume (J. C. Burns & D. S. Chamblee, 1988, personal communication). Ladino and intermediate types of white clover; crimson clover (*T. incarnatum* L.); red clover; and hairy vetch (*Vicia villosa* Roth), listed in order of importance, are their major pasture legumes. Reasons given for legumes not being grown more extensively were problems in establishment, difficulty of management, and lack of persistence. Burns and Chamblee attributed the major problems of poor persistence primarily to diseases, secondarily to insects, and subsequently to water and heat stress due to weakened plants. Over the next 5 to 10 yr, they expect legume usage in pastures to increase because of i) availability of methods for sod seeding, ii) new management techniques of shifting seeding time coupled with herbicides, iii) increased cost of N fertilizer, and iv) increased awareness by producers that legume-grass mixtures give higher animal daily gains than grass alone.

Farther south in Georgia, legumes also are not extensively grown in pastures (C. S. Hoveland, 1988, personal communication). Management difficulty, anti-quality factors (bloat), and lack of persistence are listed as reasons producers don't use more legumes. In order of importance, white clover, arrowleaf clover (*T. vesiculosum* Savi), crimson clover and sericea lespedeza [*Lespedeza cuneata* (Dumont) G. Don] are the legumes most often grown in pastures. Also, small but increasing amounts of alfalfa are used for pasture and small amounts of birdsfoot trefoil are used in the mountains. Major causes of poor legume persistence according to Hoveland are virus diseases in white clover, insects that destroy annual clover seedlings in Bahiagrass (*Paspalum notatum* Fluegge) and bermudagrass [*Cynodon dactylon* (L.) Pers.], sclerotinia (*Sclerotinia trifoliorum*) in alfalfa planted in sod, and subsoil acidity. Over the next 10 yr, more use of legumes in pastures is expected. Reasons for this increase include i) higher costs of N, ii) greater appreciation by producers of herbage quality, and iii) the development by the University of Georgia of a new grazing-type alfalfa that persists when continuously grazed.

Legumes are not grown extensively in pastures of the peninsular area of Florida (W. D. Pitman, 1988, personal communication). Difficulties in establishment, management, and lack of persistence were identified as problems facing the producer. Legumes most commonly grown include the summer annual jointvetch (*Aeschynomene americana* L.), white clover, and Florida carpon desmodium (*Desmodium heterocarpon* DC.), which is a summer perennial. A complex of nematodes, viruses, and water stress are judged responsible for declining use of white clover. Pitman believes that a lack of grazing tolerance and competitiveness with southern grasses, as well as frequent extreme dry periods have limited use of other legumes. Less white clover and about the same amount of jointvetch are expected to be grown in the future. Other tropical legumes such as 'Shaw' creeping vigna (*Vigna parkeri* Bak.) are showing promise.

SOUTHCENTRAL USA

Establishment problems, difficulty of management, bloat hazard, lack of competitiveness, improper fertilization, and lack of persistence are reasons listed in one or more of the survey questionnaires from Kentucky (W. C. Templeton, Jr., 1988, personal communication); Mississippi (V. H. Watson, 1988, personal communication); Louisiana (R. E. Joost, 1988, personal communication); and east Texas (G. E. Evers & F. M. Rouquette, 1988, personal communications) as factors limiting legume use in pastures. Most respondents listed establishment as a problem in their area and in Louisiana bloat was indicated as a factor of concern. Only in east Texas are legumes considered to be grown extensively (30% or more of the pastures). For Mississippi and Louisiana, important pasture legumes are red, white, crimson, arrowleaf, subterranean (*T. subterraneum* L.), and berseem (*T. alexandrinum* L.) clovers. Ball clover (*T. nigrescens* Viv.) is also included on the Mississippi list. White, arrowleaf, crimson, and subterranean clovers are listed for east Texas but 'Bigbee' berseem clover use is expanding in some areas. Legumes grown in Kentucky, in order of importance, are white clover, red clover, and Korean lespedeza (Lespedeza stipulacea Maxim.).

Periods with drought and heat stress are considered key factors in the lack of persistence of white and red clovers in Kentucky, Mississippi, and Louisiana. According to Joost, root and crown rot fungi limit the importance of red clover and alfalfa in Louisiana. Watson suggested that head weevil damage to crimson clover limits its production and that stand regeneration is a problem with subterranean clover. With sod seeding of subterranean clover, allelopathic effects of tall fescue and bermudagrass are believed to be responsible for the poor regeneration of stands. In Louisiana, rapid acceptance of berseem clover, and the release of 'Osceola' white clover, which has improved persistence, are expected to encourage the use of more legumes in pastures. In Mississippi, higher prices for N fertilizer, improved fencing technology, farmer realization of legume benefits, and increased cattle prices should result in greater legume usage over the next few years.

Competition from warm-season grasses (mainly bermudagrass), poor soil nutrient status, cold and water stress, and low N_2 fixation were suggested by Evers and Rouquette as factors responsible for the poor regeneration and production of winter-annual legumes. In some cases, lack of sufficient quantities of hard seed may also reduce regeneration of stands. High soil pH and low availability of P, K, and B also limit legume production in east Texas. Both Evers and Rouquette expect more legumes will be seeded into east Texas pastures in the future because of the anticipated release of improved rose clovers (T. hirtum All.) that have increased seedling vigor and hard seed. Also, more is now known about soil nutrient deficiencies in the legume growing areas.

Few legumes are grown in the semiarid pasture and rangelands of Oklahoma (S. W. Coleman & W. E. McMurphy, 1988, personal communication) and western Texas (A. G. Matches, 1988, personal communications). Alfalfa can be grown under rainfed conditions in some areas, but it is generally harvested for hay rather than grazed. Establishment is a major problem in these semiarid regions that have low and poorly distributed rainfall and high summer temperature. McMurphy indicated that some hairy vetch is grown in western Oklahoma, but wheat producers in that area consider it a serious weed problem. Research both at El Reno, OK (Coleman, 1988, personal communication) and Texas Tech University (Matches, 1988) suggests possible opportunities for use of winter-annual legumes (Trifolium and Medicago spp.) in the future. However, lack of adequate winter hardiness is a major problem with many of the winter-annual legumes. I believe that winter-annual legumes that possess improved winter hardiness and high levels of hard seed could be effectively used to enhance the quality of standing dormant warm-season grasses which are commonly grazed during the winter and early spring on the southern High Plains of Texas.

Generally, pastures in the eastern half of Oklahoma (W. E. McMurphy, 1988, personal communication) and Arkansas (J. M.

Phillips & W. P. West, 1988, personal communication) do not contain legumes. Red and white clover are listed as being grown in both states, arrowleaf clover in Oklahoma, and annual lespedeza (*Lespedeza* spp.) in Arkansas. Poor establishment and lack of persistence due to water stress, low soil pH, and low soil fertility are given as problems in both states. As a result of educational programs, legume use in pastures is expected to increase.

NORTHCENTRAL USA

In selected northcentral states, legumes are considered to be extensively grown in pastures of Missouri (P. R. Beuselinck & J. R. Forwood, 1988, personal communications) and Iowa (W. F. Wedin, 1988, personal communication), but not in Indiana (J. J. Volenec, 1988, personal communication); Nebraska (Bruce Anderson, 1988, personal communication); and Wisconsin (R. R. Smith, 1988, personal communication). Although order of importance may change among states, alfalfa, red clover, birdsfoot trefoil, and white clover were the legumes most often listed. Alfalfa is more commonly harvested as hay, but may be grazed sometime during the year. Annual lespedeza is also listed for Illinois and Missouri, and crownvetch (*Coronilla varia* L.) use is increasing in western Iowa.

Lack of legume persistence is mentioned for all states, but only in Illinois is establishment not considered to be a problem limiting legume plantings for pasture. Generally, birdsfoot trefoil is identified as the legume having the greatest establishment problems. Both Beuselink and Wedin commented that establishment of birdsfoot trefoil is a problem because of poor seedling vigor. Heaving is considered a major problem in maintaining alfalfa stands in Missouri, Illinois, Indiana, and Iowa and could also be expected to be detrimental to red clover and birdsfoot trefoil. Heat and water stress are listed as factors contributing to low persistence of legumes in Indiana, Iowa, Missouri, and Nebraska. Cold stress is listed as influencing persistence in Iowa and Wisconsin. Specific problems mentioned include diseases in red clover, birdsfoot trefoil, and annual lespedeza; lack of drought tolerance in ladino clover; lack of successful competition with tall fescue; and lack of survival under continuous grazing.

Respondents from Indiana, Iowa, Nebraska, and Wisconsin predicted that legume use in pastures would increase in the next 5 to 10 yr. Reasons for increase include i) greater emphasis on sustainable agriculture, ii) an increasing awareness by producers of the benefits of legumes, and iii) increase in animal gain on endophyte-infected tall fescue pasture through inclusion of a legume. The Illinois respondent and one Missouri respondent suggested that legume usage in pastures will not increase until more persistent legumes become available to producers.

WESTERN USA

According to R. H. Hart (1988, personal communication), alfalfa is the legume best adapted to Wyoming climate. Because it requires rotational grazing to maintain stands, its use on rangelands is expected to be limited. This might change if adapted varieties are developed that will withstand continuous grazing.

Legumes are estimated to be grown in 95% of the pastures of western Oregon (P. J. Ballerstedt, 1988, personal communication). Subterranean and white clovers are the most important legumes and some birdsfoot trefoil and big trefoil (L. uliginosis Schkuhr) are also grown. Overall, bloat risk and establishment problems are factors limiting legume use by western producers. Winter-annual legumes that have better fall and winter growth, and more drought-tolerant white clovers are needed. Ballerstedt expects usage of legumes in western Oregon pastures to remain unchanged in the future.

C. A. Raguse (1988, personal communication) estimated that 80 to 90% of the California pastures contain some legumes. Ladino clover, strawberry clover, and birdsfoot trefoil are grown in irrigated pastures. Nearly all the irrigated alfalfa grown is harvested as hay. On non-irrigated lands in the western portion of the state, rose clover and subterranean clover are the main legumes, but the annual medics (Medicago spp.) have promise. Failure of winter-annual legume stands often occur when germination is followed by extended periods of low soil moisture in the early autumn. Raguse suggested that annual legumes (particularly subterranean clovers) having greater percentages of hard seed of slow breakdown would minimize the risk of these false breaks. Also, subterranean clovers having more growing points could be expected to yield more regrowth. In general, better distribution of legume growth over the growing season is needed. Unless improved cultivars become available, usage of legumes in California pastures is not expected to change.

This survey clearly indicates that producers are facing many problems when they attempt to use legumes for pasture production in many areas of the USA. However, the development of improved management practices and release of legume cultivars have improved establishment potential, more competitiveness with grasses, and longer persistence should encourage producers to use legumes in their pastures.

REFERENCES

Burns, J. C. 1985. Environmental and management limitations of legume-based forage systems in the southern United States. p. 129-137. In R. F Barnes et al. (ed.) Forage legumes for energy-efficient animal production. Proc. Trilateral Workshop, Palmerston North, NZ. 30 Apr.- 4 May 1984. USDA-ARS, Washington, DC.

Knight, W. E. 1985. The distribution and use of forage legumes in the United States. p. 34-39. In R. F Barnes et al. (ed.) Forage legumes for energy-efficient animal production. Proc. Trilateral Workshop, Palmerston North, NZ. 30 Apr.- 4 May 1984. USDA-ARS, Washington, DC.

Matches, A. G. 1988. Forage legume research in northwest Texas. p. 27. In Proc. 10th Trifolium Conf., Corpus Christi, TX, 24-25 Mar. Texas A&M Univ., College Station.

DISCUSSION

Buxton. Much of the effort in legume improvement has been limited to alfalfa. Is the greatest need to improve present legumes for persistence or to find promising new introductions?

Matches. Many of our important breakthroughs have occurred because of new introductions. Thus, efforts are needed in both introduction and cultivar improvement of existing legumes.

Brougham. Why are viruses more important in the USA?

Irwin. The enzyme-linked immunosorbent assay (ELISA) was originally more widely used for detecting viruses in the USA than Australia. ELISA is now being used extensively in Australia.

John Hay. A virus survey was made in New Zealand in late 1970. As the survey extended further south, less viruses were present and theoretically were less important. A climate by virus interaction was assumed.

Reed. In Victoria, Australia, viruses and white clover are noticeable in stands near lucerne fields and are thought to have come from U.S. cultivars imported for their aphid resistance.

Barnes. How would you describe the availability and adequacy of statistical data on the acreage of forages grown in the U.S. compared to Australia and New Zealand?

Matches. We do not have reliable data except possibly for alfalfa and red clover hay production.

Reed. What is the most widely grown annual legume in the U.S.?

Matches. Probably crimson clover, but I have no real way of knowing for sure. We only have estimates for the hectares of clover grown, no census data.

Smith. There are no programs for development of legumes which are adapted to competition with grasses in mixtures.

Sheaffer. The priority of breeding programs is to increase persistence by breeding for disease resistance and winterhardiness. If increased persistence is obtained, plants will be more competitive or we can launch into a breeding program for competitiveness as a secondary priority.

Smith. Persistence is important to competitiveness, but I believe variability among the current germplasm does exist.

Caradus. Is the problem with the poor persistence of white clover due to the extensive use of ladino types in the USA? It is very susceptible to poor grazing management.

Matches. Probably, but other types are grown, especially intermediate in Louisiana and East Texas. The white clover cultivars are very susceptible to drought and heat which is common in many areas of the U.S.

Allen. You commented on the need for annual legumes which will persist. I suggest that good management will be a major reason for good persistence. Do you consider that American farmers will be prepared to manage annual legume pastures properly?

Matches. Yes, many will, but perhaps we need more cultivars which are "farmer resistant".

LEGUME PERSISTENCE PROBLEMS IN HAWAII: AN OVERVIEW

Peter P. Rotar

SUMMARY

Lack of legume persistence in Hawaiian pastures is a perennial problem for our ranchers. Problems of establishment, nutrition, competition, and persistence in association with aggressive grasses and weeds, as well as problems of grazing management, have caused concern about the use of legumes. We have to work with a wide range of soil and climate environments which are separated by short distances. A 1-km shift along any given elevation contour may result in as much as a 1000-mm change in rainfall, a change in the soil order with their concomitant changes in soil chemistry and a marked change in radiant energy available at the canopy level. Such rapid shifts in the environmental gradients render much of our pasture research site specific.

VEGETATION ZONES AND PASTURES

The eight major islands forming the state of Hawaii are comprised of over 16 885 km^2 of land area and lie at the southeastern end of a chain of islands and shoals more than 2567 km long. Each of the islands was originally built as one or more lava domes, and each owes its shape primarily to volcano building. Parts of the islands have been more or less modified by erosion under strongly localized conditions.

Five general types of vegetation occur in the Hawaiian Islands; shrub, forest, parkland, bog, and moss-lichen. These formations do not exist on all of the islands nor are the altitudinal limits of the formations the same on all islands. These formations tend to occur in regular patterns on the mountain slopes as one progresses from the very dry leeward side of an island to the wet windward side and, as one goes up the slopes, changing from tropical conditions at sea level to a temperate climate in the highlands and to a frigid climate at the tops of the high mountains.

Types of forages grown and their productivity are highly dependent upon the natural features of the landscape interacting with the climate and the soil. Ripperton and Hosaka (1942) incorporated climate, vegetation, soils and geographic features by classifying the lands of Hawaii into a series of Vegetation Zones which could be used to assess the potential forage productivity of various areas. Briefly, the zones and their pasture characteristics may be described as follows:

1. Zone A occurs on the lee sides of the islands or on low windward lands having no mountain background high enough or close

enough to cause cloud formation and precipitation. These lands extend from sea level to over 140 m. The climate is characterized by warm, dry summers and cool winters. Annual precipitation is less than 500 mm and occurs primarily during December to March in the form of storms from the south. Vegetation consists of lowland shrub with scattered-to-dense stands of the tree, kiawe or mesquite [Prosopis pallida (Humb. & Bonpl. ex Willd.) H.B.K.] and predominantly annual grasses (Chloris spp.). Few, if any, annual or perennial forage legumes are present. Weedy perennial grasses found in this zone include fountaingrass [Pennisetum setaceum (Forsk.) Chiov.] and sourgrass [Trichachne insularis (L.) Nees]. Buffelgrass (Cenchrus ciliaris L.) is well adapted to this zone. Pastures vary greatly as to natural cover and value for grazing. Pasture development is limited. This zone is susceptible to soil erosion due to concentrated rains and to high winds. Long periods of dry weather allow only seasonal use of pastures.

2. Zone B occurs above Zone A, if present, in areas similar to A and as narrow coastal strips in some places on the windward sides. Zone B extends to over 600 m elevation. The general climate is similar to Zone A. Rainfall varies from 500 to 1000 mm. The vegetation consists of low shrub, especially lantana (Lantana camara L.) and both annual and perennial grasses provide good ground cover. The most productive pastures are composed of guineagrass (Panicum maximum Jacq.) and koa haole [Leucaena leucocephala (Lam.) de Wit]. Green panicgrass (P. maximum var. trichoglume Eyles) and buffelgrass are also productive in this zone. The pastures of this zone are better than most parts of Zone A due to the higher rainfall and a correspondingly longer season of use. Perennial legume species adapted to the wetter parts of Zone B include various Desmodium, Stylosanthes, Centrosema and Neonotonia spp. This zone is an important part of many ranches.

3. Zone C lies above Zone B except where it reaches the ocean. It occurs more or less as a band completely encircling the lower mountains and dips to the coast on the windward sides of the islands. This zone extends from sea level to 1200 m elevation. The zone is divided into a low phase (C_1) below 760 m elevation and a high phase (C_2) from 760 to 1200 m. The low phase contains our best farm lands; the high phase has excellent pastures and farm lands. Climate at sea level is characterized by 1000 to 1500 mm rainfall, mainly of tradewind origin; summers are distinctly dry. This zone is the most desirable of all lowland zones for pastures, but most of the better lands are in cultivated crops. Pastures in Zone C produce high quality forage, have moderate carrying capacity, and have a relatively long grazing season. Summer droughts are common. Vegetation of the low phase C_1 consists of open stands of shrubs mixed with excellent ground cover of perennial grasses and forbs. Guava (Psidium guajava L.) is the predominant shrub below 600 m elevation. Grasses are particularly vigorous ranging in type from tropical to temperate. Kikuyugrass (Pennisetum clandestinum Hochst. ex Chiov.) and pangola digitgrass (Digitaria decumbens Stent) are the dominant grasses on the more fertile soils. Kikuyugrass does poorly in areas with acid soils. At low elevations in the drier regions of C_1 green panicgrass, guineagrass, buffelgrass, and bermudagrass [Cynodon dactylon (L.) Pers.] are important.

The high phase, C_2, originally forested, has a few mixed stands of trees including *Eucalyptus* and *Acacia* spp. The cleared lands are mainly grasslands. Grasses vary in type from subtropical to temperate. Kikuyugrass is the dominant grass except in areas with acid soils. Pangola digitgrass is limited to the lower parts of the C_2 as it is sensitive to low temperatures. Natal redtop [*Rhynchelytrum repens* (Willd.) C.E. Hubb.], rescuegrass or bromegrass (*Bromus unioloides* (Willd.) H.B.K.), and rattail [*Sporobolus africanus* (Poir.) Robyns et Tourn.] are prevalent at elevations above 800 m in areas not covered with kikuyugrass. Fescues (*Festuca* spp.), perennial ryegrass (*Lolium perenne* L.), and orchardgrass (*Dactylis glomerata* L.) are limited in their distribution due to low P content of soils.

Legumes in Zone C vary from subtropical *Desmodium* spp. to temperate region clovers (*Trifolium* spp.), vetches (*Vicia* spp.) and medics (*Medicago* spp.). Kaimi clover [*D. canum* (Gmell.) Schintz et Thell.], 'Greenleaf desmodium' [*D. intortum* (Mill.) Urb.], and spanish clover (*D. sandwicense* E. Mey.) do well below 900 m elevation in Zone C. Centro (*Centrosema pubescens* Benth.), perennial glycines (*Neonotonia wightii* (R. Grah. ex Wight & Arn.) Lackey), and stylo [*Stylosanthes guianensis* (Aubl.) Sw.] do well in warmer locations below 500 m in Zone C. Clovers, vetches, and medics are important temperate legumes above 300 m in Zone C. The trefoils (*Lotus* spp.) are important in the wetter regions of the high phase C_2. Distribution and abundance of legumes is limited by soil pH and P availability.

4. Zone D occurs on the windward sides on high mountains and the windward slopes, crests, and upper leeward slopes of the lower mountains. The three phases of this zone extend from sea level to 2100 m. In the low phase (D_1) the better arable lands are used for crop production and ranching is usually restricted to the poorer soils and nonarable lands. Minimum rainfall at sea level is over 1500 mm. In the middle phase (D_2) little ranching occurs except for the area between Honokaa and Kamuela on the Island of Hawaii and most of the land is in forest reserve which acts as the principal source of artesian, tunnel, and surface water for agricultural and urban use. Maximum rainfall is over 11 500 mm. The high phase, (D_3), is above the belt of maximum rainfall and varies from 1200 to 2100 m elevation. It contains some of Hawaii's most important temperate grazing lands. The climate is moderately tradewind oriented (except for the Kona District of the Island of Hawaii) and is less variable than in other vegetation Zones.

The most characteristic vegetation in the low phase is guava. Ohia [*Meterosideros collina* (Forst.) Gray subsp. *polymorpha* (Gaud.) Rock] occurs over much of the area. In cleared areas staghorn fern [*Dicranopteris linearis* (Burm.) Underw.] and Boston fern [*Nephrolepis exaltata* (L.) Schott.] are common. Carpetgrass [*Axonopus compressus* (Sw.) Beauv.], hilograss (*Paspalum conjugatum* Berg.), rattail, and ricegrass (*P. orbiculare* G. Forst.) are often dominant. The D_1 Zone is not of great importance to the grazing industry at the present time, but it offers important possibilities of extensive development through proper use of lime, fertilizers, and adapted pasture species.

The middle phase is almost entirely in forest reserve and consists mainly of subtropical rain forest. The D_2 Zone has little pasture value except as a source of reserve feed during dry periods. The high phase is principally of the open-forest type vegetation. Koa (Acacia koa Gray) is the dominant tree. The important forage species vary from kikuyugrass and pangolagrass in the low phase to the temperate region grasses and legumes in the high phase including ryegrass, fescue, clovers, and big trefoil. Some excellent pastures have been developed in Zone D_3. The most productive pastures are dominated by temperate species. Kikuyugrass, although poorly adapted, is also an important species in this zone. Management is important in maintaining the best adapted species.

5. Zone E occurs only on the Islands of Maui and Hawaii. It lies above the inversion layer where clouds form and thus is relatively dry. Zone E occurs above the high phase of Zone D in wetter localities and over the high phase of Zone C in the drier areas. The low phase (E_1) consists of a plateau or gently sloping parkland ranging from 1200 to 2100 m elevation; the middle phase consists of dwarfed trees and some grasses and forbs and varies from 2100 to 3000 m elevation; and the high phase occurs above 3000 m elevation with the vegetation being principally mosses and lichens. Rainfall is less than 1250 mm. Frosts are common as low as 1200 m elevation. Average annual temperatures are low; 15^oC at 1200 m, 10^oC at 2100 m and 4^oC at 3000 m elevation. Pastures range up to 2400 m and consists mainly of cold-tolerant temperature species including sweet vernalgrass (Anthoxanthum odoratum L.), perennial ryegrass, orchardgrass, lovegrasses (Eragrostis spp.) bromegrass, medics, and various clovers.

With the introduction of cattle by Captain George Vancouver in 1793, the search for adequate grazing lands began. The history of forage introductions is interesting, the first record being Stenotaphorum secundatum (Walt.) Kuntze in 1816. Many weedy plant species as well as useful forage species probably arrived in the feed and bedding of animals carried on board ships arriving from various parts of the world. Many of the grasses found in Hawaii at present are of little or no economic value and are only of academic interest, although they may form an appreciable component of the prevailing land cover. A few are definitely important as objectionable weeds or useful as animal forage, turf cover, ornamentals, and industrial materials. Some of the best established introductions are the least desirable and of these many are noxious weeds. Among the weedy species, notable examples are fountaingrass, broomsedge (Andropogon virginicus L.), sandbur (Cenchrus echinatus L.), torpedograss (Panicum repens L.), yellow foxtail [Setaria glauca (L.) Beauv.], rattail, sourgrass, and johnsongrass [Sorghum halepense (L.) Pers.].

Few if any, of the native legumes were useful as pasture plants and many introductions have been made, some with notable success and others failing either through the lack of understanding for the need for proper Rhizobium, or from being planted in soils which were low in available P or other plant nutrients. We estimated that several hundred species of forage or browse legumes have been introduced over the years in an attempt to find well-adapted forage types. Among the most useful are white clover (Trifolium repens L.), big trefoil (Lotus pedunculatus L.), medics and bur clovers, (Medicago spp.), vetches (Vicia spp.), kiawe or mesquite, koa haole, greenleaf desmodium, kaimi clover, red clover (Trifolium pratense L.), perennial glycines, centro, stylo and 'Siratro' [Macroptilium atropurpureum (DC.) Urb.]. A number of the introduced legumes have turned out to be pests and are of little value in pastures. Among these are Crotalaria spp., gorse (Ulex europaeus L.), and japanese tea (Cassia leschenaultiana L.).

CLIMATIC FACTORS LIMITING PRODUCTIVITY OF PASTURES

There are three dominant factors in the climate of Hawaii: the position of the islands, 19° - 22° N Lat, insular position in the Pacific ocean, and elevation and topography. Local physiographic features modify the oceanic effects to produce a wide range of microclimates.

Temperature

A major problem in pasture development in Hawaii is the fact that our climate regimes as well as our pasture species vary from tropical and subtropical to temperate in nature. For most temperate grasses and legumes, the optimum temperature for both dry matter accumulation and extension growth lies around 20-25°C. Some growth is possible at temperatures down to 5°C and at the upper extreme most species cease growth in the range from 30-35°C. In contrast, the thermal range for most subtropical and tropical grasses is at least 10°C higher, with the optimum in the vicinity of 35°C, the upper limit approximately 40-45°C and little or no growth below 15°C. Tropical legumes have not been studied to the same extent as grasses, but appear to have a slightly lower optimum temperature (25-30°C), although the minimum temperature for growth (c. 15°C) is similar for both groups (McWilliam, 1978).

In Hawaii, mean annual temperatures at sea level vary from 22 to 26°C depending more on exposure to ocean breeze than to latitude. Mean temperatures decrease at the rate of about 1.6°C per 304 m increase in elevation. At the summit of Mauna Kea, 4213 m, temperatures fall below freezing almost every night of the year. At Waimea, Hawaii at 813 m elevation, the average temperature is 17.7°C with monthly averages ranging from 12 to 24°C.

For any given location average temperatures during the winter are usually 3 to 5°C cooler than average summer temperatures. Consequently, growth is slower and recovery takes longer. Legumes which may do fairly well during the summer months slow down and cease growing during the cool season. The response, however, is variable depending upon the species as well as the location.

Low temperatures limit forage production during the winter period when most plant growth slows down or, for some species, virtually stops. Low temperature effects become more important with increasing elevation. The greater number of cloudy days during the rainy season also accounts for seasonal reduction in growth. As an example of this seasonal variation, dry matter yields of irrigated pangola digitgrass at one location were on the order of 22 kg per hectare per day in winter compared to 85 kg per hectare per day in summer (Plucknett, 1970) Most ranches (except

Zone D) are stocked according to the period of lowest forage growth (summer) and are often understocked for the rest of the year (late fall, winter, and early spring).

In some locations, we have large interactions between cloud cover and night-time temperatures. At the Volcano Experiment Farm (1188 m) average minimum temperatures during the winter season are in the range 7.2 to 10°C. However, if the night-time cloud cover disappears we can get radiation frosts.

There is a much wider change from summer to winter in air temperatures than in soil temperatures. According to U.S. Soil Taxonomy (Anonymous, 1970), all of our pastures soils have a temperature regime which varies less than 5°C from winter to summer. Temperature regimes for these soils are classified as either isomesic, isothermic, or isohyperthermic.

Precipitation

Precipitation in Hawaii is made up of three components: orographic, cyclonic, and convectional rainfall. Annual rainfall varies from less than 300 to over 11 400 mm. Large areas of arable lands in Hawaii do not regularly receive enough rainfall to support a profitable plant cover. Other areas have high rainfall but are so steep and inaccessible they cannot be cropped. The arable dry soils, the steep wet soils and the arable soils with a sufficient rainfall occur as relatively small units often all within short distances of each other.

Since most pastures depend upon rainfall, moisture is a major limiting factor in Zones A, B, C, and E, especially during the summer. Rainfall in most parts of Hawaii is seasonal, with the highest rainfall occurring during the winter period, November to March. The coastal and leeward areas receive highest solar radiation and, with irrigation, are potentially the most productive regions of the state. As elevation increases, temperatures decrease and pasture types change from tropical or subtropical to temperate. In the wetter regions (Zone D) heavy cloud cover reduces solar radiation and, consequently, reduces plant growth.

In certain sections we also get a fair (unmeasured) amount of moisture from cloud drip. The extent and amount is relatively unknown, however, from existing rainfall records it is easy to see that the amount of effective moisture for plant growth is much greater than the recorded precipitation. Cloud banks build up daily against the windward sides of the high mountains and swing around to the leesides of the mountains providing moisture to a much larger area than might be anticipated.

Sunlight

Between summer and winter there may be as much as a 40% reduction in the amount of radiant energy received. Daylength varies from a maximum of 13 h 20 min in summer to a minimum of 10 h 40 min. These values are considerably modified among locations by aspect and cloud cover.

At sea level, Waimanalo Experiment Station (windward side of Oahu) receives radiation varying from 20.50 MJ per m^2 per day (monthly average) in July and August to as low as 10.46 MJ per m^2

per day in January and February. On the leeward side at sea level it may be as much as 25.10 MJ per m^2 per day in summer and about 20% lower in winter. At Volcano Experiment Farm (Hawaii, 1122 m) solar radiation reaches 18.82 MJ per m^2 in summer and drops to as low as 7.9 MJ per m^2 per day in the winter.

Besides changes in the amount of energy received, distinct photoperiod effects occur that affect plant growth and their ability to compete. Greenleaf desmodium, for example, is a long-short-day plant (Rotar, 1970). With adequate vegetative growth in September, flower induction will occur during October to November with peak flowering during January to March. All apical and axillary buds produce flowers. Grazing during this period simply removes existing leaves and stems. There is very little if any vegetative growth in true greenleaf desmodium plants. Associated grasses, primarily kikuyugrass and pangolagrass, still put on vegetative growth although at a reduced rate. In a pasture situation the legume is put at a severe disadvantage, by the time it begins to regrow in the spring, associated grasses are much taller. The legume can recover during the summer, but if it is heavily grazed during the following winter season it is set back to the point where it doesn't recover and is lost.

PHYSIOGRAPHIC FACTORS

Most soils used for pasture lie in areas which are marginal or submarginal for cropping. Steep hilly regions present serious management problems in plant establishment, fencing, fertilization, water distribution, and weed control. Soils in low rainfall areas are often shallow and rocky; moreover, they are subject to severe erosion due to high winds during the dry season as well as to high rainfall intensities during the short rainy season. Many shallow rocky soils are covered with dense woody shrubs and trees. Steep, shallow, and rocky soils can be very difficult to productively manage.

SPECIES LIMITATIONS

All of our pasture legume species are imported from overseas; Australia, New Zealand and Continental United States being the major seed sources. Except for the Desmodium breeding program (1962-1970) we have not had a formal breeding program for pasture legumes. The clovers and trefoils which are our most important cool season legumes have been developed elsewhere in different climate--soil environments. We had a cool-season legume introduction program (1963-1978) in which we looked at a large number of clovers, trefoils, vetches, medics, and other species. Efforts were made to find suitable legumes for our diverse pasture situations (Whitney, 1982). Major white clover cultivars used in Hawaii come from either Australia, New Zealand or the United States mainland. Those areas have climatic conditions that tend to favor dormancy during the dry season or during the cold season. Perennial plants with any dormancy characteristics put much of their photosynthate into stored food reserves in their crowns and roots for use at the beginning of the next growing season. Our soil temperatures do not get cold enough in the winter to force the plants into complete dormancy and food reserves meant for next season's growth are respired away by the roots. This then puts the legume at a further competitive disadvantage with associated grasses.

Legumes have high phosphate requirements for establishment and growth as useful forage plants. Many attempts at pasture renovation in Hawaii which consisted of brush clearing, disking. and reseeding have failed for the lack of proper fertilization as well as failure to use the proper Rhizobium. Hawaii has no well-adapted legume species which are competitive with kikuyugrass or pangola digitgrass under poor management conditions. Prolonged periods of overgrazing favor grasses, and legumes die out. Tropical and subtropical legumes which are highly productive during our summer season are temperature sensitive and are relatively unproductive during the cool winter season.

With yearlong growing conditions we have not adequately matched legume species and cultivars to specific environments. For example, I have seen white clover growing from sea level (Waimanalo Experiment Station during the winter period) to as high as 2000 m on the windward slopes of Mauna Loa. It does best in a much narrower range of environments, characterized by cool, moist climates. Greenleaf desmodium will grow (with adequate moisture) from sea level to above 900 m. Tinaroo and Clarance glycine grow over the same range of environments. Tinaroo and Clarance glycine do best in limited regions in the wetter parts of the B vegetation zone under much warmer conditions than greenleaf desmodium. When grown under suboptimum environmental conditions both species either perform poorly or rapidly disappear. The Hawaiian Sugar Planters Association continually develops sugarcane varieties that are nearly site-specific for Hawaii's sugar industry, yet we have not attempted the same for pasture legumes.

One can find any number of different legume species growing along roadside banks and rights-of-way; however, if one crosses over into the adjoining pasture the legumes entirely disappear or are only occasionally present. I've examined a number of such situations and have found that there was a large amount of thatch on the ground varying from several to 15 cm or more in thickness. Kikuyugrass, if not heavily grazed, will develop a thatch which very effectively limits the growth of clover and or big trefoil. Under most conditions the green leaves above the thatch are taken off by the animals and the thatch is not removed until animals are forced to eat it; by that time the thatch will have effectively removed the legume component. The amount of thatch remaining after grazing will vary in height according to the grazing pressure. If the thatch is kept short (5 cm) clover will become dominant; however, if the thatch is allowed to develop to thicker proportions then the legume has little chance of being an effective plant in pastures. Pangolagrass behaves in much the same way although it does not develop as thick a thatch as kikuyugrass. With a shift to intensive grazing management many of our ranches are seeing an increase in clover content from less than 2% to as high as 15%. Thatch (mat) management is definitely changing species composition in our pastures.

INSECTS AND DISEASES

Several pests have severely damaged pastures in recent years. The hunting billbug (Sphenophorus venatus vestitus) has caused serious injury to kikuyugrass. The grass webworm (Herptograma licarsicalis) has also severely injured kikuyugrass, pangolagrass, and bermudagrass. The armyworm (Pseudaletia unipuncta) has occasionally injured kikuyugrass on the island of Hawaii during the spring. Koa haole (Leucaena leucocephala) has been badly damaged by the psyllid fly (Heteropsylla cubana Crawford). We have a number of pod borers which damage legume seed pods. I have not seen good documentation concerning types and amounts of insect damage on pasture legumes in Hawaii.

We have a small brown slug about 1/2-inch long which has probably caused much more damage than imagined. Some years ago I was involved with the planting of a 4 ha paddock to greenleaf desmodium; germination and emergence were excellent, however, a large infestation of slugs nearly destroyed the planting. I am certain that more than one planting has been destroyed by these slugs. Frequently, the rancher does not go back to look at the newly planted field until it is too late. Even then, there is the problem of having adequate spray equipment available for control. One of our ranches now uses a pelleted pesticide at planting with white clover; and has considerably increased white clover survival.

We had disastrous results with siratro which is well adapted in the B and C1 vegetation Zones. Siratro was widely planted during the late 1960s; however, it was quickly decimated by Rhizoctonia. Siratro can still be found in pastures but not as an effective pasture plant.

Desmodium Mosaic Virus severely damages kaimi clover; however, the legume persists even though it may not be an effective pasture legume. We are not sure about effects of other legume viruses which may be present in pastures.

Dr. Whitney (1970, personal communications) indicated that he lost many legume accessions to nematodes. At the Kula Branch Station (Maui) where Whitney did his original testing, we saw many cool-season legume accessions affected by nematodes at the end of the rainy season and plants were stressed for lack of water. Nematode susceptibility may be a much more serious problem in pastures than is realized.

There has been no concerted effort to look for chronic disease and insect problems associated with legume persistence in Hawaii. We note foliar disease and insect problems if they are sufficiently severe to come to our attention. There has been no monitoring of the root environment either summer or winter for potential parasites.

SOIL FERTILITY

Fertility of most Hawaiian soils seriously limits crop production. Of the elements essential for plant growth, N is the most deficient in our soils. Some ranchers are using fertilizer N on intensively managed pastures to raise productivity during critical seasonal periods. The increase due to N application is spectacular, however, legumes are suppressed. Pangolagrass dry-matter yields varied from 9.1 t ha^{-1} yr^{-1} in non-fertilized plots to 44.4 t ha^{-1} yr^{-1} in heavily fertilized plots (Plucknett, 1970). With current shortages and high costs of fertilizers, the use of legumes in pastures is important. Some of our legumes may provide as much as 340 kg of N ha^{-1} yr^{-1} to our pastures (Rotar et al., 1976).

Many soils in Zones C and D are acid in their natural condition and may be further acidified by use of acid-forming fertilizers without concurrent liming. As marginal pineapple [Ananas comosus (L.) Merr.] lands were abandoned, many of the areas were converted to pasture and resulted in severe management problems due to Al and Mn toxicity, Ca deficiency, and poor availability of other nutrients such as Mo in these very acid soils. Nutrient deficiencies on these soils include Mo, Ca, K, Mg, and occasionally Fe. Sulfur may be deficient in acid and neutral soils.

Nitrogen

Most of our soils are deficient in N and in highly leached soils this may be a problem in plant establishment. The use of S carrying fertilizers can accelerate acidification of soils, which, in turn, may cause a reduction in soil cation exchange capacity, thus rendering cations such as K, Ca, and Mg more readily leachable; and increases solubilities of Al and Mn which may become toxic to plants. Lowering soil pH may lead to harmful effects on root growth.

Phosphorus

Phosphorus is deficient in nearly all our highly weathered soils. These infertile soils fix large amounts of P in unavailable form. This deficiency limits forage production because of both poor root development and poor plant nutrition. Many of our pastures are deficient in P for livestock during most of the year (Younge & Otagaki, 1958). Phosphorus deficiency limits legume establishment and growth and is the major limiting factor in legume production. Use of legumes in Hawaii is especially keyed to the P status of the soil, since in many cases they are less competitive than associated grasses.

In pasture soils much of the available P is in soil organic matter and organic residues. We often have a surface layer of organic matter about 2-cm thick. Much of the available P is in this layer. This layer usually dries out before available water is exhausted from the remainder of the soil profile. Much of the P in the organic layer may be available for only brief periods. This becomes a serious problem in overgrazed areas because moisture quickly evaporates from the exposed surface. Where the surface is protected, the organic matter layer dries more slowly and P is available for a longer period of time.

Hawaiian soils vary in their phosphate sorption capacity from very low to very high (Fox, 1967). The order of fixation for the various mineralogical systems is as follows: amorphous hydrated oxides > goethite-gibbsite > kaolin > 2:1 clays.

Yost et al. (1982) presented field data from 21 transects on the Island of Hawaii where they collected 153 samples taken at 1 to 2 km intervals. Phosphate sorption curves were determined by the method of Fox and Kamprath (1970). The range in P sorption (P required to obtain 0.02 mg P/l in solution) varied from 0 to 1800 mg P/kg soil. Most of Hawaii pasture lands fall within the 0 to 200 mg P/kg soil. Some of them are in areas with much higher P sorption capacities. We have seen repeated experiments where legumes respond well to phosphate applications; however, the minimum level of P application required is variable (Tamimi & Matsuyama, 1972, 1975; Tamimi & Thompson, 1972; Tamimi et al., 1976).

For soils which are very P deficient a sigmoid-shaped yield response curve from added phosphate is obtained. This indicates that low levels of P applied are near the threshold level for uptake. Many of our fertilizer trials start with 50 kg of elemental P per ha and are incremented in multiples of 100 kg. To most ranchers these rates seem excessive; however, the benefits from higher levels of original P application may be observed for a long period of time. The original "Fixation capacity" is overcome and over time much of the originally applied P is released to the crop (Plucknett & Fox, 1965; Fox et al., 1975).

Potassium

Potassium fertilization requirements probably vary as much in Hawaii as anywhere else in the world. Yost and Fox (1981) measured exchangeable K on some Andepts of the Island of Hawaii and found values varying from 0.07 to 6.97 cmol K/100 kg with Hydrandepts (3583 mm annual rainfall) ranging from 0.07 - 0.46 cmol K/100 kg; Dystrandepts (1330 mm annual rainfall) ranging from 0.22 - 4.37 cmol K/100 kg; Vitrandepts (649 mm annual rainfall) ranging from 0.12 - 2.28 cmol K/100 kg and Eutrandepts (781 mm annual rainfall) 0.48 - 6.97 cmol K/100 kg. Most of Hawaii's pasture soils are in the Dystrandepts and Vitrandepts. Potassium is also important in maintaining a good legume/grass balance.

Sulfur

When there had been no volcanic activity for a number of years, we saw a positive response to S. Current volcanic activity (1985-1988) has probably changed S status in our soils. Several thousand tons of S are released daily into the atmosphere. We now have cumulative effects of large amounts of S being deposited over the island chain with largest accumulations on the island of Hawaii. Acid rains have affected vegetable crop and ornamentals production; however, we have no measure of the long-term effects of the S deposition on pastures.

Micronutrients

In Hawaii, micronutrient deficiency is usually associated with one or more of the following: highly leached, weathered acid soils; alkaline soils; soils which have been intensively cropped and heavily fertilized with macronutrients only; soils which have their surfaces removed through erosion; soils which have an imbalance between micronutrients or between a macronutrient and a micronutrient; and organic soils.

We look for micronutrient deficiencies in crop lands but quite frequently overlook them in pasture-range situations. The problem is passed over as not being worth the effort on range situations. The usual response is that it costs too much to apply. One might ask the question, "How much are we losing by not paying attention to micronutrient problems?"

Strongly leached acid soils tend to be low in micronutrients because acid leaching has removed much of the small quantity of these elements originally present. In soils where the surface has been removed through erosion and where the subsoil is exposed, the deficiency problem usually becomes compounded because much of the micronutrients are held in the naturally chelated form in the surface organic matter. Such soils are likely to be found in sloping lands with moderate to high rainfall in Hawaii. Zinc is the most likely micronutrient to be deficient under such conditions (Kanehiro, 1964). High pH (alkaline) and calcareous soil conditions can bring about a deficiency in five (Cu, Zn, Fe, Mn, and B) of the six micronutrients concerned. In Hawaii, a high pH is generally associated with coastal soils derived from coral parent material.

The application or the natural presence of large quantities of macronutrients, especially P or N, can reduce the availability of micronutrients, such as Fe, Zn, or Cu. Heavy N fertilization of pastures may intensify Zn and Cu deficiencies. Excess molybdenum may adversely affect copper uptake causing molybdenosis in animals. The Mo/Cu imbalance is a common problem in animals on parts of the Island of Hawaii. Boron deficiency is known to be accentuated by high K and Ca. Iron deficiency occurs even in the presence of high amounts of total Fe. This occurs where soluble Fe is rendered insoluble in the presence of higher valance oxides of Mn (Asghar and Kanehiro, 1981). With Mn, the problem in some of the Oxisols in Hawaii is one of excessive amounts (Sherman and Fujimoto, 1948). Reports on Cu deficiency have been increasingly noted (Hori, 1970; Olney, 1969; McCall, 1980, unpublished data).

Lime

The use of lime in tropical pastures in Hawaii is problematic; legumes vary widely in their lime requirements. Overliming can be highly detrimental. Munns and Fox (1976, 1977) and Munns et al. (1977) studied effects of lime on tropical and temperate legumes. Inoculated legumes representing 18 species were grown. Their lime response curves showed no distinct differences between tropical and temperate legumes. Within each group individual species varied: a) early growth or nodulation of certain species was depressed when the Oxisol was limed at rates above 6 t/ha^{-1} (pH 6); b) in some legumes the depression gave way to positive response; c) in some species the depression persisted throughout the experiment; and d) in still other species no depression was observed.

These results point out the need for careful evaluation on a species-by-species basis. In pastures a transitory depression in legume growth may lead to eventual loss of the legume due to competition. Species differences in lime response may vary among soils because lime influences growth by changing several acid-soil properties whose significance varies from one soil to another. Among the most important of these properties are acidity itself and the availability of Ca, Mn, Al, and Mo (Munns & Fox, 1977). Lime rates should be determined for individual soils.

ALLELOPATHY

Dr. Whitney (1975, personal communications) observed that greenleaf desmodium could not be established in a stand of tetraploid 'Bigalta limpograss'[Hemarthria altissima (Poir.) Stapf & Hubb.], but was readily established in the less vigorous 'Greenalta limpograss'. Young (1979) collected, isolated, and partially characterized the allelochemicals in the root exudates of bigalta and greenalta limpograss. Growth of greenleaf desmodium seedlings and cuttings were inhibited as much as 75-85 % in pots by exudates from bigalta limpograss. Root residues of bigalta limpograss, greenalta limpograss, and greenleaf desmodium added to soil increased populations of some fungi and bacteria and soil levels of the enzymes amylase, cellulase, invertase, and dehydrogenase. Lowest levels of fungi were observed in pots containing residues from bigalta limpograss. Levels of the four enzymes were highest in soil containing greenleaf desmodium root residues and lowest in soil containing bigalta limpograss and in the control. Young also observed autotoxic effects of inhibition of greenleaf desmodium seedling growth by exudates from established greenleaf desmodium cuttings.

Some allelopathic interactions among grasses and legumes include: maize--var. Minnhybrid 504 affecting growth of primary root growth of peas (Pisum sativum L.), soybeans [Glycine max (L.) Merr.], and beans (Phaseolus vulgaris L.) (LeTorneau et al., 1956); timothy (Phleum pratense L.) on soybean and alfalfa (Nielson et al., 1960). Root extracts of johnsongrass, columbus-grass (Sorghum almum Parodi.), bermudagrass, dallisgrass (Paspalum dilatatum Poir.), and bahiagrass (P. notatum Fluegge) affected germination of various clovers and crownvetch (Coronilla varia L.) (Hoveland, 1964); cogongrass [Imperata cylindrica (L.) Räusch.] inhibited Stylonsanthes guianensis (Sajise & Lales, 1975). Rice (1964) showed that fresh extracts from a number of grasses inhibited one or more species of N_2-fixing (Rhizobium leguminosarum ATC-10314 and R. sp. ATC=10703) or nitrifying bacteria. Zagallo and Bollen (1960) and Malone (1970) reported that Festuca spp. produced compounds that were toxic to soil microbes. The rhizosphere of fescues had many fewer microorganisms associated with them. A number of these grasses are grown here; one needs to ask, "What other kinds of interactions are taking place that are putting the legume at a competitive disadvantage?"

RHIZOBIUM

Although we take the need for proper Rhizobium as a matter of course, our soils do not always have the most effective strains present. Rotar et al. (1967) pointed out that the use of proper strains made the difference between successful and nonsuccessful plantings. Early on in the history of modern pasture improvement in Hawaii we had many legume establishment failures due to the fact that many introductions were tested on ranches without the realization that certain legumes were strain specific in their requirements. One of the greatest tragedies occurred when we imported large quantities of Kenya white clover (Trifolium semipilosum Fresen.) for distribution to the ranchers in the Waimea District on the Island of Hawaii. Kenya white clover was planted without the realization that it was strain specific and would not effectively nodulate with the strains effective on Trifolium repens L., consequences were disastrous. Kenya white clover never became established as a pasture plant although it is well-adapted.

Not infrequently, seeds are being imported without requesting that inoculum be supplied with them. All of our seed comes from overseas suppliers, Australia, New Zealand, The U.S. Mainland, and elsewhere. It comes by surface delivery which may take as much as 3 to 6 months. Seeds survive the journey in warm holds of the ship, but one must question the ability of the inoculum to survive such journeys. These materials are fragile and need careful handling.

MANAGEMENT

Provision of supplemental feeds during part of the year or during an extended dry period would significantly increase overall carrying capacity, but problems of feed production have never been adequately solved. Various ranches have tried growing many different animal feeds with limited success. Areas suitable for cropping are a long ways from grazing lands. Ranch operations were never oriented toward crop production, lacking equipment as well as arable land. High costs of producing, hauling, and storing locally grown forages and feed grains have led to the failure of most large-scale attempts to grow supplemental livestock feedstuffs. As a consequence many pastures have been severely overgrazed during the dry season and legumes which were formerly present have been either damaged or destroyed.

Where fertilizer N was used, ranchers frequently had an excess of forage which was either soiled by animals or became too rank and was left ungrazed. Cross-fencing for animal control to utilize benefits from N fertilizers was nonexistent. With the use of intensive grazing this situation is changing.

Much of our difficulty in maintaining legumes in pastures comes from the fact that until recently most of the ranchers involved were graziers who did not understand the need for integrating pasture management with animal management practices. Pasture management was nearly nonexistent. Paddocks were large, animals were turned loose on pasture when the grass was knee-deep and were removed after everything was gone. Cross-fencing for animal control was seldom practiced. Conservation of excess forage for silage or hay was unknown. Farm machinery for grassland management was unavailable in the Islands. When machinery was brought in, there was no readily available backup support for repairs and maintenance.

Pasture improvement consisted of replanting with improved strains of grasses and legumes imported from overseas. Some of our best successes were with buffelgrass, guineagrass, kikuyugrass and pangolagrass. Trefoils and clovers were successfully introduced in areas that had been recently cleared or where the existing sod had been turned under. Mineralization of the decaying organic residue provided needed plant nutrients for the first season; however, once these readily available supplies were exhausted the legume component disappeared.

After one or more failures, ranchers became disenchanted. A great deal of money had been invested and lost. In the ranchers defense we should point out that the necessary support system was not really in place. At the Experiment Station level we are just learning about these complex plant/soil/animal/environment interactions and have been unable to show the rancher that we could do better. Even at our main animal station we rely primarily on bagged N. Fertilizer recommendations are easy to make, but to transcribe them into actual practice and for a profit is not easy. The fiscal outlay for 100 kg P ha^{-1} on 500 or 1000 ha is large. Added to that are costs of distribution and application. The technology for aerial application of fertilizers has been well-worked-out by the sugar industry, yet we haven't been able to do the same for ranching. Up until recently, the industry itself wasn't ready to consider grassland management as part of ranch management. In the past 6 yr intensive grazing management has taken over and now we are being faced with many new and unkown plant/animal stresses.

The human factor in the plant/soil/animal/climate interaction is also often overlooked; however, if we examine it carefully we may understand why the manager does what he thinks is best when we think that he should be doing something else. In other words, it's much easier to look over the neighbor's fence and tell him how to do it, than it is for us to try and do it.

We did one study (unpublished) in which we attempted to establish a set of legume introductions into an existing kikuyugrass pasture. The work was carried out at the Mealani Experiment Station. The soil there is classified as a Maile silty clay loam and is a Typic Dystrandept. We used three fertilizer treatments plus a control; P100 applied as single superphosphate in late November 1971 plus P100 in late 1974; P100 in late 1971 only; and P100 in late 1971 and P250 in late 1974. Legumes included nine accessions of *D. intortum*, siratro, clarance and tinaroo glycine, Schofield stylo and emerald crownvetch. Existing sod was sprayed with dowpon (11 kg (a.i.) ha^{-1}) and disked before seeding. After a 6-month establishment period the paddocks were grazed at bimonthly intervals such that the animals removed about 85% (visual estimate) of the existing material on offer within a week's period. We lost all of the legume accessions in the first year except for five *Desmodium* selections. The surviving selections included greenleaf desmodium. When the trial was terminated in 1978, the five entries were still present and were contributing to the pasture. We don't know why we lost the remaining nine within the first year nor do we know why the five accessions were able to compete so well. Many of our ranchers have had the same experience. They appreciate the value of the legume contribution but haven't been able to capitalize on it.

Research Support

In looking at the pasture management literature relating to Hawaiian conditions I counted over 400 references in various categories. Although there were perhaps 40 or 50 that dealt with legume production and adaptation, I was unable to find a single reference which specifically addressed the issue of legume persistence. We have had little Experiment Station support in this area.

Part of this may be the fault of the researcher in not being able to convince his administrators that his research is worth the support. Part of the problem also lies in the way projects are funded. Three-year projects are easy to handle and there is a fairly rapid turnover of research funds. However, in looking at problems of legume persistence, one needs several years in which to get an adequate sample of environmental effects before any answers can be given. In Hawaii, much of the Experiment Station Research is commodity oriented with an industry analysis review which looks at the problems, establishes responsibilities for solving problems and sets priorities for research funding. Funding is handled through the Governor's Agricultural Coordinating Committee (GACC). The industry itself has a say in the analysis and votes on ranking of problems for support. The GACC determines which priorities are to be supported and how much money will be spent for each commodity. This is an excellent way of handling the large number of commodities, however, there are some drawbacks. In the beef cattle industry analysis program, range and pasture research is usually ranked lower than problems related to animal production, feed lots, and marketing. Monies available for any given commodity rarely reach down the list beyond the third or fourth item.

Agriculture in Hawaii is declining in importance and no longer has the influence it once did. We are also a small state with limited resources and livestock is playing a declining economic role.

Land Ownership

This is a very complicated issue but is very real. Much of our grazing land is lease-hold in which the terms of the lease are most favorable to the landlord. Some of the problems include:

a) Short-term leases in which the lessee is unable to amortize development costs (fertilizers, fences, and water supply) over a sufficiently long period of time to make a profit. Any improvements that might be made over time are carefully considered as the lease period begins to run out. Tenants are not willing to make investments which require long-term amortization if they cannot recoup their investments.

b) The way land is put out for bid; the current lessee is not given an opportunity for right of first bid. Bids are raised to the point where it is not feasible for the lessee to continue and the land changes over to a new lessee who may or may not be able to do as good a job. This then discourages the current tenant from making any improvements during the last couple of years of the lease. Needed investments in fertilizers, fences, and etc. are not made. Grazing management is nearly non-existent.

c) Lack of certification of intended lessees, i.e., that they are indeed capable of doing the job and that they have the resources needed. Bidding has been run-up to the point where it is impossible to make an adequate return on the land from grazing.

d) In some of the large land holdings, part of the land may consist of worthless pahoehoe or aa lava. No consideration, however, is given for the fact that much of the land cannot be used, yet the leased fee is the same for raw lava land or steep gullies as well as for land with usable soil; hence the effective cost of the usable land may be doubled.

e) Absentee ownership of land influences ranch managers in many ways. Stock holders are interested in short-term profits which are often inimical to sound pasture management; their perceived alternatives to long-term investments in fertilizers etc. is to develop the land for urban use.

HUMAN MOTIVATION

Many of our ranchers are actually part-time ranchers with other occupations which provide them with their needed income. For many, living on a ranch is a way of life. They may make improvements which are easy; however, they are not always willing to invest in improvements which may be risky or which they do not understand or which may require much effort. In such cases status quo is the rule.

POSSIBLE SOLUTIONS

As research faculty and administrators many of us have been (and still are) wedded to parametric statistical approaches to problem solving. Costs of such research are prohibitive; quite frequently valuable research proposals are passed over in favor of less costly proposals. We need to exploit other means of problem solving. We need to be sure that we are getting the most information possible from our research dollar. Quite often, experimental trends obtained from simple experiments are more valuable than definitive results from complicated experiments. The need to know trends in plant and animal responses over seasons is more important than to definitively determine if intensive grazing management is better than ordinary rotational or continuous grazing.

Existing information needs to be put together in a fashion such that previous work is not needlessly repeated. Existing information must also be made accessible to our clientele. How much previously gained information is being effectively used or is it collecting dust while sitting on a shelf? Responsibility for effective dissemination of information lies with researchers as well as with their administrators. We also need to be able to point out the fact that there is no information available or that what's available is not directly applicable; and having done this be able to so convince our clientele that the much needed support will be actually forthcoming.

Modeling of processes is a very quick way to determine what information is needed. Modeling procedures do not have to be very sophisticated before obvious needs become apparent. Such procedures may even tell us that we are asking the wrong questions!

REFERENCES

Anonymous. 1970. Soil taxonomy. USDA Handb. 436. U.S. Gov. Print. Office, Washington DC.

Asghar, M., and Y. Kanehiro. 1981. The fate of applied iron and manganese in an Oxisol and Ultisol from Hawaii. Soil Sci. 131:53-55.

Fox, R.L. 1967. Phosphorus fixation by Hawaiian Soils and what to do about it. p. 28-41. In Proc. First Ann. Fertilizer Conf. Coop. Ext. Serv., Univ. of Hawaii. Misc. Publ. 39.

Fox, R.L., and E.J. Kamprath. 1970. Phosphate sorption isotherms for evaluating the phosphate requirements of soils. Soil Sci. Soc. Am. Proc. 34:902-907.

Fox, R.L., R.K. Nishimoto, J.R. Thompson, and R.S. de la Pena. 1975. Comparative external phosphorus requirements of plants growing in tropical soils. p. 232-239. In X Int. Congr. Soil Sci., Trans. IV. Moscow.

Hori, T.M. 1970. "Soft rot" of dry onion caused by copper deficiency. Coll. of Trop. Agric., Univ. of Hawaii. Maui County Leaf. no. 2.

Hoveland, C.S. 1964. Germination and seedling vigor of clover as affected by grass root extracts. Crop Sci. 4:211-213.

Kanehiro. Y. 1964. Status and availability of zinc in Hawaiian soils. Ph.D. Diss. Dep. of Agronomy and Soil Science, Univ. of Hawaii. (Diss. Abstr. 25:2683).

LeTourneau, D., G.D. Failes, and H.G. Heggeness. 1956. The effects of aqueous extracts of plant tissues on germination of seeds and growth of seedlings. Weeds 4:363-368.

Malone, C.R. 1970. Short-term effects of chemical and mechanical cover management on decomposition processes in a grassland soil. J. Appl. Ecol. 7:591-601.

McWilliam, J.R. 1978. Response of pasture plants to temperature. p.17-34. In J.R. Wilson (ed.) Plant relations in pastures. CSIRO, Melbourne, Australia.

Munns, D.N., and R.L. Fox. 1976. Depression of legume growth by liming. Plant Soil 45:701-705.

Munns, D.N., and R.L. Fox. 1977. Comparative lime requirements of tropical and temperate legumes. Plant Soil 46:533-548.

Munns, D.N., R.L. Fox, and B.L. Koch. 1977. Influence of lime on nitrogen fixation by tropical and temperate legumes. Plant Soil 46:591-601.

Nielson, K.F., T. Cuddy, and W. Woods. 1960. The influence of the extracts of some crops and soil residues on germination and growth. Can. J. Plant Sci. 40:188-197.

Olney, V.W. 1969. Trace elements used in diverse agricultural crops in Hawaii. p. 55-59. In Third Ann. Fertilizer Conf. Proc. Univ. of Hawaii, Coll. of Trop. Agric., Coop. Ext. Serv. Misc. Publ. 58.

Plucknett, D.L. 1970. Productivity of tropical pastures in Hawaii. p. A38-A49. In Proceedings XI Int. Grassland Congr. Univ. Queensland Press, Surfers Paradise.

Plucknett, D.L., and R.L. Fox. 1965. Effects of phosphorus fertilization on yields and composition of pangolagrass and Desmodium intortum. v.2. p. 1525-1529. In Proc. 9th Int. Grassland. Congr., EMBRAPA, Sao Paulo.

Rice, E.L. 1964. Inhibition of nitrogen-fixing and nitrifying bacteria by seed plants I. Ecology 45:824-837.
Ripperton, J.C., and E.Y. Hosaka. 1942. Vegetation zones of Hawaii. Univ. of Hawaii, Hawaii Agric. Exp. Stn. Bull. 89.
Rotar, P.P. 1970. Variation in agronomic characteristics of Desmodium intortum (Mill.) Urb. and a related species. p. 296-300. In Proc. XI Int. Grasslands Congress, Univ. Queensland Press, Surfers Paradise.
Rotar, P.P., Y.N. Tamimi, O.R. Younge, and T. Izuno. 1976. Forage legume production trials at the Volcano Research Station, Island of Hawaii. Univ. of Hawaii, Hawaii Agric. Exp. Stn. Res. Rep. 206.
Rotar, P.P., U. Urata, and A. Bromdep. 1967. Effectiveness of nodulation on growth and nitrogen content of legumes grown on several Hawaiian soils with and without the use of proper Rhizobium strains. Univ. of Hawaii, Hawaii Agric. Exp. Stn. Tech. Prog. Rep. 158.
Sajise, P.E., and J.S. Lales. 1975. Allelopathy in a mixture of cogon (Imperata cylindrica) and Stylosanthes guianensis. Phillip. J. Biol. 4:155-164.
Sherman, D.G., and C.K. Fujimoto. 1948. The effect of the use of lime, soil fumigants, and mulch on the solubility of manganese in Hawaiian soils. Soil Sci. Soc. of Am. J. 11:206-210.
Tamimi, Y.N., and D.T. Matsuyama. 1972. The effect of fertilization on establishment and productivity of legumes. II. Effect of nitrogen, phosphorus and potassium on yield of kikuyugrass and pangolagrass in combination with big trefoil. p. 71-78. In Proc. 8th Annual Beef Cattle Field Day. Univ. of Hawaii, Coop. Ext. Serv., Misc. Publ. 97.
Tamimi, Y.N., and D.T. Matsuyama. 1975. Fertilization of grass-legume mixture. p. 30-35. In Proc. 9th Beef Cattle Field Day. Univ. of Hawaii. Coop. Ext. Serv. Misc. Publ. 128.
Tamimi, Y.N., D.T. Matsuyama, and C. Ritter. 1976. Preliminary results on fertility studies at Kohala. p. 34-38. In Proc. 10th Beef Cattle Field Day. Univ. of Hawaii, Coop. Ext. Serv. Misc. Publ. 143.
Tamimi, Y.N., and J.R. Thompsom. 1972. The effect of fertilization on establishment and productivity of legumes. I. Effect of phosphorus and lime rates on growth of ladino clover. p. 65-70. In Proc. 8th Annual Beef Cattle Field Day. Univ. of Hawaii, Coop. Ext. Serv., Misc. Publ. 97.
Whitney, A.S. 1982. Legumes for Hawaii pastures. p. 9-10. In Proc. 17th Mealani Beef Cattle Day. HITAHR. Coll. of Trop. Agric. and Hum. Res., Univ. of Hawaii, Res. Ext. Ser. 23.
Yost, R.S., and R.L. Fox. 1981. Partitioning variation in soil chemical properties of some Andepts using Soil Taxonomy. Soil Sci. Soc. Am. J. 45:373-377.
Yost, R.S., G. Uehara, and R.L. Fox. 1982. Geostatistical analysis of soil chemical properties of large land areas. II. Kriging. Soil Sci. Soc. Am. J. 46:1033-1037.
Young, C.C. 1979. Allelopathy in a grass-legume association: A case study with Hemarthria altissima and Desmodium intortum. Ph.D. Diss. Dep. of Agronomy and Soil Science, Univ. of Hawaii.(Diss. Abstr. 40:1018B).

Younge, O.R., and K.K. Otagaki. 1958. The variation in protein and Mineral composition of Hawaii Range Grasses and its potential effect on cattle nutrition. Univ. Hawaii, Hawaii Agric. Exp. Stn. Bull. 119.

Zagallo, A.C., and W.B. Bollen. 1960. Studies on the rhizosphere of tall fescue. Ecology 43:54-62.

DISCUSSION

Hay, J.: What problems are associated with tall fescue when grown with clovers (legumes) as companion species? Is it because of the associated endophytes? In New Zealand we have no problems with white clover grown in association with tall fescue which lacks the endophyte.

Barnes: What have been the effects of Festuca species toxicity to soil microbes?

Hay, J.: We have not specifically looked at the problem. Dr. Carl Hoveland is currently investigating this. At this early stage it appears that high-endophyte tall fescue affects soil microbes which indirectly affect legume growth. It may possibly be due to malic and other organic acids.

Sheath: If tall fescue has a depressive effect on associated legumes is this because the endophyte is making tall fescue more aggressive and competitive or is the endophyte directly mediating an allelopathy effect.

Hoveland: Our research shows that genetically pure clones of endophyte-infected fescues, as compared to endophyte-free clones, are higher yielding, much more drought tolerant, and more responsive to N fertilization. Other work in progress indicates that legume compatibility is poorer with endophyte-infected fescue but we do not know if it is due to competition from the grass or from allelopathy.

Matches: We have found that tall fescue exhibits allelopathic effects on birdsfoot trefoil (Lotus corniculatus L.) and that it appears to be due to certain organic acids. We don't know if it is related to the endophyte of tall fescue. Red clover does not appear to be very sensitive to these allelopathic effects. It is definitely an inhibitory factor and not just due to plant competition from tall fescue.

Reed: Was the inhibitory effect from tall fescue on birdsfoot trefoil due to a high-endophyte tall fescue?

Matches: We do not know; we think it was probably a high-endophyte line of tall fescue.

Irwin: Are the disease problems on pasture legumes such as Stylosanthes and Macroptilium, the same as those in that exist in Australia; i.e., anthracnose on Stylosanthes and rust on Macroptilium?

Rotar: Siratro's failure to persist was mainly due to Rhizoctonia which is a soil borne pathogen.

Brougham: Is the difference in Siratro persistence between Australia and Hawaii due to differences in soil type?

Rotar: Soil properties do influence the susceptibility of plants to Rhizoctonia.

Martin: Four years ago you had hopes for use of Leucaena and other shrub legumes for Hawaii; what is their latest status?

Rotar: The Psyllid wasp has become a serious problem on Leucaena leucocephala, however, there are psyllid resistant leucaenas.

Brougham: Can we use the Hawaiian Islands for collaborative research?

Rotar: Yes the opportunities are here, details need to be sorted out.

El-Swaify: Yes, there are wide opportunities for research. One has the possibility of capitalizing on the environmental and edaphic diversity which expresses itself very well over short distances. For instance adaptation studies, persistence trends alone or in competition, and effects of ecological settings on performance --, all can be investigated and systematically modeled within manageable distances and at well-documented locations. We can mimic many locations globally in "laboratory-like" settings. There are many possibilities which lend themselves well to comparative studies within a collaborative network.

GENERAL DISCUSSION OF OVERVIEW OF PROBLEMS WITH LEGUMES

The moderator, in opening the session for discussion, suggested that there were common threads running through the four overview papers presented. All authors indicated that there was a need for a wider scope of better adapted plants to fit the complex range of environments prevailing in each country. This meant the need for improved germplasm through species and variety introductions or plant breeding using both conventional and novel techniques.

The meeting participants and authors also agreed that the major factors common to all countries, with respect to the persistence of forage legume plants, and hence productivity and viability of farming were: climate, edaphic, pests and diseases, management, rhizobial presence and vitality, reseeding mechanisms, socio-economic and political. With respect to management it was agreed that some controlling factors differed in expression, depending upon established management or management of perennial pastures. Significant management components were competition between plants, severity and intensity of grazing, and fertilizer management.

Discussion then followed on legume persistence in relation to suitability and adaptability.

Sheath: For New Zealand environments, there are marginal zones where legumes fail to persist, but in most environments, persistence is not a problem. In marginal areas, other legumes should be investigated. Where persistence is satisfactory, the major future stress is fertility reduction due to economic pressure.

Scott: It should be stressed that fertilizer and management were a significant part of the environment.

Berberet: The question was asked, "How do we define expectation in terms of legume persistence?" It is suggested that various responses indicate that expectations may be quite different depending on locality and use by producers. For the USA, it is difficult to make generalizations and particularly quantification (stand longevity, as an example) as goals.

Watson: A further comment can be made concerning differing limitations and persistence during the "pioneering" introduction phase and the "maintenance" phase. During the first phase, limitations are likely to be edaphic and rhizobial, while during the second phase, pest and disease factors.

Hochman: With respect to the term economic, it should be stressed that we need to distinguish between marginal returns, cash flow, and risk factors. Any one of these factors may lead to non-adoption of research results.

Buxton: A major effort should be made among researchers from the three countries to identify the key legumes for specific environmental regimes. This approach would allow more in-depth effort on a limited, more manageable number of legumes in order to match species and environments.

Hochman: With respect to the critical legume content of swards, for subclover (*Trifolium subterraneum* L.) in Australia percentage content is important.

Curll: In determining or managing legume content in swards the key is "for what purpose is the legume required."

Marten and Reed: Some argued for a 100% legume content.

Sheath and Brougham: Others argued that this was not possible nor desirable in some environments. In most New Zealand environments for example, highest production is obtained from mixed grass-legume associations. Legume contents range from 15 to 30%.

Kretschmer: An example of where 100% legume content can be deleterious to animal production is kudzu (*Pueraria* spp.).

Rotar: I would like to emphasize the importance of monitoring the environment to establish a benchmark for studies on the soil-plant-environment interactions.

SUMMARY

This discussion concluded with the comment that percentage composition of legumes in a pasture depends on legume species and the reason for the pasture. For example, annual medic pastures in cereal rotations across South Australia were used to fix N as a break crop, and to help control cereal diseases and improve soil structure and animal production. In these situations, pastures approaching 100% annual medics are desirable.

THE ADAPTATION, REGENERATION, AND PERSISTENCE OF ANNUAL LEGUMES IN TEMPERATE PASTURE

K.F.M. Reed, M.J. Mathison, and E.J. Crawford

SUMMARY

The persistence of annual legumes in temperate pasture is reviewed with emphasis on the long-term pastures in southern Australia sown to *Trifolium subterraneum* or *Medicago* spp.

The importance of a high seed yield and hard-seededness are stressed as are the opportunities to evaluate a wide range of alternative species and suitable *Rhizobium* so that legumes for diverse environments, including problem soils, can be developed.

SIGNIFICANT GENERA AND SPECIES

The International Board for Plant Genetic Resources has established a network of centres which collects, documents, and uses plant germplasm to raise standards of living throughout the world. Some centres for genetic resources of annual legumes were established in Australia at Adelaide (for calcareous soils) and Perth (for acid soils) where there are national breeding programs; these centres have contributed greatly to our present knowledge.

Apart from their value for fixing N, recent interest in legumes has emphasized their quality for sustaining high levels of animal health, production, and reproduction (Reed, 1981). More than 4000 legume species of forage are known. Approximately 100 annual species have been listed by various reviewers as of significance for grazing by domesticated livestock. Many other species of unproven value have been extolled by various authors. Of the many annual species used for pasture, only eight have five or more cultivars registered for international trade by the OECD (Table 1).

Many annual legumes are sown as a 1- or 2-yr fodder or green manure crop in rotation with field crops. Only a few are associated with annually regenerating pasture for sustained production under close grazing [e.g., subterranean/sub. clover *(T. subterraneum*) and annual medics (*Medicago* spp.)]. There has proved to be a great need for these latter type of species in Australia, due to low summer rainfall, poor soil, and a rural economy based on low cost year-round grazing. These species were developed in Australia and development continues. Moore (1970) delineated the areas where these species grow in Australia. His area for medics is conservative (e.g., Clarkson, 1977) and, from maps of the soil groups and climatic records, we estimate the potential area suited to annual medics in Australia may approach 70 M ha (40 in the semi-arid zone and 30 in the cereal cropping zone). Interest is focussed not only on the further improvement of these species (Collins & Gladstones,

1985; Crawford *et al.*, 1988), but on the screening of a comprehensive collection of alternative legume species for pasture (Ellis *et al.*, 1985). Thus, current developments in Australia reflect an ever-increasing list of species and cultivars (Table 2).

In the USA, sub. clover is also the most important annual legume with more than 500 000 ha sown (McGuire, 1985). *Trifolium incarnatum* (120 000 ha), *T. vesiculosum* (80 000 ha), and *T. hirtum* (in California) are also of importance as are *Vicia* and *Lespedeza* spp. (Taylor, 1985). The annual *Lespedezas* also reseed naturally from hard seed resulting in good stands year after year where interspecies competition is not too severe.

In New Zealand, approximately 100 t of sub. clover seed is used annually. A cultivar of *Ornithopus sativus* has been released recently by DSIR. Other naturalized annual *Trifolium* spp. contribute in dry hill country and some annual medics (e.g., *M. polymorpha*) contribute in the Hawkes Bay region.

DISTRIBUTION

The Mediterranean and temperate species of forage legumes and their distribution have been described in some detail by Mathison (1983). They are indigenous mainly in the northern hemisphere around the Mediterranean Sea and eastwards to central Asia and/or in the temperate climatic zones of Europe, Eurasia, and eastern Asia.

Deliberate introduction of cultivated species in the new world has led to some important developments in the diversity of genotypes now available. Also, local environmental pressures have been most important for developing the unintentionally introduced species (e.g., sub. clover, Gladstones & Collins, 1983).

The geological and human influences on distribution of legumes have been reviewed by Mathison (1983) who cites evidence of *T. alexandrinum* being deliberately cultivated between 3500 and 3800 yr B.C. The common use of fertilizer, agricultural chemicals, and the sowing of monocultures is a comparatively recent phenomenon and greatly alters the environment in which legumes grow, reproduce, and perhaps eventually decline.

The need to protect refugia and to maintain genetic diversity is urgent. Sources of material tolerant to serious insect pests and, in the higher rainfall areas, to diseases, have proved vital for the breeding/selection of, for example,

* various legumes in California threatened by the Egyptian alfalfa weevil [*Hypera brunneipennis* (Boheman)]
* annual medics in Australia threatened by the spotted alfalfa aphid (*Therioaphis trifolii* fm. *maculata*) and the blue green aphid (*Acyrthosiphon kondoi*) and,
* sub. clover in Australia vulnerable to clover scorch (*Kabatiella caulivora*) and root rot (*Phytophthora clandestina*).

Within species, character-origin connections may sometimes be apparent in association with attractiveness to pollinating insects. Mathison (1983) cites examples of migration from the Mediterranean region, associated with decreased leaf-mark polymorphism, increased self-pollination and changes in corolla tube length in perennials. Further study may reveal connections in annuals, including inbreeders. For example, considerable development of new strains of sub. clover has occurred in South Australian pasture after natural hybridization (Cocks & Phillips 1979; Cocks *et al.*, 1982).

Table 1. Species of annual legumes commonly used for pasture (after
Where developed and/or cultivated: Aus, Australia;
X, extended rainfall temperate; W, winter rain;
B, brief wet season (tropical/subtrop.); I, irrigated;
S, shallow; W, well-drained; M, moist; A, alkaline;

Tribe genus and species	Habit[1]	Breeding system	No. of OECD cultivars (Anon., 1986)
GALEGEAE			
Astragalus hamosus	p	i	0
DESMODIEAE			
Lespedeza stipulacea	ep	io	2
Lespedeza striata	ep	i	0
VICIEAE			
Vicia benghalensis	e	io	0
Vicia pannonica	e	i	2
Vicia sativa	ept	io	46
Vicia villosa	ep	io	17
Lathyrus cicera[2 3]	ep	i	5
TRIFOLIEAE			
Melilotus alba[1]	e	io	2
Melilotus officinalis[1]	e	o	3
Trigonella foenum-graecum	e	o	0
Medicago arabica	ep	i	0
Medicago littoralis	ep	i	1
Medicago lupulina	ep	io	1
Medicago murex	p	i	0
Medicago polymorpha	ep	i	0
Medicago rugosa	ep	i	2
Medicago scutellata	ep	i	1
Medicago tornata	ep	i	1
Medicago truncatula	ep	i	5
Trifolium alexandrinum	e	o	10
Trifolium balansae	ep	o	0
Trifolium hirtum	e	io	0
Trifolium glomeratum	p	o	0
Trifolium incarnatum	e	io	13
Trifolium nigrescens	pt	o	0
Trifolium purpureum	e	o	0
Trifolium resupinatum	e	io	8
Trifolium subterraneum	pt	i	16
Trifolium vesiculosum	e	o	3
CORONILLEAE			
Ornithopus compressus	ep	i	0
Ornithopus pinnatus	ep	i	0
Ornithopus sativus	ep	io	1

[1]Habit: e, erect; p, prostrate; t, trailing or twining.
Breeding system: i, inbreeding; o, outbreeding.

Whyte *et al.*, 1953; Mathison, 1983).
Am, America; O, origin. Climates: P, polar and subpolar; S, summer rain; L, long wet season (tropical/subtrop.); D, drought-tolerant. Soils: L, light; H, heavy; C, calcareous; Ac, acid. Origin: Cauc, Caucasus, C, Central, Med, Mediterranean.

Where developed and cultivated	Soils	Climates	Origin of species
Aus	HCWA	XWD	E Med,W Asia,Cauc.
Am,O	Ac	WSLBD	E Asia
Aus,Am,0	Ac	WSLBD	E Asia
Aus,0	Ac	XWS	S Eur.
Eur,0	HMAc	XWS	W Asia,Med,Eur,Cauc.
Eur,Med,Am,SWAsia	WAc	XWSLB	Med,Eur,SW Asia
Eur,Med,Am,Aus	LAAc	XWS	Med,SW Asia
Eur,Med	LS	WSD	Med,SW & C Asia, S Eur,Cauc.
Am,0	HCA	XWSD	N Eur,Eurasia
Am,0	HCA	XWSD	Eurasia
O,Ind	A	WSLB	Med,Eurasia,NE Afr.
Am	CWMAAc	W	Eur,Med
Aus,Med	LWCA	W	Med,NW Asia
Eur,Asia,Am	CWA	PXWS	Eurasia,N Afr.
Aus	LHAc	W	Med
Aus,O,Am,	CWAAc	W	Med,Eurasia
Aus,0	HCWA	W	Med
Aus,0,Eur,Asia	HCWA	W	Med
Aus,0	LCWA	W	Med
Aus,0	HCWA	W	Med
Med,India	HA	WSLBI	E Med
Aus	HWMAc	W	W Asia
Aus,Am,O	Ac	XWS	SW Asia,Med
Aus	LHAc	XWS	WS Eur
Eur,Med,Am	LHAc	XWSLB	WC Eur
Am	HM	XWS	E Med,W Asia
Aus	M	XW	E Med,SW Asia
Eur,Med,0,Aus,Am	HMAAc	WSLBI	W Asia,E Med
Aus,0,Am,NZ	HMAc	WBI	Med
Am,0	W	XW	SCE Eur,Cauc
Aus	LWAc	XW	Med
Aus	LWAc	W	N Afr
Eur,Aus,O,NZ	LWAc	XW	SW Eur

[2]Maybe biennial. [3]Maybe perennial.

Table 2. Annual pasture legumes in Australia.

Species	Certified seed produced 1986/87 (t)	No. of cv. certified in 1986/87	No. of new cv. to be registered in 1988
T. subterraneum			
ssp. subterraneum	3230	15	0
ssp. yanninicum	1453	3	0
ssp. brachycalycinum	327	1	1
M. truncatula	1273	7	0
M. polymorpha	583	2	1
T. resupinatum	450[1]	1	0
T. balansae	427	1	0
M. scutellata	218	1	1
M. littoralis	118	1	1
M. rugosa	40	3	0
V. benghalensis	5	1	0
O. compressus	3	1	4
M. tornata	-	1	1
O. pinnatus	0	-	1
M. murex	0	-	1
TOTAL	8127	38	11

[1]Mainly uncertified seed.

ADAPTATION TO SOILS

Adaptation to general categories of climates and soils are given in Table 1. The differences in adaptation by *Trifolium* and *Medicago* spp. in their general preference for acidic and alkaline soils, respectively, is well recognized. Recent rhizobial studies, however, have re-emphasized the acid-soil tolerance of *M. polymorpha* and *M. murex* (Howieson & Ewing, 1986; Gillespie, 1987). Examples of nutritional adaptation involving a succession of species of *Ornithopus* and *Trifolium* with increasing levels of soil fertility are given by Hely *et al.* (1976). Differences in the depth and development of root systems commonly underlie differences between legume species in their requirement for specific nutrients (Williams & Andrew, 1970). In his review, Mathison (1983) refers to work on Ca absorption by the fruit of various legumes, especially sub.clover (Robson & Loneragan, 1978). It is most important to consider the broad soil groups (e.g., Hubble, 1970) to appreciate the geographical distribution likely with increasing rhizobia specificity, and the further development of alternative legumes.

Tolerance to low levels of Mo and high levels of Mn at low pH also characterizes clovers, while ability to absorb Zn at high pH may characterize medics; tolerance to Al may also be involved (Robson, 1969). Among sub. clovers, subspecies *T. brachycalcycinum* has demonstrated better adaptation to heavy textured alkaline soils than has subspecies *T. subterraneum* ; this too may be associated with the ability to absorb Zn at high pH (Millikan, 1963).

Drainage affects the distribution of species. The adaptation of the subspecies *T. yanninicum* cv. Yarloop to poorly drained sites is well known. Its success is associated with high activity of the enzyme alcohol dehydrogenase (which prevents the accumulation of ethanol) and with shallow distribution of roots (Francis *et al.*, 1974). Yarloop is capable of greater substrate oxidization by roots (Katznelson, 1970). Because of a high oestrogen content in cv. Yarloop (Beck, 1964) aerial-seeding species of *Trifolium* were examined as alternatives on waterlogged lateritic podsols (Beale and Crawford, 1972). Subsequent work identified 19 *Trifolium* spp. with good winter growth and resistance to the clover scorch fungus (Beale & Crawford, 1975). Of these, *T. resupinatum* and *T. balansae* have been commercialized and a wide collection of species for four major soil types is being assessed (Ellis *et al.*, 1985). More than 100 *Trifolium* species were also grown at Grafton in New South Wales; 12 of the annuals listed for further evaluation included three of the species noted by Beale and Crawford (1975), viz. *T. isthmocarpum, T. constantinopolitanum,* and *T. resupinatum* (Wilson, 1979).

Medicago polymorpha tolerates poor drainage better than other medic species (Andrews & Hely, 1960). Francis and Poole (1973) found the most tolerant medics to flooding were *M. polymorpha, M. arabica,* and *M. intertexta*. *Trifolium clusii* and *T. ornithopodioides* appear to be adapted to even wetter soils than cv. Yarloop, and tolerance of *T. dubium* lies between that of cv. Yarloop and other subterranean clovers (P.S. Cocks, 1982, pers. comm.).

A number of legumes are capable of colonizing deep siliceous sand, one of the more important being yellow serradella (*Ornithopus compressus*). This annual Mediterranean species has deep roots which greatly extend the growing season compared with sub. clover; it is noted for its ability to grow well in low-fertility situations (Gladstones *et al.* 1975; Bolland & Gladstones, 1987). Other legumes to grow well in these soils are (*Vicia villosa* ssp. *dasycarpa* (Archer, 1981), *M. tornata,* and several annual species of *Trifolium*.

Micro-environment

A study of a 40 yr-old mixture of sub. clover cv. Mt. Barker and Dwalganup, found that 60% of the individuals differed genetically from the original strains; late-flowering strains colonized the wettest sites and early strains the driest sites (P.S. Cocks, 1982, pers. comm.). The linking of polymorphism in flowering time with site heterogeneity is probably the main reason for the poor relationship, at the regional level, between flowering time and length of growing season, referred to earlier; sometimes however, a reasonable association has been observed (Cocks, 1985). Adaptation to aspect and micro-environment has not been studied in depth in Australia, as it has elsewhere (e.g., Antonovics, 1978).

PHYSIOLOGICAL ASPECTS OF ADAPTATION

Time to Maturity

The development of a range of commercial cultivars which can set seed over a long period has been fundamental in Australian improvement of annual legumes. Generally, the annual medics flower earlier than sub. clover when they are grown at the same centre (Table 3). Flowering time shows considerable evidence of a genotype by environment (temperature; latitude/daylength) interaction (Aitken & Drake, 1941; Devitt *et al.*, 1978). Selection for flowering time has, more than any other characteristic, enabled us to extend areas of adaptation.

Sub. clover may form seed in Australia from late September to mid-December, depending on cultivar and location. Geraldton sub. clover is a cultivar noted for fast seed development. Although flowering a week later than Dwalganup, it produces mature seed about 12 d earlier.

Flowering in sub. clover and annual medics occurs in response to daylength and temperature (Evans, 1959), but cultivars vary considerably in their response to each factor (e.g., Clarkson & Russell 1975; Hochman, 1987). Grazing can delay floral initiation and flowering by up to 30 d (Collins & Aitken, 1970).

Seed Yield

The germination, establishment, and natural regeneration of annual medics and sub. clover in Australia has been reviewed by Carter (1987). The importance of an adequate seed set is paramount (Crawford & Nankivell, 1984; Carter, 1987). Carter claimed that establishment is far more difficult on old farming land compared to newly developed land. He emphasized the role of management in (i) preparing the green pasture for an adequate seed yield, and (ii) minimizing the depletion of the seed bank through sensible management of the dry pasture.

The annual medics are early flowering relative to sub. clover and are used in more drought prone areas in ley farming systems. They, therefore, suffer from a greater risk of low or zero seed yields. Seed yields vary between species and interact with the phase of the ley system. In Syria, Cocks (1985) sowed mixtures and found *M. orbicularis* set the most seed in the year of sowing (light grazing). In Year 2 under crop, *M. polymorpha* set the most seed. When continuous heavy grazing was applied (Year 3), *M. truncatula* had the greatest seed yield.

Seed yield is subject to great environmental influence. Frost and drought can interfere so that the later-maturing strains may set more or less seed, respectively, than earlier ones - a reversal of what may have been expected with average seasonal conditions. In the absence of frost and drought, Francis and Gladstones (1974) found no correlation between time of flowering and seed yield of sub. clover strains at Perth. However, in a survey of colder regions (viz. Victoria) mid- and late-maturing strains of sub. clover were found to be the most common strains persisting (Aitken & Drake, 1941). New Zealand workers have similarly noted better persistence with the later strains (MacFarlane & Sheath, 1984; Chapman *et al.*, 1986). Mid and late strains have also set more

Table 3. Maturity range and other characteristics of Australian cultivars of some Medicago, Ornithopus and Trifolium species.

Medicago and *Ornithopus* species	Days to flowering[1] Adelaide	*Trifolium* species	Days to flowering[1] Adelaide	Relative hard-seededness[2]	Oestrogenic activity
		T. subterraneum			
M. polymorpha		ssp. *subterraneum*			
Serena	75	Nungarin	102	10	Low
Circle Valley	110	Northam	103	8	Low
		Dwalganup	106	7	V. High
M. scutellata		Geraldton	118	8	High
Sava	83	Daliak	120	6	Low
Robinson	83	Dalkeith	122	8-9	Low
		Seaton Park	123	5	Low
M. truncatula		Enfield	123	1-2	Low
Parabinga	97	Junee	123	7	Low
Cyprus	100	Dinninup	124	7	V. High
Borung	110	Green Range	127	5	Low
Sephi	112	Woogenellup	127	3	Low
Paraggio	114	Esperance	128	5	Low-Med
Ascot	115	Karridale	134	3	Low
Jemalong	115	Mt. Barker	134	1	Low
Hannaford	115	Nangeela	138	1	Low
		Tallarook	158	1	High
M. littoralis		ssp. *yanninicum*			
Harbinger	99	Yarloop	122	4	V. High
		Trikkala	126	3	Low
M. rugosa		Larisa	137	2	Low
Paraponto	101	Meteora	144	8	Med.
Paragosa	112				
Sapo	112				
M. tornata		ssp. *brachycalycinum*			
Tornafield	115	Rosedale	113	4-5	Low
Rivoli[3]	126	Clare	130	3	Low
		T. balansae			
		Paradana	138	High	Low
O. compressus		*T. hirtum*			
Pitman	150	Hykon	124	Med	Low
		Kondinin	134	Med	Low

[1] After sowing in early May at Parafield Plant Introduction Centre, S.A.
[2] Scale of 1 (low) to 10 at the break of season in Autumn, at Perth, W.A.
[3] Registration pending.

seed than early/early midseason cultivars in the higher rainfall areas of Tasmania (Evans & Carpenter, 1987) and under heavy grazing pressure in South Australia (de Koning & Carter, 1987).

The seed yield of sub. clover can be considerably influenced by soil Ca which is absorbed directly by the fruit (Ozanne & Howes, 1974). Some strains have a propensity to bury burrs (e.g., Dwalganup). Such strains may develop greater seed reserves under grazing than do "non-buriers" (e.g., Woogenellup) (Rossiter, 1972). Indeed Rossiter attributes the success of sub. clover as a pasture species mainly to its ability to bury seed and to its genetic diversity (Rossiter *et al.*, 1985).

Seed Quality

Hard-seededness or seedcoat impermeability helps to ensure the persistence of annual legumes intermittently under conditions of low or zero seed set. Thus after a drought year, hard seed (seed set in previous years) whose testae had hitherto resisted the penetration of water can re-establish the species (Donald, 1959; Quinlivan, 1971). Some hard seeds form in individual fruits when their water content declines to a low level and when seed maturation is slow (Aitken, 1939). Hard-seededness breaks down in the field when diurnal summer temperature fluctuations cause the testa to crack (Quinlivan & Millington, 1961). Consequently, the hard seeds in buried burrs break down slowly compared to those in surface burrs. Crawford and Nankivell (1984) found that the regeneration of annual medics was better in a permanent pasture system, than in cropping rotations where seed was buried over a 5-yr period. The rate of breakdown of hardseededness in sub. clover may fall dramatically if seed is buried 10 cm (Taylor & Ewing, 1988).

Within sub. clover, there is a wide range in hard-seededness between cultivars. Generally, earlier cultivars produce the hardest seeds. Meteora is a notable exception to this association (Table 3). *Ornithopus compressus* and *T. vesiculosum* have high levels of hard seed and require more scarification than most other annual legumes. More than 90 and 70% of seed of these respective species may be hard at seed harvest. *Trifolium nigrescens* also has a high proportion of hard seed. The release of *T. resupinatum*, cv. Kyambro in Australia is the result of a deliberate search for hardseededness in this species.

Seeds of annual medics are much harder than those of most *Trifolium* spp. More importantly, they are less inclined to "break down" between maturity and the break-of-season. There is some variability within and between species in this rate of break down (Fig. 1) and Paraggio was developed because of its more desirable behavior. Even after 6 months of fluctuating temperature approximately 70% of hard seed of various annual medics may remain hard (Quinlivan, 1968).

Hard-seededness would seem to be of far greater significance to the persistence of annual legumes in a ley-farming system, than in a grazing-only situation. Important questions for some ley farming systems include: What is the rate of field breakdown? What effect has tillage on hard-seededness? How important is natural seed burial in a ley-farming system? From what depth can the buried seeds emerge? In view of the relationship between seed size and hypocotyl length (Black, 1957), possession of large seeds by a

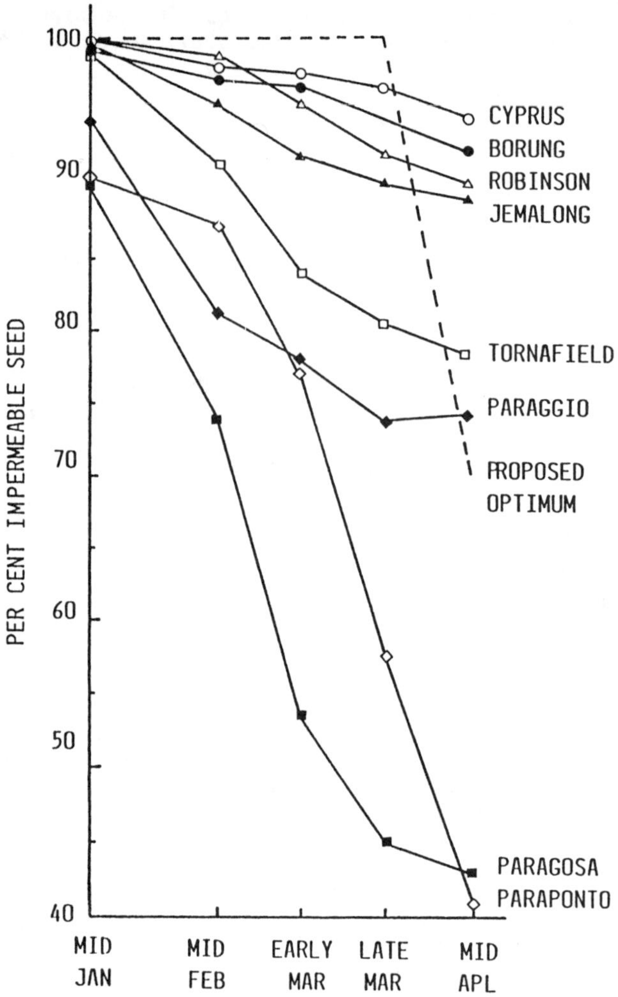

Fig. 1. (From Crawford, *et al.*, 1988)
Seasonal changes in seedcoat permeability in eight commercial annual medic cultivars at Parafield Plant Introduction Centre.

species may be more important than ability to bury seeds. No matter how effective hard-seededness is in protecting seed populations, there can be heavy losses during tillage (Quigley *et al.*, 1987; Heida & Jones, 1988). The ability to produce large quantities of seed may be far more important in cropping systems than in systems of continuous grazing. While much has been written about empirical relationships with environmental and management factors, greater knowledge of the physiological development of hard-seededness is needed to help predict the subsequent behavior of seeds.

Embryo dormancy - a transient character that can delay germination immediately after maturity - can also assist in the survival of annual legumes. Embryo dormancy can be broken by CO_2 enrichment and cold temperature (Ballard, 1961) and by the leaching of seedcoat inhibitors (Taylor, 1970). Prolonged warm temperatures relieve dormancy in the field (Quinlivan & Nichol, 1971). Embryo dormancy varies between strains. It is rare in medics and more pronounced among the mid- and late-maturing cultivars of sub. clover than among the early maturing cultivars (Taylor, 1970). It may represent an adaptive mechanism that has evolved in strains originating from moister/cooler areas of the Mediterranean region where hard-seededness usually fails to develop (Morley, 1958).

Persistence of *T. vesiculosum* is aided by seed germinating over a prolonged period extending into the cold days of winter when temperatures are too low for the germination of e.g., *T. incarnatum* (Hoveland & Elkins, 1965). *Trifolium nigrescens* also has the ability to germinate in cold conditions following a dry autumn.

SYMBIOSIS

Rhizobial specificity is common among temperate and Mediterranean legumes. Knowledge and research on the genetic diversity and ecology of *Rhizobium* is limited, yet the subject is complex. The ultimate usefulness of particular species in areas with appropriate soils and climate cannot be forecast accurately without supporting evidence that suitable Rhizobium are present or available. In Australia, inoculants are derived from pure identified sources so strain x host x environment interactions can be monitored. Recent expansion of areas of *M. rugosa, M. murex, M. polymorpha,* and *T. resupinatum* in Australia has been aided by the development of new strains of appropriate *Rhizobium* (Brockwell & Hely, 1966; Howieson & Ewing, 1986; Cunningham *et al.*, 1986). The deliberate introduction and subsequent isolation of strain CC 280 specifically for *Tetragonobolus purpureus* from Jordan, has greatly altered the potential of this species in current research (E.J. Crawford and J. Brockwell, 1988, pers. comm.). In Syria, endemic strains of *M. rigidula,* inoculated with endemic strains of rhizobia, have shown considerable potential over Australian cultivars and strains (Cocks, 1985; Cocks & Ehrman, 1987). There is a need to improve the supply of N to *T. subterraneum* in some regions of Australia during winter (Cocks, 1980). Strains of cold and/or waterlogging tolerant rhizobia for *Medicago, Ornithopus,* and *Trifolium* spp. amongst others, have been proposed to improve species adaptation in South Australia (M.J. Mathison, 1988, unpubl. data). Cocks (1985) has drawn attention to effective cold-tolerant strains endemic to the highlands of West Asia where a wide range of *rhizobia* have been collected specific to a wide range of genera.

The matching of "alternative" legumes with effective rhizobia may lead to considerable expansion of legume in problem areas. For example, in the upper north of the South Australian Wheat Belt, hard-setting red brown earths are located on elevated land that enhances the continentality of the climate (cold winters, heavy frosts). *Medicago rigidula* and West Asian rhizobia may suit this situation (Abd El-Moneim & Cocks, 1986).

PESTS AND DISEASES

The most serious of the recorded fungal diseases of annual legumes in Australia are root rot and clover scorch, both of which affect sub. clover (Johnstone & Barbetti, 1987). Annual medics are susceptible to spring blackstem (*Phoma medicagenis*). Both native and introduced insects pose serious threats (Allen, 1987). In annual legume breeding programs in Australia, screening is carried out for resistance to up to four diseases and seven insect/mite pests including the redlegged earthmite (*Halotydeus destructor)* and lucerne flea (*Sminthurus viridis*). The significance of seed-borne viruses is under examination.

Under conditions of low soil fertility, annual legumes have often dominated pastures. Their decline in dominance over time may not only be caused by inappropriate management (see Seed Yield section above). Temporary isolation from various diseases and pests may have also been a factor. Research with sub. clover has demonstrated the seriousness of hitherto unknown fungi which are, however, not considered to be recent introductions to Australia viz. *Phytophthora clandestina* and *Aphanomyces euteiches*. It is of interest that most Australian cultivars developed from naturalized ecotypes of all three subspecies of sub. clover are severely affected by *P. clandestina,* whereas new cultivars based on recent introductions of ssp. *T. yanninicum* and some newly bred ssp. *T. subterraneum* (e.g., Karridale) have fortuitously proved relatively resistant (Taylor & Greenhalgh, 1987; F.C. Greenhalgh, 1988, pers. comm.).

PASTURE-ANIMAL INTERFACE

Plant Habit

Perhaps the most relevant adaptation in the context of long-term legume persistence in pasture is the adaptation to grazing. Under intensive grazing, the natural occurrence of *T. cherleri* is common in the Mediterranean region but that of *T. hirtum* is rare (Gladstones, 1973). Heavily grazed sub. clover ssp. *subterraneum* is more common than ssp. *brachycalycinum* on alkaline soils in Turkey, but the reverse occurs under light grazing (Francis *et al.*, 1975). In their Australian review, Reed and Cocks (1982) cite several examples where under a heavy grazing pressure, the density of sub. clover has declined, although the contribution of *T. cernuum, T. glomeratum,* and *T. dubium* has increased. Success in adaptation to grazing is likely to result from complex interactions between plant characters and various environmental stresses.

Table 4. Nutritive value of red *(T. pratense)* and annual clovers at Hamilton, Victoria: Sown Sept. 1982; harvested at flowering, 18 January.

Species of Trifolium	No. of lines	N (%DM)	Neutral det. fibre (%DM)	*In vitro* digestibility of DM (%)
T. aintabense	1	2.3	-	67
T. alexandrinum	1	2.6	35	72
T. balansae	2	1.9,2.1	41,44	55,56
T. brachycalycinum	1	1.8	44	59
T. cherleri	2	2.1,2.5	47,57	49,58
T. globosum	1	2.7	46	58
T. hirtum	2	2.2	46,48	59,60
T. isthmocarpum	1	2.6	36	63
T. pilulare	2	2.1,2.2	45,47	54,56
T. pratense	1	2.9	24	81
T. purpureum	2	1.6,2.3	51,55	51,53
T. resupinatum	2	2.9,3.1	23,25	82,84
T. subterraneum	1	2.3	41	62
T. yanninicum	1	2.6	40	57

In much of Australia, sub. clover is sown in permanent pasture with perennial grasses. Under heavy grazing pressure, the decline in clover density in the first 6 wk of the season can vary with the species of sown grass (Reed, 1974). It may be important that the habit of companion grasses is such that they can maximize moisture in the vicinity of the clover seed. The importance of allelopathy in grass-clover relationships is yet to be clarified.

Nutritive Value and Oestrogenicity

Sub. clover is not a particularly palatable species (Kenny & Black, 1983) and has a low-feeding value when mature (Fels *et al.*, 1959; Kenny & Reed, 1984). Palatability may be affected by isoflavone glycosides (Francis, 1973). In the long-term, strains with a high formonenetin content may persist well under grazing (Cocks *et al.*, 1972; Rossiter *et al.*, 1985).

Although mainly used as a fodder crop, *T. resupinatum* has been shown to have a particularly low fibre content (Table 4) and is both extremely palatable and nutritious (Flinn *et al.*, 1985). A hard-seeded cultivar (cv. Kyambro) has recently been released by the South Australian Department of Agriculture. *Trifolium hirtum* may also have some nutritional advantages over sub. clover (Rossiter *et al.*, 1972). Some alternative genera (e.g., *Astragalus, Lathyrus*) contain otherwise promising pasture legumes which produce toxins. Modern breeding techniques may help to eliminate these bad traits and so develop such species.

Seed Distribution

Animals can greatly affect the distribution of plant species, particularly via spreading seed in their feces. Small-seeded clovers (e.g. *Trifolium glomeratum, Trifolium balansae*) largely survive digestion by sheep (*Ovis aries*) but for sub. clover and the annual medics, seed is nearly all destroyed by digestion (Carter, 1987). The importance of the burr burial/grazing escape mechanism of the former species and *T. israeliticum* can thus be appreciated. The extreme hairiness of surface-laid burr, characteristic of some annual clovers (viz. *Trifolium batmanicum, Trifolium globosum,* and *Trifolium pauciflorum*) may make the burr unpalatable. Bolland (1987) found their seed banks were reduced much less by grazing than those of other annual legumes. His work also confirmed the need for grazing, during the period of vegetative growth, in order to obtain high seed yields from sub. clover and annual medics. Pod spininess can be quite variable within annual medics and spiny burrs aid seedling persistence via anchorage to the soil, and via deterring grazing animals. They also adhere to animal coat fibres thereby aiding distribution. They remain, however, a major source of vegetable fault in wool and so the length, and the angle of insertion of the spines in pods, are assessed in breeding programs (Crawford *et al.*, 1988).

FUTURE NEEDS

Our ability to increase the contribution of annual legumes will require concerted efforts by multi-disciplinary teams. Equal emphasis should be given to pasture management and plant improvement. Early flowering, seed yield, hard-seed, and pest resistance will be important for low rainfall areas; disease resistance for high rainfall areas. Cold tolerance is required in some parts of the world but to date this characteristic has received only slight attention.

The development of lesser known species may help to improve neglected niches. e.g., *Trifolium* spp. are needed for high rainfall calcareous soils and we await with interest the performance of the newly released hardseeded *T. resupinatum*, and, on neutral to acidic soil, *T. balansae*. Later-flowering cultivars of these species with hard-seededness are needed.

A problem in much of Australia has been the lack of a perennial legume to raise the quality of feed for the dry months. In many cases, the limitation for *M. sativa* has been the water-logging of shallow soils in winter. This problem has been offset to some useful extent by the use of the productive annual, *T. resupinatum*. Similar lateral approaches may be successful with other problem environments. The development of low coumarin *Melilotus* spp. could have an impact in saline areas.

Rhizobium strain development studies are vital to the assessment of alternative species and to maximize legume persistence in acid soils. Acid tolerant *Rhizobium* may extend the area of adaptation of annual medics into areas too dry for sub. clover. Compatible and nutritious grasses may reduce nitrate leaching and so aid legume persistence by slowing the soil acidification process.

Management studies must define the critical range of grazing pressure beyond which persistence is threatened. Persistence is sensitive to both under-grazing and over-grazing. Prior to flowering, grazing must control weeds. After flowering, grazing management must ensure some seed reserve and allow for a late break of season or a drought. Studies of whole-farm systems should consider alternative strategies for feeding so that the vital legume component can be preserved at critical times rather than sacrificed.

REFERENCES

Abd El-Moneim, A.M., and P.S. Cocks. 1986. Adaptation of *Medicago rigidula* to a cereal-pasture rotation in north-west Syria. J. Agric. Sci. 107:179-186.

Aitken, Y. 1939. The problem of hard seeds in subterranean clover. Proc. R. Soc. Victoria. 51:187-213.

Aitken, Y., and F. Drake. 1941. Studies of the varieties of sub. clover. Proc. R. Soc. Victoria. 53:342-393.

Allen, P.G. 1987. Insect pests of pasture in perspective. p. 211-225. In J.L. Wheeler et al. (ed.) Temperate pastures. CSIRO, Melbourne.

Andrews, W.D., and F.W. Hely. 1960. Frequency of annual species of *Medicago* on the major soil groups of the Macquarie region of NSW. Aust. J. Agric. Res. 11:705-714.

Antonovics, J. 1978. The population genetics of mixtures. p. 233-252. In J.R. Wilson (ed.) Plant relations in pastures. CSIRO, Melbourne.

Archer, K.A. 1981. Evaluation of legumes for use in short term leys and natural pastures on the Northern slopes of NSW. Aust. J. Exp. Agric. Anim. Husb. 21:485-490.

Ballard, L.A.T. 1961. Studies of dormancy in the seeds of subterranean clover. Aust. J. Biol. Sci. 14:173-186.

Beale, P.E., and E.J. Crawford. 1972. Preliminary comparisons of several annual *Trifolium* species on Kangaroo Island, South Australia. Aust. J. Exp. Agric. Anim. Husb. 12:634-637.

Beale, P.E., and E.J. Crawford. 1975. Assessment of *Trifolium* species and other pasture legumes with particular reference to tolerance to *Kabatiella caulivora*. S.A. Dept. Agric., Agric. Record. 2:54-59.

Beck, A.B. 1964. The oestrogenic isoflavones of subterranean clover. Aust. J. Agric. Res. 15:223-230.

Black, J.N. 1957. Early vegetative growth of 3 strains of sub. clover in relation to seed size. Aust. J. Agric. Res. 8:1-14.

Bolland, M.D.A. 1987. Effect of grazing and soil type on seed production of several *Trifolium* species compared with sub. clover & annual medics. J. Aust. Inst. Agric. Sci. 53:293-295.

Bolland, M.D.A., and J.S. Gladstones. 1987. Serradella (*Ornithopus* spp.) as a pasture legume in Australia. J. Aust. Inst. Agric. Sci. 53:5-10.

Brockwell, J., and F.W. Hely. 1966. Symbiotic characteristics of *Rhizobium meliloti*. Aust. J. Agric. Res. 17:855-899.

Carter, E.D. 1987. Establishment and regeneration of annual pastures. p. 35-51. In J.L. Wheeler et al. (ed.) Temperate pastures. CSIRO, Melbourne.

Chapman, D.F., G.W. Sheath, M.J. Macfarlane, P.J. Rumball, B.M. Cooper, G. Crouchley, J.H. Hoglund, and K.H. Widdup. 1986. Performance of sub. and white clover varieties in dry hill country. Proc. N.Z. Grassld. Assoc. 47:53-62.

Clarkson, N.M. 1977. Annual medics in Queensland. Queensland Agric. J. 105:39-48.

Clarkson, N.M., and J.S. Russell. 1975. Flowering responses to vernalization and photoperiod in annual medics. Aust. J. Agric. Res. 26:831-838.

Cocks, P.S. 1980. Limitations imposed by N deficiency on the productivity of sub. clover. Aust. J. Agric. Res. 31:95-107.

Cocks, P.S. 1985. Utilization of local resources for pasture plant cultivars. p. 99-107. In Trop. Agric. Series No. 18, Ministry of Agric., Forestry and Fisheries, Tsukuba, Japan.

Cocks, P.S., A.D. Craig, and R.V. Kenyon. 1982. Evolution of sub. clover in S.Australia: Genetic composition of a mixed population after 19 years under grazing. Aust. J. Agric. Res. 33:679-695.

Cocks, P.S., and T.A.M. Ehrman. 1987. Geographic origin of frost tolerance in Syrian pasture legumes. J. Appl. Ecol. 24:673-683.

Cocks, P.S., and J.R. Phillips. 1979. Evaluation of sub. clover in South Australia. I. The strains and their distribution. Aust. J. Agric. Res. 30:1035-1052.

Collins, W.J., and Y. Aitken. 1970. Effect of leaf removal on flowering in sub. clover. Aust. J. Agric. Res. 21:893-903.

Collins, W.J., and J.S. Gladstones. 1985. Breeding to improve Subterranean Clover in Australia. p. 308-316. In R.F Barnes et al. (ed.) Forage legumes for energy-efficient animal production. USDA-CSIRO-DSIR, Palmerston North.

Crawford, E.J., A.W.H. Lake, and K.G. Boyce. 1988. Breeding annual *Medicago* species for semi-arid conditions in southern Australia. Adv. Agron 42: (in press).

Crawford, E.J., and B.G. Nankivell. 1984. The effect of rotations on annual medic species seed and seedling populations. p. 155-164. J.E. Butler (ed.) Proc. Aust. Seeds Research Conf., Lawes, Queensland. Sept. 10-13th 1984. CSIRO.

Cunningham, P.J., R.R. Gault, and J. Brockwell. 1986. Field responses of Persian clover to inoculation with Rhizobium trifolii. p. 149-150. In W. Wallace and S. Smith (ed.) Proc. 8th Aust. N. Fixation. Conf.; Occas. Publ. 25. Aust. Inst. Agric. Sci., Adelaide.

de Koning, C.T., and E.D. Carter. 1987. Impact of grazing on sub. clover cultivars. p. 169. In T.G. Reeves (ed.) Proc. 4th Aust. Agron. Conf. Melbourne, Victoria. 24-27 Aug. 1987. Aust. Soc. of Agron.

Devitt, A.C., B.J. Quinlivan, and C.M. Francis. 1978. The flowering of annual legumes in Western Australia. Aust. J. Exp. Agric. Anim. Husb. 18:75-80.

Donald, C.M. 1959. The production and life span of seed of subterranean clover. Aust. J. Agric. Res. 10:771-787.

Ellis, R.W., A.D. Craig, and R.S. Martyn. 1985. Alternative Pasture legumes for Southern Australia. Dept. Agric., S. Aust., Tech. Rep. No.85.

Evans, L.T. 1959. Flower initiation in *Trifolium subterraneum* L. Aust. J. Agric. Res. 10:1-15.

Evans, P.M., and J.A. Carpenter. 1987. Sub clovers for Tasmania: Production of early, mid-season and late groups. p. 162. In T.G. Reeves (ed.) Proc. 4th Aust. Agron. Conf., Melbourne. Victoria. 24-27 Aug. 1987. Aust. Soc. of Agron.

Fels, H.E., J.J. Moir, and R.C. Rossiter. 1959. Herbage intake of sheep in S.W. Australia. Aust. J. Agric. Res. 10:237-247.

Flinn, P.C., K.F.M. Reed, G.R. Saul, G.N. Ward, and J.F. Graham. 1985. The cultivation and feeding value of Persian clover. p. 961-962. In T. Okubo (ed.) Proc. 15th Int. Grassl. Congr., Kyoto, Japan. 24-31 Aug. 1985. Sci. Council of Japan and Jap. Soc. Grssl. Sci., Japan.

Francis, C.M. 1973. Influence of isoflavone glycosides on the taste of sub. clover leaves. J. Sci. Food Agric. 24:1235-1240.

Francis, C.M., A.C. Devitt, and C. Steele. 1974. Influence of flooding on the alcohol dehydrogenase activity of roots of *Trifolium subterraneum* L. Aust. J. Plant Physiol. 1:9-13.

Francis, C.M., and J.S. Gladstones. 1974. Relationships among rate and duration of flowering and seed yield components in subterranean clover. Aust. J. Agric. Res. 25:435-442.

Francis, C.M., J. Katznelson, and W.J. Collins. 1975. Legume seed collection tour of Turkey. Aust. Plant Introd. Rev. 10(3):1-10.

Francis, C.M., and M.L. Poole. 1973. Effect of waterlogging on annual medics. Aust. J. Exp. Agric. Anim. Husb. 13:711-713.

Gillespie, D.J. 1987. Murex, a new medic for acid soils. p. 172-175. In J.L. Wheeler et al. (ed.) Temperate pastures. CSIRO, Melbourne.

Gladstones, J.S. 1973. Observations on the environment and ecology of some annual legumes in Southern Italy. Aust. Plant Introd. Rev. 9:11-40.

Gladstones, J.S., and W.J. Collins. 1983. Sub. clover in Australia. J. Aust. Inst. Agric. Sci. 49:191-202.

Gladstones, J.S., J.F. Loneragan, and W.F. Simmons. 1975. Mineral elements in temperate crop and pasture plants. Aust. J. Agric. Res. 26:113-126.

Heida, R., and R.M. Jones. 1988. Seed reserves of Barrel medic and snail medic in the topsoil on pastures in a brigalow soil in southern Queensland. Trop. Grassl. 22:16-21.

Hely, F.W., C.A. Neal-Smith, T.A. Salguerio, and N.F. Pereira. 1976. Volunteer leguminous species on low fertility grazing lands of S. Portugal. Aust. Plant Introd. Rev. 11:22-48.

Hochman, Z. 1987. Quantifying vernalization and temperature effects on time of flowering in *Medicago truncatula*. Aust. J. Agric. Res. 38:279-86.

Hoveland, C.S., and D.M. Elkins. 1965. Germination response of arrowleaf, ball and crimson clover varieties to temperature. Crop Sci. 5:244-46.

Howieson, J.G., and M.A. Ewing. 1986. Acid tolerance in the *Rhizobium meliloti-Medicago* symbiosis. Aust. J. Agric. Res. 37:55-64.

Hubble, G.D. 1970. Soils of Australia. p. 44-58. In R.M. Moore (ed.) Australian Grasslands. ANU Press, Canberra.

Johnstone, G.R., and M.J. Barbetti. 1987. Impact of fungal and viral diseases on pasture. p. 235-248. In J.L. Wheeler et al. (ed.) Temperate pastures. CSIRO, Melbourne.

Katznelson, J. 1970. Edaphic factors in the distribution of subterranean clover in the Mediterranean region. p. 192-196. In M.J.T. Norman (ed.) Proc. 11th Int. Grassld. Congr., Surfers Paradise. Apr. 1970. Univ. of Qld. Press, Queensland.

Kenny, P.A., and J.C. Black. 1983. Factors affecting diet selection by sheep. p. 249-253. In G.B. Robards and R.G. Packham (eds.) Feed information and animal production. Comm. Ag. Bureaux, Slough.

Kenny, P.T., and K.F.M. Reed. 1984. Effects of pasture type on the growth and wool production of weaner sheep during summer and autumn. Aust. J. Exp. Agric. Anim. Husb. 24 :322-331.

MacFarlane, M.J., and G.W. Sheath. 1984. Clover for dry hill country. Proc. N.Z. Grassl. Assoc. 45:140-150.

Mathison, M.J. 1983. Mediterranean and temperate forage legumes. p. 63-81. In McIvor, J.G. and R.A. Bray (ed.) Genetic resources of forage plants. CSIRO, Melbourne.

McGuire, W.S. 1985. Sub. clover. p. 515-535. In N. Taylor (ed.) Clover science and technology. Amer. Soc. Agron., Madison.

Millikan, C.R. 1963. Effects of different levels of zinc and phosphorus on the growth of subterranean clover. Aust. J. Agric. Res. 14:180-205.

Moore, R.M. 1970. Pastures of Australia (Map 5). Opp. p. 102. In R.M. Moore (ed.) Australian grasslands. ANU Press, Canberra.

Morley, F.H.W. 1958. Inheritance and ecological significance of seed dormancy in sub. clover. Aust. J. Biol. Sci. 11:261-274.

OECD. 1986. OECD schemes for the varietal certification of seed moving in international trade: List of cultivars eligible for certification. OECD, Paris.

Ozanne, P.G., and K.M.W. Howes. 1974. Increased seed production of sub. clover pastures in response to fertilizers supplying calcium. Aust. J. Exp. Agric. Anim. Husb. 14:749-757.

Quigley, P.E., E.D. Carter, and R.C. Knowles. 1987. Medic seed distribution in soil. p. 193. In T.G. Reeves (ed.) Proc. 4th Aust. Agron. Conf., Melbourne, Vic. 24-27 Aug. 1987. Aust. Soc. of Agron.

Quinlivan, B.J. 1968. Seed coat impermeability in the common annual legume pasture species of Western Australia. Aust. J. Exp. Agric. Anim. Husb. 8:695-701.

Quinlivan, B.J. 1971. Seed coat impermeability in legumes. J. Aust. Inst. Agric. Sci. 37:283-295.

Quinlivan, B.J., and A.J. Millington. 1961. The effect of a mediterranean summer environment on the permeability of hard seeds of subterranean clover. Aust. J. Agric. Res. 13:377-387.

Quinlivan, B.J., and H.I. Nichol. 1971. Embryo dormancy in subterranean clover seeds. Aust. J. Agric. Res. 22:599-614.

Reed, K.F.M. 1974. Production of *Phalaris tuberosa* and *Lolium perenne* pastures. Aust. J. Exp. Agric. Anim. Husb. 14:640-648.

Reed, K.F.M. 1981. A review of legume-based vs nitrogen fertilized pasture systems for sheep and beef cattle. p. 401-417. In J.L. Wheeler and R.D. Mochrie (ed.) Forage evaluation CSIRO and American Forage and Grassland Council, Melbourne,

Reed, K.F.M., and P.S. Cocks. 1982. Limitations of pasture spp. in Southern Australia. p. 142-60. In M.J.T. Norman (ed.) Proc. 2nd Aust. Agron. Conf., Wagga, NSW. July 1982. Aust. Soc. of Agron.

Robson, A.D. 1969. Soil factors affecting the distribution of annual Medicago species. J. Aust. Inst. Agric. Sci. 35:154-167.
Robson, A.D., and J.F. Loneragan. 1978. Responses of pasture plants to soil chemical factors other than N and P, with emphasis on the legume symbiosis. p. 128-142. In J.R. Wilson (ed.) Plant relations in pastures. CSIRO, Melbourne.
Rossiter, R.C. 1972. The effect of defoliation on flower production in sub. clover. Aust. J. Agric. Res. 23:427-435.
Rossiter, R.C., R.A. Maller, and A.G. Parkes. 1985. A model of changes in the composition of binary mixtures of subterranean clover strains. Aust. J. Agric. Res. 36:119-143.
Rossiter, R.C., G.B. Taylor, and G.W. Anderson. 1972. The performance of subterranean, rose, and cupped clovers under set-stocking. Aust. J. Exp. Agric. Anim. Husb. 12:608-613.
Taylor, G.B. 1970. The germinability of soft seed in strains of sub. clover. Aust. J. Exp. Agric. Anim. Husb. 10:293-297.
Taylor, G.B., and M.A. Ewing. 1988. Effect of burial on the longevity of hard seeds of subterranean clover and annual medics. Aust. J. Exp. Agric. 28:77-81.
Taylor, N.L. 1985. Clovers around the world. p. 1-6. In N. Taylor (ed.) Clover science and technology. Amer. Soc. Agron., Madison.
Taylor, P.A., and G.C. Greenhalgh. 1987. Significance, causes and control of root rots of subterranean clover. p. 249-252. In J.L. Wheeler et al. (ed.) Temperate pastures. CSIRO, Melbourne.
Whyte, R.O., G. Nilsson-Leissner, and H.C. Trumble. 1953. Legumes in agriculture. FAO, Rome.
Williams, C.H., and C.S. Andrew. 1970. Mineral nutrition of pastures. p. 321-338. In R.M. Moore (ed.) Australian grasslands. ANU Press, Canberra.
Wilson, G.P.M. 1979. *Trifolium* species evaluation. Dep. Agric. N.S.W., Grafton Agric. Stn. Ann. Rep. 1978-79.

DISCUSSION

Sheaffer: Are there any native rhizobia in Australia suitable for the annual medics? How well do introduced rhizobia persist?

Reed: There are no native rhizobia for annual medics. We have introduced *Rhizobium* strains that nodulate annual medics effectively. These introduced rhizobia can persist in the soil, but may not survive in adequate numbers when diluted by tillage, long cropping phases (particularly when herbicides are used to destroy the host), long hot summers, and increasing soil acidity.

Woomer: What concentration of rhizobia in the soil must be achieved before rhizobia are considered to be persistent? Can you successfully introduce a new strain in the presence of such numbers of resident bacteria?

Reed: Numbers of rhizobia in the soil at locations such as Wagga Wagga (NSW) and Rutherglen (Victoria) can be as low as 100 per gram of soil at the end of summer, and may build up to 10^6 per gram in the spring. We have inoculated clover seeds with CC2483g (the new inoculant for *T. resupinatum*), both as a single strain or in mixtures with other commercial strains. We have found CC2483g to be very competitive with other strains in the year after sowing, at sites with both high and low resident populations of *Rhizobium trifolii*.

J. Hay: In the work you mentioned on persistence of the Persian clover *Rhizobium* strain, did you inoculate the seed in the second year?

Reed: No. The original strain was identified in nodules of a reseeded, but non-inoculated, second-year stand.

Watson: Would a cold-tolerant *Rhizobium* strain boost winter production of white clover in New Zealand conditions?

Reed: I don't know. I expect that some *R. trifolii* strains may be more effective than others on white clover.

Clements: Maintenance of a seed reserve is essential for the persistence of an annual legume. We once had almost half a million hectares of rose clover in Australia; now there's none. Under the heavy grazing pressures in southern Australia, how well is the maintenance of a seed bank correlated with the height at which seed pods are borne? Also, I was interested to hear you say that scientists are advising farmers to control their grazing management to retain Balansa and hard-seeded Persian clover in pastures. This is a change in traditional Australian attitudes to annual legumes. Please comment.

Reed: Australia is not alone in experiencing difficulties in maintaining clovers with elevated seedheads. The area of land containing such clovers has declined in the USA also. Many farmers are probably not receptive to management-demanding species at present.

Sheath: Do terminal-flowering annuals fit into efficient grazing systems?

Reed: They tend to have a specific rather than a "general" role. For example, Persian clover is not a general alternative to subterranean clover. It can, however, provide excellent hay or standing feed to fill late summer feed gaps when the general or conventional pasture has minimal quality.

Jones: The general principle is that plants dependent on seedling regeneration must be allowed to re-seed. Legumes with terminal inflorescences (such as Persian clover) often are not able to re-seed without careful grazing management.

Fisher: With regard to morphological variation in *Stylosanthes, S. humalis* has the ability to modify its reproductive morphology. When left undefoliated, it has terminal infloresccences, but when grazed from an early age it produces a dense herbage mat yet still produces a great deal of seed. Perhaps we need plants with this facultative characteristic.

Irwin: How important is *Phytophthora clandestina* in reducing subterranean clover establishment?

Reed: There are large genotype environment interactions in response to *P. clandestina*. The relative ranking of the yields of subterranean clover cultivars may change during the season in the presence of the fungus. The impact of the disease during establishment can sometimes be dramatic where seasonal conditions add further stress to the plant.

Hoveland: No mention was made in your paper of bloat in animals grazing Persian clover. In the southern USA, Persian clover has earned a reputation for being a bad "bloater" of cattle. Have you had any difficulties with this in Australia?

Reed: Bloat has occurred in dairy cows, but few complaints have been received. Persian clover is commonly used by dairy farmers in irrigation districts.

DEVELOPMENT AND GROWTH CHARACTERISTICS OF TEMPERATE PERENNIAL LEGUMES

Margot B. Forde, M.J.M. Hay and J.L. Brock

SUMMARY

The development and growth form of the major temperate perennial forage legumes are briefly described and compared, and their ecological adaptation and response to defoliation related to their morphological characteristics. Because of the importance of white clover as a pasture legume this species is considered in more detail.

INTRODUCTION

Management of perennial forage legume crops or pastures must relate to the growth characteristics of the species concerned, which govern their suitability for grazing or cutting (Haynes, 1980; Marten, 1985). As establishment of new plants from seed occurs infrequently in stable swards with closed canopies, mechanisms of perennation and vegetative reproduction under such conditions are of great significance (Snaydon, 1985).

This paper will briefly review the growth characteristics of the major temperate perennial legumes and the various perennation strategies they serve, and discuss how these reflect climatic and edaphic adaptation and interact with periodic defoliation. In view of the importance of white clover as a pasture legume and the increasing volume of work on its growth (Frame & Newbould, 1986), this species will be considered in more detail. No attempt will be made to cover the large literature available on well-known species, and only certain relevant references will be cited. Use has also been made of a number of standard agronomy texts and reference books including Heath et al. (1973), Gill et al. (1980), Duke (1981), Langer (1973a), Martin and Leonard (1967), Whyte et al. (1953), Walton (1983), Zohary and Heller (1984), and these will not be cited again as individual references. Hays (1888) gives detailed descriptions of the root systems of several major species.

GROWTH TYPES IN TEMPERATE PERENNIAL LEGUMES

For the purposes of this review, temperate perennial legumes will be grouped into four major classes:

1. Subshrubs or shrubs that perennate by forming a more or less permanent woody or semi-woody framework aboveground.

2. Herbs perennating by formation of a deep taproot system and

woody crown. This group includes most of the major temperate forages including lucerne (Medicago sativa L.), red clover (Trifolium pratense L.) and birdsfoot trefoil (Lotus corniculatus L.)

3. Herbs that both perennate and spread by the formation of rhizomes [L. pedunculatus Cav., zigzag clover (T. medium L.)], or adventitious shoots arising from the root system [creeping lucernes, crownvetch (Coronilla varia L.)], in addition to an initial crown.

4. Herbs such as white clover (T. repens L.) and strawberry clover (T. fragiferum L.) that perennate by forming clonal patches of rooted stolons which can persist after the original plant has died.

The perennial vetches and vetchlings such as flatpea (Lathyrus sylvestris L.) that produce annual climbing stems from rhizomes or tuberous roots form a fifth group which will not be considered here, as none are forages of major importance.

SHRUBS AND SUBSHRUBS

Leguminous browse shrubs are principally a feature of the semi-arid tropics and subtropics, but also occur in dry warm-temperate regions. The species discussed here are currently the subject of research in N.Z. and Australia (Van Kraayenoord & Hathaway, 1986; Logan & Radcliffe, 1985).

Tagasaste, Tree Lucerne [Chamaecytisus palmensis (Christ) Bisby et K. Nicholls] (Logan, 1982; Woodfield & Forde, 1987). Tagasaste is a fast-growing shrub or small tree from the island of La Palma which flowers in winter and spring and produces its main growth in spring and summer. Different genotypes vary in the degree of branching and the height at which apical dominance is lost. Types with dense branching close to the ground are the most productive and best suited for browsing, and make fast regrowth. However defoliation by livestock must be controlled to prevent overgrazing causing death. Tagasaste has a deep taproot system which confers considerable drought resistance, but is highly susceptible to pathogens in wet soils.

Tree medic (M. arborea L.), a dense, leafy, much-branched Mediterranean shrub 1-3 m tall, is suitable for warmer drier areas than tagasaste but is less vigorous in growth.

Garcia [Teline stenopetala (Webb et Berth.)] is a fast-growing leafy broom from the Canary Islands with soft ribbed stems and silvery green leaves. Plants are densely branched from the base and produce fast regrowth after grazing but are not as long-lived as tagasaste.

Canary clover [Dorycnium hirsutum (L.) Ser.] (Wills et al., 1989). A low-growing Mediterranean subshrub up to 50 cm tall with a deep taproot and soft hairy grey-green foliage. Vegetative growth begins in early spring from the base of the plant, slows down during flowering, and resumes in autumn. Branches may die back after pods have ripened, especially in summer-dry regions. The species tolerates drought and severe frosts but is very susceptible to root rots in heavy wet soils. A related species with less hairy leaves, and more prostrate open growth, D. pentaphyllum, is more palatable.

Lucerne (M. sativa, M. falcata, M. sativa x falcata) (Iversen & Meijer, 1967; Palmer, 1967; Langer, 1973b; Marten, 1985). Medicago sativa is believed to have originated in arid continental regions characterised by low humidity and high day temperatures, cool nights, and deep soil moisture. This led to the development of an erect, highly productive plant recovering quickly after defoliation, with a deep, almost unbranched taproot and variable winter hardiness based on dormancy. By contrast, M. falcata is a plant of colder, more humid upland areas. Consequently it is a less productive but much more cold-tolerant species with a deep-set crown, prostrate stems and a widely branching root system. Hybrids between these species are involved in the origin of almost all modern cultivated forms, with resulting variation in growth characteristics and ecological tolerances, but sativa characteristics predominate in the most widely grown, high-producing cultivars.

The young lucerne plant is generally erect and does not form an initial rosette. Hence it is especially vulnerable to grazing damage in the first year before a crown has formed. Defoliation must also be correctly timed to allow development of the deep taproot system on which the drought tolerance of the species depends. A seedling 10 cm tall may have a taproot 30 cm deep, growing to 10 m in mature plants in favorable soil.

The crown is formed by branching at the base of the plant, starting from buds in the axils of the cotyledons and first leaves. Low axillary buds on these stems in turn produce further basal shoots, forming a branched crown of short perennial stems just above the ground which is the main source of regeneration after defoliation. Up to 20 erect stems may arise from the crown in the course of a season as the older stems mature and are cut. These produce axillary racemes of flowers near the tips.

Active extension growth of new shoots at the base of the plant occurs after cutting or when the previous crop of shoots has reached a certain stage of maturity, usually the production of flower buds. Regrowth can also occur from axillary buds on stubble, especially if defoliation has been premature, but although these stubble shoots appear more quickly, their contribution to the recovery of the plant is soon superseded by regrowth from the base.

Because the bulk of leaves and growing points are borne on upright stems lucerne is very vulnerable to excessive or badly timed defoliation, and yield and vigour are much reduced by continuous grazing. Although defoliation is followed by relatively rapid regrowth, subsequent build-up of carbohydrate reserves is rather slow, and frequent defoliation results in depletion of root reserves and eventual death. Cutting in late autumn is especially critical as it may reduce root reserves at a time when they cannot be restored before winter. Winter-hardy cultivars have a progressively slower rate of regrowth after harvest and a more decumbent growth habit in autumn than do non-hardy cultivars.

Red Clover (T. pratense L.) (Fergus & Hollowell, 1960; Bowley et al., 1984; Taylor, 1985). Red clover is characteristic of more humid regions with moderate or cool temperatures, and is not as deep-rooted or tolerant of drought as lucerne. Its growth characteristics also show some significant differences. The primary shoot has little internode elongation but forms a rosette of leaves with the growing points at ground level. At this stage it is more tolerant of grazing than lucerne. Growth of axillary buds from the cotyledons, primary leaves and early branches forms a crown at the top of a branched taproot system which can penetrate to 3 m in favorable situations. Adventitious roots from the crown region often form the major part of the root system in 3 to 4 yr-old plants. Prostrate stems can also produce adventitious roots from the nodes in damp soil.

Erect, leafy, branched stems bearing terminal flower heads begin to appear in spring. These are strictly annual, and the plant reverts to the rosette form again in winter. Early-flowering strains with a short photoperiod requirement produce relatively few stems at the first flowering, but a succession of flowering stems is produced throughout the season, resulting in rapid regrowth after cutting and two or more cuts per season. However as few dormant buds are left to continue growth in the following year, such varieties tend to be short-lived. Late-flowering varieties with a longer photoperiod requirement produce a single heavy flush of flowering shoots and are slow to recover from cutting, leaving many dormant buds to continue growth for next season. Hence they tend to live longer and tolerate more cold. As with lucerne, late defoliation in cold regions may result in premature death because of depletion of root reserves.

Alsike Clover (*T. hybridum* L.) is adapted to colder moister climates than red clover and its root system does not penetrate as deeply as that of more drought-tolerant legumes. However it will tolerate poorly drained soils better. The well-developed crown produces many prostrate or scrambling hollow stems up to a metre long which do not root at the nodes. As the growing points are held above the ground they are vulnerable to defoliation. In contrast to red clover, alsike stems are indeterminate and bear the long-stalked flower heads in the leaf axils, as does white clover. Generally the species gives one good flush of growth with little aftermath, and is not long-lived.

Birdsfoot Trefoil (*L. corniculatus* L.) [McDonald, 1944; Seaney & Henson, 1970]. Like lucerne and red clover, birdsfoot trefoil has a well-developed taproot, not as long as that of lucerne but with a better developed thick fibrous root system in the upper soil. The species will tolerate a wide range of environmental conditions and is more successful than lucerne on shallow and poorly drained soils.

Stems and branches arise from a crown formed by branching from buds in the axils of the cotyledons and early seedling leaves of the short primary stem, which soon ceases growth. In prostrate forms stems may arise from crown buds below ground level and grow for several inches as short rhizomes before emerging, but do not generally root. Shoots can also develop from root tissue close to the crown. The stems are indeterminate, and bear flowers in stalked axillary clusters towards the tips.

During the growing season regrowth arises from stubble buds, and root reserves are maintained at a low level until autumn storage occurs. The remaining stubble must carry green leaves to support the regrowth of the upper axillary buds. Where winters are very cold the plant may die back to the crown.

Sainfoin (Onobrychis viciifolia Scop.), a species adapted to continental climates, forms a deep taproot from which arises a rosette of large pinnate leaves followed by numerous hollow leafy stems bearing axillary lupin-like racemes of flowers. A dense crown of decumbent branches is formed close to the ground, and the upper parts of the rootstock become dark brown and gnarled with age.

Giant sainfoin has erect growth taller than lucerne and flowers in the first year, providing two or three cuts. Common sainfoin is prostrate and does not flower in the first year. It also returns to the prostrate form after the first cut, and the aftermath is normally grazed. Because of its deep rooting system the species is very drought resistant and withstands cold winters, but will not tolerate wet or saline soils.

Sulla (Hedysarum coronarium L.) (Watson, 1982; Van Kraayenoord & Hathaway, 1986). This Mediterranean relative of sainfoin is a shorter-lived perennial herb with a deep branching taproot and decumbent to erect, succulent stems up to 1.5m. It will not tolerate heavy grazing and withstands only moderate frosts, but has considerable drought tolerance. Sulla grows actively in spring/early summer and autumn/early winter, whereas sainfoin is primarily a summer grower.

Sericea Lespedeza (Lespedeza cuneata L.) (Henson, 1957). A perennial herb tolerant of drought and low fertility, which has erect branched stems and an extensively branched deep taproot. New stems arise from the crown in spring and grow throughout the summer until they flower and set seed. After cutting recovery growth develops from axillary buds on the stubble, so the crop must not be cut too low or too frequently, generally only twice a season. When grazed the pasture is best kept short so that the plants are tender and growing vigorously. As with lucerne, late defoliation prevents the build-up of root reserves necessary for survival. Topgrowth is killed by frost in temperate climates and renewed from crown buds in the spring.

RHIZOME-FORMERS

Big Trefoil (L. pedunculatus Cav.) (Sheath, 1980a, 1980b, 1981). Lotus pedunculatus initially forms a crown and taproot which are later supplemented by a shallow network of rhizomes and associated fibrous roots in the upper 5-10 cm of soil. Air spaces in the cortex of these shallow or superficial rhizomes assist in enabling the species to tolerate waterlogged conditions.

Rhizome growth chiefly occurs in late summer and autumn and is encouraged by lenient defoliation. In winter and spring, old rhizomes break down and new discrete plant units are formed. New growth from the crown is generally in the form of rhizomes, and leafy aerial shoots are chiefly initiated from rhizome nodes. As in lucerne, most active growing shoot tips are removed during defoliation. The stems are indeterminate and prostrate or semi-scandent, bearing long-stalked axillary flower clusters at the tips.

As defoliation becomes more lenient and herbage accumulates, axillary growth from rhizome nodes tends to be laterally growing rhizomes rather than leafy shoots. Following defoliation of large canopies there is a release of leafy axillary buds and a transition from rhizome formation to aerial shoots, which involves delays in regrowth. New shoot growth can also develop from aboveground nodes on both intact and stubble stems but this is a transient phase and rhizome shoots eventually dominate regrowth (just as crown shoots dominate stubble shoots in lucerne). More frequent defoliation increases stubble shoot regrowth at the expense of rhizome shoot regrowth. Thus management seems to have little positive influence in hastening initiation of new shoots because of the demands of the indeterminate growth habit of the species.

Kura Clover, Caucasian Clover (*T. ambiguum* M. Bieb.,) (Speer & Allinso 1985; Bryant, 1974). **Zigzag clover** (*T. medium* L.). Though not closely related, these species are very similar in growth habit. The initial crown and deep branched root system are soon supplemented by extensive deep rhizomatous growth resulting in a multi-tufted habit, with new plants arising progressively farther out from the parent until the whole soil mass is occupied with roots and rhizomes. Flower heads are terminal on determinate stems in zigzag clover, but axillary in Kura clover.

These species are very slow to establish, and die back to rosette leaves or completely to the ground in winter. Regrowth after cutting is also slow, and total production much less than that of red clover. However because of their protected underground rhizomes they are extremely persistent and will tolerate considerable cold and drought.

Cicer milkvetch (*Astragalus cicer* L.) is a vigorous long-lived rhizomatous perennial which can spread up to 70 cm a year. It is extremely drought-resistant and winter-hardy and tolerates unfavourable soil conditions, but because seedling growth is slow, good stands may not be obtained in the first season.

Crownvetch (*Coronilla varia* L.) (Henson, 1963; Lambrechtsen, 1974). This hardy long-lived perennial has a deep taproot system and spreads by underground creeping roots which produce new crowns from adventitious buds. The weak hollow stems are prostrate to semi-erect and up to 1.5 m long, forming a dense tangled mass and bearing flowers in long-stalked axillary clusters. The species is slow to establish and does not recover from defoliation as well as lucerne, but the plants are drought-tolerant and revert to a dormant rosette in winter, tolerating severe frosts.

Creeping Lucernes (Iversen & Meijer, 1967; Palmer, 1967). A similar growth habit is seen in forms of *Medicago falcata* and the hybrid creeping lucernes such as 'Rambler' and 'Cancreep', which have widely branching roots from which adventitious stems and new crowns can develop. One 3-year-old plant of Rambler was reported to be 9 ft across with 286 separate crowns. Lucernes of this type would be expected to be more persistent and tolerant of mismanagement and harsh climatic conditions than conventional types, but a price is paid in lower production.

Rhizomatous strains of lucerne and birdsfoot trefoil occur in which short rhizomes grow out horizontally from crown buds below the soil before emerging, effectively enlarging the size of the crown but not resulting in great spread or the formation of independent new plants.

White Clover (Trifolium repens L.), the most important perennial legume for grazed pastures in humid temperate regions, is now described in more detail.

Seedling Growth

The process of germination and seedling development has been described in detail elsewhere (Erith, 1924; Thomas, 1987b). The seedling has a single primary stem with no apparent internode elongation, and over the first weeks of growth the rate of leaf emergence is low (one per week) although buds are initiated in the axils of the older leaves. A distinct second phase of growth is marked by an increase in the rate of leaf emergence as the leaf primordia on the axillary buds appear. This is accompanied by elongation of internodes on the axillary shoots and the formation of secondary lateral stolons of which there can be up to five per plant by the seventh week. The primary stem either begins to elongate or commonly ceases growing (Thomas, 1987b).

Growth of secondary lateral stolons involves adventitious root development at nodes which eventually allows these stolons an individual existence when the seminal root system dies (Erith, 1924). The stolon is the basic structural unit of the plant and consists of a series of internodes separated by nodes which form as a result of growth at the apical bud. Each node bears a trifoliate leaf with an erect petiole, two root primordia and an axillary bud which is capable of growing into a lateral stolon or into an inflorescence. Production of lateral stolons allows vegetative spread of the plant. Growth of the secondary and all subsequent stolons is negatively geotropic, the strength of the geotropic response decreasing as light intensity increases (Thomas, 1987b). Under natural conditions this results in stolon growth closely adpressed to the ground except in tall pasture where stolons tend to grow upwards. Where stolons are buried growth of both branches and the stolon tip is vertical (Hay, 1983).

Plant Habit

The response of the plant to gravity and light gives rise to the prostrate creeping growth habit which greatly assists the plant to tolerate defoliation by grazing. Most stolon, which harbors current and potential apices, is found on or below the soil surface (Hay et al., 1987) and is thus generally below grazing height. However as grazing intensity increases, the amount of stolon consumed increases (Curll & Wilkins, 1982) and this adversely affects the subsequent presence of white clover in such pastures (Curll & Wilkins, 1982, Hay et al., 1987). Another characteristic which contributes significantly to the success of white clover as a forage legume is the ability to recover from defoliation of stolon apices. Removal of a stolon apex stimulates the development of in particular the youngest one, two or three lateral axillary buds and often one such bud rapidly assumes the characteristics of the previous apical bud of the stolon (Curll, 1980; Thomas, 1987b). The strongly unidirectional acropetal movement of nutrients and carbohydrates towards the production of new leaves at the stolon apex assists the plant to tolerate defoliation of older leaves (King et al., 1978).

The ability to rapidly vary the elongation of petioles in order to position the laminae close to the top of the sward, so ensuring they are well illuminated, contributes much to the success of white clover as a forage species. However this positioning of the laminae makes them particularly susceptible to complete defoliation, especially under systems of infrequent but intensive grazing (rotational grazing). Following defoliation white clover has to rely on the rapid production of new leaves to restore photosynthetic area. At this stage where sward Leaf Area Index is <2.0 the horizontal positioning of laminae confers an advantage to white clover over grass in light interception (Hart, 1987).

Each node has two root primordia which are situated above and below the lateral bud. The upper primordium usually remains inactive, but under moist conditions (Stevenson & Laidlaw, 1985) the lower primordium develops into an adventitious nodal root (Thomas, 1987a). Nodal roots can grow to 15 cm long (Erith, 1924) before lateral branching and secondary thickening may begin in the normal manner. The number and proportion of nodal roots that show secondary development varies with cultivar (Caradus, 1977) and probably with environmental conditions. The occurrence of secondary thickening may ultimately result in the formation of a secondary taproot [usually defined by a diameter of >1.5 mm at its proximal end (Caradus, 1977)]. The presence of thickened roots influences branch development at the node (Chapman, 1983), longevity of the node and those acropetal to it (Wilman & Asiegbu, 1982; Chapman, 1983; Hay, 1983), plant survival over dry periods (Smith & Morrison, 1983; Stevenson & Laidlaw, 1985) and probably size of individual plants. The ability to develop independent root systems from each node as well as a branch favours the performance of white clover in intensively grazed systems as it enables the species to tolerate and even spread following fragmentation of plants by defoliation or treading.

Growth of the Mature Plant

The description of growth of mature plants in grazed swards will include published (Brock, 1988; Brock et al., 1988; Hay, 1983; Hay et al., 1988) and as yet unpublished data from work at Palmerston North. They examined aspects of white clover growth in three differently managed Huia white clover/Ruanui perennial ryegrass swards all stocked at 22.4 ewes plus lambs per hectare.

Growth of the mature plant involves growth at stolon apices and death of stolon at the basal end of the oldest (or primary) stolon of the plant (Kershaw, 1959). Thus size of individual plants at any time is a function of both the rate of stolon formation at the apices and the rate of death at the base of the primary stolon. In contrast with glasshouse grown or spaced plants in the field, the size of most white clover plants in grazed swards is small (mean dry weight of stolon per plant, 80 mg, Hay et al., 1988) and the distribution of the population by branching structure favors simpler structures (20% of plants were unbranched, 45% secondarily branched, 27% tertiarily branched, 7% quaternarily branched and less than 2% more complexly branched). The distribution of plants by their shoot dry weight in these white clover populations was highly positively skewed indicating there were many small plants and few large individuals in the population. The overall mean Gini coefficient was 0.51 and 68% of the population was smaller than the mean plant. However, the few large plants in the population are important as they contribute disproportionately to the population of growing points as the first, second, third and fourth branching orders contribute 3, 31, 44 and 21%, respectively. Thus larger plants have the greatest potential to increase the number of nodes of the population.

The size hierarchy in these populations was maintained over the 2-yr period of study by variability of both the growth rate of individuals (Burdon & Harper, 1980) and the rate of death of older stolon. Death of older stolon within individual plants has variable effects on plant size as the rate of progression along stolons is inconsistent in that death proceeds towards the stolon apex in stages, pausing at nodes with strong root development (Curll & Wilkins, 1982; Hay, 1983; Sackville-Hamilton, 1987, personal communication), and the number and size of branches released upon death of parent nodes is highly variable.

The clonal growth form of white clover just described has an important ramification as regards root growth. The constant turnover of nodes [maximum longevity, 2-yr (Chapman, 1983)] appears to prohibit a large investment of carbon into root systems. White clover is considered a species with a superficial root system (Turkington & Burdon, 1983) and a low root/shoot ratio (Mouat, 1983) [approximately 1:4 (Young, 1958)]. A direct result of this morphological characteristic is the often noted correlation between summer rainfall and clover growth, both in New Zealand (Hoglund et al., 1979) and in Australia (Lazenby & Swain, 1969). This low investment of the species in root growth seriously affects the persistence and productivity and hence usefulness of white clover in summer-dry regions.

Seasonal Growth

White clover has a particularly strong seasonal pattern of growth as regards the vertical distribution of stolons in the sward (Hay et al., 1987), mean plant size (Brock et al., 1988, Hay et al., 1988) leaf size (Brougham, 1962; Davies & Evans, 1982), dry matter allocation to plant organs (Davies & Evans, 1982; Brock et al., 1988) stolon extension rate (Chapman, 1983) branching and flowering frequency (Chapman, 1983; Chapman & Anderson, 1987) and root initiation (Chapman, 1983; Brock et al., 1988) and growth (Caradus & Evans, 1977). Under New Zealand conditions white clover has by late winter about 80% of stolon buried, a low leaf to stolon ratio (0.58) and few unbranched plants (14%). During spring the buried stolon pool decreases by 70% as older basal stolon dies and this is associated with morphological changes in plants with decreases in dry weight, branching complexity, stolon weight per unit length and stolon/leaf ratio and increases in root number per plant and per unit length of stolon. Over this period, growth of branches vertically to the soil surface from buried stolon, and subsequent rapid development, gives these branches the appearance of new plants (Hay, 1983). Such branch growth often contributes significantly to the persistence of white clover as such stolons have strong root development (Brock et al., 1988). However small size of individuals in many plant communities is associated with decreased survival and productivity (Harper, 1977). In spring when white clover plants are of minimum size and simplest structure they may be more vulnerable to stress (environmental and managerial) and this might be related to the often noted variable performance of white clover in mixed swards (Frame & Newbould, 1986) especially in the more marginal of environments. For instance, a high loss (75%) of white clover in swards has been reported following two successive spring droughts (Hoglund, 1985). In late spring in response to increasing photoperiod and temperature the rate of growth of plants increases and the mean dry weight per plant increases, reflecting the new equilibrium between rates of formation and death of stolon, and the branching structure of the population changes back to the pre-spring structure and remains stable until the following spring (Brock et al., 1988; Hay et al., 1988).

The difficulty of measuring the density of white clover plants in swards has meant that this measurement is not generally made (Snaydon, 1984). However plant density has been assessed in the Palmerston North swards once a month for a year although no significant seasonal effects were observed. Mean values of plant density for set stocked and rotationally grazed swards containing 14 and 46 g stolon DW m^{-2} were 330 and 550 plants m^{-2}, respectively. Individual clones (genotypes) of white clover do not form local exclusive zones of occupancy but form intimately intermingled mixtures (Harper, 1985). For instance, Trathan (1983) has reported the presence of between 44-49 genotypes m^{-2} in an old Welsh pasture. If similar densities of genotypes occurred in Palmerston North swards then each genotype would be represented by 5-20 independent plants m^{-2}. This production of a number of physiologically independent plants of a genotype favors survival of the genotype and acts to preserve the genetic diversity of the population (Snaydon, 1985).

As a node may produce either an inflorescence or a vegetative branch but not both, each inflorescence precludes a potential branch. In USA environments, profuse flowering in spring negatively affects white clover persistence by reducing the number of stolons going into the hot dry summer (Gibson, 1957). Transplant experiments show white clover to display great phenotypic plasticity in relation to resource allocation to reproductive effort (Turkington & Maze, 1982; Horikawa, 1986a). Plant structures related to the survival of the individual constitute a primary sink and reproductive structures a secondary sink to which resource is allocated (Turkington & Maze, 1982). Defoliation stress reduces both reproductive effort and number of seed heads m^{-2} (Horikawa, 1986b, Chapman & Anderson, 1987) and under reasonably intensive grazing systems the maximum percentage of nodes flowering in summer (5.7%) compares with a maximum of 43% of nodes which branched and 51% of nodes with untapped potential to branch or flower (Chapman, 1983). Thus under such grazing conditions flowering must be considered unlikely to influence the vegetative spread of white clover.

On the other hand persistence and spread of white clover is of course possible via establishment from seed. Suckling (1959) advocated the grazing management strategy of summer spelling of pastures to promote seed head development and grazing by stock and transfer of seed to pastures of low white clover content via dung.

However recent studies in moist temperate environments in pastures with closed canopies have found that successful establishment of white clover from seed is a rare event (Turkington et al., 1979; Chapman, 1987) indicating that recruitment from seed plays a minor role in white clover persistence in sheep-grazed pastures in such environments. However where plant canopies are more open, white clover may persist as an annual (Hollowell, 1966) especially in subtropical environments (Jones, 1982). As seedling development in white clover is slow in the early stages (Thomas, 1987b) this militates against white clover successfully persisting in this manner and insufficiently advanced seedling development has been cited as a cause of low white clover content in swards in Queensland following dry spring conditions (Jones, 1982). This would suggest that perhaps white clover is not as well adapted for an annual life habit as was previously suggested (Hollowell, 1966) and that other legumes should be considered for such environments.

All the same, efforts have been made recently to improve the persistence of white clover in dryland environments by breeding for re-seeding potential (Macfarlane & Sheath, 1984), increased vegetative survival by increasing the degree of taprootedness (Woodfield & Caradus, 1987) or root shoot ratio (Smith & Morrison, 1983) and improved moisture conservation through reduced leaf size, stomatal closure and high cuticular resistance (Parsons, 1982).

Phenotypic Plasticity

Clonal species of indeterminate life span often display high levels of phenotypic plasticity and this is true of white clover (Hill, 1977; Brougham et al., 1978; Horikawa, 1986a,b). The primary method by which white clover adapts to any defoliation regime is by a plastic response which changes the size of plant organs such that, where possible, it is in an equilibrium with the defoliation regime (Briseno de la Hoz & Wilman, 1981; Wilman & Asiegbu, 1982; Hay & Baxter, 1984; Horikawa, 1986b). Plastic changes in the partitioning of dry matter to different organs has been recorded in response to season (Brougham, 1962; Davies & Evans, 1982; Brock et al., 1988), drought stress (Thomas, 1984), defoliation (Horikawa, 1986a) and nutrients (Harvey, 1979). Plastic responses are rapid, changes of 50% in size of plant organs occur within a month in the field (Briseno de la Hoz & Wilman, 1981; Brock et al., 1988). Phenotypic plasticity is under genetic control, can be subject to selection, may vary within and among populations and each trait may vary in the level of plasticity within and among individuals (Silander, 1985). Should plasticity be desired for a particular characteristic then it appears that scope exists for improving plasticity by selection for it. In view of the significance of phenotypic plasticity in conferring white clover with a mechanism to survive adverse conditions and to exploit favorable conditions the topic has attracted surprisingly little research.

The phenotypic response of white clover to nutrient stress is to restrict branch development by aborting young leafy buds (Wong, 1988, personal communication), to inhibit lateral bud development in subsequently produced nodes (Harvey, 1979) and to reduce internode length and leaf size (Harvey, 1979). These factors all negatively affect the vegetative spread of the species and hence persistence. Therefore to assist the establishment of white clover-based pastures, the correction of all nutrient deficiencies by fertilizer application at sowing has long been advocated (Levy, 1970). Although pot trial work has consistently found white clover to compete poorly with grass for nutrients (Haynes, 1980), in established pastures varying the phosphate inputs has inconsistent effects on the white clover component of the sward (Jackman & Mouat, 1972; Feyter et al., 1988). These inconsistencies may arise as sometimes other factors (e.g., drought, disease) have an overriding influence in determining persistence. However it is also highly likely that in grazed swards the wandering guerilla habit (Harper, 1985) of white clover allows many white clover plants to obtain nutrients from the many small pockets of soil enriched by earthworm castings, urine and dung returns and herbage decay, without intense competition from grass. Further work is required in this area.

Strawberry Clover (*T. fragiferum* L.) is unrelated to white clover but has essentially the same growth habit with less extreme stolon development. The initial crown and taproot system are stronger and longer-lived and the stolons do not root so freely. This species tolerates wet saline soils.

Kenya White Clover (*T. semipilosum* Fresen., *T. africanum* Ser., *T. burchellianum* Ser.) (Mannetje 1966; Jones et al., 1974). This group of African clovers are similar in habit to white clover, but the growing points of the stolons are not restricted to the soil surface and may become semi-erect under lax defoliation. This makes them more resistant to shading, but also more susceptible to grazing. *Trifolium burchellianum* also produces subsurface stolons, sometimes referred to as rhizomes. A number of native American clovers including *T. polymorphum* Poir. are also stoloniferous.

Adesmia bicolor, a little-known Uruguayan legume, forms a turf of prostrate rooted stolons bearing pinnate leaves and erect racemes of flowers on short axillary branches. Preliminary observations suggest it may be more drought tolerant than white clover but less cold tolerant.

DISCUSSION

Each perennial growth form meets the challenge of perennation and/or spread in the face of climatic stresses and defoliation in a different way, and poses its particular problems for agronomists who wish to maximize yield while maintaining persistence.

Persistence of the original plant is achieved most successfully by the tap-rooted crown-formers, which are classed as "k-strategists" or "phalanx" species and are successful competitors in stable environments because they monopolize environmental resources in their near neighborhood by their large, long-lived structures (Snaydon, 1985). Because of the depth to which the taproot penetrates and the protection of the crown by partial burial and/or shelter by stubble, these species will tolerate considerable drought and cold and show good regrowth characteristics, with topgrowth suppression of lateral buds. However those like lucerne where the growing points are elevated and exposed to defoliation cannot be cut or grazed too often, and root reserves must not be unduly depleted (especially before winter) if the plant is to survive. Species with deep taproots are generally very intolerant of wet soils.

These species remain spot-bound however, and although the size of the crown increases slowly and may be augmented in some species by peripheral growth of short buried rhizomes or casual adventitious rooting from prostrate stems, new independent plants (ramets) are not formed except where adventitious shoots are produced from the roots, as in the creeping lucernes.

Shrubs and subshrubs with their woody or semiwoody superstructure show another form of k-strategy, though most of the species discussed here have the r-selection traits of rapid early growth, high seed output and relatively short life (Williams, 1981). They also have deep taproot systems and are drought tolerant, but are the most vulnerable to excessive defoliation and physical damage because their stems and growing points are constantly exposed. Consequently they must be protected from overgrazing.

The clone-formers, which propagate themselves by rhizomes or stolons or by adventitious shoots from the root system, form ramets in new sites which can persist after the original plant has died. These are sometimes termed "vegetative r-strategists" or "guerilla" species, and can vary the proportion of resources invested in stolons or rhizomes (as opposed to investment in the parent plant) both seasonally and in different environments (Snaydon, 1985), thus achieving demographic flexibility in unstable environments. They also draw upon the resources of a wider area than the species relying only on a deep taproot system.

Species that send out underground rhizomes or suckering roots show great resistance to climatic extremes and also to defoliation, but are slow to establish and have slower regrowth after cutting because of the competing demands of other growth centres. The stolon-formers, which root relatively superficially from prostrate stems, have the least resistance to climatic adversity because they lack deep roots and dormant growing points, but the greatest adaptation to frequent defoliation because the stems and growing points are mostly at ground level and regrowth is rapid.

REFERENCES

Bowley, S.R., N.L. Taylor, and C.T. Dougherty. 1984. Physiology and morphology of red clover. Adv. Agron. 37: 318-341.

Briseno de la Hoz, V.M., and D. Wilman. 1981. Effects of cattle grazing, sheep grazing, cutting and sward height on a grass-white clover sward. J. Agric. Sci. 97: 699-706.

Brock, J.L. 1988. Evaluation of New Zealand bred white clover cultivars under rotational grazing and set stocking with sheep. Proc. N.Z. Grassl. Assoc. 49: 203-206.

Brock, J.L., M.J.M. Hay, V.J. Thomas, and J.R. Sedcole. 1988. Morphology of white clover (*Trifolium repens* L.) plants under intensive sheep grazing. J. Agric. Sci. 111: 273-283.

Brougham, R.W. 1962. The leaf growth of *Trifolium repens* as influenced by seasonal changes in the light environment. J. Ecol. 50: 449-459.

Brougham, R.W., P.R. Ball, and W.M. Williams. 1978. The ecology and management of white clover-based pastures. p.309-324. *In* J.R. Wilson (ed.) Plant relations in pastures. CSIRO, East Melbourne.

Bryant, W.G. 1974. Caucasian clover (Trifolium ambiguum - a review. J. Aust. Inst. Agric. Sci. 40: 11-18.

Burdon, J.J., and J.L. Harper. 1980. Relative growth rates of individual members of a plant population. J. Ecol. 68: 953-957.

Caradus, J.R. 1977. Structural variation of white clover root systems. N.Z. J. Agric. Res. 20: 213-219.

Caradus, J.R., and P.S. Evans. 1977. Seasonal root formation of white clover, ryegrass and cocksfoot in New Zealand. N.Z. J. Agric. Res. 20: 337-342.

Chapman, D.F. 1983. Growth and demography of Trifolium repens stolons in grazed hill country pastures. J. Appl. Ecol. 20: 597-608.

Chapman, D.F. 1987. Natural re-seeding and Trifolium repens demography in grazed hill pastures. II. Seedling appearance and survival. J. Appl. Ecol. 24: 1037-1043.

Chapman, D.F., and C.B. Anderson. 1987. Natural re-seeding and Trifolium repens demography in grazed hill pastures. 1. Flower-head appearance and fate, and seed dynamics. J. Appl. Ecol. 24: 1025-1035.

Curll, M.L. 1980. The effects of grazing by set-stocked sheep on a perennial ryegrass/white clover pasture. Ph.D. Thesis, Reading University.

Curll, M.L., and R.J. Wilkins. 1982. Frequency and severity of defoliation of grass and clover by sheep at different stocking rates. Grass Forage Sci. 37: 291-297.

Davies, A., and M.E. Evans. 1982. The pattern of growth in swards of two contrasting varieties of white clover in winter and spring. Grass Forage Sci. 37: 199-207.

Duke, J.A. (ed.) 1981. Handbook of legumes of world economic importance. Plenum Press, New York.

Erith, A.G. 1924. White clover (Trifolium repens L.). A monograph. Duckworth, London.

Fergus, E.N., and E.A. Hollowell. 1960. Red clover. Adv. Agron. 12: 365-436.

Feyter, C., M.B. O'Connor, R.J. Fris, and B. Addison. 1988. The effects of restricting or stopping fertilizer application to Waikato dairy pastures. Proc. N.Z. Grassl. Assoc. 49: 157-160.

Frame, J., and P. Newbould. 1986. Agronomy of white clover. Adv. Agron. 40: 1-88.

Gibson, P.B. 1957. Effect of flowering on the persistence of white clover. Agron. J. 49: 213-215.

Gill, N.T., K.C. Vear, and D.J. Barnard. 1980. Agricultural botany I. Dicotyledonous crops. Ed. 3. Gerald Duckworth & Co. Ltd, London.

Harper, J.L. 1977. Population biology of plants. Academic Press, London.

Harper, J.L. 1985. Modules, branches and the capture of resources. p.1-33. In J.B.C. Jackson et al. (ed.). Population biology and evolution of clonal organisms. Yale University Press, New Haven.

Hart, A.L. 1987. Physiology. p.125-147. In M.J. Baker and W.M. Williams (ed.) White clover. CAB International, Wallingford, Oxfordshire.

Harvey, H.J. 1979. The regulation of vegetative reproduction Ph.D. thesis, University of Wales.

Hay, M.J.M. 1983. Seasonal variation in the distribution of white clover (Trifolium repens L.) stolons among 3 horizontal strata in

2 grazed swards. N.Z. J. Agric. Res. 26: 29-34.
Hay, M.J.M., J.L. Brock, V.J. Thomas, and M.V. Knighton. 1988. Seasonal and sheep grazing management effects on branching structure and dry weight of white clover plants in mixed swards. Proc. N.Z. Grassl. Assoc. 49: 197-201.
Hay, M.J.M., D.F. Chapman, R.J.M. Hay, C.G.L. Pennell, P.W. Woods, and R.H. Fletcher. 1987. Seasonal variation in the vertical distribution of white clover stolons in grazed swards. N.Z. J. Agric. Res. 30: 1-8.
Hay, R.J.M., and G.S. Baxter. 1984. Spring management of pasture to increase summer white clover growth. p.132-137. In Proc. of the 34th Lincoln College Farmers conference, Gore, Southland, 6-7 June 1984. Rural Development and Extension Centre, Lincoln College, N.Z.
Haynes, R.J. 1980. Competitive aspects of the grass-legume association. Adv. Agron. 33: 227-256.
Hays, W.M. 1888. Roots of clovers. Rep. Dep. Agric. Univ. Minn. 5, 1887-1888: 18-197.
Heath, M.E., D.S. Metcalf, and R.E. Barnes. 1973. Forages. Ed.3. Iowa State Univ. Press, Ames.
Henson, P.R. (ed). 1957. The lespedezas. Adv. Agron. 9: 113-22.
Henson, P.R. 1963. Crownvetch. A soil conserving legume and a potential pasture and hay plant. Crops Res. ARS 34-53, USDA, Washington, DC.
Hill, J. 1977. Plasticity of white clover grown in competition with perennial ryegrass. p.24-25. In Welsh Plant Breed. Stn. Annu. Rep. 1976. University College of Wales, Aberystwyth, U.K.
Hoglund, J.H. 1985. Grazing intensity and soil nitrogen accumulation. Proc. N.Z. Grassl. Assoc. 46: 65-69.
Hoglund, J.H., J.R. Crush, J.L. Brock, P.R. Ball, and R.A. Carran. 1979. Nitrogen fixation in pasture XII. General discussion. N.Z. J. Exp. Agric. 7: 45-51.
Hollowell, E.A. 1966. White clover Trifolium repens L., annual or perennial? p.184-187. In Proc. 10th Int. Grassl. Cong., Helsinki, 7-16 July 1966. 10th Int. Grassl. Cong., Helsinki, Finland.
Horikawa, Y. 1986a. Plastic allocation of photosynthetic product in white clover (Trifolium repens L.). J. Jpn. Soc. Grassl. Sci. 32: 225-234.
Horikawa, Y. 1986b. Reproductive strategy in white clover (Trifolium repens L.) of different habitats. J. Jpn. Soc. Grassl. Sci. 32: 235-242.
Iversen, C.E., and G. Meijer. 1967. Types and varieties of lucerne. p.74-84. In R.H.M. Langer (ed.) The lucerne crop. Reed, Wellington.
Jackman, R.H., and M.C.H. Mouat. 1972. Competition between grass and clover for phosphate. 1. Effect of browntop (Agrostis tenuis Sibth.) on white clover (Trifolium repens L.) growth and nitrogen fixation. N.Z. J. Agric. Res. 15: 653-666.
Jones, R.M. 1982. White clover (Trifolium repens) in subtropical South-East Queensland. 1. Some effects of site season and management on the population dynamics of white clover. Trop. Grassl. 16: 118-127.
Jones, R.M., B.W. Strijdom, and E.P. Theron. 1974. The indigenous South African clovers (T. africanum Ser. and T. burchellianum Ser. and their potential as pasture legumes. Trop. Grassl. 8: 7-16.

Kershaw, K.A. 1959. An investigation of the structure of a grassland community. 11. The pattern of Dactylis glomerata, Lolium perenne and Trifolium repens. III. Discussion and conclusions. J. Ecol. 47: 31-53.

King, J., W.I.C. Lamb, and M.T. McGregor. 1978. Effect of partial and complete defoliation on regrowth of white clover plants. J. Brit. Grassl. Soc. 33: 49-55.

Lambrechtsen, N.C. 1974. Crownvetch - a legume with soil conservation potential. Soil and Water, March 1974: 48-50.

Langer, R.H.M. (ed.) 1973a. Pastures and pasture plants. Reed, Wellington.

Langer, R.H.M. 1973b. Lucerne. p.347-364. In R.H.M. Langer (ed.) Pastures and pasture plants. Reed, Wellington.

Lazenby, A., and F.G. Swain. 1969. Pasture species. p.67-97. In Intensive pasture production. Angus and Robertson Pty Ltd., Sydney.

Levy, E.B. 1970. Grasslands of New Zealand. Ed.3. Government Printer, Wellington.

Logan, L.A. 1982. Tree lucerne in New Zealand. Crop Research Division DSIR. Christchurch, N.Z.

Logan, L.A., and J.E. Radcliffe. 1985. Fodder trees. A summary of current research in New Zealand. Crop Res. Div. Rep. 106. Christchurch, N.Z.

Macfarlane, M.J., and G.W. Sheath. 1984. Clover - What types for dry hill country. Proc. N.Z. Grassl. Assoc. 45: 140-150.

McDonald, H.A. 1944. Birdsfoot trefoil (Lotus corniculatus L.) Its characteristics and potentialities as a forage legume. Cornell Univ. Agric. Exp. Stn. Mem. 261.

Mannetje, L.t' 1966. A punched card key to species of Trifolium L. in Africa south of the Sahara excluding Ethiopia. East Afr. Agric. For. J. 31: 261-270.

Marten, G.C. 1985. Environmental and management limitations of legume-based forage systems in the northern United States. In Forage legumes for energy-efficient animal production. Proc. Trilateral Workshop 1984, Palmerston North, N.Z. USDA, Washington, DC.

Martin, J.H., and W.H. Leonard. Principles of field crop production. Ed. 2. Macmillan, New York.

Mouat, M.C.H. 1983. Competitive adaptation by plants to nutrient shortage through modification of root growth and surface charge. N.Z. J. Agric. Res. 26: 327-332.

Palmer, T.P. 1967. Lucerne breeding in New Zealand. p.85-93. In R.H.M. Langer (ed.) The lucerne crop. Reed, Wellington.

Parsons, L.R. 1982. Plant responses to water stress. p.175-192. In M.N. Christiansen and C.F. Lewis (ed.) Breeding plants for less favourable environments. John Wiley and Sons, New York.

Seaney, R.R., and P.R. Henson. 1970. Birdsfoot trefoil. Adv. Agron. 22: 120-153.

Sheath, G.W. 1980a. Effect of season and defoliation on growth habit of Lotus pedunculatus Cav. cv. 'Grasslands Maku'. N.Z. J. Agric. Res. 23: 191-200.

Sheath, G.W. 1980b. Production and regrowth characteristics of Lotus pedunculatus Cav. cv 'Grasslands Maku'. N.Z. J. Agric. Res. 23: 201-209.

Sheath, G.W. 1981. Lotus pedunculatus - an agricultural plant? Proc. N.Z. Grassl. Assoc. 4: 160-167.

Silander, J.A. 1985. Microevolution in clonal plants. p.107-155. In J.B.C. Jackson et al. (ed.) Population biology and evolution of clonal organisms. Yale University Press, New Haven.
Smith, A., and A.R.J. Morrison. 1983. A deep rooted white clover for South African conditions. Proc. Grassl. Soc. South Africa 18: 50-52.
Snaydon, R.W. 1984. Plant demography in an agricultural context. p.389-407. In R. Dirzo and J. Sarukhan (ed.) Perspective on plant population ecology. Sinauer Assoc. Inc., Sunderland, Mass.
Snaydon, R.W. 1985. Aspects of the ecological genetics of pasture species. In J. Haeck and J.W. Woldendorp (eds.) Structure and functioning of plant populations. North Holland Publ. Co., Amsterdam.
Speer, G.S., and D.W. Allinson. 1985. Kura clover (Trifolium ambiguum). Legume for forage and soil conservation. Econ. Bot. 39: 165-176.
Stevenson, C.A., and A.S. Laidlaw. 1985. The effect of moisture stress on stolon and adventitious root development in white clover (Trifolium repens L.). Plant and Soil 85: 249-257.
Suckling, F.E.T. 1959. Pasture management trials on unploughable hill country at Te Awa. 11. Results for 1951-57. N.Z. J. Agric. Res. 2: 488-543.
Taylor, N.L. (ed.) 1985. Clover science and technology. ASA Madison, Wisconsin.
Thomas, H. 1984. Effects of drought on the growth and competitive ability of perennial ryegrass and white clover. J. Appl. Ecol. 231: 591-602.
Thomas, R.G. 1987a. The structure of the mature plant. p.1-29. In M.J. Baker and W.M. Williams (ed.) White Clover. C.A.B. International, Wallingford, Oxfordshire.
Thomas, R.G. 1987b. Vegetative growth and development. p.31-62. In M.J. Baker and W.M. Williams (ed.) White Clover. C.A.B. International, Wallingford, Oxfordshire.
Trathan, P. 1983. Clonal interactions of Trifolium repens and Lolium perenne. Ph.D. thesis, University of Wales.
Turkington, R., and J.J. Burdon. 1983. The biology of Canadian weeds. 57. Trifolium repens L. Cand. J. Plant Sci. 63: 243-266.
Turkington, R., M.A. Cahn, A. Vardy, and J.L. Harper. 1979. The growth, distribution and neighbour relationships of Trifolium repens in permanent pasture. 111. The establishment and growth of Trifolium repens in natural and perturbed sites. J. Ecol. 67: 231-243.
Turkington, R., and J. Maze. 1982. Patterns of dry-matter distribution in transplanted populations of Trifolium repens and its bearing on ecological relationships. Cand. J. Bot. 60: 2014-2018.
Van Kraayenoord, C.W.S., and R.L. Hathaway (ed.) 1986. Plant materials handbook for soil conservation, Vol. 2. Introduced plants. Water & Soil Misc. Publ. 94. Nat. Water & Soil Conservation Authority, Wellington.
Walton, P.D. 1983. Production and management of cultivated forages. Reston Publ. Co., Reston, VA.
Watson, M.J. 1982. Hedysarum coronarium - a legume with potential for soil conservation and forage. N.Z. Agric. Sci. 16: 189-93.
Whyte, R.O., C. Milsson-Leissner, and H.C. Trumble. 1953. Legumes

in agriculture. FAO Agricultural Studies No. 21.
Williams, P.A. 1981. Aspects of the ecology of broom (*Cytisus scoparius*) in Canterbury, New Zealand. N.Z. J. Bot. 19: 31-43.
Wilman, D., and J.E. Asiegbu. 1982. The effects of variety, cutting interval and nitrogen application on the morphology and development of stolons and leaves of white clover. Grass Forage Sci. 37: 15-27.
Wills, B.J., J.S.C. Begg, and A.G. Foote. 1989. *Dorycnium* species - new legumes with potential for dryland pasture improvement and soil resource conservation. Proc. N.Z. Grassl. Assoc. 50: 169-174.
Woodfield, D.R., and J.R. Caradus. 1987. Adaptation of white clover to moisture stress. Proc. N.Z. Grassl. Assoc. 48: 143-149.
Woodfield, D.R., and M.B. Forde. 1987. Genetic variability within tagasaste. Proc. N.Z. Grassl. Assoc. 48: 103-8.
Young, D.J.B. 1958. A study of the influence of nitrogen on the root weight and nodulation of white clover in a mixed sward. J. Brit. Grassl. Soc. 13: 106-114.
Zohary, M., and D. Heller. 1984. The genus *Trifolium*. Israel Academy of Science and Humanities, Jerusalem.

DISCUSSION

Kretschmer: Are there non-toxic *Dorycnium* spp?

Forde: They are non-toxic as far as I know, certainly the two species being studied in New Zealand (*D. hirsutum* and *D. pentaphyllum*).

Lowther: Does damage caused by grazing predispose shrubs to disease?

Forde: There is no evidence of this as yet. However ring barking by goats can be a problem.

Leath: Is slow establishment related to high levels of persistence?

Forde: I am not aware that it is directly related. However, from an evolutionary point of view perhaps there has not been selection pressure for strongly perennial species to establish rapidly.

Sheath: *Lotus pedunculatus* is a rhizome former and invests a lot of energy in underground growth, but it is this same growth that is very sensitive to management and/or competition stress. Underground root and rhizome size decreases markedly under intensive grazing and becomes intolerant of drought. Ungrazed *L. pedunculatus* can be quite drought tolerant.

Jones: In the subtropics, the seasonal pattern of stolon density is similar to Mike Hay's story, but 6 months out of sequence. Stolon branching peaks in spring, plants fragment over summer, and stolon length and branching are least in autumn. In some years stolons are even eliminated by autumn.

Gramshaw: Could stolon mortality in spring be due to pathogens?

Hay: The specific causes of stolon loss have not been identified.

Woomer: Please elaborate on the function of "buried stolons".

Hay: Buried stolon is important in the regeneration of white clover in spring as it provides sites for branches that grow to the soil surface. Another function often served by buried stolon is to connect the younger more actively growing portion of the plant with a node with a well established root system.

Sheaffer: Is there a threshold level for stolon carbohydrates, below which persistence is affected?

Hay: We do not know the answer to that question as yet, but work is already under way to investigate carbohydrate levels in stolons in relation to branching.

Watson: Is the white clover plant under a canopy before grazing, and therefore more subject to stress than after grazing?

Hay: I doubt it. There is recent evidence suggesting that white clover competes effectively for light in most situations. Post grazing recovery after defoliation is very energy demanding and can drop carbohydrates in stolons to very low levels so any additional stress at this time can adversely affect persistence.

Brougham: This work of Mike Hay's is giving us the reasons why (mechanisms) white clover shows such wide fluctuations in persistence in relation to yield from year to year in New Zealand farming.

SELECTION FOR ROOT TYPE IN RED CLOVER

R.R. Smith

SUMMARY

Red clover (Trifolium pratense L.) is generally characterized as being adapted to cool, temperate climates, adapted to wet, acid soils, and having well-developed adventitious (fibrous) roots. The objectives of this study were to characterize six cultivars of red clover for root type and to divergently select for tap and adventitious root types. Six cultivars of red clover (Lakeland, Arlington, Marathon, Florex, Chesapeake, and Pennscott) were grown in a silt loam soil and the roots were subsequently classified for root type. Both Arlington and Marathon, selected for persistence subsequent to Lakeland, had a higher frequency of plants classified as fibrous rooted types. This was reflected in greater persistence of the cultivars in 3-yr-old stands. The distribution of root types in Florex was similar to Lakeland. Chesapeake and Pennscott, adapted to high clay soils of eastern USA, had a high frequency of fibrous-rooted plants. Two cycles of divergent selection in Arlington for tap and fibrous-rooted types were effective. Forage yield and persistence of the second cycle taproot selection was similar to Arlington in performance when tested in well-drained silt loam soils.

INTRODUCTION

Red clover (Trifolium pratense L.) is best adapted to cool, temperate climates. It is known to tolerate moist wet soils and is poorly adapted to drought conditions. Persistent plants have well-developed adventitious root systems in lieu of a taproot (Kendall & Stringer, 1985). Early in the life of a red clover plant, the primary axis begins to deteriorate with an eventual loss of the initial taproot. However, adventitious roots develop simultaneously to support further growth of the plant. Ryle et al. (1981) provide evidence that the main axis of red clover is programmed genetically to be susceptible to physiological stresses. Using ^{14}C, they report that most (78%) of the photosynthate from mature leaves was translocated to the branches of red clover and only 22% to the primary root.

Root morphology in red clover can be influenced by soil type and conditions. Weaver (1926) reported that red clover roots can penetrate to 3 m in prairie soils while Farris (1934) observed that they only penetrated to 1 m in heavy clay soils. In Wisconsin field tests, roots of red clover penetrated to a depth of 86 cm in a Miami silt loam, 66 cm in a Plainfield sand, and 41 cm in a Spencer silt loam (Lamba et al., 1949). The ability to penetrate deep into the soil and to produce adventitious roots would suggest that red clover has the plasticity to adapt to varying growing conditions and/or cultivars are composed of a group of genotypes, each with different genetic potential for a different root morphology.

Both Cressman (1967) and Taylor et al. (1962) suggest that the most persistent red clover cultivars could be developed by selecting genotypes with the greatest potential of producing adventitious roots.

The objectives of this study were to examine the variability of root type in select cultivars of red clover and to divergently select plants for tap and adventitious (fibrous) root types.

MATERIALS AND METHODS

Six red clover cultivars, Lakeland, Arlington, Marathon, Florex, Chesapeake, and Pennscott were seeded in the spring in six rows each spaced 50 cm apart in a Parr silt loam soil at the University of Wisconsin Experimental Station, Arlington, WI. After emergence, plants were thinned to 2.5 cm apart to provide approximately 100 plants per row (total 600). Herbage growth was removed twice during the year. The following spring the 1-yr-old plants were undercut to a depth of 17.5 cm and lifted from the soil. Roots of each cultivar were scored for root type (RT) based on the following scale: 1=primarily tap, only an occasional adventitious or secondary root visible; 2=predominately tap with some secondary branching and adventitious roots; 3=moderate taproot with moderate secondary branching and adventitious roots; 4=slight to no taproot with little branching but profuse adventitious roots; 5=no taproot or secondary branching but profuse adventitious roots.

The cultivars Lakeland, Arlington, and Marathon were developed in Wisconsin cooperatively by the USDA, ARS and the University of Wisconsin and were released in 1961, 1973 and 1987, respectively. Florex was developed in northcentral USA by the Northrup-King Company and released in 1973. Chesapeake was selected in Maryland and released in 1952 and Pennscott was selected in Pennsylvania and released about 1954.

Thirty (5%) plants with tap or near taproot types (classes 1 and 2) and 30 plants with fibrous or near fibrous root types (classes 4 and 5) were selected from Arlington to constitute the parents of the divergent selected populations. Plants of each set of Co parents were intercrossed in isolation cages to provide seed of the first cycle of selection. Honeybees (<u>Aphis</u> spp.) were used as pollinating agents. In subsequent selections and for final testing, plants were established as described above in August in Plainfield sand at the University of Wisconsin Experiment Station, Hancock, WI. Plants were undercut, lifted and scored for RT in the spring following the August seeding. Selection intensity in the second cycle was 2.5%.

Forage yield (dry matter) and stand assessment of cultivars and experimental populations were evaluated in yield plots (0.9 X 7.5 m) established at Arlington, WI in Parr silt loam soil. Distributions were compared using the Chi-square statistic and assuming either Lakeland or Arlington distributions as expected distributions. Means were compared using the protected LSD.

RESULTS AND DISCUSSION

Distribution of plants for root type between the cultivars Lakeland, Arlington, and Marathon was significantly different (Table 1). Both Arlington and Marathon had a higher frequency of plants in the more fibrous RT classes (scores 3 and 4) than Lakeland. Plants in Marathon were also significantly more fibrous than the plants in Arlington. Some germplasm in Arlington was selected for persistence in old stands, but approximately 35 percent of the germplasm in Marathon was selected for persistence to wet, acid soils of central Wisconsin. While no conscious selection was applied for root type in either cultivar, it would appear that the distribution of root type has shifted towards more adventitious roots in these two cultivars. This may have contributed to the improved stands in yield trials after 3 yr. Plants in Florex were distributed across RT classes similar to Lakeland, but both Pennscott and Chesapeake, adapted to eastern USA had more fibrous types. Neither of the latter two cultivars persist well in the northcentral USA suggesting that other factors in addition to root type are involved with persistence.

Divergent recurrent phenotypic selection for RT in the cultivar Arlington was effective (Table 2). Both cycles of selection for tap and fibrous roots were significantly different from Arlington. Selection for the taprooted type was more effective in that 24 and 16 % of the plants were characterized as taprooted type (class 1) in cycles 1 and 2, respectively. Only 6 and 2% of the plants were in the fibrous class (class 5) cycles 1 and 2%, respectively, in the fibrous rooted selections. Progeny test of cycle 2 plants suggests that sufficient genetic variability remains in both populations to warrant further selection (data not presented). In the second cycle taprooted

Table. 1. Characterization of red clover cultivars for root type (RT)[1].

Cultivar	Year Release	Percent plants with RT of 1	2	3	4	5	Mean RT	Stand-fall 3rd yr
		%						%
Lakeland	1961	1	60	36	3	-	2.40	32
Arlington	1973	2	52	40	6	-	2.51*	48
Marathon	1987	3	25	59	14	-	2.53**	67
Florex	1974	8	52	37	3	-	2.38	41
Pennscott	1952	4	30	64	2	-	2.65	10
Chesapeake	1954	3	41	53	3	-	2.56	15

*RT mean and distribution significantly different from Lakeland at 5% level.
**RT distribution significantly different from Lakeland and Arlington at 1% level.
[1]RT score 1=taproot 5=no tap (fibrous).

Table 2. Response to two cycles of recurrent phenotypic selections for root type (RT)[1]

Germplasm	Cycle	Percent plants with RT of 1	2	3	4	5	Mean RT	Performance in 3rd yr[2] Yield	Stand
								% Arl	%
Tap Sel.	2	16	48	35	1	0	2.21*	100	55
Tap Sel.	1	24	29	28	17	2	2.43*	-	-
Arlington	0	2	52	40	6	0	2.50	100	52
Fibrous Sel.	1	3	25	42	24	6	3.05*	-	-
Fibrous Sel.	2	1	22	55	20	2	3.00*	28	15

*Both mean RT and distribution of RT different from Arlington at the 1% level of significance.
[1]RT score 1=taproot 5=no tap (fibrous).
[2]Evaluated at Arlington, WI in well drained silt loam soil.

population means ranged in root type score from 1.98 to 2.27 and in the fibrous rooted population from 2.53 to 3.67. This would suggest that an exclusively taprooted population would be achieved earlier than a fibrous rooted population.

The shift in soil type between cycles of selection did not appear to influence selection since both populations moved in the direction of applied selection. Forage yield and persistence of the taproot selected populations was similar to Arlington when tested in a well drained silt loam soils.

Cultivars developed by selecting in old stands (Arlington) or in wet, acid soils (Marathon) support the concept that more persistent cultivars have a higher proportion of plants with more adventitious root systems. However, when selecting directly for root type, the taprooted type population was more persistent than the fibrous rooted population. Additional studies are required to ascertain the responses of these selected populations to varying moisture conditions.

REFERENCES

Cressman, R.M. 1967. Internal breakdown and persistence of red clover. Crop Sci. 7:357-361.

Farris, N.F. 1934. Root habits of certain crop plants as observedin the humid soils of New Jersey. Soil Sci. 38:87-111.

Kendall, W.A. and W.C. Stringer. 1985. Physiological aspects of clover. In N.L. Taylor (ed.) Clover science and technology. Agronomy 25:111-159.

Lamba, P.S., H.L. Ahlgren, and R.J. Muchenhern. 1949. Root growth of alfalfa, medium red clover, bromegrass, and timothy under various soil conditions. Agron. J. 41:451-458.

Ryle, G.J.A., C.E. Powell, and A.J. Gordon. 1981. Patterns of ^{14}C labelled assimilate partitioning in red and white clover during vegetative growth. Ann. Bot. 47:505-514.

Taylor, N.L., W.H. Strobe, W.A. Kendall, and E.N. Fergus. 1962. Variation and relation of colonal persistence and seed production in red clover. Crop Sci. 2:203-305.

Weaver, J.E. 1926. Root development of field crops. McGraw-Hill,New York.

DISCUSSION

Clements: What were the management strategies applied to the cultivars and/or selected material to estimate third year stand?

R. Smith: Germplasm was evaluated in well drained Plainfield sand subjected to two harvests over each of two years, i.e. simulated hay conditions.

Caradus: Are there any shoot characters in red clover that change with selection for root type that confound results regarding shoot growth? In white clover increased taproot diameter is related to large leaf size and upright habit.

R. Smith: This did not appear to happen with red clover. There did not appear to be correlated changed in shoot characteristics with selection for root type.

Buxton: If red clover is selected for more fibrous roots, will it be more drought susceptible?

R. Smith: It hasn't been tested as yet, but it may.

Leath: The primary benefit from the program to selected fibrous versus taprooted red clover might be adapation rather than persistence. Under Pennsylvania conditions, there generally is not enough of the root system after two growing seasons to carry the plant through the next winter.

Gramshaw: Taprooted habit does not always confer drought tolerance. In some soils the taproot length may be constrained by soil properties or disease factors.

M.J. Hay: New Zealand work has shown a general trend from taproot to fibrous over time. This occurs more rapidly in early flowering cultivars, than late flowering cultivars. The most persistent plants tend to be the ones which retain their taproot characters.

R. Smith: We found this to be true in our germplasm but red clover only persists for two to three years. I believe that ideally it would be best to have plants which retain the taproot but also have the potential to produce numerous fibrous or adventitious roots.

Scott: In this and other papers, red clover is described as a species best suited to high fertility and moist conditions. However, red clover, along with timothy at Yorkshire as a group of species, is adapted to low fertility clay soils.

Brougham: Would you use a similar selection approach if you were developing plants in an environment where there is some winter growth of red clover, such as in New Zealand? This, as opposed to your environment where true dormancy occurs.

R. Smith: Yes, I think I would, the procedure seems applicable irrespective of the growing conditions. The root type desired may change depending upon selection objectives.

Brougham: Would this type of study i.e., selection of root types in different environments, be conducive to a collaborative research project?

R. Smith: Yes, I believe it would develop into an excellent cooperative project, especially if practiced on one set of germplasm under two or three different environmental conditions and soil types.

TROPICAL FORAGE LEGUME DEVELOPMENT, DIVERSITY, AND METHODOLOGY FOR DETERMINING PERSISTENCE

Albert E. Kretschmer, Jr.

SUMMARY

Of the 17 250 species of Leguminosae, all but three cultivars occur in the subfamily Papilionoideae; and most are classified as tropical. Morphological diversity among tropical legumes is large compared with that among temperate species. The diversity provides germplasm for most edaphic and climatic environments in the tropics and subtropics with annual rainfall above about 500 mm. Although more than 40 cultivars have been released for commercial use, many are no longer planted and most others are restricted to relatively small environmental niches. This may be a result of faulty evaluation techniques when predictions of legume persistence under grazing are based in large part on cutting regimes persistence. Because of this possibility, evaluation methods should include reducing the time from initial introduction of germplasm to measuring its reaction to grazing. Modifications of conventional cut and weigh procedures are suggested that would include grazing and ratings of legumes populations. Strip seeding and site method with grazing are discussed.

INTRODUCTION

Research with tropical forage legumes for pasture, stored feed, and special uses is recent compared with that of temperate species. Research methodology used with temperate legumes generally has not been successful with tropical legumes. Lack of knowledge of the diversity, adaptability, and reasons for persistence in tropical species were the primary reasons for the limited success in developing persistent cultivars for grazing.

The object of this presentation is to review the historical and ecological development, diversity, and adaptability of presently used and experimental cultivars, and to suggest the need for alternate and shorter methodologies for determining persistence under grazing.

HISTORY AND DEVELOPMENT

The Leguminosae family is comprised of three subfamilies, Caesalpinioideae, Mimosoideae, and Papilionoideae. It is one of the three largest families among the angiosperms (Polhill et al., 1981). The emergence and speciation (Stebbins, 1982) of legumes probably occurred during the upper Cretaceous era in west

Gondwanaland with the emergence of the Caesalpinioideae about 60 to 70 million years ago (m.y. BP) (Crepet & Taylor, 1985; Raven & Polhill, 1981). Its large divergence probably began during the Eocene (38 to 54 m.y. BP) with diversification of the Caesalpiniodeae into the Mimosoideae by the middle Eocene, and into the Papilionoideae by the Pliocene Epoch. Because of the geographic configurations among continents at that time, there is evidence that Africa was the primary site of legume evolution. Furthermore, the evidence suggests that the evolution of legumes in Africa, Madagascar, and Latin American took place during most of the Tertiary. Some of the dispersal has been long-range across large water barriers. The present-day distribution of the legumes is from the polar ice areas to the equator.

There are 42 legume tribes, 651 genera, and about 17 250 species (Polhill & Raven, 1981; Kretschmer, 1985). Most of the herbaceous forage legumes are included in the Papilionoideae which is comprised of about 440 genera and 12 000 species. The 7 genera and 477 species of the Trifolieae tribe contain the majority of the well-known temperate forage legumes including the clovers and alfalfa. Smaller temperate tribes include Loteae (the trefoils with 4 genera and 127 species), and Vicieae (vetches and peas with 1 genus and 298 species).

TROPICAL LEGUMES

The tropical forage legumes that have been released for commercial use are included in 6 tribes with 168 genera and about 4 300 species. Many species are not suitable as forage because of toxicity, antiquality, or woody characteristics. The remaining, largely untested herbaceous species, belong in the Aeschynomeneae, Desmodieae, and Phaseoloeae tribes (Williams & Clements, 1985). Except for leucaena (Leucaena leucocephala deWit) and desmanthus (Desmanthus virgatus Willd.) in Mimosoideae, and Cassia rotundifolia Pers. in Caesalpinioideae, all others that have been released or have exhibited potential are found in the Papilionoideae. The most economically important tribe in Papilionoideae is Phaseoleae, species of which are found worldwide. It contains the edible soybean [Glycine max (L.) Merr.], cowpeas [Vigna unguiculata (L.) Walp], edible beans (Phaseolus vulgaris L.) and lima bean (P. lunatus L.), pigeon pea [Cajanus cajan (L.) Millsp.], lablab bean [Lablab purpureus (L.) Sweet], yam bean (Pachyrhizus DC.), and winged bean [Psophocarpus tetragonolobus (L.) DC.]. Other genera in Phaseoleae that include forages (about 680 spp.) are velvet bean (Mucuna Adan.), Canavalia DC., Galactia P. Brown, calopo (Calopogonium mucunoides Desv.), tropical kudzu, Teramnus P. Br., perennial soybean (Neonotonia wightii Lackey), Centrosema Benth., Clitoria L., Dolichos L., Macrotyloma (Wight & Arn.) Verdc., and Macroptilium (Benth.) Urban. Aeschynomeneae (Benth.) Hutch. is another large tribe that has added to the available commercial cultivars. Primarily found in the tropics and subtropics it includes: Aeschynomene L., Zornia J. F. Gmel., Stylosanthes Swartz (Stace & Edye, 1984), and Arachis L., and totals about 330 species (Kretschmer, 1985). Examples of commercial cultivars in this tribe are American jointvetch (Aeschynomene americana L.) (Kretschmer & Bullock, 1979; Hodges et al., 1982), Australian jointvetch (A. falcata DC.)

(Wilson et al., 1982), stylo (Stylosanthes guianensis Sw.), Townsville style (S. humilis HBK.), Caribbean stylo (S. hamata Taub.), Shrubby stylo (S. scabra Vog.), S. capitata Vog. (Stace & Edye, 1984) Arachis glabrata Benth. (Prine et al., 1981), and A. pintoi (Grof, 1985).

DIVERSITY

Species Distribution

The main centers of evolution of the tropical legumes are in Brazil, Mexico, eastern Africa, and the Sino-Himalayan region. Most of the tropical legume species, such as Centrosema (Clements & Williams, 1980) that have been used or that have "potential" come from tropical America. Notable exceptions are alyce clover, tropical kudzu, carpon desmodium (Desmodium heterocarpon DC.) and hairy indigo from Asia, and perennial soybean and lotononis (Lotononis bainesii Baker) from Africa (Williams, 1983a).

Morphology

Morphological diversity of tropical legumes is much greater than that among prominent temperate legumes (Kretschmer, 1978; Williams, 1983a). They range from rather insignificant herbaceous species such as lotononis, Vigna parkeri Bak., Desmodium heterophyllum DC., and prostrate forms of alyce clover, to the vines, centro, and Macroptilium Urb. and to shrubs and trees such as desmanthus and leucaena. The variation among species can be high, as evidenced by that found in Desmodium (Ohashi, 1973) and Aeschynomene (Kretschmer & Bullock, 1979); or within species such as Stylosanthes (Stace & Edye, 1984). Generalized gross morphological diversity and extent of use of important legumes are presented in Table 1.

Edaphic Adaptation

Tropical legumes differ in their tolerance to soil acidity and Al. For example, Centrosema macrocarpum Benth. (Schultze-Kraft, 1986), C. acutifolium (Schultze-Kraft et al., 1987), and C. brasilianum (Belalcazar & Schultze-Kraft, 1986) grow satisfactorily in soil with a pH value of 4.3 and an 80% Al saturation, while common centro, Siratro, American jointvetch, and carpon desmodium do not (Snyder et al.,1978). Legumes adapted to soil pH's below about 5.0 may reach maximum productivity with lower rates of lime (about 500 kg ha^{-1}) than those requiring higher pH values (Snyder & Kretschmer, 1983).

Although Stylosanthes spp. are known to persist at very low levels of soil P, they respond on these soils to P fertilization (Probert, 1984; Kretschmer & Snyder, 1979).

Aeschynomene nivea Brandeg. survives with an annual rainfall of 250 mm while A. fluminensis Vell. and others have flooded or waterlogged habitats (Kretschmer & Bullock, 1979; Rudd, 1955).

Growth habit and adaptability to various single factor stresses are summarized in Table 2 for the most prominent tropical forage legumes.

PRESENT KNOWLEDGE

Available Germplasm

Twenty years ago the knowledge of germplasm that might have a potential was limited to less than 10 (Williams, 1964). Presently, more than 30 Australian and several tropical American cultivars have been released (Kretschmer, 1985). Many of these, unfortunately, have narrow adaptability. Tropical legume researchers must utilize the information developed over the intervening 25 yr, to prevent duplication of efforts.

Much of the tropical legume germplasm in resource banks probably has not been adequately evaluated (Williams & Clements, 1985). Also, there is an expanded knowledge of growth habits and adaptations, at least with many of the genera such as Stylosanthes (Stace & Edye, 1984), Centrosema (Schultze-Kraft & Clements, 1988), Aeschynomene (Kretschmer & Bullock, 1979; Bishop et al., 1985; Wilson et al., 1982) and Macroptilium (Kretschmer et al., 1985b).

Although some species of Aeschynomene are well known, only about 20 species out of 160 can be found in germplasm banks. Because of its proven potential and large species diversity, this genus needs to be collected more thoroughly in isolated areas of South America but particularly in Africa. Even A. americana is quite diverse and contains plants that appear to be, at least, short-lived perennials (A. E. Kretschmer, Jr., 1983, unpublished).

Quality

Cutting and feeding experiments have provided data on N, digestibility and mineral contents of most tropical legumes over a range of environmental conditions (Minson, 1982; Minson, 1985; Skerman, 1977). Further analyses in a similar environment would be redundant until the legume is known to persist. Contents of tannin, cyanoglucosides, alkaloids, mimosine , and other anti-quality or toxic factors in legumes also can be found in the literature (Barry & Reid, 1985, Minson & Hegarty, 1985; Hegarty, 1982; Duke, 1981) They vary only slightly from site to site depending on plant age and plant fraction. In any event, tropical legume protein content generally is higher than that for grasses; and digestibility does not decrease over time as fast as that in tropical grasses (Minson, 1985).

Insects and Diseases

The literature contains many references from different countries concerning pests. Descriptive papers on diseases (Lenne, 1982) and insects (Calderon, 1982), and a 35 mm slide package and guide (Calderon et al., 1982) summarize the potential pest problems that might arise with tropical legumes. Salinas et al, (1982) discuss and describe foliar mineral deficiencies and toxicity symptoms.

Table 1. Range of tropical forgae legume diversity, extent of evaluation and use, and persistence in the subtropics.

Genus	NS[1]	D[2]	Species	D[2]	T[3]	CU[4]	ST[5]
Arachis	60	M	glabrata	S	S	S	+
Aeschynomene	160	VL	americana	M	M	M	+
			falcata	S	M	S	+
Alysicarpus	30	M	vaginalis	M	L	S	+
Calopogonium	8	S	mucunoides	S	L	L	-
Centrosema	33	L	pubescens	S	L	L	-
			virginianum	M	M	N	+
			pascuorum	S	S	S	-
Clitoria	70	M	ternatea	S	M	S	+
Desmanthus	25	M	virgatus	M	S	N	+
Desmodium	300	VL	heterocarpon	M	L	S	+
			intortum	S	L	S	+
			ovalifolium	S	L	N	+
			uncinatum	S	L	S	+
Indigofera	700	L	hirsuta	S	L	S	+
Lablab	1	S	purpureus	S	L	L	+
Leucaena	40	M	leucocephala	S	L	L	+
Lotononis	100	M	bainesii	S	M	S	+
Macroptilium	20	M	atropurpureum	S	L	M	+
			lathyroides	S	L	S	+
Macrotyloma	24	M	axillare	S	M	S	-
			uniflorum	S	S	N	-
Neonotonia	1	S	wightii	S	L	M	+
Pueraria	20	M	phaseoloides	S	L	L	-
Stylosanthes	34	L	capitata	S	M	S	-
			guianensis	M	L	M	+
			hamata	M	L	M	+
			humilis	S	L	S	+
			scabra	S	S	M	+
Vigna	150	L	adenantha	S	S	N	+
			luteola	S	M	N	+
			parkeri	S	S	S	+
Zornia	80	M	latifolia	S	M	N	+

[1]NS = Approximate number of species.

[2]D = Diversity: VL = very large, L = large, M = moderate, S = small area of testing.

[3]T = Extent of evaluation: L= tested worldwide, M= tested in many areas with varying environments within a continent, S = testing just begun or limited to specific environment.

[4]CU = Present commercial use: L= many countries, M= limited to a few countries or widely sown in one country, S= limited to a specific environment within a country, N= not used

[5]ST = persistence in subtropics: + = persists, - = does not persist

Table 2. Growth habit and environmental adaptation of selected tropical forage legumes.

Legume	1	2	3	4	5	6	7	8	9	10	11
Aeschynomene americana L.	E-A	E	2	L-M	1,2,3	M	M	H	L	M	H
Aeschynomene falcata DC.	D-P	D	2	L	2,3	M-H	H	H	L	L	H
Alysicarpus vaginalis DC.	E-A	ED	2	L	1,2,3,4	L	M	H	L*	M	H
Arachis glabrata Benth.	R-P	R	2	M	1,2,3	M	H	H	L	L	L#
Cajanus cajan DC.	E-PB	E	3	M	2,3	M	M	M	M*	M	H
Calopogonium mucunoides Desv.	TD-ABP	D	2	M	1,2,3	L	H	VL	L	L	M
Cassia rotundifolia Pers.	D-A	D	2	L	3,4,5	L-M	H	L	L	L	H
Centrosema acutifolium	TS-P	D	1	L	2,3	M	M	H	M	L	M
Centrosema brasilianum Benth.	DS-P	D	1	L	3,4,5	M	M	M	M	L	M
Centrosema macrocarpum Benth.	TS-P	TS	1	L	2,3,4	M	M	H	L	L	M
Centrosema pascuorum Mart. ex Benth.	TD-A	TD	2	M	1,3,4,5	L	M	M	L	L	M
Centrosema pubescens Benth.	TS-P	D	2	M-H	2,3	M	M	H	L	M	M
Centrosema virginianum Benth.	TS-AP	D	3	M	2,3	M	M	M	L	L	M
Clitoria ternatea L.	T-P	ED	2	M-H	2,3,4	M	M	M	L	L	M
Codariocalyx gyroides Hassk.	E-PB	E	2	M	1,2,3	M	M	M	L*	L	L
Desmanthus virgatus Willd.	E-P	E	3	M	2,3,4,5	M	H	H	L	L	H
Desmodium adscendens DC.	D-P	D	2	U	2	L	M	M	L	L	M
Desmodium barbatum Benth.	E-PA	D	2	M	2,3	M	H	M	L	M	H
Desmodium heterocarpon DC.	ED-P	DE	2	M	1,2	M	H	M	L*	M	H
Desmodium heterophyllum DC.	DS-P	DS	2	M	1,2	L	H	M	L	L	M
Desmodium incanum DC.	EDR-P	ER	3	M	1,2,3	M	H	L	L	L	H
Desmodium intortum Urb.	ET-P	E-TS	2	H	2	H	M	M	L	M	M
Desmodium ovalifolium	DS-P	DS	1,2	L	1,2,3	L	H	L	H*	L	H
Desmodium sandwicense E.Mey	E-P	E	2	U	2	M	M	M	L	M	M
Desmodium uncinatum DC.	ET-P	E-TS	2	H	2	H	L	M	L	M	M
Galactia striata Urb.	T-P	D	2	M	2,3	M	M	M	L	L	M
Indigofera hirsuta Linn.	E-A	E	2	L	2,3,4	M	H	VL	L	L	H
Lablab purpureus SW.	T-AB	E	3	M	2,3	M	L	M	M*	M	H

Legume	1	2	3	4	5	6	7	8	9	10	11
Leucaena leucocephala de Wit	E-P	E	3	M-H	2,3,4	M	H	H	L	H	L
Lotononis bainesii Baker	DS-P	DS	2	L-M	2,3	M	H	H	M	L	L
Macroptilium atropurpureum (DC.) Urb.	T-P	D	3	L-M	2,3	M	M	H	H*	M	H
Macrotyloma axillare (E.Mey)Verdc.	T-P	E	3	M	3,4	M	M	L	M*	L	H
Macrotyloma uniflorum (Lam.)Verdc.	T-A	D	3	L-M	3,4	M	L	L	L	L	H
Mucuna pruriens DC.	T-A	D	2	M	2	L	H	VL	M	L	M
Neonotonia wightii Lakey	T-P	TD	3	H	2,3	H	M	M	M	L	L
Pueraria phaseoloides Benth.	T-P	TD	2	M	1,2	L	H	VL	L	L	H
Stylosanthes capitata Vog.	E-P	E	1	L	2,3,4	L	H	H	M	M	H
Stylosanthes guianensis Sw.	E-P	ED	1,2	L	2,3	M	M	M	H	L	H
Stylosanthes hamata Taub.	E-AB	D	3	L	3,4,5	L	H	M	H	L	H
Stylosanthes humilis HBK	ED-A	D	2	L	2,3,4	L	H	M	H	L	H
Stylosanthes scabra Vog.	E-P	E	1,2	L	3,4,5	M	H	L	M	L	M
Stylosanthes viscosa Vog.	E-P	ED	1,2	L	3,4,5	L	H	VL	L	L	H
Teramnus labialis Spreng	TDS-AP	D	3	M	1,2,3,4	L	H	M	L	L	M
Teramnus uncinatus Sw.	T-P	T	3	M	2,3	M	L	H	L	L	H
Trifolium semipilosum Fresen.	DS-P	D	2	M-H	2,3	H	H	H	L	L	M
Vigna adenantha M.M.& S.	TS-P	D	2	M	1,2	L	M	M	M	M	H
Vigna luteola Benth.	T-AB	D	3	M-H	1,2	L	L	H	M	M	H
Vigna parkeri Bak.	DS-P	D	2	M	1,2	M	H	H	L	L	H
Vigna vexillata A. Rich.	T-P	TD	2	M-H	1,2,3	M	L	H	M	M	H
Zornia latifolia Sm.	E-P	ED	1,2	M	2,3	L	H	H	L	L	H

[1]Normal growth habit without influence of grazing: D= decumbent, procumbent; E= erect, ascending; T= twining, climbing; S= stoloniferous or adventitious rooting; R= rhizomatous;--- A= annual; B= biennial, or short-lived perennial; P= perennial.

[2]Normal growth habit under grazing: same as "1".

[3]General soil pH adaptability range : 1= 4.0-5.5; 2= 5.0-6.5; 3= 5.0-8.3.

[4]Soil fertility requirements: L= low, M= medium, H= high, U= unknown

[5]Conditions favorable for 80% of normal growth: 1= Periodic 2 to 3 wk waterlogging or flooding; 2= Annual rainfall >1500 mm, < 3 months effective drought; 3= Distinct 3-6 month dry season; 4= Distinct dry season > 6 months; 5= Semiarid, 500-750 mm rainfall.

[6]Near-equatorial altitude adaptability for 80% of normal growth: L= 0-500; M= 0-1000; H= > 1000 m.

[7]Grazing pressure constraints after the seedling stage of growth assuming annual plants are permitted to produce seeds: H=survives with continuous, high (3-4 A.U. ha^{-1}) grazing pressures without rest period; M=survives with continuous moderate grazing (1.0-1.5 A.U. ha^{-1}), or high grazing pressures for 1-2 months with infrequent 5 to 6-wk rest periods; L= survives under intermittent grazing pressure with 5 to 6-wk rest periods. Plants resistant to grazing in low rainfall environments may not tolerate heavier stocking in higher rainfall environments.

[8]Cattle acceptability or palatability: VL= very low; L= low; M= moderate; H= high.

[9]Soil borne or aerial disease (* nematodes) on some ecotypes: L= no reduction in productivity; M= possible reduction in productivity but no permanent effect; H= loss of productivity and survival.

[10]Insect damage on some ecotypes: L= no reduction in productivity; M= possible reduction in productivity but no permanent effect; H= loss of productivity and survival.

[11]Ease of establishment : L= difficult; M= moderate; H= easily established; #established vegetatively.

Note: I wish to thank Dr. R. J. Clements and Mr. R. M. Jones, CSIRO, Cunningham Lab., Brisbane, Australia, and Dr. R. Schultze-Kraft, CIAT, Cali, Colombia for their help in preparing this table.

Table 3. Some factors affecting tropical legume persistence.

Category	Factor
Soils:	acidity, fertility, texture, drainage
Rainfall:	quantity, distribution
Grass competition:	height, sunlight,nutrients, water
Grass growth form:	bunch, stolons, rhizomes
Legume growth form:	prostrate, erect, vine, crown or bud area height
Legume attributes:	annual, perennial, quality, antiquality, palatability
Harmful factors:	pathogens, insects, nematodes, mineral deficiencies, weeds
Grazing:	continuous, rotation, trampling, preferential grazing,
Seed production:	quantity, determinant-non-determinant, hardseededness

Factors That Affect Persistence

Many scientific papers are available from most environments of the tropics and subtropics that deal with persistence of tropical legumes in mixtures with many grasses. Factors presented in Table 3 can have an effect on persistence.

Cutting Versus Grazing

Jones et al. (1980) found that mowing gave a good indication of Siratro dry matter production under a set-stocked, rotational grazing system, but did not want to generalize on this comparison, Hodgkinson and Williams (1983), Jones and Walker (1983), and Grof (1986) discuss the modification of legume growth form resulting from grazing compared to that from cutting. Bryan et al. (1964) indicated that grazing and cutting results were not always the same. Staples et al. (1985) although not comparing cutting vs. grazing on persistence, noted that the predicted range of adaptability of legumes developed under evaluation was much greater than that experienced after the legume was tried commercially. Some obvious differences have been noted by the author with 'Siratro' and commercial centro which survives well under cutting but fail to persist as well under grazing. In contrast, some cultivars of S. guianensis do not survive under lenient cutting but do survive under grazing. What is even more perturbing is that even 3 to 5 yr grazing for persistence evaluation may not predict the most persistent legume after 15 yr. Jones (1988) found this to be true with Siratro which did not persist as well as predicted from results in trials established in the early 1970s (Jones & Rees, 1972; Rees et al., 1976). The success of conventional methods of evaluation have not been very successful in predicting persistence. Furthermore, several widely used legumes (Townsville stylo, tropical kudzu, calopo, hairy indigo, alyce clover, and American jointvetch) have been accepted by growers without any formal evaluation, because they persisted under their grazing management systems.

In summary, there is a vast literature base dealing with all phases of tropical legume growth and persistency under cutting. This can be used to reduce the time required from introduction to grazing evaluation. Results of persistence of tropical legumes under cutting and conventional evaluation methods, however, have not been very successful in predicting persistence under grazing. New, innovative techniques are needed to speed up the evaluation process, through earlier use of grazing.

EVALUATION METHODOLOGY FOR DETERMINING PERSISTENCE

The primary aim of tropical forage legume introduction and evaluation should be to identify those legumes which will persist under grazing; with higher yield, quality, and/or higher seed production secondary considerations. Which is more important, a high-yielding, nonpersistent legume under grazing, or a lower yielding, long-lived legume?

Conventional Methodology

Many research organizations have been involved in evaluation for persistence and have tried to quantify the general procedures for evaluations; and flow charts are often used to diagram these procedures (Jones & Walker, 1983; Edye & Grof, 1983; Rees et al., 1976; Williams, 1964; Bryan et al., 1964; Jones & Rees, 1972; Toledo, 1982; Toledo & Schultze-Kraft, 1982; Cameron & McIvor, 1980; Kretschmer, 1979; Lascano & Pizarro, 1986; Spain & Pereira, 1986; Jones, 1984; Staples et al., 1985, Jones et al., 1980; Symons & Jones, 1971; Thomas & Andrade, 1984; Goncalves et al., 1982; Paladines, 1982; Kitamura, 1985; Cameron et al., 1984; Thomas & Rocha, 1984; Hodgkinson & Williams, 1983; Clements et al., 1983; Shaw & Bryan, 1976, Barnes et al., 1970; Mott & Jimenez, 1979; Jones et al., 1985; Ivory et al., 1985; McCosker, 1987; Anning, 1982; Myers et al., 1974). Based on work by Toledo and Schultze-Kraft (1982), and Jones and Walker (1983), most strategies require 5 or 6 yr from plant introduction before cattle are used to determine grazing effects on persistence (Table 4).

There is general agreement on the initial or stage one testing that normally takes a minimum of one growing season. Two growing seasons are preferred in order to determine survival over a stress period. Main objectives of this stage are to increase seed supplies, categorize plant growth form, and obtain flowering data. Commercial legumes from other countries or areas that are believed to have potential in the area of evaluation should be included in all initial testing. If one or more of these entries prove successful, their commercial use would be much faster than that for a new ecotype, simply because adequate seed supplies would be available immediately.

Secondary testing (stage 2, Table 4) normally includes small-plot clipping experiments with or without grass, and almost never includes grazing (Jones & Walker, 1983; Bryan et al., 1964; Toledo, 1982; Toledo & Schultze-Kraft, 1982). Spending the suggested 3 to 4 yr on clipping experiments is expensive and does not determine persistence under grazing. In addition to the formalized stages listed in Table 4, there are times when large numbers of breeding lines or introductions of the same species or genus need to be tested concurrently. Single plant (Kretschmer et al., 1985b) or two to three plants per plot (Brolman, 1979) may have to be used because of seed shortages and to reduce cost and space requirements. This type of test lends itself to morphologic-agronomic characterization through numerical classification which categorizes plants into groups based on the attribute studied (Burt & Williams, 1979a; Burt & Williams, 1979b; Burt et al., 1979; Burt et al., 1980; Edye & Grof, 1983).

Statistical (Haydock, 1964) and data storage and retrieval methods (Clements & Bray, 1980; Haydock, 1964; Williams, 1983b; Ivory et al., 1985) are well documented.

After about 5 yr of conventional testing, animals (stage 3) are used to test persistence of selected legume entries obtained from stage 2. Finally, after another several years, animal performance experiments begin using one or several legumes chosen from stage 3. These expensive tests may include different stocking rates and grazing methods (Shaw & Bryan, 1976; Haydock, 1964; Lascano & Pizarro, 1986; Spain & Pereira, 1986; Pizarro & Toledo, 1986).

Table 4. Idealized format for conventional and alternate systems of evaluation for persistence of tropical legumes.

System	Plot[1] type	Reps[2]	Mixed[3] with grass	Cattle[3] used	Yr[4]	Method[5]	Data[6]
Conventional							
Stage 1	Row	1,2	N	N	2	R	M, A, F
Stage 2	SP	3,4	N,Y	N	3	C	Y, N, IV
Stage 3	SP,LP	3,4	Y	Y	3	G,C	P
Revised							
Stage 1	Row	1,2	N	N	2	R	M, A, F
Stage 2	SP	4	Y	N	3	R,SH	N,IV,SY
	Strip	4	Y	Y	3	R	P
	Site	10	Y	Y	3	R	P

[1]SP = small plot, LP = large plot.
[2]Replications.
[3]Y = yes, N = no.
[4]Minimum number of years.
[5]Method of evaluation: R = rating, C = cutting for yield, G = grazing, R,SH = primarily rating with selective harvest for yield.
[6]Data obtained: M = morphologic, A = agronomic, F = flowering, N = nitrogen content, IV = digestibility, Y = yield, P = persistence, under grazing, PP = population density (legume coverage) , SY = selective yield.

Proposed Alternate Methods

I propose that a rating system for yield and persistence and earlier use of grazing be used in place of the conventional system. In spite of numerous evaluations, the use of ratings for yield, plant populations, or for persistence, has been limited (Bryan et al., 1964; Toledo & Schultze-Kraft, 1982; Clements & Bray, 1980; Cameron & McIvor, 1980), and almost always in conjunction with conventional cutting methods. The use of a rating system for persistence, with limited, selective harvesting for yields would provide a faster, and I believe equally effective method for evaluation of small and especially massive numbers of entries.

The following suggestions are based on results obtained at the ARECFP (Agricultural Research and Education Center, Ft. Pierce) and in commercial pastures in Florida and in several foreign countries. These methods are based on the assumption that the primary interest is to predict legume survival under grazing. No change is proposed for the initial evaluation stage.

Small Plots

Our work began in 1979 in a small-plot experiment of more than 100 entries and four replications that were seeded in an establishing Pangola area (Kretschmer et al., 1985a). Included were 36 Aeschynomene entries (9 spp.) and 11 other genera (25 spp.). Most were selected from the initial evaluation stages after two growing seasons, however, there were several cultivars included. Because of limited manpower it was impossible to harvest all plots on a consistent basis; and it was decided to use a rating system for legume plant populations (plot coverage by legumes) and legume vigor. Selected plots (highest ratings for different growth forms) were cut and components separated for yields and quality assessment. The same plots were not necessarily harvested each time since highest ratings varied from period to period. It was assumed that legume yields of lower rated plots would be less than higher rated plots of legumes with similar growth habits. The sliding scale rating system used was: 1=no plants, very poor vigor; 3=25% plant coverage, poor vigor; 5=50% , moderate; 7=75%, good; and 9=100%, excellent. During subsequent field strip tests, a rating system of 0=no plants and 1=1 to 3 plants present, was substituted for those of the small-plot test. After rating and harvesting the selected plots, remaining foliage from the area was removed with a flail-type harvester. There was a considerable savings in time and labor. Vigor ratings were difficult to make because of the diversity of ecotypes used, i.e., from prostrate to erect growth habits and appeared to be positively related to plant population. With a group of like entries, the use of vigor ratings could be an asset. The flexibility of this method would permit animals to graze the area in place of cutting (Table 4) and still obtain selective yield and quality data.

Strip or Row Method

Although this method was designed to replace the conventional stage 3, it could be used under the small-plot, selective harvest design or to replace the conventional stage 2 test. Several of the more persistent entries from the small plot test were selected and included with commercial cultivars in several area strip tests (Pitman & Kretschmer, 1984; Pitman et al., 1988). These and additional testing in Costa Rica, Mexico, and the Virgin Islands indicated that the strip testing technique was a viable less expensive alternative to the second and third stage conventional methods. Persistence under the strip method depended on the use of grazing animals in place of mechanical harvesting. The legumes are planted in rows 5 to 15-m long with 2 to 4 m between rows. The sliding scale rating system used for the small-plot method works well, i.e., 0 = no plants, 1 = 1 to 3 plants, 5 = 50% of plant cover, and 9 = 100% plant coverage of the strip. The shorter row length is more practical because the whole row can be seen and evaluated by one technician when passing across the center of the rows. In addition, less seeds are required than for longer rows. The object, although not always achieved, was to establish at least one seedling every 10 cm in the row. No attempt was made to count the number of plants or number of seedlings. Seeding rate was not uniform, but it provided enough seed per plot for at least 1 seed per 2 cm of row. When insufficient seeds are available, plants may be transplanted at 0.5 to 0.75-m intervals. The most important rating is that obtained several weeks after germination in order to have a basis of comparison for later plant populations ratings.

Strip tests can be monitored with more precision at research centers, however, with cooperative growers there has not been a problem as long as the grower's grazing management is not altered greatly. As a matter of fact, within a similar environmental zone, the grower's management probably represents that of the surrounding growers as well, if not better than that suggested by the researcher. We have recommended no grazing during a 2 to 3 month period during maximum seed production (fall). If the cooperator will fence small half-ha areas within a larger fenced area, then a system of opening gates provides one or more grazing pressures. Flexibility of procedures are much better than that of conventional stages.

In 1982 in Queensland, Australia, a strip method test was initiated to evaluate 32 accessions of temperate and tropical legumes under heavy grazing pressure (R.M. Jones, CSIRO Cunningham Lab., Brisbane, Australia, 1988, personal communication). The nine sites ranged from 21 to 29°S Lat. Evaluation consisted of rating presence or absence in 10, 20 by 20 cm quadrates per 4 m seeded row. Results are nearly ready to publish. Recently, similar strip experiments were begun in Queensland, Australia. They include important climatic zones and the same legumes are being compared at each site (D. Gramshaw, Pasture Management Branch, Queensland Dep. Primary Industries, G.P.O. Box 46, Brisbane 4001 QLD., Australia, 1988, personal communication). Results from these experiments will help to assess the value of the strip test method in predicting legume persistence.

Site Method

A modification of the strip method, which is now under examination at the ARECFP, is named the site-spot or simply the site method. This method has two advantages compared with the strip method. One, it reduces the number of seeds needed, and two, it permits seeding in established grass pastures without soil or grass disturbance. Although the statistical ramifications have not been evaluated, it is believed that the following explanation will help those who may want to try the procedure. The object is to seed sites, 4 to 5 cm in diameter, of each entry spaced at convenient intervals in a gridwork or rectangular style. We have tried a spacing of 1 m between sites in a heavily stocked establishing bahiagrass (Paspalum notatum Flugge) pasture, when most of the 27 entries were easily differentiated and no two similar entries were planted next to each other. With a group of the same species, probably 2 to 3 m would be appropriate because smaller spacings may hinder attempts to determine spread of an individual entry after the establishment year. Five to 10 seeds of each replication (we used 10 replications) are placed in a separate envelope for later distribution to appropriate sites. The object is to assure at least one germinating plant at each site. Seeding can be done by hand, burying the seeds 1/2 to 1 cm deep. We used a "plug mix" seeder designed for plugging a mixture of peat soil and vegetable seeds for more rapid establishment and seedling growth (Hayslip, 1974). We eliminated the peat and used pure seed. The implement can be adjusted to plant the seeds at any reasonable depth. After staking out the area to be seeded, the plugger method requires less than 1 min per site (two technicians). Germination of plants in our initial test was disappointing in that we had only 4 to 10 replications of an entry germinating; a severe drought occurred, which caused death of many seedlings. Rating of the 270 sites (0=no plant, 1=plant present) required about 1 to 1.5 hr. Another test has been seeded in the same area but in an established sod of bahiagrass. In this instance, the sites were sprayed (approximate 0.5-m circles) with glyphosate (Roundup) about 4 wk prior to seeding.

The advantage of requiring less seed and less area than the strip method can decrease the time from the initial seed collection evaluation stage to the grazing stage. No sophisticated equipment is needed; the grower is not required to use his equipment nor is an established grass pasture disturbed. Time saved over the conventional evaluation method could be used to plant more tests in other nearby ranches; or in other environmental zones. The procedure is rapid and simple enough for nontechnical personnel. Questions as to the effect of self-competition among legume plants in a given site or long-term persistence at sites is not known at this time, however it seems logical that establishment year plants must be permitted to achieve near maximum seed production as recommended for the strip planting method.

The small space required for 200, 1 by 1 m sites compared with the strip or conventional methods, would permit a larger number of smaller fenced areas (the use of electric fencing should be considered) for different management treatments; or a much greater number of entries could be used. The question of how many good entries or species may be rejected because of poor methodology is not known, but most likely some are lost using the conventional method, also. The important positive consideration is to find a more persistent tropical legume as rapidly as possible. It is believed that the small-plot selective harvesting (optional grazing), strip, and site method are superior in obtaining the end result more rapidly than the conventional methods presently employed.

ACKNOWLEDGMENT

I wish to thank Dr. R. J. Clements, and Mr. R. M. Jones, Principal Research Scientists, CSIRO Cunningham Laboratory, 306 Carmody Rd., Brisbane, Queensland, Australia 4067, for helping to develop Tables 1, 2, and 3, and to Dr. R. Schultze-Kraft, Germplasm Agronomist, CIAT, Apartado aereo 67-13, Cali, Colombia for his help in developing Table 2.

The work was supported in part by the USDA under CSRS Spec. Grant Nos. 83-CRSR-2-2134 and 86-CRSR-2-2846 managed by the Caribbean Basin Advisory Group (CBAG).

REFERENCES

Anning G.P. 1982. Evaluation of introduced legumes for pasture in the dry tropics of north Queensland. Trop. Grassl. 16:146-155.

Barnes, R.F., DC. Clanton, C.H Gordon, T.J. Klopfenstein, and D.R. Waldo. 1970. Proceedings of the national conference on forage quality evaluation and utilization. Nebraska Center Continuing Educ. Lincoln, NE.

Barry, T.N., and C.S.W. Reid. 1985. Nutritional effects attributable to condensed tannins, cyanogenic glycosides and oestrogenic compounds in New Zealand forages. p. 251-259. In R.F Barnes et al. (ed.) Forage legumes for energy efficient animal production. Proc. Trilateral Workshop, Palmerston, North, NZ. 30 Apr.- 4 May 1984. USDA-ARS, Washington, D.C.

Belalcazar, J., and R. Schultze-Kraft. 1986. Centrosema brasilianum (L.) Benth.: descripcion de la especie y evaluacion agronomica de siete ecotipos. Pasturas Tropicales 8:14-19. CIAT, Apartado 6713, Cali, Colombia.

Bishop, B. G., B. Walker, D. H. Ludke, and M. T. Rutherford. 1985. Aeschynomene-a legume genus with potential for the Australian tropics. p. 160-161. In Proc. XV Int. Grassl. Congr., Kyoto. 24-31 Aug. Iroha Insatsu Kogei Co., Nagoya, Japan.

Brolmann, J. B. 1979. Evaluation of various Stylosanthes accessions in south Florida. Soil Crop Sci. Soc. Fla. Proc. 39:102-104.

Bryan, W.W., N.H. Shaw, L.A. Eyde, R.J. Jones, L.'t Mannetje, and J.J. Yates 1964. The development of pastures. In Cunningham Lab., CSIRO, Brisbane, Australia (ed.) Some concepts and methods in sub-tropical pasture research. Bull. 27:123-143.

Burt, R.L., R.F. Isbell, and W.T. Williams. 1979. Strategy of evaluation of a collection of tropical herbaceous legumes from Brazil and Venezuela. l. Ecological evaluation at the point of collection. Agro-Ecosystems 5:99-117.

Burt, R.L., and W.T. Williams. 1979a. Strategy of evaluation of a collection of tropical herbaceous legumes from Brazil and Venezuela. ll. Evaluation in the quarantine glasshouse. Agro-Ecosystems 5:119-134.

Burt, R.L., and W.T. Williams. 1979b. Strategy of evaluation of a collection of tropical herbaceous legumes from Brazil and Venezuela. lll. The use of ordination techniques in evaluation. Agro-Ecosystems 5:135-148.

Burt, R.L., R.J. Williams, and W.T. Williams. 1980. Observation, description and classification of plant collections. p. 40-51. In R.J. Clements and D.G. Cameron (ed.) Collecting and testing tropical forage plants. CSIRO, Melbourne.

Calderon, M. 1982. Evaluacion del dano causado por insectos. p. 57-71. In J.M. Toledo (ed.) Manual para la evaluacion agronomica. Red internacional de evaluacion de pastos tropical. Serie CIAT 07SG-1(82). CIAT, Apartado 6713, Cali, Colombia.

Calderon, M., F. A. Varela, and C. A. Valencia. 1982. Guia de estudio. Descripcion de los plagas que atacan los pastos tropicales y caracteristicas de sus danos. Serie 04SP-03.01. CIAT, Apartado 6713, Cali, Colombia.

Cameron, D.G., and J.G. McIvor. 1980. Evaluation. p.71-87. In R. J. Clements and D. G. Cameron (ed.) Collecting and testing tropical forage plants. CSIRO, Melbourne.

Cameron, D.G., I.L. Miller, P.G. Harrison, and R.J. Fritz. 1984. A review of pasture plant introduction in the 600-1500 mm rainfall zone of the Northern Territory. Dep. Primary Production, G.P.O. Box 4160, Darwin N.T. 5794 Australia.

Clements, R.J., and R.A. Bray. 1980. Information storage and retrieval. p. 52-70. In R.J. Clements and D.G. Cameron (ed.) Collecting and testing tropical forage plants. CSIRO, Melbourne.

Clements, R.J., M.D. Howard, and D.E. Byth. 1983. Genetic adaptation in pasture plants. p. 101-115. In J.G. McIvor and R.A. Bray (ed.) Genetic resources of forage plants. CSIRO, East Melbourne, Australia.

Clements, R.J., and R.J. Williams. 1980. Genetic diversity in Centrosema. p. 559-567. In R.J. Summerfield and A.H.Bunting (ed.) Advances in legume science. Royal Botanic Gardens, Kew, England.

Crepet, W.L., and D.W. Taylor. 1985. The diversification of the Leguminosae: first fossil evidence of the Mimosoideae and Papilionoideae. Science 228:1087-1089.

Duke, J.A. 1981. Handbook of legumes of world economic importance. Plenum Press, New York.

Edye, L.A., and B. Grof. 1983. Selecting cultivars from naturally occurring genotypes: evaluating Stylosanthes species. p. 217-232. In J.G. McIvor and R.A. Bray (ed.) Genetic resources of forage plants. CSIRO, East Melbourne, Australia.

Goncalves, C.A., J. de Cunha Medeiros, and J.R. da Cruz Oliveira. 1982. Introducao e avaliacao de gramineas e leguminosas forrageiras em Rondonia. Bol. de Pesquisa 1. EMBRAPA, UEPAE, Porto Velho, Brazil.

Grof, B. 1985. Forage attributes of the perennial groundnut, Arachis pintoi, in a tropical savanna environment in Colombia. p. 168-170. In Proc. XV Int. Grassl. Congr., Kyoto. 24-31 Aug. Iroha Insatsu Kogei Co., Nagoya, Japan.

Grof, B. 1986. Forage potential of some Centrosema species in llanos Orientales of Colombia. Trop. Grassl. 20:107-112.

Haydock, K.P. 1964. Statistics in pasture research. In Cunningham Lab., CSIRO, Brisbane, Australia (ed.) Some concepts and methods in sub-tropical pasture research. Bull. 27:159-169.

Hayslip, N.C. 1974. A "plug-mix" seeding method for field planting tomatoes and other small seeded hill crops. Ft. Pierce ARC Res. Rep. RL1974-3.

Hegarty, M.P. 1982. Deleterious factors in forages affecting animal production. p. 133-150. In J.B. Hacker (ed.) Nutritional limits to animal production from pastures. CSIRO, Div. Trop. Crops Past. Commonwealth Agric. Bur. Farnham Royal, England.

Hodges, E. M., A. E. Kretschmer, Jr., P. Mislevy, R. D. Roush, O. C. Ruelke, and G. H. Snyder. 1982. Production and utilization of the tropical legume Aeschynomene. Florida Agric. Exp. Stn. Circ. S-290.

Hodgkinson, K.C., and O.B. Williams. 1983. Adaptation to grazing in forage plants. p. 85-100. In J.G. McIvor and R.A. Bray (ed.) Genetic resources of forage plants. CSIRO, East Melbourne, Australia.

Ivory, D.A., A. Ella, J. Nulik, R.Salem, M.E. Siregar, and S. Yuhaeni. 1985. Assessment of agronomic performance of large numbers of plant accessions using a computerized data management system. p. 171-173. In Proc. XV Int. Grassl. Congr., Kyoto. 24-31 Aug. Iroha Insatsu Kogei Co., Nagoya, Japan.

Jones, R.J., and B. Walker. 1983. Strategies for evaluating forage plants. p. 185-201. In J.G. McIvor, and R.A. Bray (ed.) Genetic resources of forage plants. CSIRO, East Melbourne, Australia.

Jones, R.M. 1984. Persistencia de las especies forrajeras bajo pastoreo. p. 167-199. In E.A. Pizarro and J.M. Toledo (ed.) Evaluacion de pasturas con animales-alternativas metodologicas. Red internacional de evaluacion de pastos tropicales. CIAT, Apartado 6713, Cali, Colombia.

Jones, R.M. 1988. Inspection of old species evaluation trials after twenty years of farm grazing. Trop. Grassl. News 4(1):4-9.

Jones, R.M., R.J. Jones, and E.M. Hutton. 1980. A method for advanced stage evaluation of pasture species: a case study with bred lines of Macroptilium atropurpureum. Aust. J. Exp. Agric. Anim. Husb. 20:703-709.

Jones, R.M., and M.C. Rees. 1972. Persistence and productivity of pasture species at three locations in subtropical south east Queensland. Trop. Grassl. 6:119-134.

Jones, R.M., R.J. Jones, and D.F. Sinclair. 1985. New methods and current needs in pasture research. p. 80-91. In G. J. Murtagh and R.M. Jones (ed.) Proc. Third Australian Conf. on Tropical Pastures, Rockhampton. 8-12 July 1985. Trop. Grassl. Occas. Publ. 3.

Kitamura, Y. 1985. Introduction of tropical legumes and development of legume-based pastures in subtropical Japan. Trop. Agric. Res. Series 18. Trop. Agric. Res. Center, Ishigaki, Japan.

Kretschmer, A.E., Jr., 1978. Tropical forage and green manure legumes. p. 97-122. In G.A. Jung (ed.) Crop tolerance to suboptimal land conditions. ASA Spec. Publ. 32. ASA, CSSA, and SSSA, Madison, WI.

Kretschmer, A.E., Jr. 1979. Characterization and preliminary evaluation. p. 33-43. In G.O. Mott and A. Jimenez C. (ed.) Handbook for the collection, preservation and characterization of tropical forage germplasm resources. CIAT, Apartado 6713, Cali, Colombia.

Kretschmer, A.E., Jr. 1985. A list of positions of subfamilies and tribes of the family Leguminosae with a brief description and approximate numbers of genera and species. Ft. Pierce AREC Res. Rep FTP-85-3.

Kretschmer, A.E., Jr., and R.C. Bullock. 1979. Aeschynomene spp.: distribution and potential use. Soil Crop Sci. Soc. Fla. Proc. 39:145-152.

Kretschmer, A.E., Jr., and G.H. Snyder. 1979. Forage production on acid infertile soils of subtropical Florida. p.227-258. In P.A. Sanchez and L.E. Tergas (ed.) Pasture production in acid soils of the tropics. CIAT, Apartado 6713, Cali, Colombia.

Kretschmer, A.E., Jr., G.H. Snyder, and T.C. Wilson. 1985a. Productivity and persistence of selected Aeschynomene spp. Soil Crop Sci. Soc. Fla. Proc. 45:174-178.

Kretschmer, A. E., Jr., R. M. Sonoda, R. C. Bullock, G. H. Snyder, T. C. Wilson, R. Reid, and J. B. Brolmann. 1985b. Diversity in Macroptilium atropurpureum (DC.) Urb. p. 155-157. In Proc. XV Int. Grassl. Congr., Kyoto. 24-31 Aug. Iroha Insatsu Kogei Co., Nagoya, Japan.

Lascano, C., and E. Pizarro. 1986. Evaluacion de pasturas con animales-alternativas metodologicas. Red internacional de evaluacion de pastos tropicales. CIAT, Apartado 6713, Cali, Colombia.

Lenne, J.M. 1982. Evaluacion de enfermedades en pastos tropicales en el area de actuacion. p. 45-55. In J.M. Toledo (ed.) Manual para la evaluacion agronomica. Red internacional de evaluacion de pastos tropical. Serie 07SG-1(82), CIAT, Apartado 6713, Cali, Colombia.

McCosker, T.H. 1987. Establishment and persistence of introduced grasses on the Marrakai land system, in the Northern Territory semi-arid tropics. Trop. Grassl. 21:28-41.

Minson, D.J. 1982. Effects of chemical and physical composition of herbage eaten upon intake. p. 167-182. In J.B. Hacker (ed.) Nutritional limits to animal production from pastures. CSIRO, Div Trop. Crops Past. Commonwealth Agric. Bur. Farnham Royal, England.

Minson, D.J. 1985. Nutritional value of tropical legumes in grazing and feeding systems. p. 192-196. In R.F Barnes et al. (ed.) Forage legumes for energy efficient animal production. Proc. Trilateral Workshop, Palmerston North, NZ. 30 Apr.- 4 May 1984. USDA-ARS, Washington, DC.

Minson, D.J., and M.P. Hegarty. 1985. Toxic factors in tropical legumes. p. 246-250. In R.F Barnes et al. (ed.) Forage legumes for energy efficient animal production. Proc. Trilateral Workshop, Palmerston North, NZ. 30 Apr.- 4 May 1984. USDA-ARS, Washington, DC.

Mott, G.O., and A. Jimenez C. 1979. Handbook for the collection, preservation and characterization of tropical forage germplasm resources. CIAT, Apartado 6713, Cali, Colombia.

Myers, L.F., J.V. Lovett, and M.H. Walker. 1974. Screening of pasture plants: a proposal for standardizing procedures. J. Aust. Inst. Agric. Sci. 40:283-289.

Ohashi, H. 1973. The Asiatic species of Desmodium and its allied genera (Leguminosae). No. 1. Ginkgoana. Contributions to the flora of Asia and Pacific region. Academic Sci. Book Inc., Tokyo.

Paladines, M. O. 1982. Evaluacion y seleccion de germoplasma forrajero. Dept. Zootecnia, Pontifica Univ. Catolica de Chile.

Pitman, W. D., C. G. Chambliss, and A. E. Kretschmer, Jr. 1988. Persistence of tropical legumes on peninsular Florida flatwoods (Spodosols) at two stocking rates. Trop. Grassl. 22:27-33.

Pitman, W.D., and A.E. Kretschmer, Jr. 1984. Persistence of selected tropical pasture legumes in peninsular Florida. Agron. J. 76:993-996.

Pizarro, E.A., and J.M. Toledo. 1986. La evaluacion de pastures con animales: consideration para los ensayos regionales (ERD). p. 1-11. In E.A. Pizarro and J.M. Toledo (ed.) Evaluacion de pasturas con animales-alternativas metodologicas. Red internacional de evaluacion de pastos tropicales. CIAT, Apartado 6713, Cali, Colombia.

Polhill, R.M., and P.H. Raven. 1981. Advances in legume systematics. Part 1. Royal Botanic Gardens, Kew, Surrey, England.

Polhill, R.M. P.H. Raven, and C.H. Stirton. 1981. Evolution and systematics of the Leguminosae. p. 1-26. In R.M. Polhill and P.H. Raven (ed.) Advances in legume systematics. Part 1. Royal Botanic Gardens, Kew, Surrey, England.

Prine, G.M., L.S. Dunavin, J.E. Moore, and R.D.Roush. 1981. 'Florigraze' rhizoma peanut. A perennial forage legume. Florida Agric. Exp. Stn. Circ S-275.

Probert, M.E. 1984. The mineral nutrition of Stylosanthes. p. 203-226. In H.M. Stace and L.A. Edye (ed.) The biology and agronomy of Stylosanthes. Academic Press, Australia.

Raven, P.H., and R.M. Polhill. 1981. Biogeography of the Leguminosae. p. 27-34. In R.M. Polhill and P.H. Raven (ed.) Advances in legume systematics. Part 1. Royal Botanic Gardens, Kew, Surrey, England.

Rees, M.C., R.M. Jones, and R. Roe. 1976. Evaluation of pasture grasses and legumes grown in mixtures in south-east Queensland. Trop. Grassl. 10:65-78.

Rudd, V. E. 1955. American species of Aeschynomene. U. S. Natl. Herb. 32:1-172. Smithsonian Inst., Washington, DC.

Salinas, J.G., J.I. Sanz, and R. Garcia. 1982. Sintomas foliares de deficiencias y toxicidades minerales en pastos tropicales. p. 73-81. In J.M. Toledo (ed.) Manual para la evaluacion agronomica. Red internacional de evaluacion de pastos tropicales. Serie 07SG-1(82). CIAT, Apartado 6713, Cali, Colombia.

Schultze-Kraft, R. 1986. Natural distribution and germplasm collection of the tropical pasture legume Centrosema macrocarpum Benth. Angew. Botanik 60:407-419.

Schultze-Kraft, R., G. Benavides, and A. Aris. 1987. Recoleccion de germoplasma y evaluacion preliminar de Centrosema acutifolium. Pasturas Tropicales 9:12-20. CIAT, Apartado 6713, Cali, Colombia.

Schultze-Kraft, R., and R. J. Clements. 1988. Centrosema: biology, agronomy, and utilization. CIAT, Apartado 6713, Cali, Colombia.

Shaw, N.H., and W.W. Bryan 1976. Tropical pasture research principles and methods. Bull. 51. Commonwealth Agric. Bur. Hurley, England.

Skerman, P.J. 1977. Tropical forage legumes. FAO, Rome.

Snyder, G.H., and A.E. Kretschmer, Jr. 1983. Liming for tropical legumes for establishment and production. p. 302-305. In Proc. XIV Int. Grassl. Congr., Lexington, 15-24 June, 1981. Westview Press, Boulder, CO.

Snyder, G.H., A.E. Kretschmer, Jr., and J.B. Sartain. 1978. Field response of four tropical legumes to lime and superphosphate. Agron. J. 70:269-273.

Spain, J.M., and J.M. Pereira. 1986. Sistemas de manejo flexible para evaluar germoplasma bajo pastoreo: una proposita. p. 85-97. In E.A. Pizarro and J.M. Toledo (ed.) Evaluacion de pasturas con animales-alternativas metodologicas. Red internacional de evaluacion de pastos tropicales. CIAT, Apartado 6713, Cali, Colombia.

Stace, H. M., and L. A. Edye. 1984. The biology and agronomy of Stylosanthes. Academic Press, Australia

Staples, I.B., R. Reid, and G.P.M. Wilson. 1985. Plant introduction for specific needs in northern Australia. p. 29-37. Proc. Third Australian Conf. on Tropical Pastures, Rockhampton. 8-12 July 1985. Trop. Grassl. Occas. Publ. 3.

Stebbins, G. L. 1982. Plant speciation. p. 21-39. In Claudio Barigozzi (ed.) Mechanisms of speciation. Alan R. Liss, Inc., NY.

Symons, L.B., and R.I. Jones. 1971. An analysis of available techniques for estimating production of pastures without clipping. Proc. Grassl. Soc. South Africa 6:185-190.

Thomas, D., and R.P. de Andrade. 1984. The persistence of tropical grass-legume associations under grazing in Brazil. J. Agric. Sci. 102:257-263.

Thomas, D., C.M.C. de Rocha. 1984. Manejo de pasturas y evaluacion de la produccion animal. p. 43-84. In E.A. Pizarro and J.M. Toledo (ed.) Evaluacion de pasturas con animales-alternativas metodologicas. Red internacional de evaluacion de pastos tropicales. CIAT, Apartado 6713, Cali, Colombia.

Toledo, J.M. 1982. Objectivos y organizacion de la red internacional de evaluacion de pastos tropicales. p. 13-21. In J.M. Toledo (ed.) Manual para la evaluacion agronomica. Red internacional de evaluacion de pastos tropical. Serie 07SG-1(82). CIAT, Apartado 6713, Cali, Colombia.

Toledo, J.M., and R. Schultze-Kraft. 1982. Metodologia para evaluacion agronomica de pastos tropicales. p. 91-110. In J.M. Toledo (ed.) Manual para la evaluacion agronomica. Red internacional de evaluacion de pastos tropical. Serie 07SG-1(82). CIAT, Apartado 6713, Cali, Colombia.

Williams, R.J. 1964. Plant introduction. In Cunningham Lab., CSIRO, Brisbane, Australia (ed) Some concepts and methods in sub-tropical pasture research. Bull. 27:60-78.

Williams, R.J. 1983a. Tropical legumes. p. 17-37. In J.G. McIvor and R.A. Bray (ed.) Genetic resources of forage plants. CSIRO, East Melbourne, Australia.

Williams, R.J., and R.J. Clements. 1985. The future role of plant introduction in the development of tropical pastures in Australia. p. 20-28. In G. J. Murtagh and R. M. Jones (ed.) Proc. Third Australian Conf. on Tropical Pastures, Rockhampton. 8-12 July 1985. Trop. Grassl. Occas. Publ. 3.

Williams, W.T. 1983b. Analysis of plant evaluation data. p. 293-298. In J.G. McIvor and R.A. Bray (ed.) Genetic resources of forage plants. CSIRO, East Melbourne, Australia.

Wilson, G.P.M., R.M. Jones, and B.G. Cook. 1982. Persistence of jointvetch (*Aeschynomene falcata*) in experimental sowings in the Australian subtropics. Trop. Grassl. 16:155-156.

DISCUSSION

Brougham: Explain your strip test procedure of evaluation.

Kretschmer: The details of the procedure are in the paper. Briefly, the sliding rating system of 0 = no plants; 1 = 1 to 3 plants; 2, 3, 4 = intermediate low; 5 = 50% of row with plants; 6,7,8 = high intermediate; and 9 = 100% of row covered with plants provides a workable scale that is easily handled in computers. A portable hand-held computer is used to record data which are later transferred to a microcomputer. The recorder walks across (perpendicularly) the rows to make recordings.

Marten: Do an especially high proportion of tropical legumes contain antiquality and palatability-inhibiting components?

Kretschmer: Aside from the possible physical properties, particularly leaf and stem "hairiness", high tannin contents up to 10 to 12% in certain species can reduce palatability and protein digestibility. Tannin can be found in almost all tropical legumes.

Marten: At what point in your evaluation scheme do you test for animal response?

Kretschmer: If you mean legume response to grazing, then we would be able to do this from the first rating after grazing and ensuing grazings. We have a fair idea from other research which legume species are eaten well and avoid suspect species or those known to be toxic. Animal response would be tested in about 4 to 5 yr after plant introduction in an idealized situation.

Fisher: CIAT is involved in similar tests in most countries from Mexico to Brazil where small plots are used under cutting for several years and then selected entries are tested for response to grazing (see paper by Fisher & Thornton).

Fisher: I don't feel too worried if the legumes are somewhat unpalatable because they are easier to manage under grazing in association with vigorous grasses.

Kretschmer: I agree, generally; but sometimes if the legume is very unpalatable, i.e, tropical kudzu (Pueraria phaseoloides (Roxb.) Benth.) and particularly Calopogonium caeruleum Sauv., the grass may be temporarily eliminated as a component, and animal production suffers.

Gramshaw: We have tests in several areas, in Queensland using a similar rating procedure. The sites correspond to the important climate zones and the same legumes are being compared at each site under grazing.

Hochman: What are the merits of extrapolation of evaluation tests applied to various legumes?

Kretschmer: There is a poor relationship between clipping and grazing. Thus, grazing should be started as early as possible in the evaluation scheme.

Clements: In the temperate areas there is primary interest in only two or three genera (Trifolium, Medicago, Lotus). What will be the situation in the tropics.

Kretschmer: There are about 18 genera and 28 species most of which have at least one released cultivar and the remainder are in the last stages of testing. Because there are genera with large diversity that have not been fully tested (i.e., about 20 of 160 species in Aeschynomene and less than 20 of more than 300 species in Desmodium), I believe that there will be at least 10 genera from which species will have a significant and permanent contribution to animal production during the next decade. The climatic diversity (500 to 4000 mm annual rainfall) necessitates a broader diversity of legume species than in temperate regions where legume diversity is much less.

Rotar: Breeding in tropical legumes is still in its infancy.

DEMOGRAPHY OF PASTURE LEGUMES

R.M. Jones and E.D. Carter

SUMMARY

Pasture legumes can be grouped into three demographic groups: annuals, short-lived herbaceous perennials, and long-lived trees and shrubs. The short-lived perennials may replace themselves by sexual or asexual mechanisms, or a combination of both pathways. The major factors limiting successful population persistence in each category have been considered. Demographic understanding can help to improve legume persistence by defining better management practices and objectives for breeding and evaluation programs. Modelling of persistence, and ways of improving extension advice on legume persistence, are discussed.

INTRODUCTION

In this contribution we shall outline the various demographic pathways by which forage legumes can persist. Many authors have listed plant attributes which are important for survival of plants under grazing. Hodgkinson and Williams (1983), for example, listed 13 adaptations but recognized that some of these (e.g., "rapid reproduction") related primarily to annuals while others (e.g., "firm anchorage by roots") related mainly to long-term perennials. For this reason, we will consider the major demographic pathways to legume persistence separately and conclude by discussing how demographic studies can be used to improve legume persistence.

PATHWAYS TO PERSISTENCE

Let us assume a legume population is obtained by sowing a new legume-based pasture or by oversowing legume seed into an existing pasture. The individual plants resulting from this initial sowing will eventually die, and, for all but extremely long-lived plants, must be replaced by new plants if the population is to be maintained. New plants can develop from both sexual and asexual reproduction. Maintenance of adequate density depends on the survival rate of existing individuals and on rates of recruitment of new individuals. These rates are affected by a wide range of factors, but we can broadly group them into "environmental" and "management" factors. Environmental factors include soil moisture and temperature stress, waterlogging, frosting and soil type. Most of these factors are largely or completely out of the farmer's control and farmers can only react by choosing the most appropriate cultivars. Management factors, such as stocking rate, stock movement and fertilizer application are under more control by the farmer. Management can affect legume persistence by a direct effect on the legume (e.g., grazing of seed heads or pods and burrs) or indirectly through either weakening or enhancing the competitive effect of the associated grasses. Choice of the companion grasses grown with a legume can also affect legume persistence. Usually effective legume persistence is poorer with a more vigorous or denser companion grass (Hay & Hunt, this book).

For convenience we will group legumes into three classes: annuals, herbaceous perennials, and trees or shrubs. The herbaceous perennials will be further subdivided into legumes that do not recruit new plants under grazing, and those which recruit sexually, asexually, or by a combination of both pathways. The ecological theory of these pathways in terms of "r" and "k" reproductive strategies has been briefly discussed by Forde et al. (this book).

ANNUALS

This section primarily deals with obligate annuals but we will also consider instances where legumes may behave as facultative annuals, as in the case of Caribbean stylo (Stylosanthes hamata (L) Taub. cv. Verano) in northern Australia (Gardener, 1984) and white clover (Trifolium repens L.) at its extreme end of subtropical adaptation (Blaser & Killinger, 1950; Jones, 1982; Dzowela et al., 1986).

Annual legumes tend to be more important in areas with a distinct and relatively reliable wet season followed by an equally reliable dry season, as exemplified by the use of subterranean clover (Trifolium subterraneum L.) and annual medic (Medicago spp.) pastures in southern Australia and of Townsville stylo (S. humilis Kunth) and Verano stylo in northern Australia (Gramshaw et al., this book).

The key requirements for persistence of an annual are that sufficient seed must be produced and survive under grazing, that breakdown of hardseededness should ensure adequate seedling numbers at the onset of the following wet season, and that these seedlings survive to maturity and set seed. However, a reasonable proportion of seed must survive for more than 1 yr because poor rainfall, or management interventions such as a cropping phase, may prevent seed set in a year or run of years.

Seed Production

Seed production reflects the genetic potential of the plant, environmental conditions (particularly rainfall) and the grazing management imposed. Days to flowering has been very important in the selection of appropriate annual legumes for reliable seed production with the different rainfall zones with a Mediterranean environment in southern Australia (Reed et al., this book). An example of the effect of grazing on seed set is the finding that the proportion of rose clover (T. hirtum All.) in mixtures with subterranean clover declined in grazed pastures compared with ungrazed pastures, partly because the erect seed heads of rose clover were more readily grazed (Taylor & Rossiter, 1974). Heavier grazing pressures during flowering and seeding reduced seed production of barrel medic (Tow & Hodgkins, 1982). A moderate grazing pressure imposed on a range of annual legumes had no effect on seed production of some species (e.g., *T. globosum* L.) as compared with ungrazed swards, but reduced seed production of other species (e.g., *T. balansae* Boiss) by up to 50% (Bolland, 1987). Insects can also reduce seed set (Tow & Hodgkins, 1982). Once seed is set, adequate amounts must be left after grazing during the dry season. Carter (1983) documented how sheep select for medic pods during the dry season, and McKeon and Mott (1984) found that approximately 40% of seed set by three *Stylosanthes* spp. was removed during the dry season by cattle grazing and by invertebrate predators. Haymaking can drastically reduce seed production by medics but if correctly timed has little effect on seed production by subterranean clover (Carter et al., 1988a).

Hardseededness

During the dry season, there is a decrease in the hardseededness of the soil seed reserve. The higher the soil temperature and longer the dry season, the greater the amount of breakdown. Rates of breakdown of hardseededness in four *Stylosanthes* species at three widely separated sites in northern Australia were correlated with the number of days during which maximum soil temperatures exceeded 50-55°C, and from this relationship soil temperature data could be used to predict the rate of hardseed breakdown in different environments (Mott et al., 1981). If hardseededness is low or is broken down too readily there is a greater likelihood of germination or potential seed loss following "out of season" rains, as found for Mt. Barker and Woogenellup subterranean clover in a marginal area of New South Wales (Hagon, 1974). In some areas of southern Australia, with less extreme summer temperatures and drought, seedlings resulting from these summer rains can persist and make a major contribution to later production. Hence cultivars with lower hardseededness may be more productive (Fitzgerald & Archer, 1987). In contrast to these Australian experiences, Pavone and Reader (1982) found that in Canada, the major breakdown of hardseededness in *Medicago lupulina* L. occurred during winter. In several species, e.g., subterranean clover (Francis & Gladstones, 1983); serradella (*Ornithopus* spp.) (Bolland, 1985); and crimson clover (*T. incarnatum* L.) (Knight & Hollowell, 1973), selection for hardseededness has been of high priority. However, despite successful breeding for hardseededness in subterranean clover to give cv. Nungarin, Bolland (1986) found that cv. Nungarin does not have the same degree of hardseededness as the annual medics. In southern Australia, hardseededness is of greater significance in persistence of annual legumes in ley-farming systems than under permanent pasture (Reed et al., this book).

Cultivation may also affect hardseededness. Taylor (1985) recorded greater emergence of subterranean clover in the year following minimum tillage as compared with conventional tillage, but also found that greater burial of seed by conventional cultivation resulted in better-long term survival of seed.

Emergence

Depending on the cultivation equipment used, periodic cropping phases in an annual legume pasture can bury seed below 5 cm, at least until further cultivation (Quigley et al., 1987; Heida & Jones, 1988), and 5 cm is probably the limit for successful seedling emergence in all but very sandy soils (Carter & Challis, 1987; Heida & Jones, 1988). After deep mouldboard plowing, Carter et al. (1988b) found that only 15% of medic seed was in the top 5 cm of soil, compared with 90% following scarifying. However, seedling emergence as a percentage of the viable seed bank in the top 5 cm of soil can still be low, e.g., ca. 10% for annual medics (Carter, 1983; Heida & Jones, 1988), largely as a result of high levels of hardseededness. Furthermore, there are often considerable losses of soil seed reserves that cannot be accounted for as soil seed or seedlings (e.g. Gardener, 1981). Based on data for barrel medic (Carter, 1983), subterranean clover (Collins et al., 1983; Taylor et al., 1984) and Verano stylo (Gardener, 1982) a reasonable conceptual model of a successful self-regenerating annual pasture legume is one with $\geq$ 5000 seeds/m^2 and $\geq$ 10% emergence resulting in at least 500 seedlings/m^2 at the start of the wet season.

High stocking rates not only reduce seed reserves, as recorded for barrel medic by Carter (1983) and for subterranean clover by Carter and Lake (1985), but presumably increase the percentage emergence of the soil seed bank due to higher temperatures experienced on the more exposed soil surface. Likewise in southern Queensland, where summer rains result in growth from summer-growing grasses, experience suggests that emergence of cool-season annual medics is encouraged by close grazing over summer (E.J. Weston, 1988 pers. comm.). Close grazing of grasses over summer in the wetter subtropics also increases emergence of white clover seedlings in autumn/winter, attributed to to enhanced breakdown of hardseededness and to better seedling survival (Jones, 1982). Grazing pressure can also affect seedling survival. Fitzgerald and Archer (1987), for example, found that leaving some protective cover enhanced survival of seedlings of subterranean clover that emerged in summer and early autumn.

Seedlings that emerge at the beginning of the wet season, and then survive, usually make a proportionally larger contribution to legume yield than seedlings which emerge later (e.g., Massa & Mannetje, 1982; Jones, 1984a). Moisture stress is frequently nominated as the main cause of seedling death (e.g., Jones, 1980; Massa & Mannetje, 1982). In areas with reliable rainfall during the growing season, plants usually survive well over the growing season. For example, J.J. Mott (1988, pers. comm.) found that 60% of *Stylosanthes* seedlings emerging at the start of the wet season survived through the rest of the wet season.

Seed in Feces

A desirable characteristic of a pasture legume is that it spreads through the feces of grazing livestock, enabling colonization of unsown areas. A higher percentage of ingested seed will be excreted intact when legumes are hardseeded, and in feces of cattle rather than sheep or goats (Simao Neto et al., 1987; C.J. Gardener, 1988, pers. comm.).

The percentage of ingested seed passed in sheep feces can vary widely, e.g., from 2% for annual medics (Carter, 1980; Carter et al., 1988c), ca. 5% for subterranean clover (Wilson & Hindley, 1968; de Koning & Carter, 1988), 38% for *Trifolium balansae* (Carter et al., 1988c) and 57% for *T. glomeratum* L. (Franklin & Powning, 1942). In general, percentage passage increases with decreasing seed size (Playne, 1969) and increasing level of hardseededness. However, the low recovery of medic seed may well be explained by the chewing required to break down the enclosing pods so that they can be passed through the rumen. However, even with only 2% passage of ingested barrel medic seed, sheep can still excrete up to 10 viable seeds/g dry feces or 6000 seeds/ sheep per day (Carter & Lake, 1985). Jones (1982) recorded a peak concentration of 100 white clover seeds/g dry cattle feces. The importance of seed dissemination in feces has been briefly discussed by Curll and Jones (this book).

Improving Persistence of Annual Legumes

Persistence of annuals can be improved by selecting better genotypes and by improving pasture management (Carter, 1987). Important characters that may be sought in improved genotypes are disease and pest resistance, more reliable seed production under grazing, higher seed yield, and higher levels of hardseededness. These criteria, in the case of subterranean clover, have been discussed by Rossiter (1978), Carter et al. (1982), and Francis and Gladstones (1983).

Management can improve persistence by allowing adequate seed set and avoiding excessive removal of seed during the dry season, as discussed by Carter and Lake (1985) for annual medics grazed by sheep. Walton (1975) found that defoliation of subterranean clover during the growing season increased the proportion of clover seed which was buried, and which presumably would be protected from grazing and have a slower breakdown of hardseededness. Management may also aid seedling emergence and survival, as in the instance of close grazing of summer-growing pastures in south-east Queensland discussed earlier. Similarly, after oversowing *Aeschynomene americana* L. into grass pastures, Sollenberger et al. (1986) suggested grazing the sward until the legumes were at the "two-leaf stage". However, treading associated with very high stocking rates has the potential to reduce legume density (Carter & Sivalingam, 1977). There may also be scope for improving persistence of seed banks of annual legumes through a cropping phase by choosing appropriate tillage techniques (Taylor, 1985; Carter et al., 1988a; Taylor & Ewing, 1988; Gramshaw & Gilbert, this book) and by choosing appropriate timing of cultivation to favor the legumes rather than the associated species (Forcella & Gill, 1986).

There is little information on the effect of fertilizer on the demography of annual legumes as the direct effect of fertilizer is confounded with changes in grazing pressure, animal selectivity and other factors (e.g., Gardener, 1984). Fertilizer can increase legume yield and hence seed production and and soil seed reserves, as found for Verano stylo fertilized with superphosphate by J.J. Mott (1988, pers. comm.). However, if the growth of companion species is enhanced by extra N previously fixed by the fertilized legume, this may in turn reduce legume yield and seed set. Also if there is water stress during flowering and seeding, seed set could be reduced more on a higher-yielding fertilized sward than on a lower-fertility sward.

Although many of the factors limiting persistence of annual legumes are outside management control (e.g., low rainfall) there is nevertheless considerable scope for using management practices, such as grazing pressure and altered crop/pasture rotations, for improving persistence. Smith (1977) noted that farmers who had less success with medic pastures emphasized rainfall conditions as being more of a limitation than did successful farmers who presumably had higher management skills. A greater use of legume mixtures may assist in reducing the impact of variable environment and management on persistence. Whatever the causal factor(s), it is reasonable to conclude that inadequate seed reserves are frequently a major limitation to commercial success with annual legumes (e.g., Carter et al., 1982; Carter & Lake, 1985; Dear & Loveland, 1985; Carter & Cochrane, 1985; Carter, 1987).

PERENNIAL HERBACEOUS LEGUMES

Persistence of perennial herbaceous legumes can be improved by enhancing the survival of the original plants and/or improving the sexual or asexual recruitment of new plants.

Species with No Recruitment

Lucerne (*Medicago sativa* L.) is a good example of a plant that almost always relies on persistence of the originally established plants. Lucerne survival is strongly influenced by environment: persistence of lucerne in Australia decreases from southern Australia (half life ca. 3 yr) to subtropical Australia (half life ca. 1 yr) (Leach, 1978). Management cannot override this environmental gradient, but can improve persistence at any point in the gradient, e.g., by rotational grazing as distinct from continuous grazing (Leach, 1978) and, in the subtropics, by ensuring that C_4 grasses growing with lucerne are well grazed (Leach & Ratcliff, 1979). Persistence of red clover (*T. pratense* L.) is also dependent on the life span of individual plants, and survival is generally better under cutting than under grazing (e.g., McBratney, 1984). Breeding of lucerne and red clover varieties resistant to pests and diseases will also improve persistence (e.g., cv. Trifecta, Clements et al., 1984).

Perennials with Asexual Recruitment

In milder temperate regions, white clover is the classic example of a plant that persists through vegetative reproduction (Turkington et al., 1979; Chapman, 1987). In this sense we will consider white clover as a perennial although it can also be described as a "vegetative annual" (Sheath & Hay, this book). The primary taproots only live for ca. 2 yr (Westbrooks & Tesar, 1955) and clover subsequently persists by a dynamic network of stolons. "Senescence and death of the distal ends of stolons proceeds as growth progresses and, except for the early stages of growth from seed, no part of the plant lives for more than twelve months" (Spedding & Diekmahns, 1972). Chapman (1983) found that only 10 to 20% of nodes on clover stolons survived for more than 1 yr, and Hay et al. (1987) have shown that, for most of the year in New Zealand, there are more stolons buried below the soil surface than there are above it. The seasonal changes in the rates of extension, fragmentation and death of white clover stolons in New Zealand pastures have been described by Forde et al. (this book).

Stolon survival and stolon density can be enhanced by grazing with cattle rather than sheep (Briseno de la Hoz & Wilman, 1981; Chapman, 1983), by avoiding excessively close grazing (Briseno de la Hoz & Wilman, 1981; Curll & Wilkins, 1983), and, by inference from yield data, through seasonal manipulation of grazing pressure (Jones, 1933). Treading can also reduce the stolon density of white clover (Curll & Jones, this book). As a general rule, good white clover swards can have 40-100 m of stolon/m^2 (e.g., Curll & Wilkins, 1983). Genotypes with a high frequency of branch and root formation at the same node persist better than other genotypes under intensive grazing (Chapman, 1983). In old clover swards, containing a range of grass species, there is genetic differentiation between stolons growing in association with the different species of companion grass (Turkington & Harper, 1979).

Other species, such as _Lotus pedunculatus_ Cav. cv. Grasslands Maku also persist by rhizomes and stolons. In New Zealand, rhizomes and stolons of Maku extend in late summer and autumn and fragment in winter (Sheath, 1980). Rhizome development is reduced by grazing during late summer/early autumn (Wedderburn & Lowther, 1985). In subtropical Queensland, the stolon/rhizome length of Maku and a perennial non-seeding _Arachis_ sp. (_glabrata_ type) was compared 12 months apart, before and after a very dry summer. The stolon/rhizome length of Maku decreased from 160 to 5 m/m^2 whereas that of _Arachis_ sp. remained constant at ca. 190 m/m^2 (Vos & Jones, 1986). This highlights the important point that if a pasture legume persists solely, or nearly so, by vegetative means, the rhizome/stolon network must be hardy enough to persist through any likely stress periods (e.g., drought, freezing, waterlogging, fire).

This point is further illustrated by the frequent failure of _Desmodium intortum_ (Mill.) Urb. and _D. uncinatum_ (Jacq.) DC. to persist in subtropical Queensland, even in well-managed stands. Although the primary taproots of these _Desmodium_ spp. last far longer than the primary taproots of white clover, persistence of older desmodium relies to a very great extent on adventitious rooting from stolons. Good legume-grass pastures of these species contained 30-50 m of legume stolon/m^2 with ca. 200-300 adventitious roots of >1 mm diam/m^2 (Jones, 1988a). However, as larger adventitious roots were damaged by weevil larvae, the capacity of the sward to survive through the inevitable dry period was severely restricted (Jones, 1988a).

Many attributes could be involved when selecting genotypes for improved vegetative persistence, e.g., stolon density, stolon branching, roots per unit stolon length, stolon survival, deeper rooting systems, growth habit, drought tolerance and resistance to pests and diseases. In most cases, direct field selection for improved persistence may be the most appropriate route to take unless it is possible to clearly define the limiting factor - and this may be easiest with pests and diseases.

In this instance, the inevitable death of older plants must be compensated for by the successful recruitment of new plants from seedlings. Tropical legumes appear to have more plants in this category, e.g., shrubby stylo (Stylosanthes scabra Vog.), than do temperate legumes (Forde et al., this book). Most of the comments made about seedling recruitment of annual species also apply here. However, seedlings of perennial legumes will often be under greater competitive stress as they are competing with an established sward, whereas in annual pastures, particularly in a Mediterranean environment, seedlings are often only competing with other seedlings. Hence, compared with a new sowing into a cultivated seedbed, we can expect that in an established pasture there will be more seedling death, and the plants which develop from these seedlings will often have poorer survival (Gardener, 1981). This is illustrated for siratro (Macroptilium atropurpureum (D.C.) Urb.) in Fig. 1 (Jones & Mannetje, 1986).

While it is easy to measure the soil seed bank of a perennial legume, it is more difficult to know what the results mean than it is in the case of an annual legume. This is partly because the need for successful recruitment depends on the survival rate of established plants, and partly because seedling survival may be poorer. In some cases perennial legumes may have substantial seed reserves but these may be of little consequence for persistence because of extremely poor seedling survival, e.g., D. intortum (Jones, 1988a). Also, it is possible to build up substantial reserves of seed which are of minimal consequence because of the extremely slow rate of breakdown of hardseed in that environment, e.g., Lotononis bainesii Baker in moist soils (Jones, 1982).

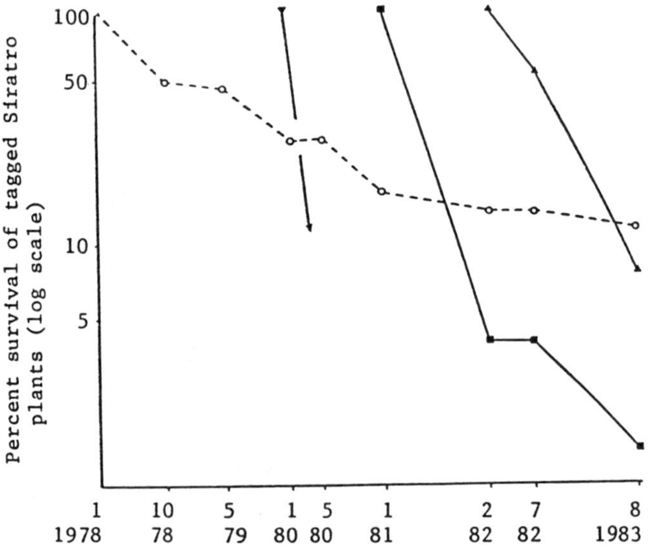

Figure 1: A comparison of the decline in the initial population of siratro (broken line) and subsequent cohorts (solid lines) in 1980, 1981 and 1982

However, there is also a minimum desirable level for a seed bank. For example, if we assume a half life of 2 yrs for a perennial legume, and a desired density of 10 plants/m^2, five plants are needed every 2 yrs, or 2.5 plants per year, to maintain this density. If we assume 10% emergence of the soil seed bank per year, and 10% survival of emerging seedlings (before they develop into an established plant), then we need a soil seed bank of 250 seeds/m^2 to produce 2.5 plants/m^2 per year. All these assumptions are quite reasonable (e.g. Gardener, 1984; Jones & Bunch, 1988a, 1988b; J.J. Mott, 1988 pers. comm.). However, a larger seed bank is highly desirable to allow for one or more unfavorable years with below average seed set and seedling survival. Thus, the good persistence and spread of Aeschynomene falcata (Poir) D.C. Prodr. cv. Bargoo is in part attributed to its ability to maintain high seed reserves (1000 - 7000 seeds/m^2) even under heavy grazing (Wilson et al., 1982). On the other hand, extremely high levels of soil seed may be of no extra benefit (e.g., 100 000 seeds of lotononis/m^2; Jones & Evans, 1977). Measurements of soil seed reserves can usually only be interpreted adequately when the other components of the demographic pathway (e.g., seedling emergence and plant survival) are known. Thus Suckling and Charlton's excellent 1978 review of legume seed reserves in New Zealand pastures could say very little about the role of these reserves in legume persistence but much more about problems of buried seeds producing contamination in commercial seed production of legumes.

Fire not only affects persistence through its effect on the established plant (Gardener, 1980) but affects numbers and hardseededness of the soil seed store (Gardener, 1980; Mannetje et al., 1983). Seed banks can also be increased by appropriate grazing management, e.g., reducing grazing pressure or spelling during flowering and seeding of siratro (Jones, 1988b).

An established stand of perennial legumes that persist by seedling recruitment will inevitably contain plants of different ages, and the age structure will vary with species, as shown for Stylosanthes by Gardener (1984), and often with management. As an example, the age structure of siratro plants in a heavily stocked and a lightly stocked pasture, both 10 yrs old, in southeast Queensland was as follows (taken from life-tables in Jones & Bunch, 1988a):

	Plants/m^2 of different ages (in years)								
	>8	7	6	5	4	3	2	1	Total
heavy stocking	-	-	-	-	-	0.1	-	0.8	0.9
light stocking	0.3	0.2	0.1	0.8	0.7	1.1	0.9	2.4	6.5

As both pastures had similar siratro densities to start with, these data show that, after eight years, both survival and recruitment of siratro were depressed at the heavy stocking rate.

There is very little information on the effect of fertilizer on the demography of perennial legumes forming new plants from seedlings. Preliminary data from siratro and shrubby stylo on a soil with 5 ppm available P suggest that superphosphate application may have a greater effect on seed set, soil seed reserves and seedling regeneration than it does on survival of original plants (R.M. Jones, unpublished data).

Many plants in natural communities require periodic disturbance, such as occurs with treefall, for successful seedling recruitment to occur e.g., Viola fimbriatula Sm. (Cook & Lyons, 1983). Such disturbance aids germination, emergence, seedling survival and plant survival whereas little or no successful recruitment may occur without disturbance. Likewise, some legumes, e.g., siratro, may persist better with periodic rough cultivation (= disturbance) (Bishop et al., 1983). Furthermore, a grain or fodder crop could be grown after the cultivation to help recover cultivation and maintenance fertilizer costs, as used by Hurwood (1979) for siratro.

As with annuals, the ability of a perennial legume to colonize new areas through seed spread in feces is a most useful attribute. Suckling (1952) suggested that farmers could deliberately manage pastures to maximize dissemination of white clover seed in sheep feces. The maximum expression of this character is found in legumes, such as Stylosanthes spp., which produce large quantities of seed which are retained on the plant after maturity, enabling animals to ingest them, and which have sufficient hardseededness to survive passage through the digestive tract (C.J. Gardener, 1988. pers. comm.).

Perennials with Both Sexual and Asexual Recruitment

Many pasture legumes replace themselves vegetatively and also by seedlings. For example, seedling recruitment of white clover is of minimal consequence in favored temperate areas but can be the main persistence pathway in subtropical environments (Blaser & Killinger, 1950; Jones, 1982; Dzowela et al., 1986). This change is attributed to the poor survival of white clover stolons in the hot, humid summers of the subtropics. The balance between sexual and asexual persistence can also be affected by management. For example, Lotononis bainesii can persist by both mechanisms, but in one study persistence by stolons was relatively more important at higher stocking rates (Pott & Humphreys, 1983). In contrast, vegetative persistence of siratro was only of consequence at a very light stocking rate (Jones & Bunch, 1988a). Vigna parkeri Bak. and, in some situations, Centrosema virginianum (L.) Benth. can also persist by both stolons and seedling recruitment (Jones & Clements, 1987) and it is likely that other tropical legumes, particularly those suited to higher rainfall environments, behave similarly.

TREES AND SHRUBS

There has recently been increasing interest in using leguminous trees and shrubs for fodder. In many cases, once shrubs are established they live for such a long time that there is no need for replacement. For example, even under regular "hedgerow grazing" some leucaena (*Leucaena leucocephala* (Lam.) de Wit) stands in southeast Queensland have a half life of over 50 yrs (Jones & Harrison, 1980; author's unpublished data). Some other shrubs are less persistent. Snook (1986) implies that, given correct grazing management, tagasaste (*Chamaecytisus palmensis* (Christ) F.A. Bisby & K.W. Nicholls) will survive for many years but acknowledges that survival will otherwise be poor. There are some instances, such as for *Codariocalyx gyroides* (Roxb. ex Link) Hassk., where seedling regeneration may be critical for long-term stand persistence, as in Belize (Lazier, 1981). Thus although some lines of *C. gyroides* grew well in southeast Queensland, the accessions used were unable to set seed in the cool winter and so stand persistence was poor (Jones, 1984b).

WHAT IS THE USE OF DEMOGRAPHIC STUDIES ON PASTURE LEGUMES?

The potential benefits, of demographic studies (Jones & Mott, 1980; Jones, 1986) are as follows:

Predicting the Effect of Management and Climate on Persistence

Reports of legume evaluation frequently have little predictive role and merely describe the success or failure of persistence under one or more treatments at one site. However, if the basic demography of a species is reasonably well known, there is a greater likelihood of predicting how changed management or environment will affect persistence, e.g., how changes in soil temperature will affect breakdown of hardseededness and longevity of a soil seed reserve. Likewise, demographic studies can assist in better defining good and bad management, e.g., it is bad management to regularly heavily graze barrel medic stands in summer and autumn so that low soil seed reserves limit stand density and productivity (Carter, 1982, 1983; Carter & Lake, 1985).

Aiding Breeding and Selection

Demographic studies can highlight breeding and selection objectives that are directly related to population biology, e.g., the need for a higher (or lower) level of hardseededness. They can also highlight factors indirectly related to persistence, e.g., poor persistence resulting from disease or insect damage that is not readily obvious. The recognition that many perennial tropical legumes must replace themselves for long-term population persistence has, in recent years, led to greater attention being paid to measurements of soil seed reserves in plant evaluation.

Modelling - Quantitative or Qualitative

Pasture legumes are inevitably exposed to a wide range of soil, rainfall, temperature, fertilizer, and grazing environments. It is of little value to have a quantitative model restricted to one specific situation, and it will be unusual to get enough information to devise a quantitative model that will accurately predict persistence in all likely situations. The chances of achieving this are much greater with annual legumes growing in a predictable environment with a reliable wet and dry season. This is evidenced by the models produced for Townsville stylo (e.g., Torssell & McKeon, 1976; Torssell & Nicholls, 1978) and subterranean clover (Galbraith et al., 1980; Rossiter et al., 1985) where equations describe the progressive movement between the different biological phases of germination, establishment, growth, seed production and break of hardseededness in annual legumes. In other situations the best we may be able to hope for is to have an adequate conceptual model, based on quantitative data, which can suggest how a legume is likely to respond to changes in its environment, management and genotype. A critical problem in modelling demography of perennial legumes is that of quantifying the overriding effect that competition from the established sward can have on all phases of plant demography, particularly on survival of legume seedlings emerging in undisturbed pasture (Cook & Ratcliff, 1985; Keating & Mott, 1987; Fisher & Thornton, this book). We support the suggestion of Torssell and Nicholls (1978) that modelling of perennial legumes could be split into a first year (or juvenile) phase and a 'mature plant' stage.

How can Demographic Information be used on the Farm?

Extension advisers can obviously incorporate demographic information and conceptual and quantitative models into their advice. Good progress has been made in developing an analytical service (or do-it-yourself kit?) for measuring seed reserves on farmers' paddocks (Carter & Le Leu, 1988), just as is done for available P. It may be possible to develop simple rating scales for seedling, plant, or stolon density that would enable farmers or graziers to adequately assess the population of legumes in their own pastures.

WHEN TO UNDERTAKE DEMOGRAPHIC STUDIES

Demographic studies should not be automatically undertaken as a routine activity in research on legume-based pastures. Some measurements, particularly those involving survival of plants and seedlings can be very time consuming, and hence are time wasting if done inappropriately. If a pasture legume is persistent, or is only used in short-term, high-input pastures where resowing costs are of minimal importance, demographic studies of legume persistence have little or no role.

Demographic studies may be of most practical value in lower-cost systems (where costs of resowing may be relatively higher), where pasture species show considerable promise but persistence is variable, or where persistence appears to be unduly restricted by site or management requirements. These aspects are considered in more detail by Jones and Mott (1980) and Jones (1986). Perhaps the main point of this paper is to encourage readers to think of legume persistence from a demographic viewpoint. This will usually indicate if demographic measurements are required.

REFERENCES

Bishop, H.G., B. Walker, and M.T. Rutherford. 1983. Renovation of tropical legume-grass pastures in northern Australia. p. 555-559. In J.A. Smith and V.W. Hays (ed.) Proc. 14th Int. Grassl.

Congr., Lexington. 15 Jun. - 24 Jun. 1981. Westview Press, Boulder, USA.

Blaser, R.E., and G.W. Killinger. 1950. Life history studies of Lousiana white clover (Trifolium repens L.). 1. Seed germination related to temperature, pasture management and adaptation. Agron. J. 42:215-220.

Bolland, M.D.A. 1985. Serradella (Ornithopus sp.): maturity range and hard seed studies of some strains of five species. Aust. J. Exp. Agric. 25:580-587.

Bolland, M.D.A. 1986. A laboratory assessment of seed softening patterns for hard seeds of Trifolium subterraneum subspp. subterraneum and brachycalycinum, and of annual medics. J. Aust. Inst. Agric. Sci. 52:91-94.

Bolland, M.D.A. 1987. The effect of grazing and soil type on the seed production of several Trifolium species compared with subterranean clover and annual medics. J. Aust. Inst. Agric. Sci. 53:293-295.

Briseno de la Hoz, V.M., and D. Wilman. 1981. Effects of cattle grazing, sheep grazing, cutting and sward height on a grass-clover sward. J. Agric. Sci. 97:699-706.

Carter, E.D. 1980. The survival of medic seeds following ingestion of intact pods by sheep. p. 178. In Proc. 1st Aust. Agron. Conf., Lawes. p.178.

Carter, E.D. 1982. The need for change in making the best use of medics in the cereal-livestock farming systems of South Australia. p.180. In Proc. 2nd Aust. Agron. Conf., Wagga Wagga. Aust. Soc. Agron., Parkville, Victoria.

Carter, E.D. 1983. Seed and seedling dynamics of annual medic pastures in South Australia. p. 447-450. In J.A. Smith and V.W. Hays (ed.) Proc. 14th Int. Grassl. Congr., Lexington, 15 Jun. - 24 Jun. 1981. Westview Press, Boulder, USA.

Carter, E.D. 1987. Establishment and natural regeneration of annual pastures. p. 35-51. In J.L. Wheeler, et al. (ed.) Temperate pastures: their production, use and management. Aust. Wool Corp. & CSIRO, Melbourne.

Carter, E.D., and S. Challis. 1987. Effects of depth of sowing medic seeds on emergence of seedlings. p.192. In Proc. 4th Aust Agron. Conf., Melbourne. Aust. Soc. Agron., Parkville, Victoria.

Carter, E.D., and M.J. Cochrane. 1985. The poor subterranean clover status of dairy pastures in the Adelaide Hills. p.217. In Proc. 3rd Aust. Agron. Conf., Hobart. Aust. Soc. Agron., Parkville, Victoria.

Carter, E.D., and A. Lake. 1985. Seed, seedling and species dynamics of grazed annual pastures in South Australia. p.654-656. In Proc. 15th Int. Grassl. Congr., Kyoto, 24 Aug. - 31 Aug. 1985. Jap. Soc. Grassl. Sci., Tochigi - Ken, Japan

Carter, E.D., and K. Le Leu. 1988. Predicting emergence of annual pasture legumes. The University of Adelaide, Waite Agricultural Research Institute, 1986-87 Biennial Report, in press.

Carter, E.D., and T. Sivalingam. 1977. Some effects of treading by sheep on pastures of the Mediterranean climatic zone of South Australia. p.307-312, Section 5. In Proc. 13th Int. Grassl. Congr. Leipzig.

Carter, E.D., E.C. Wolfe, and C.M. Francis. 1982. Problems of maintaining pastures in the cereal-livestock areas of Southern

Australia. p.68-82. In Proc. 2nd Aust. Agron. Conf., Wagga Wagga. Aust Soc. Agron., Parkville, Victoria.

Carter, E.D., M. Armstrong, and K.J. Sommers. 1988a. Effects of tillage and haymaking practices on survival and productivity of annual pasture legumes. p N4, N5. In Workshop on tillage systems, rotations, nitrogen and cereal root diseases, Adelaide, South Australia.

Carter, E.D, P. Thomas, E. Fletcher, and E. Cotze. 1988b. Effects of tillage practices on annual medics. The University of Adelaide, Waite Agricultual Research Institute, 1986-87 Biennial Report, in press.

Carter, E.D., S. Challis, and R.C. Knowles. 1988c. Legume seed survival following ingestion by sheep. The University of Adelaide, Waite Agricultural Research Institute, 1986-87 Biennial Report, in press.

Chapman, D.F. 1983. Growth and demography of *Trifolium repens* stolons in grazed hill pastures. J. Appl. Ecol 20:597-608.

Chapman, D.F. 1987. Natural re-seeding and *Trifolium repens* demography in grazed hill pastures. J. Appl. Ecol. 24:1037-1043.

Clements, R.J., et al. 1984. Breeding disease resistant, aphid resistant lucerne for subtropical Queensland. Aust. J. Exp. Agric. Anim. Husb. 24:178-188.

Collins, W.J., R.C. Rossiter, and E.C. Wolfe. 1983. The winter production of some strains of subterranean clover grown in defoliated swards. Aust. J. Exp. Agric. Anim. Husb. 23:140-145.

Cook, R.E., and E.E. Lyons. 1983. The biology of *Viola fimbriatula* in a natural disturbance. Ecology 64:654-660.

Cook, S.J., and D. Ratcliff. 1985. Effect of fertilizer, root and shoot competition on the growth of Siratro (*Macroptilium atropurpureum*) and green panic (*Panicum maximum* var. *trichoglume*) seedlings in a native speargrass (*Heteropogon contortus*) sward. Aust. J. Agric. Res. 36:233-245.

Curll, M.L., and R.J. Wilkins. 1983. The comparative effects of defoliation, treading and excreta on a *Lolium perenne* - *Trifolium repens* pasture grazed by sheep. J. Agric. Sci. 100: 451-460.

Dear, B.S., and B. Loveland. 1985. A survey of seed reserves of subterranean clover pastures on southern Tablelands of New South Wales. p.214. In Proc. 3rd Aust. Agron. Conf., Hobart. Aust. Soc. Agron., Parkville, Victoria.

de Koning, C.T., and E.D. Carter. 1988. The ecology and productivity of subterranean clover. The University of Adelaide, Waite Agricultural Research Institute, 1986-87 Biennial Report, in press.

Dzowela, B.H., G.O. Mott, and W.R. Occumpaugh. 1986. Summer grazing management of a white clover-bahia grass pasture: legume survival. Expl. Agric. 22:363-371.

Fitzgerald, R.D. and K.A. Archer. 1987. Improvement of natural pastures on the Northern Slopes of New South Wales. Final report to the Australian Wool Corporation, Dan 10P.

Forcella, F., and A.M. Gill. 1986. Manipulation of buried seed reserves by timing of soil tillage in Mediterranean-type pastures. Aust. J. Exp. Agric. 26:71-77.

Francis, C.M., and J.S. Gladstones. 1983. Exploitation of the genetic resource through breeding: *Trifolium subterraneum*. p.

251-260. In J.G. McIvor and R.A. Bray (eds) Genetic resources of forage plants. CSIRO, Melbourne.

Franklin, M.C., and R.F. Powning. 1942. Studies on the chemical composition of pods and seeds of certain species of Medicago, with a note on the apparent digestibility of cluster clover (Trifolium glomeratum L.) seed. J. Council Sci. Ind. Res. Aust. 15:190-200.

Galbraith, K.A., G.W. Arnold, and B.A. Carbon. 1980. Dynamics of plant and animal production of a subterranean clover pasture grazed by sheep. 2. Structure and validation of the pasture growth model. Agric. Syst. 6:23-44.

Gardener, C.J. 1980. Tolerance of perennating Stylosanthes plants to fire. Aust. J. Exp. Agric. Anim. Husb. 20:587-593.

Gardener, C.J. 1982. Population dynamics and stability of Stylosanthes hamata cv. Verano in grazed pastures. Aust. J. Agric. Res. 33:63-74.

Gardener, C.J. 1984. The dynamics of Stylosanthes pastures. p. 333-357. In H.M. Stace and L.A. Edye (ed.) The biology and agronomy of Stylosanthes. Academic Press, Sydney.

Hagon, M.W. 1974. Regeneration of annual winter legumes at Tamworth, New South Wales. Aust. J. Exp. Agric. Anim. Husb. 14:57-64.

Hay, M.J.M. et al. 1987. Seasonal variation in the vertical distribution of white clover stolons in grazed swards. N.Z. J. Agric. Res. 30:1-8.

Heida, R., and R.M. Jones. 1988. Seed reserves of barrel medic (Medicago truncatula) and snail medic (M. scutellata) in the topsoil of pastures on a brigalow soil in southern Queensland. Trop. Grassl. 22:16-21.

Hodgkinson, K.C., and O.B. Williams. 1983. Adaptation to grazing in forage plants. p. 85-100. In J.G. McIvor and R.A. Bray (ed.) Genetic resources of forage plants. CSIRO, Melbourne.

Hurwood, R. 1979. Deterioration and renovation of pastures at "Bungawatta". Trop. Grassl. 13:181-182.

Jones, M.G. 1933. Grassland management and its influence on the sward. II. The management of a clovery sward and its effects. Emp. J. Exp. Agric. 1:122-127.

Jones, R.M. 1980. Survival of seedlings and primary taproots of white clover (Trifolium repens) in subtropical pastures in south-east Queensland. Trop. Grassl. 14:19-22.

Jones, R.M. 1982. White clover (Trifolium repens) in subtropical southeast Quensland. 1. Some effects of site, season and management practices on the population dynamics of white clover. Trop. Grassl. 16:118-127.

Jones, R.M. 1984a. White clover (Trifolium repens) in subtropical south-east Queensland. III Increasing clover and animal production by use of lime and flexible stocking rates. Trop. Grassl. 18:186-194.

Jones, R.M. 1984b. Yield and persistence of the shrub legumes Codariocalyx gyroides and Leucaena leucocephala on the coastal lowlands of southeast Queensland. CSIRO Australia Div. Trop. Crops and Pastures, Tech. Mem. No. 38.

Jones, R.M. 1986. Persistencia de las especies forrajeras bajo pastoreo p. 167-200. In C. Lascano and E. Pizarro (ed.) Evaluacion de pasturas con animales. CIAT, Cali, Colombia.

Jones, R.M. 1988a. Productivity and population dynamics of Silverleaf desmodium (Desmodium uncinatum), greenleaf desmodium (D. intortum) and two D. intortum x D. sandwicense hybrids at two stocking rates in coastal southeast Queensland. Trop. Grassl. 22: in press.

Jones, R.M. 1988b. The effect of stocking rate on the population dynamics of Siratro in Siratro (Macroptilium atropurpeum)/ setaria (Setaria sphacelata) pastures in south-east Queensland. III. Effects of spelling on restoration of Siratro in overgrazed pastures. Trop. Grassl. 22:5-11.

Jones, R.M., and G.A. Bunch. 1988a. The effect of stocking rate on the population dynamics of Siratro in Siratro (Macroptilium atropurpureum) - setaria (Setaria sphacelata) pastures in southeast Queensland. I. Survival of plants and stolons. Aust. J. Agric. Res. 39:209-219.

Jones, R.M., and G.A. Bunch. 1988b. The effect of stocking rate on the population dynamics of Siratro in Siratro (Macroptilium atropurpureum) - setaria (Setaria sphacelata) pastures in southeast Queensland. II. Seed set, soil seed reserves, seedling recruitment and seedling survival. Aust. J. Agric. Res. 39:221-234.

Jones, R.M., and R.J. Clements. 1987. Persistence and productivity of Centrosema virginianum and Vigna parkeri cv. Shaw under grazing on the coastal lowlands of south-east Queensland. Trop. Grassl. 21:55-64.

Jones, R.M., and T.R. Evans. 1977. Soil seed levels of Lotononis bainesii, Desmodium intortum and Trifolium repens in subtropical pastures. J. Aust. Inst. Agric. Sci. 43:164-166.

Jones, R.M., and R.E. Harrison. 1980. Note on the survival of individual plants of Leucaena leucocephala in grazed stands. Trop. Agric. (Trinidad) 57:265-266.

Jones, R.M., and L.'t Mannetje. 1986. A comparison of bred Macroptilium atropurpureum lines and cv. Siratro in subcoastal southeast Queensland with particular reference to legume persistence. CSIRO Australia Div. Trop. Crops and Pastures, Tech. Mem. No. 47.

Jones, R.M., and J.J. Mott. 1980. Population dynamics in grazed pastures. Trop. Grassl. 14:218-224.

Keating, B.A., and J.J. Mott. 1987. Growth and regeneration of summer-growing pasture legumes on a heavy clay soil in south-eastern Queensland. Aust. J. Exp. Agric. 27:633-641.

Knight, W.E., and E.A. Hollowell. 1973. Crimson Clover. Adv. Agron. 25:47-76.

Lazier, J.R. 1981. Effect of cutting height and frequency on dry matter production of Codariocalyx gyroides (syn. Desmodium gyroides) in Belize, Central America. Trop. Grassl. 15:10-16.

Leach, G.J. 1978. The ecology of lucerne pastures. p. 290-308. In J.R. Wilson (ed.) Plant relations in pastures. CSIRO, Melbourne.

Leach, G.J., and D. Ratcliff. 1979. Lucerne survival in relation to grass management on a brigalow soil in south-east Queensland. Aust. J. Exp. Agric. Anim. Husb. 19:198-207.

Mannetje, L.'t, S.J. Cook, and J.H. Wildin. 1983. The effects of fire on a buffel grass and Siratro pasture. Trop. Grassl. 17:30-39.

Massa, F.E., and L.'t Mannetje. 1982. The behaviour of Townsville stylo (Stylosanthes humilis) in a native pasture at Narayen Research Station in south-east Queensland. Trop. Grassl. 16:186-196.

McBratney, J.M. 1984. Productivity of red clover grown alone and with companion grasses; further studies. Grass Forage Sci. 39:167-75.

McKeon, G.M., and J.J. Mott. 1984. Seed biology of Stylosanthes. p. 311-332. In H.M. Stace and L.A. Edye (ed.) The biology and agronomy of Stylosanthes. Academic Press, Sydney.

Mott, J.J., G.M. McKeon, C.J. Gardener, and L.'t Mannetje. 1981. Geographic variation in the reduction of hard seed content of Stylosanthes seeds in the tropics and subtropics of northern Australia. Aust. J. Agric. Res. 32:861-869.

Pavone, L.V., and R.J. Reader. 1982. The dynamics of seed bank size and seed state of Medicago lupulina. J. Ecol. 70:537-547.

Playne, M.J. 1969. The nutritional value of intact pods of Townsville lucerne (Stylosanthes humilis). Aust. J. Exp. Agric. Anim. Husb. 9:502-507.

Pott, A., and L.R. Humphreys. 1983. Persistence and growth of Lotononis bainesii - Digitaria decumbens pastures. 1. Sheep stocking rate. J. Agric. Sci. 101:1-7.

Quigley, P.E., E.D. Carter, and R.C. Knowles. 1987. Medic seed distribution in soil profiles. p. 193. In Proc. 4th Aust. Agron. Conf., Melbourne. Aust. Soc. Agron., Parkville, Victoria.

Rossiter, R.C. 1978. The ecology of subterranean clover-based pastures p. 325-339. In J.R. Wilson (ed.) Plant relations in pastures. CSIRO, Melbourne.

Rossiter, R.C., R.A. Maller, and A.G. Pakes. 1985. A model of changes in the composition of binary mixtures of subterranean clover strains. Aust. J. Agric. Res. 36:119-143.

Sheath, G.W. 1980. Effects of season and defoliation on the growth habit of Lotus pedunculatus Cav. cv. Grasslands Maku. N.Z. J. Agric. Res. 23:191-200.

Simao Neto, M., R.M. Jones, and D. Ratcliff. 1987. Recovery of pasture seed ingested by ruminants. 1. Seed of tropical pasture species fed to cattle, sheep and goats. Aust. J. Exp. Agric. 27:239-246.

Smith, M.V. 1977. Farmer usage of annual medics in a ley farming area in South Australia. p. 301-305, Section 5. In Proc. XIII Int. Grassl. Congr., Leipzig.

Snook, L.C. 1986. Tagasaste-tree lucerne. Night Owl Publishers, Shepparton, Australia.

Sollenberger, L.E., K.H. Quesenberry, and J.E. Moore. 1986. Effect of grazing management on establishment and productivity of Aeschynomene overseeded in limpograss pastures. Agron. J. 79:78-82.

Spedding, C.R.W., and E.C. Diekmahns. 1972. White clover (Trifolium repens). p. 347-369. In Grasses and legumes in British Agriculture. Bulletin 49. Commonwealth Agricultural Bureaux, Farnham Royal, U.K.

Suckling, F.E.T. 1952. Dissemination of white clover (Trifolium repens) by sheep. N.Z. J. Sci. Technol. A, 33:64-77.

Suckling, F.E.T., and J.F.L. Charlton. 1978. A review of the significance of buried legume seeds with particular reference to New Zealand agriculture. N.Z. J. Exp. Agric. 6:211-215.

Taylor, G.B. 1985. Effect of tillage practices on the fate of hard seeds of subterranean clover in a ley farming system. Aust. J. Exp. Agric. 25:568-573.

Taylor, G.B., and M.A. Ewing. 1988. Effect of depth of burial on the longevity of hard seeds of subterranean clover and annual medics. Aust. J. Exp. Agric. 28:77-81.

Taylor, G.B., and R.C. Rossiter. 1974. Persistence of several annual legumes in mixtures under continuous grazing in the south west of Western Australia. Aust. J. Exp. Agric. Anim. Husb. 14:632-639.

Taylor, G.B., R.C. Rossiter, and M.J. Palmer. 1984. Long term patterns of seed softening and seedling establishment from single crops of subterranean clover. Aust. J. Exp. Agric. 24:200-212.

Torssell, B.W.R., and G.M. McKeon. 1976. Germination effects on pasture composition in a dry monsoonal climate. J. Appl. Ecol. 13:593-603.

Torssell, B.W.R., and A.O. Nicholls. 1978. Population dynamics in species mixtures. p. 217-232. In J.R. Wilson (ed.) Plant relations in pastures. CSIRO, Melbourne.

Tow, P.G., and D. Hodgkins. 1982. Factors affecting seed production in Jemalong barrel medic based pastures at Roseworthy Agricultural College. p. 179. In Proc. 2nd Aust. Agron. Conf., Wagga Wagga. Aust. Soc. Agron., Parkville, Victoria.

Turkington, R., M.A. Cahn, A. Vardy, and J.L. Harper. 1979. The growth, distribution and neighbour relationships of Trifolium repens in a permanent pasture. III. The establishment and growth of Trifolium repens in natural and perturbed sites. J. Ecol. 67:231-243.

Turkington, R., and J.L. Harper. 1979. The growth, distribution and neighbour relationships of Trifolium repens in a permanent pasture. IV. Fine-scale biotic differentiation. J. Ecol. 67:245-254.

Vos, G., and R.M. Jones. 1986. The role of stolons and rhizomes in legume persistence. CSIRO, Australia, Div. Trop. Crops and Pastures, Annu. Rep. 1985-86, 70-71.

Walton, G.H. 1975. Response of burr burial in subterranean clover (Trifolium subterraneum) to defoliation. Aust. J. Exp. Agric. Anim. Husb. 15:69-73.

Wedderburn, M.E., and W.L. Lowther. 1985. Factors affecting establishment and spread of "Grasslands Maku" lotus in tussock grasslands. Proc. N.Z. Grassl. Assoc. 46:97-101.

Westbrooks, F.E., and M.B. Tesar. 1955. Taproot survival of Ladino clover. Agron. J. 47:403-410.

Wilson, A.D., and N.L. Hindley. 1968. The value of seeds, pods and dry tops of subterranean clover (Trifolium subterraneum) in the summer nutrition of sheep. Aust. J. Exp. Agric. Anim. Husb. 8:168-176.

Wilson, G.P.M., R.M. Jones, and B.G. Cook. 1982. Persistence of jointvetch (Aeschynomene falcata) in experimental sowings in the Australian subtropics. Trop. Grassl. 16:155-156.

DISCUSSION

B. Smith: Can you look at new cohorts of seedlings and determine the overall health of the legume as far as persistence is concerned?

Jones: It is easier to use seedling density as a measure of the population health of an annual than it is for a perennial, but you have to know your species.

Kretschmer: As long as we see some seedling recruitment in an annual plant such as *Aeschynomene americana*, do we need to know more than this?

Jones: Yes, we need to know more. We need to know how many seedlings are necessary to ensure that yield is not limited by plant numbers and to ensure long-term persistence. For example, I do not know how many seedlings per square meter are needed for *A. americana* because I have little knowledge of that species.

Fisher: How many adult plants comprise a good sward of subclover?

Jones: About 500 (or more) plants per square meter, but the figure would vary with environment and grazing pressure.

Clements: What is an acceptable seed bank for subclover or an annual medic?

Jones: As a broad generalization, it would be good to have over 5000 seeds per square meter, but his figure also varies with environment and management.

Sheath: What is the relative merit of annual legumes vs. free seeding perennial legumes as the regularity of "disturbance" (e.g., drought) increases?

Jones: As "disturbance" becomes more regular, annuals tend to become more important.

Allen: You have clearly presented a demographic concept on a whole-system basis, including a range of biotic and abiotic factors affecting plant persistence. Do you see a need for interdisciplinary and integrated projects to provide sufficient understanding of a system?

Jones: The interdisciplinary and integrated approach is desirable. However, it may not be practical in all situations.

Sheath: If it is necessary for plants to set seed in order to persist, is there a cost to the adult plants in terms of plant vigor and yield?

R. Smith: With red clover and similar forage plants our concern is to delay or reduce the reproductive phase in order to maximize yield, unless we are after the "reseeding" characteristic.

Management can be used to promote the particular activity (vegetative or reproductive growth) that we are seeking.

Hochman: In annuals, the "cost" of seed production in terms of herbage yield is probably not important, because feed demand at the time of seeding will be well below feed supply.

Clements: Would one of the factors which limit demographic research be that the work is long term, and consequently not suitable for higher degree studies?

Jones: To some extent this is true, but some aspects of demography are well suited to such studies; for example the work on white clover in the USA by Fred Westbrook and Roy Blaser some 40 yr ago was in this category.

ROOTING CHARACTERISTICS OF LEGUMES

Arthur G. Matches

SUMMARY

Rooting characteristics of forage legumes have not been investigated extensively. Little is known about how or if rate of root elongation, rooting depth, root distribution and density, number of rooting points, and root/shoot ratio may influence legume persistence across different soils and environments. Genera, species and cultivars of temperate and tropical legumes show marked variability in rooting characteristics. Variation in soil temperature influence root growth and development, and N_2 fixation. Water stress, low soil pH and aluminum toxicity, high soil bulk-density and strain of rhizobia also are reported to affect root development. Variability among legumes may offer opportunities for breeding and selecting more persistent legumes for specific soil and climatic conditions, and for particular uses under grazing and cutting. Improved methods for conducting long-term research on legume roots are needed. Future research should focus on how rooting characteristics affect legume persistence.

INTRODUCTION

About half the growth of pasture plants is not available to livestock because it is partitioned into roots (Davidson, 1978). Compared to top-growth, much less is known about the roots of legumes. Therefore, greater attention should be directed towards understanding the root-growth of legumes. For example, rate of root elongation, rooting depth, root density, and number of rooting points all may contribute to the persistence of legumes. These and other characteristics influence water uptake and/or drought tolerance (Asher & Ozanne, 1966; Caradus, 1981a; Carter et al., 1982; Davidson, 1978; Hamblin, 1985; Hamblin & Hamblin, 1985, Hamblin & Tennant, 1987; Humphries & Bailey, 1961; Kramer, 1983; Ozanne et al., 1965; Stevenson & Laidlaw, 1985; Thomas, 1984). Nutrient uptake (Asher & Ozanne, 1966; Barley, 1970), and competition with other plants (Cook & Ratcliff, 1985; Jackman & Mouat, 1972; Lee & Cho, 1985) are also influenced by root characteristics. Likewise, the presence or absence of rhizomes or stolons may influence rooting characteristics and survival under divergent intensities of defoliation (Sheath, 1980).

TECHNIQUES FOR ROOT INVESTIGATIONS

Three references may be useful to those initiating root development investigations. Bohm (1979) indicated that nearly 10 000 papers on root ecology had been published on methods of studying root systems. Direct methods for studying roots that he reported included excavation, monolith, auger, profile wall, and glass wall. Thirteen indirect methods included determination of soil water, staining techniques, uptake of non-radioactive and radioactive tracers, and container methods (including tubes, flexible tubes, boxes, and pots). A growth pouch method was described by McMichael et al. (1985). The clear polyethylene growth pouches (164 x 174 mm) containing a nutrient solution and an absorbent paper and trough for seed placement are useful in studying both root elongation and lateral root development of seedlings over short periods of time. Perhaps the most exciting new development is the mini-rhizotron system involving clear acrylic tubes (51 mm id x 3 m length) placed in the soil. A battery-operated color video camera is used to observe root development around the outside of the tube (Upchurch & Ritchie, 1984). This technique allows following root development of undisturbed plants in the field.

ROOT CHARACTERISTICS

Temperate Legumes

In comparisons of white clover (*Trifolium repens* L.) lines grown in the greenhouse or field, Caradus (1977, 1981a, 1981b) reported differences in root dry weight and shoot dry weight per plant, root/stem ratio, basal width of tap root, length of first primary lateral root, number of nodal roots, number of stolon tips, number of root tips at different layers (depths), and size of tap root in the field. For example, in one test of 10 lines, 6-wk-old seedlings differed by 50% for root dry weight, 39% for mean shoot dry weight, 27% for root/shoot ratio, and 30% for basal tap root diameter. Plants collected from sunny aspects in the field had more fibrous roots and fewer large tap roots than those collected from shady aspects. Caradus and Evans (1977) found that white clover root development was highly seasonal, with most new nodal roots being produced during the autumn, winter and spring. Seasonal changes in the vertical distribution of stolons of white clover at each of seven sites were observed in New Zealand by Hay et al. (1987). They found that approximately 80% of the stolons were buried in winter and 40% in summer and autumn. They concluded that white clover follows an annual cycle consisting of burial of stolons in winter and re-emergence of growing points over summer, providing that harsh environmental conditions do not interrupt this cycle. Earthworm biomass (which affects the quantity of surface castings) and effective rainfall were equally important in the burial of stolons. Stolon mass (kg ha^{-1}) decreased with increasing water stress, but individual stressed stolons were thicker and heavier than stolons at non-stressed sites. In trials in Japan with red clover (T. pratense L.), alfalfa (Medicago sativa L.), and white clover, Ueno (1982) reported complete loss of red clover stands, no thinning of alfalfa, and taproot decay for ladino clover. However, the thinning of ladino was temporary and new stolons replaced thinned areas. Ability of white clover nodal roots to compensate for loss of taproots depended on their position and size, and on the cultivars.

Ennos (1985) reported that among 272 progeny plants from a natural white clover population, genetic variation in root length was significant and estimated narrow sense heritability was high (from 0.42 to 0.84). Long-rooted plants produced 22% more stolons than short-rooted types and during a period of drought, dry matter yield was higher for plants with long roots. However, Stevenson and Laidlaw (1985) found that root initiation for even well-watered stolons was inhibited at a relative humidity of 85% or less. Thomas (1984) suggested that the competitive ability of white clover during drought could be improved by breeding for greater root density and for the ability to adopt a low defoliation-avoiding habit during dry weather.

Dry weights of tops, roots, and total plants and the root/shoot ratio of four subterranean clover (T. subterraneum L.) cultivars grown in the field differed by 66, 55, 64, and 34%, respectively (Humphries & Bailey, 1961). Pearson and Jacobs (1985) reported that the rooting depth of 'Northam' and 'Nungarin' subterranean clover plateaued in about 90 days after planting, but the difference between cultivars was small. For plant populations ranging from 2 to 1510 plants m^{-2}, they concluded that the amount of root per unit area increased with population, but plants appeared to adjust (lower leaf area per root length) while being unable to sustain dry matter partitioning (reducing burr/top yields) at the highest population studied.

Hamblin and Hamblin (1985) conducted field evaluations which included Lupinus angustifolius L., L. cosentinii Guss., T. subterraneum L., T. hirtum All., Pisum sativum L., Vicia benghalensis L., Medicago truncatula Gaertner, M. polymorpha L., M. tornata (L.) Miller, M. littoralis Rhode ex Loisel, and M. scutellata (L.) Miller, grown on a sandy soil in Western Australia. Species and cultivars showed a wide range in rooting depth, root length and root density (Table 1). Lupins had less than 50% of their total root length in the top 20 cm of soil, whereas other species had over 70% in the top 20 cm. Pasture legumes with high root density in the top soil caused drying of the top soil and a subsequent reduction in nutrient availability and vegetative growth. Ozanne et al. (1965), studied the root distribution of some pasture plants in a deep sandy soil and found large differences in root distribution and rooting depth among species. For example, 'Yarloop' subterranean clover had no roots below 40 cm, and 95% of its total roots were present within the upper 10 cm of soil. In contrast, roots of lupins (L. digitatus Forsk.) were well developed down to at least 100 cm, and only 25% of roots occurred in the upper 10 cm of soil. Although the highest concentration of roots was near the surface in all species, there

Table 1. Range in root measurements for different legumes under 298-mm winter rainfall in Western Australia (adapted from Hamblin and Hamblin, 1985).

Legumes (No. of entries)	Maximum root depth	Root length/ ground area	Average root density
	cm	cm cm^{-2}	cm cm^{-3}
Lupins(4)	185 - 220	7 - 10	0.03 - 0.05
Clovers (4)	40 - 105	37 - 86	0.82 - 1.13
Peas (2)	50 - 90	9 - 13	0.16 - 0.18
Medics (8 and 9)1/	65 - 130	10 - 138	0.10 - 1.06

1/Eight entries for depth, nine for other measurements.

were large differences in root distribution and effective rooting depth among species. Asher and Ozanne (1966) found that barrel medic (M. tribuloides Desr.) had a deeper root system than either Yarloop or 'Mount Barker' subclover.

In semiarid regions with minimal winter precipitation, legume seedlings having rapid root elongation might possibly be able to maintain roots in moist soil as the soil moisture front moves downward over winter. With this in mind, three experiments with 14 winter-annual legumes (Medicago and Trifolium spp.) were conducted in the greenhouse by Reyes (1986) at Lubbock, TX. Seedlings were grown in clear acrylic tubes which were filled with washed sand, and the tubes were slanted 15° from vertical. The range of differences for root measurements from the first experiment is shown in Table 2. Among entries, maximum depth of root penetration, elongation rate, dry weight, and root/shoot ratio differed by 40, 38, 68, and 62%, respectively. Again, considerable genetic variability existed among winter-annual legumes for certain root characteristics.

Working with alfalfa, Simpson et al. (1977) reported differences in root growth among 40 genotypes grown in soils low in calcium. From a second phase of this research, they concluded that improved genotypes could be selected from Australian alfalfas for establishment in areas with acidic subsoils, but that selections based on root penetration alone would not necessarily lead to increased herbage yields. Alfalfa genotypes selected for tolerance to soil acidity, and an unselected control, were grown on a clay loam soil with a subsoil pH of 4.2 (Bouton et al., 1982). Root growth was compared in areas where the subsoil was limed and not limed. On the unlimed subsoil, alfalfas selected for acid tolerance had greater depth of root penetration than the unselected control; however, where the subsoil was limed, the reverse occurred. Matches et al. (1962) found that 'African' alfalfa had

Table 2. Range in root measurements for different winter-annual legumes grown in slanted tubes in the greenhouse (adapted from Reyes, 1986). 1/ 2/

Legumes (No. of entries)	Maximum depth	Seedling root Elongation rate	Dry wt./ plant	Root/shoot ratio
	cm	$cm\ d^{-1}$	mg	
Medics (5)	53 - 64	0.88 - 1.07	74 - 160	0.57 - 1.42
Sub. clovers (5)	50 - 65	0.84 - 1.09	69 - 180	1.14 - 1.49
Other clovers (4)	39 - 63	0.66 - 1.05	57 - 116	0.76 - 1.41

1/ A 60-day trial conducted in the greenhouse from 17 Nov. '84 to 17 Jan. '85.

2/ Mean of three N levels in nutrient solution (0, 105, and 210 mg ammonium nitrate/L Hoagland's solution).

lower seedling root dry weights than either 'Vernal' or 'DuPuits' at 100 d after planting. Also, African, which has low cold tolerance, had higher root concentration of total available carbohydrates in the fall than the other cultivars which are medium to high in winter hardiness (Matches et al., 1963). By spring, all African plants were dead (not reported in paper), but the other cultivars survived.

Effects of season and defoliation on the growth habit of Lotus pedunculatus Cav. were investigated by Sheath (1980). The Grasslands Maku cultivar was dominated by aerial shoot growth during spring and early summer and by expansion of underground organs during late summer and autumn. Of the underground components, rhizomes were the most affected by season and defoliation, as well as being the principal sites for shoot initiation. Few new shoots were produced on primary crowns. Limiting intensive defoliation during the late summer and autumn was suggested for encouraging rhizome spread.

Another interesting aspect of root-growth is the finding of Sawai et al. (1986). They examined a population of 3888 plants derived from crossing 198 plants of 10 cultivars of red clover. They found that nonflowering plants which persisted longer than flowering plants had a greater portion of roots arising from the crown.

Tropical Legumes

Like temperate legumes, tropical and subtropical legumes may also differ in rooting characteristics. Jones and Clements (1987)

Table 3. Stolon length, rooted points, and number of adventitious roots greater than 1 mm diam in the winter of 1982 and 1986 (adapted from Jones & Clements, 1987).

Species	Stolon length 1982	Stolon length 1986	Rooted points 1982	Rooted points 1986	Roots > 1 mm diam 1982	Roots > 1 mm diam 1986
	$m\ m^{-2}$		No. m^{-2}		No. m^{-2}	
C. virginianum						
Line 1	13.6	3.7	194	88	36	24
Line 2	25.4	9.1	475	366	40	68
CP 140057	26.2	2.3	460	48	82	12
Shaw creeping vigna	23.2	0.5	433	4	116	0

compared three lines of Centrosema virginianum (L.) Benth., and 'Shaw' creeping vigna (Vigna parkeri Bak.) under grazing when grown with Setaria sphacelata (Schum.) Stapf and Hubb. Grazing began in 1978 and root and stolon measurements for 1982 and 1986 showed considerable change over years (Table 3). Species differed significantly in 1982 for stolon length, number of rooting points, and root diameter, but were not different in 1986. Centrosema virginianum did not persist when grazed at 2.3 steers ha^{-1} but did persist when grazed at 1.5 steers ha^{-1}. Siratro did not persist after 4 yr. The lower stolon density in 1986 was attributed to very dry conditions during the previous summer which affected the shallower rooting creeping vigna relatively more than the C. virginianum. In another stocking rate trial, Jones and Bunch (1988) found that Siratro was unable to persist under heavy grazing. Heavy grazing had detrimental effects on recruitment and on the development and survival of Siratro crowns and stolons.

In another study conducted in 1983 and 1984, Tsugawa et al. (1988) reported that centro (Centrosema pubescens Benth.) had more rooted points per unit area (1300 to 2100 m^{-2}) than desmodium [Desmodium intortum (Mill.) Urb.] (600 to 1200 m^{-2}). This was believed due to centro's greater total length of stolons (c. 300 m m^{-2}) compared to those of desmodium (c. 90 m m^{-2}). Desmodium stolons were thicker and heavier than those of centro. By the end of the growing season, desmodium had slightly greater weight of roots in the surface 5 cm of soil (72 g m^{-2}) than did centro (62 g m^{-2}).

Bray et al. (1969) reported that Glycine javanica L. developed lateral roots faster and to a greater extent than alfalfa, but alfalfa had a more rapid and early development of vertical roots. Alfalfa roots reached a depth of 53 cm in 4 to 6 wk as compared to 6 to 7 wk for Glycine.

Four desmodium lines (D. intortum cv. Greenleaf, D. uncinatum (Jack) DC. cv. Silverleaf, and two D. intortum x D. sandwicense E. Meyer hybrids) were evaluated at two stocking rates by Jones (1988) for 7 yr. Stolon lengths ($m\ m^{-2}$) were significantly different only in the first year. Within 3 yr after initiation of the experiment, only 10% of the original taproots of Greenleaf and the two hybrids were alive compared to 50% survival for Silverleaf. All entries relied on adventitious rooting from stolons for long-term persistence. Damage of larger roots by weevil (Amnemus quadrituberculatus) larvae were believed responsible for decreased plant persistence during dry periods. Also, stolon length was depressed at the higher stocking rate (1.9 vs. 1.1 steers ha^{-1}). Where regeneration of seedlings from self-reseeding is not dependable, maintaining a good density of stolons and adventitious roots might help ensure the continuation of good stands.

Keating and Mott (1987) found that Siratro, 'Murray' phasey bean [Macroptilium lathyroides (L.) Urb.], Rhynchosia minima (L.) DC., and a line of Stylosanthes scabra Vog. were productive over 3 yr of investigation. Siratro, a perennial, had the deepest root system of the legumes investigated. Irrigation did not improve the root length density profile of any legume. The grass Panicum coloratum var. makarikariense L. cv. Bambatsi was also investigated. Because of its more extensive root system and greater extraction of water during dry periods, the authors suggested that the legumes tested would be at a serious disadvantage during dry periods if grown with this grass.

FACTORS INFLUENCING ROOT DEVELOPMENT

Temperature

Root growth is often limited or stopped by low or high temperature (Kramer, 1983). The optimum temperature varies with species and stage of development. For example, Cohen and Tadmor (1969) reported that root elongation rates of small- and large-seeded range plants (including grasses and legumes) were increased two-to threefold over a temperature range from 10 to 20° C. Over this temperature range, they found that the rate of increase was faster in the upper soil (2 to 12 cm depth) than in a deeper layer (12 to 22 cm). Between 20 to 25° C, rate of root elongation varied with species and soil depth. The optimal temperature for root elongation in the 2 to 12 cm depth was generally higher than in the 12 to 22 cm depth. Also, the positive influence of seed size on root elongation decreased as the temperature increased from 10 to 25° C.

Summer et al. (1972) concluded that air ambient temperature appeared to be less important than root ambient temperature as a determinant of the growth rate of subterranean clover. Silsbury et al. (1984) reported that percent germination and emergence of Mt. Barker subterranean clover were hardly affected at a temperature range of 10 to 20° C, but at 25° C germination was reduced 50% and at 30° C to about 10%. Rates of germination and emergence were slowest at 10° C. However, Fukai & Silsbury (1976) reported that the optimum temperature for growth of subterranean clover was 20 to 25° C when plants were young (under 100 days old), but decreased during growth so that after 100 days total dry matter was inversely related to temperature over the range of 15 to 30° C. Davidson (1978) found that root/shoot ratios of white clover increased with increasing temperature above its optimum temperature for growth. For example, the root/shoot ratio increased from about 0.25 to 0.70 with an increase of 15° C above its optimum temperature. Also, Gibson (1965) stated that at root temperatures above 20° C, changes in dry weight increase of subterranean clover cultivars and its distribution between the root and shoot were largely controlled by the effect of root temperature on symbiotic N_2 fixation. There was a root temperature by bacterial strain interaction, and cultivar effects were also evident. Thus, different morphogenic processes can have different temperature optima (Fukai & Silsbury, 1976).

Water Stress

Water stress may also influence root and shoot growth. In glasshouse experiments, Jodari-Karimi et al. (1983) found that deep irrigation did not affect total root production of alfalfa, but it did affect root distribution at the irrigation zone. Deeply irrigated plants produced slightly more roots at lower depths than plants receiving a shallow irrigation. Also, under high evaporative demand, root production of nonirrigated alfalfa was lower than for irrigated alfalfa. The root/shoot ratios for shallow, deep, and nonirrigated plants were 0.28, 0.22, and 0.62, respectively, for plants grown from September to November. The authors contended that this higher root/shoot ratio for nonirrigated alfalfa indicated that the rate of root-growth was increased in nonirrigated alfalfa as a result of limited water stress and low evapotranspiration demand during the autumn period. In other glasshouse trials, Carter et al. (1982) observed a water regime by cultivar interaction for root weight of alfalfa. Cultivar root lengths did not change significantly as water level decreased.

Hamblin and Tennant (1987) examined the relationship of root length per unit ground area (La) and growing season water loss from the soil profile (WL) in cereals and grain legumes. They found in all cases that maximum rooting depth was better correlated with WL than with La ($r^2 = 0.58$ vs. 0.22). There was also a higher association of WL with maximum rooting depth than with total root density.

Aluminum Toxicity

Low pH and high Al content of soils are detrimental to many legumes. On soils high in Al, Joost and Hoveland (1986) reported that in limed surface soil, alfalfa taproots elongated at twice the rate of serecia lespedeza [Lespedeza cuneata (Dum.) G. Don], but in unlimed soil, alfalfa root-growth was reduced 69%. Three of the five lespedeza entries showed no reduction in root-growth rate on unlimed soil. Similar reduced root and shoot growth from Al for alfalfa (Rechcigl et al., 1986; Simpson et al., 1977), subterranean clover (Wright & Wright, 1987), and various tropical legumes (Ogata et al., 1986) have also been reported. Thus, in many areas of the world, low pH and high Al content minimize the number of legumes which may be successfully grown. However, the research of Joost and Hoveland (1986) is but one example suggesting that legume cultivars may be bred for improved Al tolerance.

Soil Bulk Density and Competition for Nutrients

Bulk density of soil can be a factor limiting root-growth of plants. For example, Shierlaw and Alston (1984) reported that a bulk density greater than or equal to 1550 kg m^{-3} restricted the growth of some grasses. Jones (1983) stated that bulk densities may affect root-growth even with optimum soil moisture. Eavis (1972) observed that in some cases pea roots were shorter and thicker as soil impedence increased, but in other cases roots became shorter and thicker as soil impedence decreased. The latter was attributed to poor soil aeration.

Competition for nutrients may also influence root development (Barley, 1970; Hart and Jessop, 1984; Jackman and Mouat, 1972; Mengel and Steffens, 1985). For example, deep-rooted legumes may be less susceptible to potassium (Ozanne et al., 1965) and sulphur (Barley, 1970) deficiency because roots can draw from greater depths and explore greater soil volume.

Influence of Rhizobia

Some strains of rhizobium may have an influence on root development of legumes. Cremers (1986) reported that certain rhizobia strains capable of inducing nodulation on clover may produce heat-stable, water soluble substances that can induce such root responses as a shortening of roots, an increase in number of root hairs, and distortion or curling of root hairs.

FUTURE FOCUS

The main emphasis of this review has been directed to the extent of variability which exists among legumes for various rooting parameters and factors which may affect root development. Certainly the high variability in rooting depth, rate of root elongation, root density, root/shoot ratio, and stolon and rhizome production among and within temperate and tropical legume species is good news. Such variability offers opportunities to possibly breed and select new legumes for specific soil and climatic conditions, and for particular uses in grazing and under cutting management.

Too little research has been directed towards studying the whole plant in clarifying cause-and-effect relationships of poor legume persistence. Increased attention should be given to understanding the role roots play in legume persistence and in herbage and seed production. Also, the literature search revealed that rooting of only a few legumes has been investigated relative to the number of pasture legumes grown throughout the world. Root investigations should be expanded to include more temperate and tropical legume species, as well as cultivars and lines within species.

More research is needed where legumes are grown in mixtures as is usually the case in pastures. Root responses might be quite different when a legume is grown alone or in a mixture with grasses and/or other legumes.

I believe our future research focus should also answer the following questions:

1. What is the optimum root/shoot ratio for sustained herbage production and legume persistence?
2. Under what conditions (soil, environmental, management schemes) are the various rooting characteristics desirable? For example, should we look for increased rooting depth, for increased root density ($m\ m^{-2}$ or m^{-3}), or both?
3. When and where are the presence or absence of rhizomes or stolons most desirable? What configuration is advantageous when grazed with cattle or sheep, or under cutting management?
4. How do rooting characteristics change with differences in the nutrient status of soil, and what does this mean in respect to legume persistence, herbage production, and drought and cold tolerance?
5. Can root production during the summer be increased through breeding or management manipulations? Would this be desirable or detrimental to legume persistence and/or herbage production?
6. What roles do root age and rooting dynamics play in legume persistence? Would an increase in root replacement offer any advantages?
7. Do roots differ in their rate of recovery from stresses such as drought, heat, cold, or severe defoliation?
8. Under what conditions are deep, shallow, highly fibrous, or less fibrous roots desirable?
9. How important are the effects of different strains of rhizobia on legume rooting as related to legume persistence?
10. What do we really know about root physiology as related to legume persistence? For example, do roots of all legumes have similar rates of respiration?

Some of these questions may be addressed at this workshop, but there is another paramount question which must be considered in the future. It is, how should root investigations be conducted? Gomez et al. (1986, 1987) and Ozanne et al. (1965) reported that rooting depths measured by glasshouse and laboratory techniques are well correlated with rooting depth in the field. However, Caradus (1981a) showed that vertical penetration of white clover roots did not stabilize until 12 to 18 months after seed germination. Therefore, he expressed concerns about the usefulness of short-term studies, especially those conducted in containers. Also, he questioned whether results are meaningful where grazing has not been imposed when experimenting with pasture legumes. Certainly, his concerns are valid. Perhaps further development of mini-rhizotron systems which allow root examination with a video camera will allow more comprehensive investigation of roots under field conditions.

REFERENCES

Asher, C. J., and P. G. Ozanne. 1966. Root growth in seedlings of annual pasture species. Plant Soil 24:423-426.

Barley, K. P. 1970. The configuration of the root system in relation to nutrient uptake. Adv. Agron. 22:159-201.

Bohm, W. 1979. Methods of studying root systems. Springer-Verlag, Berlin, Heidelberg, New York.

Bojorquez, C. L., A. G. Matches, and H. M. Taylor. 1986. Root growth of forage legumes using the slant-tube technique. p. 107. In Agronomy abstract. ASA, Madison, WI.

Bouton, J. H., J. F. Hammul, and M. E. Sumner. 1982. Alfalfa, Medicago sativa L., in highly weathered, acid soils IV. Root growth into acid subsoil of plants selected for acid tolerance. Plant Soil 65:187-192.

Bray, R. A., J. B. Hacker, and D. E. Byth. 1969. Root mapping of three tropical pasture species using 32P. Aust. J. Exp. Agric. Anim. Husb. 9:445-448.

Caradus, J. R. 1977. Structural variation of white clover root systems. N. Z. J. Agric. Res. 20:213-219.

Caradus, J. R. 1981a. Root growth of white clover (Trifolium repens L.) lines in glass-fronted containers. N. Z. J. Agric. Res. 24:43-54.

Caradus, J. R. 1981b. Root morphology of some white clovers from New Zealand hill country. N. Z. J. Agric. Res. 24:349-351.

Caradus, J. R., and P. S. Evans. 1977. Seasonal root formation of white clover, ryegrass and cocksfoot in New Zealand. N. Z. J. Agric. Res. 20:337-342.

Carter, P. R., C. G. Sheaffer, and W. B. Voorhees. 1982. Root growth, herbage yield and plant water status of alfalfa cultivars. Crop Sci. 22:425-427.

Cohen, Y., and N. H. Tadmor. 1969. Effects of temperature on the elongation of seedling roots of some grasses and legumes. Crop Sci. 9:189-192.

Cook, S. J., and D. Ratcliff. 1985. Effect of fertilizer, root and shoot competition on the growth of Siratro (Macroptilium atropurpureum) and green panic (Panicum maximum var. trichoglume) seedlings in a native speargrass (Heteropogon contortus) sward. Aust. J. Agric. Res. 36:233-245.

Cremers, H. C. J. C. 1986. Symplasmid and chromosomal gene products of Rhizobium trifolii elicit developmental responses on various legume roots. J. Plant. Physiol. 122:25-40.

Davidson, R. L. 1978. Root systems - the forgotten component of pastures. p. 86-94. In J. R. Wilson (ed.) Plant relations in pastures. CSIRO, Melbourne.

Eavis, B. W. 1972. Soil physical conditions affecting seedling root-growth. I. Mechanical impedance, aeration and moisture availability as influenced by bulk density and moisture levels in a sandy loam soil. Plant Soil 36:613-622.

Ennos, R. A. 1985. The significance of genetic variation for root growth within a natural population of white clover (Trifolium repens). J. Ecol. 73:615-624.

Fukai, S., and J. H. Silsbury. 1976. Response of subterranean clover to temperature. I. Dry matter production and plant morphogenesis. Aust. J. Plant. Physiol. 3:527-543.

Gibson, A. H. 1965. Physical environment and symbiotic nitrogen fixation III. Root temperature effects on shoot and root development and nitrogen distribution in Trifolium subterraneum. Aust. J. Biol. Sci. 19:219-232.

Gomez, J. F., A. G. Matches, B. L. McMichael, and H. M. Taylor. 1987. Root development of forages using the pouch technique. p. 112. In Agronomy abstract. ASA, Madison, WI.

Gomez, J. F., J. C. Shin, C. L. Bojorquez, A. G. Matches, B. L. McMichael, and H. M. Taylor. 1986. Using the slant tube technique to predict rooting depth of forage legumes in the field. In Agronomy abstracts. ASA, Madison, WI.

Hamblin, A. P. 1985. The influence of soil structure on water movement, crop root growth, and water uptake. p. 95-158. In N. C. Brady (ed.) Adv. Agron. Academic Press, Inc., Orlando, FL.

Hamblin, A. P., and J. Hamblin. 1985. Root characteristics of some temperate legume species and varieties on deep, free-draining Entisols. Aust. J. Agric. Res. 36:63-72.

Hamblin, A., and D. Tennant. 1987. Root length density and water uptake in cereals and grain legumes: how well are they correlated? Aust. J. Agric. Res. 38:513-527.

Hart, A. L., and D. Jessop. 1984. Leaf phosphorus fractionation and growth responses to phosphorus of the forage legumes Trifolium repens, T. dubium and Lotus pedunculatus. Physiol. Plant. 61:435-440.

Hay, M. J. M., D. F. Chapman, R. J. M. Hay, C. G. L. Pennell, P. W. Woods, and R. H. Fletcher. 1987. Season variation in the vertical distribution of white clover stolons in grazed swards. N. Z. J. Agric. Res. 30:1-8.

Humphries, W. A., and E. T. Bailey. 1961. Root weight profiles of eight species of trifolium grown in swards. Aust. J. Exp. Agric. Anim. Husb. 1:150-152.

Jackman, R. H., and M. C. H. Mouat. 1972. Competition between grass and clover for phosphate. II. Effect of root activity, efficiency of response to phosphate, and soil moisture. N. Z. J. Agric. Res. 15:667-675.

Jodari-Karimi, F., V. Watson, H. Hodges, and F. Whisler. 1983. Root distribution and water use efficiency of alfalfa as influenced by depth of irrigaton. Agron. J. 75:207-211.

Jones, C. A. 1983. Effect of soil texture on critical bulk densities for root growth. Soil Sci. Soc. Am. J. 47:1208-1211.

Jones, R. M. 1988. Productivity and population dynamics of silverleaf desmodium (Desmodium uncinatum), greenleaf desmodium (D. intortum) and two D. intortum x D. sandwichense hybrids in coastal south-east Queensland. Trop. Grassl. (in press).

Jones, R. M., and G. A. Bunch. 1988. The effect of stocking rate on the population dynamics of siratro in siratro (Macroptilium atropurpureum) - setaria (Setaria sphacelata) pastures in south-east Queensland. I. Survival of plants and stolons. Aust. J. Agric. Res. 39:209-219.

Jones, R. M., and R. J. Clements. 1987. Persistence and productivity of Centrosema virginianum and Vigna parkeri cv. Shaw under grazing on coastal lowlands of south-east Queensland. Trop. Grassl. 21:55-64.

Joost, R. E., and C. S. Hoveland. 1986. Root development of serecia lespedeza and alfalfa in acid soils. Agron. J. 78:711-714.

Keating, B. A., and J. J. Mott. 1987. Growth and regeneration of summer-growing pasture legumes on a heavy clay soil in south-eastern Queensland. Aust. J. Exp. Agric. 27:633-641.

Kramer, P. J. 1983. Water relations of plants. Academic Press, New York.

Lee, H. J., and M. J. Cho. 1985. Root competition and productivity in mono- and binary association of four forage species. p. 663-665. In Proc. Int. Grassl. Cong. 15th, Kyoto, Japan. 24-31 Aug. The Nat. Grassl. Res. Inst., Nishi-Narino, Japan.

Matches, A. G., G. O. Mott, and R. J. Bula. 1962. Vegetative development of alfalfa seedlings under varying levels of shading and potassium fertilization. Agron. J. 54:541-543.

Matches, A. G., G. O. Mott, and R. J. Bula. 1963. The development of carbohydrate reserves in alfalfa seedlings under various levels of shading and potassium fertilization. Agron. J. 55:185-188.

McMichael, B. L., J. J. Burke, T. D. Berlin, J. L. Hatfield, and J. E. Quisenberry. 1985. Root vascular bundle arrangements among cotton strains and cultivars. Environ. Exp. Bot. 25:23-30.

Mengel, K., and D. Steffens. 1985. Potassium uptake of ryegrass (Lolium perenne) and red clover (Trifolium pratense) as related to root parameters. Biol. Fertil. Soils 1:53-58.

Ogata, S., K. Fujita, and K. Moroshima. 1986. Effect of Al concentration in culture solutions on the growth and N_2 fixation of some tropical pasture legumes. Soil Sci. Plant Nutr. 32:27-35.

Ozanne, P. G., C. J. Asher, and D. J. Kirton. 1965. Root distribution in a deep sand and its relationship to the uptake of added potassium by pasture plants. Aust. J. Agric. Res. 16:785-800.

Pearson, C. J., and B. C. Jacobs. 1985. Root distribution in space and time in Trifolium subterraneum. Aust. J. Agric. Res. 36:601-614.

Rechcigl, J. E., R. B. Reneau, Jr., D. D. Wolf, W. Kroontje, and S. W. van Scoyoc. 1986. Alfalfa seedling growth in nutrient solution as influenced by aluminum, calcium and pH. Commun. Soil Sci. Plant Anal. 17:27-44.

Reyes, Custodio Lucio Bojorquez. 1986. Root and top growth of annual legume forages using the slant-tube technique. M. S. thesis. Texas Tech Univ., Lubbock.

Sawai, A., M. Gau, and S. Ueda. 1986. Differences in root systems among growth types of red clover. J. Jpn. Soc. Grassl. Sci. 32:164-166.

Sheath, G. W. 1980. Effects of season and defoliation on the growth habit of Lotus pedunculatus Cav. cv. 'Grasslands Maku'. N. Z. J. Agric. Res. 23:191-200.

Shierlaw, J., and A. M. Alston. 1984. Effects of soil compaction on root growth and uptake of phosphorus. Plant Soil 77:15-28.

Silsbury, J. H., D. Zuill, and P. H. Brown. 1984. Effects of temperature on germination emergence and early seedling growth of swards of Mt. Barker subterranean clover plants grown with and without nitrate. Aust. J. Agric. Res. 35:539-549.

Simpson, J. R., A. Pinkerton, and J. Lazdouskis. 1977. Effects of subsoil calcium on the root growth of some lucerne genotypes (*Medicago sativa* L.) in acidic soil profiles. Aust. J. Agric. Res. 28:629-638.

Stevenson, C. A., and A. S. Laidlaw. 1985. The effect of water stress on stolon and adventitious root development in white clover (*Trifolium repens* L.). Plant Soil 85:249-257.

Summer, D. C., C. A. Raguse, and K. L. Taggard. 1972. Effects of varying root/shoot temperatures on early growth of subterranean clover. Crop Sci. 12:517-520.

Thomas, H. 1984. Effects of drought on growth and competitive ability of perennial ryegrass and white clover. J. Appl. Ecol. 21:591-602.

Tsugawa, Hyoe, R. M. Jones and R. J. Clements. 1988. Growth of plant tops, stolons and roots in first year swards of *Centrosema pubescens* and *Desmodium intortum* cv. Greenleaf. II. Development of stolons and adventitious roots. Jpn. J. Grassl. Sci. (submitted).

Ueno, M. 1982. Root development and function of white clover in warmer region of Japan. Jpn. Agric. Res. Q. 16:198-201.

Upchurch, D. R., and J. T. Ritchie. 1984. Battery-operated color video camera for root observations in mini-rhizotrons. Agron. J. 76:1015-1017.

Wright, R. J., and S. F. Wright. 1987. Effects of aluminum and calcium on the growth of subterranean clover in Appalachian soils. Soil Sci. 143:341-348.

DISCUSSION

Buxton. What is the relationship between root elongation rate and final footing depth with species?

Matches. I would expect that rate of root elongation would change over time. Secondly, changes in soil density at different depths might influence the rate of root elongation. Actually, the answer to your question is not known.

Fisher. Seedlings of *Stylosanthes humilis* H.B.K. (Townsville stylo) grow deep roots very quickly at the expense of top growth. In contrast, the seedlings of competing annual grasses do not grow deep roots but make more top growth. Thus, in the wet-dry tropic, drought after the opening rains favor the grasses.

Fisher. *Macroptilium atropurpureum* (DC.) Urb. (siratro) and *Cenchrus ciliaris* L. (bufflegrass) have contrasting physiological reactions to drought, yet they coexist successfully in a dry environment. Siratro is deep rooted and doesn't compete with bufflegrass during drought.

B. Smith. There are different kinds of drought.

GENERAL DISCUSSION OF DEVELOPMENT AND GROWTH CHARACTERISTICS OF LEGUMES

Clements: Let me introduce the general discussion concerning the relationship of persistence to development and growth characteristics of legumes like this: What generalizations can we make about persistence and development/growth characteristics? Can we develop any principles? Can we extrapolate from temperate models to the tropics?

Reed: You seem to be making the assumption that relationships between persistence and development/growth characteristics are better understood in temperate regions than in the tropics, perhaps because tropical pasture legumes have been more recently developed for agriculture. True, temperate pastures are largely based on a relatively small number of species by comparison with tropical pastures, but it is also a fact that the number is increasing. Many temperate species have yet to be classified with respect to their appropriate ecological niche and appropriate role and management. This work needs considerable resources, and progress therefore is likely to be slow.

Clements: That's correct. My point is that the widely sown species of Trifolium and Medicago are comparatively well-developed, with relatively long histories of domestication. With the tropical legumes we are struggling to develop species from a whole range of genera, as Al Kretschmer has pointed out, with a comparatively limited understanding of each one.

Kretschmer: We can't do it alone. We need to collaborate with organizations such as International Board for Plant Genetic Resources (IBPGR), Centro Internacional de Agricultura Tropical (CIAT), and International Livestock Centre for Africa (ILCA). There is scope for collaboration in developing Aeschynomene in Malawi and surrounding countries, Lotononis in and from South Africa, Zornia in Africa, and Desmodium in Asia, to name just a few.

Clements: Margot Forde listed good and bad points in her description of four general classes of temperate perennial legumes; for example, stolon-formers tend to be drought-susceptible, while shrubs tend to have elevated growing points and are susceptible to overgrazing. Are we always going to be in a "trade-off" situation, sacrificing benefits in one character to achieve gains in another?

Forde: I believe that no one plant form can have all the advantages for all types of farming situations; the trick is to match the needs of the site with the characteristics of the plant. That being said, it is possible within limits to try to amend specific disadvantages by breeding, e.g., to breed white clover with deeper nodal roots. Some natural solutions are not acceptable, e.g., spines to protect edible shrubs.

Sheaffer: In the future, we will make progress in persistence by breeding for improved root morphology and disease resistance.

Fisher: Root distribution patterns may be key factors in the success of some tropical legumes in drought-prone climates. Leaf characteristics such as stomatal control, desiccation tolerance, and sun-tracking are also important.

Clements: Myles, you might comment on the rapid root elongation of Townsville stylo seedlings.

Fisher: Townsville stylo seedlings after emergence seem to grow very slowly for some time, but the apparent lack of above ground action contrasts with a rapid elongation of roots. This has been suggested as an important adaptation for annual legumes growing in a semiarid tropical environment.

Caradus: The success of white clover in grazed mixed swards is due largely to its ability to regenerate from grazing and treading. This is related to its stoloniferous habit and its ability to root at the nodes.

(Several brief comments on specific adaptations were made at this point.)

Brougham: Most of the tropical grazing zones of the world are grass dominant (e.g., Serengeti in Africa, Great Plains of North America). There is (or was) good animal production from these regions (e.g., wild animals in the Serengeti region, buffalo on native blue stems etc. in North America). These regions do not have many productive legumes. Can we expect successful introduction of persistent legumes into these kinds of regions?

Hoveland: Legumes are unnatural plant species in grassland. Where in the world does one find solid stands <u>in nature</u> where legumes are the dominant species? Legumes have to be cultured as a crop. They lack the tolerance to adversity that is characteristic of grasses. Thus, it should be expected that it is difficult to grow legumes in a pasture grazed by animals.

Fisher: Annual legumes can become dominant in Australian pastures. Why are they not dominant in the Mediterranean region where so many occur naturally?

Reed: John Brockwell (CSIRO, Canberra) has suggested that these legumes may grow so well in Australia because they are temporarily free from pests and diseases that are common in their centres of origin or diversity.

Sheath: Persistence of legumes also depends on flexibility within the plant (both phenotypic and genotypic) and on a mixture of species making up the legume component.

Clements: Let's pursue this point. Is this a generalization we can make across the range of legumes described in this series of papers?

Jones: It can be useful to have a mixture of genotypes, and/or plasticity in morphology or in persistence mechanisms within a species. For example, white clover and *Vigna parkeri* Bak. can persist both by stolon rooting (vegetative reproduction) and by regeneration from seed. Generally, mixtures and plasticity will be more useful as the environment becomes more variable. Variability in this context includes not just rainfall and soil variation, but also grazing management.

Forde: When we speak of the need for plasticity in legume populations, we need to distinguish between the phenotypic plasticity of individual plants, as found in white clover, and the high genetic variability in legume cultivars which will allow local genetic adaptation in response to heterogeneous spatial and temporal selection pressures. The two are somewhat opposed strategies.

CLIMATIC AND EDAPHIC CONSTRAINTS TO THE PERSISTENCE OF LEGUMES IN PASTURES

Z. Hochman and K. R. Helyar

SUMMARY

Climatic constraints to legume persistence are discussed with emphasis on the effects of cyclical, as well as long term, climatic changes. We illustrate the use of modelling methods for developing hypotheses on the effects of past and putative future changes in temperature and rainfall patterns on persistence of legumes.

Edaphic factors reviewed include competition for nutrients, soil salinity, compaction, waterlogging, and soil acidity. Emphasis is given to the implications of soil acidification processes for long term viability of legumes in pastures. Liming and plant tolerance strategies are evaluated in that context.

The importance of interactions between climatic and edaphic factors is stressed and illustrated. We conclude that in the Australian environment legume persistence is becoming more difficult to manage and suggest priorities for research.

INTRODUCTION

Legumes seldom dominate natural ecosystems. Maintenance of an agriculturally desirable legume content in pastures may therefore be viewed as an attempt to sustain a non equilibrium condition. Neither the climate nor the soil can be considered as static factors in the legume maintenance equation. We will show that climatic changes taking place as a result of the "greenhouse effect" will be sufficient to cause profound changes in the flowering behavior of annual legumes and how changes in soil moisture availability have already influenced the persistence of annual and perennial legumes in Australia. Soils are also subject to changes in nutrient status, and in levels of salinity and acidity. We will review the data to show that management or reversal of these processes is required for the persistence of legumes and other desirable species in pastures.

CLIMATIC CONSTRAINTS

Temperature

The responses of pasture plants to temperature were comprehensively reviewed by McWilliam (1978). Temperature influences growth rate, morphology, photosynthesis, respiration, nodulation and N_2-fixation rate. Initiation of germination and reproduction in response to temperature and the ability to survive temperature extremes are critical factors in the adaptation of species.

McWilliam found that a wide variation in temperature tolerance exists between species and between ecotypes within species. Knowledge of temperature adaptation and the thermal requirements of pasture plants can therefore assist in the selection of successful combinations of species to provide more stable and productive pasture systems.

The factors outlined above have a significant role on the adaptation and selection of legumes. We are therefore concerned with the growing scientific consensus that in the first half of the next century, as a result of the increasing concentration of greenhouse gases, a rise of global mean temperatures could occur which is greater than any in man's history ("The Villach Statement" as reported by Zillman, 1986). Recently published data on long-term global temperature variation tends to support such predictions (Jones *et al.*, 1986). It is anticipated that by 2030 AD. temperatures in the inland will increase by about 2°C in northern Australia and up to 3°C or 4°C further south. It is also predicted that winter and overnight minimum temperatures will warm more than summer and daily maximum temperatures (Pittock, 1987).

We will not attempt a review of the very complex effects which temperature and other anticipated environmental changes are likely to have on legume persistence in pastures. However, an example of how process based models could be used to help us understand these implications may indicate the kind of work which is required: The roles of vernalization and subsequent high temperature promotion on flowering of annual legumes is well understood (Aitken, 1955; Evans, 1959; Morely & Evans, 1959; Clarkson & Russell, 1975). Recently, a model which quantified these processes was developed and validated for three cultivars of *Medicago truncatula* Gaertn. (Hochman, 1987). On the basis of the predictions reported by Pittock above, the temperature data for the seven validation sites were modified initially by adding 1°C to the maximum daily temperature and 2°C to the minimum daily temperature, and in a subsequent run, by increasing temperatures by 2°C and 4°C, respectively. In the first run the effect on predicted flowering dates for the cultivar Sephi ranged from a 23-day delay at one site to a seven-d hastening at another. The second run further increased the maximum delay to 32 d while the maximum hastening remained unchanged (Table 1). Clearly the range of adaptation of this newly released cultivar is likely to change considerably between now and 2030 A. D.

Water

Donald (1963) observed that success of any plant or species will depend on the rate and completeness with which it can make use of the soil water supply. These, in turn, depend on relative growth rate and corresponding earliness of water demand and on the rate of root extension. In terms of their ability to exploit stored soil moisture, legumes are unlikely to fare very well in competition with grasses. Hamblin & Hamblin (1985) noted that while wheat (*Triticum aestivum* L.) had less than 50% of its total root length in the top 20 cm of a deep free-draining Entisol, pasture legumes had more than 70% in that layer. This tends to make legumes more dependent

Table 1. Predicted changes in the duration of the vegetative period of *M. truncatula* cv. Sephi due to temperature changes anticipated as a consequence of the "glasshouse effect".

Site	Latitude	Sowing date	Predicted duration	Predicted change in duration +1.5°C[1]	+3°C[2]
	degree		days	± days	± days
Salmon Gums	33	30 May 1980	84	+ 8	+17
		4 Jun 1980	79	+10	+20
Esperance	34	6 Jun 1980	88	+16	+28
Downs		30 May 1981	87	+10	+30
Walpeup	35	17 Apr 1984	89	+23	+32
Leeton	34	22 Apr 1983	123	− 3	+ 2
		28 Jun 1983	85	− 7	− 7
Condobolin	33	1 May 1983	119	− 5	0
		14 May 1985	89	+ 5	+ 5
		26 Jun 1985	80	− 4	− 5
Tamworth	31	7 Aug 1980	57	+ 4	+ 8
		23 Jun 1983	79	− 4	− 2

1 Each daily minimum is raised by 2° and daily maximum by 1°C.

2 Each daily minimum if raised by 4° and daily maximum by 2°C.

on frequent wetting of the topsoil. Similarly, the extensive deep rooting ability of *Panicum coloratum* var. *makarikariense* Goosens cv. Bambatsi, enabled it to extract soil water more effectively during a dry period than various legumes growing in pure swards on a heavy clay soil (Keating & Mott, 1987).

Based on the water-use efficiency (WUE) concept developed by de Wit (1958), the most efficient plants are those with crassulacean acid metabolism (CAM), followed by C_4 grasses, C_3 grasses, and finally legumes (Fischer & Turner, 1978; Tanner & Sinclair, 1983). However, persistence in a moisture limited environment is not determined by WUE alone. Plant responses to drought can also be classified in terms of savers and spenders (de Wit, 1978): Savers of water react sensitively to water stress by reducing leaf expansion and closing stomates. Spenders continue to transpire at near potential rates until their water supply is exhausted. In the absence of competition savers can mature and form seed within a normal developmental period and may profit from late showers. Spenders must be able to complete development in order to set seed before limited soil moisture is depleted. In competition with each other, spenders can use up the water saved by the savers and, assuming small differences in WUE, are at a distinct competitive advantage.

The available data suggest that legumes tend to fall in the category of savers rather than spenders. In a study of irrigated pastures at Kyabram, the productivity of white clover (*Trifolium repens* L.) was reduced by approximately 50% as a result of water shortage during an irrigation cycle. In contrast, paspalum (*Paspalum dilatatum* Poir.) was insensitive while ryegrass (*Lolium perenne* L.) had an intermediate response. Stomata of white clover and ryegrass closed when leaf water potential (LWP) was about -2.0 MPa while those of paspalum closed at -2.5 MPa, thus allowing paspalum to continue using the soil moisture "saved" by its companion species. These results are consistent with field observations of paspalum dominating ryegrass and clover in mixed swards (Blaikie & Martin, 1987). Similarly, in a semiarid subtropical environment, data presented by Peak *et al.* (1975) suggests that the competitive advantage of buffel grass (*Cenchrus ciliaris* L.) over siratro (*Macroptilium atropurpureum* (DC.) Urban) can be attributed to the grass being a spender while the legume is a saver.

Although there is little data on the competitive ability of legumes in the presence of dicotyledonous weeds, the information at hand is not encouraging. In a study of capeweed in pastures of south-west Western Australia, capeweed content was favored in the lower-rainfall areas and in seasons when germination was followed by a 4-5 wk dry period (Arnold *et al.*, 1985).

Modelling the Effects of Soil Moisture on Persistence

One explanation offered for the poor persistence of legumes in recent years is that changes in rainfall patterns have occurred (Russell, 1981; Pittock & Salinger, 1982). McCaskill (1987) used a water balance model to study the effects of such changes in the Armidale area on the cumulative duration of water stress endured by white clover in each season over the period from 1858 to 1986. His simulation results show that rainfall patterns during the 1950s, 1960s (apart from 1965), and 1970s were extremely favorable (averaging 6.3 critical clover days per year), while the 1980s (with an average of 29 critical clover days per year) appear to approximate the longer term pattern. The successful adaptation of white clover in the New England may have been a relatively short lived phenomenon.

Murray & Hochman (1988, unpublished data) are further developing this concept to determine whether such changes have also played a role in the decline of subterranean clover (*Trifolium subterraneum* L.) on the southern slopes of New South Wales. In that study, a soil moisture submodel is being used in a model describing seed production and survival assuming ideal grazing management and no competition from other species. The assumptions made about subterranean clover seed dynamics are summarized below:

1. Germination can occur after 15 February when available soil moisture (ASM) in the top 5 cm is greater than 0.3 of maximum ASM and if soil moisture is sufficient to sustain growth for a period which would allow roots to grow 7 cm [root growth rates are based on measurements reported by Pearson & Jacobs (1985)]. As the soil dries roots die until there are no living roots and

a "false break" is flagged. Subsequent germinations can occur.
2. Flowering occurs on 15 September unless germination occurred later than 1 May. The subsequent delay in flowering date is 1/4 day for each day's delay in germination after 1 May. [Based on Archer *et al.* (1987) data for cv. Woogenellup.]
3. Maturity will occur when either:
   a. 86 days elapse after commencement of flowering (Collins & Quinlivan, 1980). or,
   b. moisture stress after the commencement of flowering exceeds 14 "day units". Day units are 1 if ASM in the root zone is 0 and 0.5 if ASM is less than 0.3 of maximum ASM. The effect of withholding water on length of the growing season is documented (Collins, 1981). The day units system is an estimate.
4. Seed yield and viability are calculated from estimated ASM in the postflowering period. The underlying seed development pattern is based on data of Collins & Quinlivan (1980) and Rossiter (1977) as follows:
   a. Seed set requires a minimum of five wk from the onset of flowering.
   b. Between five and six wk postflowering, seed set may increase linearly to 30 g m^{-2} and viability to 10%.
   c. Between 6 and 8 wk seed can gain weight at 5 g m^{-2} d^{-1}.
   d. Between 8 and 12 wk seed can gain weight at 3.6 g m^{-2} d^{-1}.
   e. If ASM < 0.3 of maximum ASM from weeks 5 to 12, the above rates of gain are halved.
   f. Viability of seed, determined one wk after maturity is as follows for different times between flowering and maturity:

Weeks:	6	8	9	10	12
Percent viable:	10	20	50	70	90

   Linear interpolation is assumed.
5. Studies of the persistence of seeds suggest that the contribution to the germinable seed pool of seed set in any year is spread over the next three yr in the approximate ratio of 6:3:1 (Taylor *et al.*, 1984; Bolland & Collins, 1987). 60% of viable seed is assumed lost by predation and other causes between maturity and the following season (Carter, 1981; Carter, 1988 pers. comm.).

Results of the simulation, using daily rainfall and pan evaporation data for Wagga Wagga from 1948 to 1987, are summarized in Table 2. The model predicted highly variable seed production ranging from 0 in 1977 & 1982 to over 175 g m^{-2} in 7 yr. The mean yield in the first decade was 110 g m^{-2} compared with 59 g m^{-2} for the following three decades. The pool of viable soft seed each autumn also showed considerable variability between years but never declined to less than 3 g m^{-2}, equivalent to a sowing rate of 30 kg seed/ha. Moisture stress after early germination is associated with increased weed competition (Arnold *et al.*, 1985). It may also have implications for the severity of yield reduction due to root rot diseases (Taylor & Greenhalgh, 1987). This condition was simulated in about 40% of years.

Using Dear & Loveland's (1985) classification of seed pool data and assuming 60% of the seed pool was soft and viable; seed pool results were only fair to poor in 63% of seasons. Seed set and seed pool results show clearly that the 50s were favorable years. These were also the years of high wool prices in which most of the pasture development in the Wagga Wagga area occurred. The

Table 2. Simulated seed bank (gm^{-2}), seed production (gm^{-2}), and weeks of seedling moisture stress of subterranean clover at Wagga Wagga, 1948–1987.

Year	Dry matter	Germ. seed pool	Seed production total	Seed production viable seed		Early stress
	$t\ ha^{-1}$	$g\ m^{-2}$	$g\ m^{-2}$	%	$g\ m^{-2}$	weeks
1948	7.5		30.0	10	3	4
1949	14.5		172.0	90	155	9
1950	16.3		197.2	90	177	3
1951	8.3	61	128.6	90	116	4
1952	11.9	55	159.1	90	143	2
1953	11.4	55	179.2	90	161	4
1954	3.8	61	45.0	12	5	6
1955	16.1	26	175.6	90	158	7
1956	15.4	45	182.8	90	165	0
1957	3.0	59	72.5	18	13	4
1958	11.7	29	132.9	90	120	7
1959	7.7	37	85.0	18	15	10
1960	8.0	19	78.4	71	56	0
1961	6.6	20	61.1	61	37	3
1962	7.2	16	32.5	19	6	0
1963	8.6	8	77.7	56	44	1
1964	9.4	13	72.2	79	57	0
1965	7.6	19	139.5	90	126	0
1966	9.5	39	77.2	29	22	6
1967	1.9	23	30.0	10	3	0
1968	7.2	8	40.0	13	5	1
1969	10.7	3	70.0	19	13	0
1970	12.9	4	209.8	90	189	2
1971	5.7	47	37.5	14	5	1
1972	4.3	24	40.0	13	5	0
1973	12.0	9	146.2	90	132	2
1974	15.9	32	146.8	90	132	0
1975	12.7	48	119.8	77	92	10
1976	8.4	43	177.4	90	160	10
1977	9.7	55	0.0	0	0	8
1978	11.9	23	109.6	73	80	6
1979	9.5	26	93.0	59	55	3
1980	5.6	23	62.5	41	26	4
1981	6.0	16	61.2	41	25	0
1982	4.4	11	0.0	0	0	3
1983	12.0	4	134.4	80	108	1
1984	8.0	27	69.0	41	28	0
1985	12.4	20	85.2	59	50	4
1986	12.8	20	194.8	90	175	0
1987	6.3	49	30.0	19	6	0
1988		24				
Mean	9.4	29	98.9	55	72	

simulation indicates that while we may not expect subterranean clover performance to be as consistent as it was in the 1950s, any decline in its performance over the last 25 yr is probably due to factors other than soil moisture availability.

Light

In a review of light relations of pasture plants Ludlow (1978) argued that: "Shade tolerance plays only a minor role in competitiveness for light compared with the ability to gain preferential access to incident radiation, which is determined primarily by height. A combination of legumes with more-or-less horizontal or diaphotonistic leaves and grasses with erect leaves would appear to provide a pasture canopy that minimizes the difference in competitiveness between grasses and legumes and promotes stability in botanical composition."

Subsequent work (Wong & Wilson, 1980; Nicholls *et al.*, 1987; Broom & Arnold, 1986; Arnold & Anderson, 1987) has tended to support Ludlow's arguments. Is is worth noting however that while shading of *Trifolium* species from the beginning of flowering reduces the number of infloresences produced per unit area (Collins *et al.*, 1978), at the stage of seed and burr development light is an inhibiting factor (Taylor, 1979). The management of light relations in a pasture can be achieved primarily through choice of companion species but also through grazing management (see Curll this book) and forage conservation strategy (Kaiser & Curll, 1987).

EDAPHIC CONSTRAINTS

Nutrient Supply

Satisfaction of the nutrient requirement of plants depends on a number of factors which vary with genotype and soil-type. The availability of nutrients in the soil solution varies with the total amount of available nutrient present, and with distribution of the available nutrient between the soil solution and the adsorbed forms. In this review available nutrient is regarded as any form of the nutrient that can be absorbed by the roots and that is rapidly exchangeable (in seconds to hours rather than days to weeks) with the soil solution form. The ratio of solution to adsorbed forms varies greatly between nutrients, and particularly for more strongly adsorbed ions such as phosphate, Cu, and Zn, with soil adsorption characteristics. Differences between nutrients in their adsorption ratios result in different leaching rates and different rooting patterns required to maximize their adsorption from the soil. Plants vary in their internal requirement for each nutrient, that is the concentration of the nutrient in the plant, or specific parts of the plant, at which yield is near optimal. Once nutrients are taken up, there are further differences in the efficiency with which genotypes transport nutrients to their active sites (Nye & Tinker, 1977). With these factors in mind we review recent literature to determine the limitations of major nutrients to legume persistence . Two major reviews (Nicholas & Egan, 1975; Hannam & Reuter, 1987) of the trace element nutrition of pastures have more than adequately covered this subject which is of critical importance to pastoral development in Australia. Legume growth may be limited by deficiencies of Mo, Co, Zn, Cu, B, Mn, and Fe. In particular, Mo and Co are required in N-fixation processes.

Calcium

The responses of grasses and legumes to Ca supply were reviewed by Robson & Loneragan (1978). Further developments in this area have not been extensive, reflecting perhaps the impression that Ca deficiency of sown pastures does not appear to be widespread, due partially to the application of Ca in superphosphate.

In essence, Robson and Loneragan describe legumes as able to absorb Ca faster than grasses but having a much greater internal requirement for Ca. This together with the relative immobility of Ca within plants, means that we can expect legumes to be disadvantaged by low Ca availability. Because subterranean clover buries its seeds and the flow of Ca to organs is dependent on transpiration, Ca has been found to increase seed yield of subterranean clover in cases where it does not increase herbage production.

Similarly, plant roots growing in topsoils where Ca may be abundant cannot transport Ca to roots growing in Ca-deficient subsoils. Such a situation may limit the ultimate usefulness of Al tolerance in leached acid subsoils (Adams & Moore, 1983).

Nitrogen

Since the legume is normally part of a N-supplying symbiosis, N availability in the whole system is affected by and in turn affects, the competition between the legume and the grass components. The competitive ability of legumes is always reduced by increasing N-supply which favors the grass and thus suppresses the legume to the point where the legume becomes a minor component of the pasture as a maximum organic N content is achieved (Simpson, 1987). The general experience of legume dynamics in temperate Australian pastures is broadly consistent with the N-driven regeneration cycle postulated by Turkington & Harper (1979) and illustrated in Fig. 1. It is reasonable to expect that older permanent pasture areas in Australia which are not rotated with cereal crops or subject to heavy leaching and other losses of N could now be experiencing the factors which operate on the downward side of this cycle.

Phosphorus

Ryegrass (*L. rigidum* Gaudin) requires less soil P than subterranean clover for near maximum yield of young plants (Ozanne *et al.*, 1969; Barrow, 1975a) and differences are greater in soils which adsorb larger amounts of P (Barrow, 1975a). These and other authors (Jackman & Mouat, 1972; Barrow, 1975b) point out that differences in root system geometry account for the superior performance of grasses at low levels of soil P. Recently Bolan, *et al.* (1987) showed that while non-mycorrhizal ryegrass took up more P than non-mycorrhizal subterranean clover at all levels of application, mycorrhizal infection only increased P uptake by subterranean clover. Furthermore, there was no difference in P

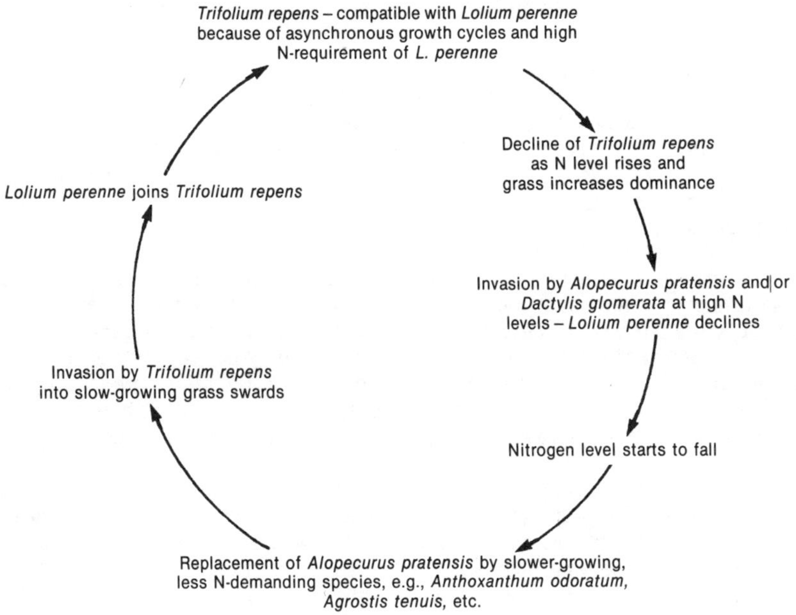

Fig. 1. Postulated regeneration cycle within an area of permanent pasture in Wales (from Turkington & Harper 1979).

uptake between mycorrhizally infected ryegrass and subterranean clover at low levels of P application.

In a competition study the response curve of subterranean clover to applied P in the presence of annual ryegrass was very similar to that of the pure subterranean clover sward, but the grass component responded to a much higher level of applied P in the presence of the legume (Ozanne *et al.*, 1976). A contrasting result was obtained by Dahmane & Graham (1981) who found that while ryegrass produced maximum shoot dry matter at 10 mg/Kg P, annual medic required 160 mg/Kg P. Ryegrass in mixture with medic produced equal dry matter to ryegrass alone, except at high P rates where the medic was better able to compete. The difference between the two studies may well be due to the N status of the pastures. In the earlier study P applied to the mixed sward led to higher N content in the grass while in the later study N content of ryegrass was similar in pure culture and in mixture.

Comparisons of the P requirements of annual legume species, have shown that serradellas (*Ornithopus* spp.) absorb more P from added superphosphate than subterranean clover, while the internal P efficiency of the two species was similar (Bolland, 1985a). Seed yields of subterranean clover, serradella and annual medics responded markedly to applied P on newly cleared soils (Bolland, 1985b, 1986).

In contrast with temperate pastures, the most important introduced grasses in the seasonally dry tropical areas have high P requirements whereas the *Stylothenthes* can be expected to dominate swards at low P levels (McIvor, 1984).

Sulfur

Sulfur deficiency is most likely to restrict pasture growth in areas with soils of low S sorption capacity, with low S inputs in rainfall or fertilizers, and where organic S mineralization rates are insufficient to meet plant requirements (McLachlan, 1975). In a recent review on the S nutrition of temperate pastures, Holford (1987) noted that there is little evidence of any increase in the occurrence or severity of S deficiency in temperate pastures during the past 12 yr.

It appears likely that competition for S in ryegrass/subterranean clover mixtures is of little significance except in cases of very severe S deficiency combined with an adequate N supply. S application increased the relative yields (yield in mixture as a percent of yield in monoculture) of subterranean clover and reduced the relative yield of annual ryegrass when sufficient time was allowed for expression of competitive effects and with applied N (Gilbert & Robson, 1984a).

The external requirement for S is of the order medics > brome grass and wimmera ryegrass > subterranean clovers. A low external requirement is not necessarily associated with a low internal requirement because species vary in their ability to absorb and transfer S from root to shoot. Although ryegrass has longer roots and root hairs and larger root surface area than subterranean clover roots, they absorb much less S per unit of root parameter (Gilbert & Robson, 1984b, c, d).

Another important consideration is the role played by rooting depth; White *et al.* (1981) and others, have demonstrated the importance of subsoil sulphate (40-80 cm) in certain soils. The deeper rooting tendency of grasses may give them an advantage in obtaining sulphate from such soils.

Adding S to a *Stylosanthes guianensis* (Aubl.) Sw./*Themeda australis* (R.Br.) Stapf pasture resulted in a strong response by the legume component. *T. australis* was found to be much better adapted to low S soils (Gilbert & Shaw, 1981).

Potassium

A number of studies show that correction of K deficiency should favor legume persistence and increase the proportion of legume in the sward. For example, K addition increased pasture yield in 20 of 27 subterranean clover/ryegrass pastures in south-eastern South Australia (when exchangeable K in the soil was less than 0.2 cmol(+)/kg). When K increased yield, the proportion of clover in the sward was either maintained or increased (Meissner & Clark, 1977). In southeastern New South Wales critical soil K levels were higher for white clover in competition with grasses than for pure clover swards (Spencer & Govaars, 1982).

A native pasture (*Heteropogon contortus* (L.) Roem. & Schult. dominant) oversown with Townsville stylo (*S. humilis* Kunth) was found, after 6 yr of annual application of superphosphate to have developed a severe K-deficiency. Correction of this deficiency increased legume yield two-to threefold (Shaw & Andrew, 1979).

Using an adaptation of the de Wit (1960) procedure, Hall (1974) found that at low K, the grass *Setaria anceps* Massey severely restricted the growth of the legume *Desmodium intortum* (Miller) Fawc. & Randle by competing for K.

Soil Salinity

The salinization of irrigated and dryland soils is a serious problem in some pasture areas (Noble & West, 1987). Clovers are the most salt sensitive component of both annual and perennial pastures (Maas & Hoffman, 1977). The response of legumes to salt is primarily dependent on the sensitivity of the host rather than on the response of the rhizobial symbiosis (Wilson, 1970, 1985; Balasubramanian & Sinha, 1976).

In a study of eleven tropical legumes, ten temperate legumes and eleven tropical grasses, Russell (1976) found that grasses were more persistent at high salt levels than legumes; lucerne *M. sativa* L. being the most tolerant legume. Keating *et al.* (1986) found that despite varying degrees of salt tolerance among tropical pasture legumes (siratro was the best) the grass *P. coloratum* L. was markedly more tolerant.

Following five yr of irrigation with saline water on a perennial pasture, Mount & Schuppan (1978) observed that lucerne and perennial ryegrass had valuable tolerance of saline and sodic conditions. They also found that while the contribution of tall wheat grass (*Agropyron elongatum* (Host) Beauv.) increased with salinity, that of *Trifolium* species declined. Current research in Victoria indicates scope for identifying variation in the tolerance of clover to salt, and for utilizing such variation in a breeding program (Noble & West, 1987).

In the long-term, salinity itself must be tackled. Reclamation of a saline/sodic soil by aquifer pumping transformed an unproductive site carrying salt tolerant weeds into a weed-free pasture consisting of white clover (70%) and grasses (30%) over two yr (Mehanni, 1987).

Soil Compaction and Waterlogging

Soil compaction eliminates many of the large 'transmission pores' that serve simultaneously as the major pathways for the drainage of water, the exchange of gasses between the atmosphere and the soil, and the penetration of roots (Drew, 1983). The effect of treading by sheep of an irrigated annual ryegrass/subterranean clover pasture was to increase soil bulk density by 40% and reduce winter production by one-third, due mainly to a reduction in the contribution of ryegrass (Witschi & Michalk, 1979). Artificial compaction of a soil was also found to be more damaging to wheat than to subterranean clover (Reeves *et al.*, 1984). The limited evidence seems to suggest that grasses may be more sensitive to soil compaction than legumes.

Interference in root growth and function in flooded soil leads to either an insufficient supply of essential substances to the shoots or to an excess of toxins and other substances originating in the anaerobic soil or in the roots themselves (Drew, 1983).

Studies in North America have shown that grasses have greater tolerance of waterlogging than legumes (Bolton & McKenzie, 1946; Finn *et al.*, 1961). In Australia, a comparison of the effects of ponding after irrigation, revealed that white clover was sensitive to a 6-h irrigation treatment, while there was no effect on paspalum even when the duration was 24 h (Blaikie & Martin, 1987). While *Medicago* species are generally susceptible to waterlogging (Robson, 1969), a comparison with subterranean clovers (Francis & Pool, 1973) also indicated that large differences in sensitivities exist between lines of these species. Selection and breeding of legumes tolerant to waterlogging is probably the most practical way of dealing with this problem.

SOIL ACIDITY

In southern Australia and to a lesser extent in northern Australia, soil acidification processes and the effects of soil acidity on plant growth, have been of interest because of widespread observations of accelerated soil acidification rates under P, S, and Mo fertilized legume based pastures (Helyar *et al.*, 1988). There is a significant amount of data describing the responses of various pasture legumes to the acidity of the main root zone, and to the complete root zone in pot experiments. However there are few examples of studies on the persistence of legumes in competition with other species. Furthermore, our understanding of subsurface acidity effect on growth and persistence is imprecise (Helyar *et al.*, 1988).

In this section, we briefly outline our understanding of soil acidification processes and the soil pH profiles that result from the different soil acidification regimes. This discussion leads to a description of the way the growth and persistence of pasture legumes may be affected by the different soil pH profiles.

Sources of Acid and Acid Addition Rates

Understanding the mechanisms of soil acidification assists us to manage soil acidity, and hence to manage legume growth and persistence. A model that defines the sources of acids causing soil acidification has been described by Helyar and Porter (1988). The model and the suggestions for soil acidity management that arise from it are outlined below.

Acids are produced in natural and agricultural ecosystems mainly in the N, C, S, Fe, and Mn cycles. The amount of acid contributed by a given nutrient cycle can be assessed by first defining a reference state for the element in question. The reference state is chosen to minimize the difficulty of the measurements required to define acid additions, and such that transformations between any forms of an element in the reference state do not involve acid or alkaline production or consumption. Second, any acid or alkaline products are defined. These are forms of the element associated with acid production (acid product, AcP) or alkali production (alkaline product, AlP), when transformed from a reference state form.

The amount of acid contributed by a given nutrient cycle can be assessed by measuring the net transport to and from the soil of AcP and AlP compounds, and by measuring their depletion or accumulation in the soil. Helyar and Porter (1988) applied this model to a grazed pasture ecosystem. The results indicated the major sources of acid were from the the C and N cycles. The acid added was accounted for by organic anion accumulation in soil organic matter (30-40%) and removal in products and waste products (12-15%) and by the leaching plus run-off of nitrate following N_2 fixation (40-50%).

This description of acid sources highlights the processes that can be manipulated to reduce acid addition. For example in legume based pasture systems N accumulates dominantly as organic nitrogen (RNH_2) and may be lost in this form (product removal and erosion), as nitrate or as ammonium (leaching, run-off, erosion) or as gasses (denitrification, ammonia volatilization). Action which would result in reducing losses of nitrate is therefore consistent with both reducing soil acidification and with increasing the efficiency of N utilization. It may be possible to halve current acid addition rates (Cregan & Helyar, 1986) through better soil nitrate management (minimum tillage, early sowing of crops, perennial rather than annual pastures) and by managing soil organic matter and product removal to minimize the contribution of carbon cycle acids to the system (feeding hay on hay cutting paddocks, minimizing removal of waste product to yards and camp areas).

The Distribution of Acidity in the Soil Profile

The soil profile usually does not acidify uniformly with depth because the acid (H^+) and alkali (OH^- or HCO_3^-) are added to the soil in particular locations (e.g. nitrification dominantly in the surface organic-rich layer, and root excretion of H^+ or HCO_3^- at the root surface). Furthermore transport of H^+, HCO_3^-, and mobile ions such as nitrate in the mass flow of water is responsible for transfer of acid and alkali within the soil profile. Recently a dynamic model has been developed (Soil Profile Acidification Model - SPAM) that describes the acid balance of a series of soil layers in a root zone given acid and alkali inputs as described by the Helyar-Porter model. Helyar *et al.* (1988) includes a preliminary description of this model, its initial fitting to data, verification against an independent data set, and an outline of the major simplifying assumptions used.

In the SPAM model acid distribution in the soil profile is affected by H^+, HCO_3^-, and Al^{3+} mass flow in water, and $CaCO_3$ and $Al(OH)_3$ precipitation/dissolution reactions dependent on a standard solubility product equation. A pH buffer capacity (pHBC) term accounts for other reactions involving H^+- mainly pH dependent H^+ association/disassociation reactions.

The locations of the acid and alkali additions to the soil are controlled by the ion intake balance of the roots, the site of lime additions, and estimates of where other acid/alkaline reactions occur. For example, organic matter humification, oxidation and nitrification occur mainly in the surface horizons. The main features of the model are illustrated in the flow diagram (Fig. 2).

The verification runs of SPAM show that a reasonable description of the development of soil pH profiles is achieved (Fig. 3 from Helyar *et al.*, 1988). Some discrepancies may be attributed to the fact that diffusion processes are not included in the model.

Given a stable plant production system and acid addition rate, and given a positive and stable leaching regime, the SPAM model shows that the soil pH profile will approach steady-state with time. Once each horizon or layer is at steady state, net acid

SPAM Flow-Chart

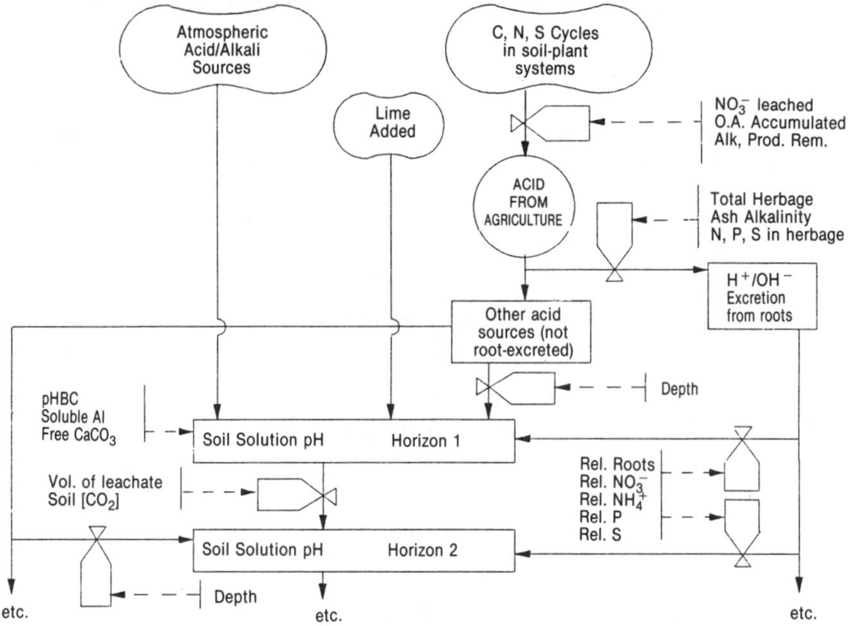

Fig. 2. A simplified relational diagram for the soil profile acidification model SPAM.

losses in the leachate balance acid gains from other sources. Several steady-state soil pH profiles have been generated by SPAM for a range of acid addition rates (Fig. 4). The different rates represent differing degrees of neutralization by lime of the acid added in typical southern Australian annual pasture ecosystems. The model inputs other than the lime rate were the same as for the model verification run.

Addition of acid at a rate typical of southern Australian agricultural ecosystems (3.46 Kmol H^+/ha per yr) and similar to acid addition rates in the polluted rainfall of much of Europe and north eastern USA (Likens *et al.*, 1979), results in a deep highly acid profile with a slightly higher pH in the surface and 5 to 10 cm layers (Fig. 4). The higher surface pH results mainly from the strong alkaline effect of mass flow once steady state is achieved. Use of lime to reduce acid addition to 2 Kmol H^+/ha per yr increases the surface soil pH slightly, but has a minor effect on the subsurface pH. Reduction of acid addition to 0.4 Kmol H^+/ha per yr (equivalent to only 20 Kg $CaCO_3$/ha per yr) produces an adequate pH for plant growth to about 30 cm. Below this depth however subsoil pH values are still low enough (pH 4.1 to 4.3) to inhibit the root growth of acid sensitive species (Fig. 4).

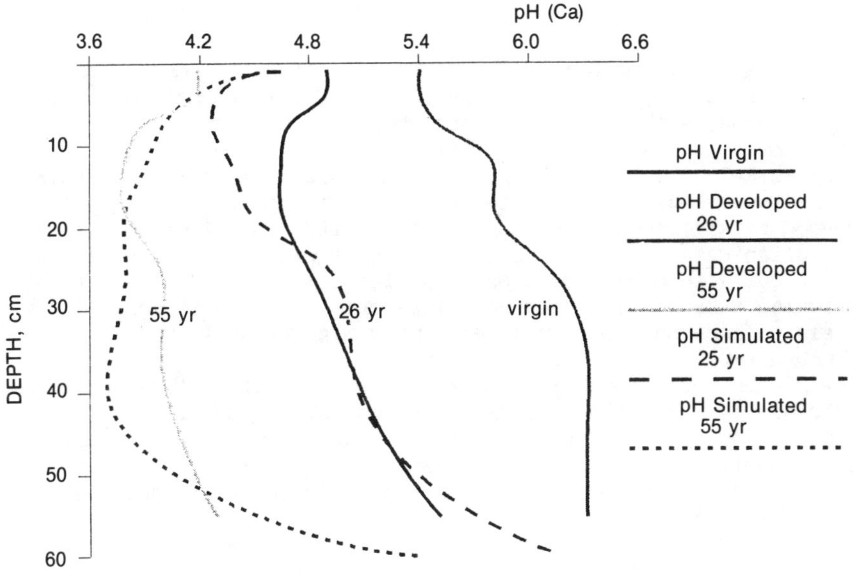

Fig. 3. Comparison of measured and simulated data for an acidifying soil profile.

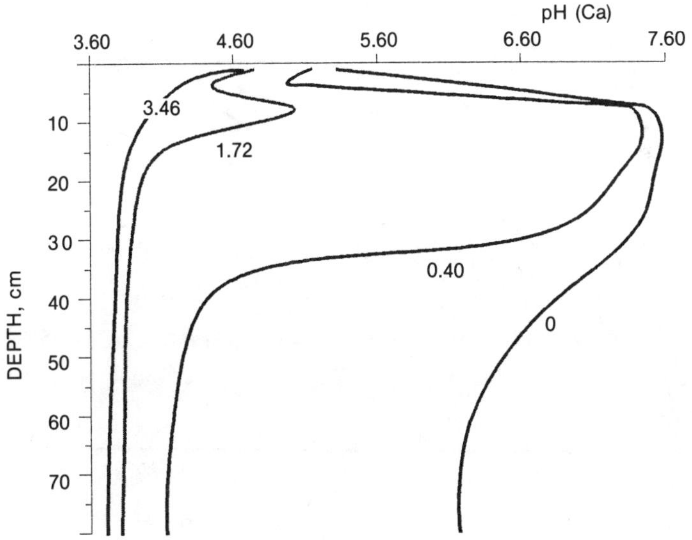

Fig. 4. Simulated steady-state curves for various levels of net acid addition (Kmol/ha per year).

Surface soil acidity often reduces plant yields more than subsurface acidity, presumably reflecting the fact that the surface soil is usually the most important source of nutrients and water for plants. Two examples are provided by Adams *et al.* (1967) and Gonzalez-Erico *et al.* (1979), where the effect of toxic Al levels in the topsoil caused about twice the decline in yield that was induced by similar levels of subsoil acidity. Some limited Australian data from lime rate experiments with lucerne on similar soils, but with differing subsoil Al levels (Mahoney *et al.*, 1981), indicated subsoil Al levels were less critical than topsoil Al levels to lucerne yield in a relatively high rainfall zone. Although topsoil conditions are usually more critical to plant growth than subsoil conditions, there are many situations where the yield of tops is critically dependent on subsoil supplies of nutrients, water, or both. Supplies of subsoil water and/or nitrate may be particularly important, as are supplies of other nutrients such as K (Spencer & Govaars, 1982) and S (Probert & Jones, 1977), in some situations.

The effects of pH on the growth of plants with varying degrees of tolerance to acidity is represented in Fig. 5. When we wish to account for the effect of subsurface acidity we must also consider the importance of subsurface soil for the species growing in the environment in question. This can range from plants highly dependent on subsurface water for survival (e. g. lucerne growing through a drying summer) to much less dependent situations such as subterranean clover, drawing on subsurface water only during intermittent dry periods.

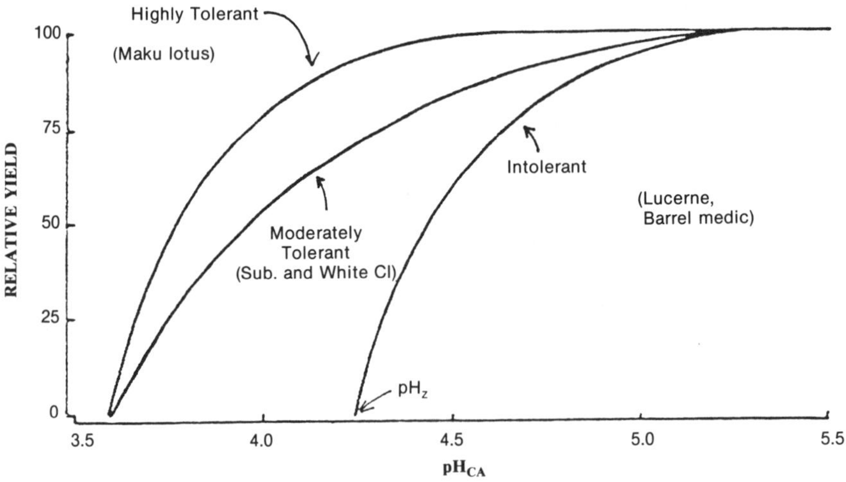

Fig. 5. Response curves for the effects of soil acidity on the relative yields of species with varying degrees of tolerance to soil acidity.

The importance of subsurface acidity to a given variety-environment combination can be represented as follows:

$$A = A_{max}(1 - Be^{-c(pH_{10-30} - pH_z)})$$

where: A_{max} = Potential yield of the species where there are no acidity related restrictions to root growth at any depth.

A = Yield potential corrected for subsurface acidity effects.

B = Maximum yield with no roots below 10 cm.

c = Coefficient dependent on the plant's tolerance to acidity.

pH_{10-30} = pH in the 10 to 30 cm soil horizon.

pH_Z = Critical pH below which the yield of this species is zero (Fig. 5).

The acidity of the surface soil (the 0-10 cm layer) restricts the yield further according to the following equation:

$$Y = A(1 - e^{-c(pH_{0-10} - pH_z)})$$

where:

Y = predicted yield

pH_{0-10} = pH in the top 10 cm of the soil

All other symbols are as in the previous formula.

Figure 6 shows the results of applying this model to three soils with pH profiles in steady state with different levels of net acid addition, for species with different levels of dependence on the subsoil and with different tolerance to soil acidity. The results indicate that in the absence of any reduction in current rates of acid addition to southern Australian pastures, highly acid tolerant shallow rooting annuals will eventually dominate these pastures. With partial compensation for acid addition (such as surface liming at 100 kg lime/year) the range of adapted species will broaden somewhat to allow for species like clovers and ryegrass although these would still be at a competitive disadvantage compared with highly tolerant deep rooting weeds, especially in situations where subsoil moisture supply is important.

It is clearly desirable in the long term that liming rates should match the rate of acid addition. Unfortunately, bio-economic analysis of alternative liming strategies (Hochman, 1984; Godyn *et al.*, 1987; Hochman *et al.*, 1988) show that economically rational decisions made by farmers would commonly result in the surface soil pH being maintained in the range of 4.6 to 5.0 in many Australian pasture and crop-pasture systems. This is a pH consistent with

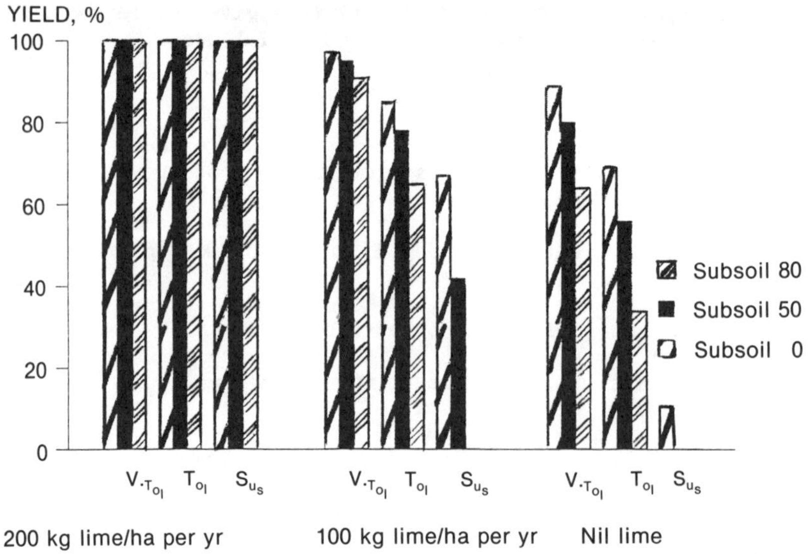

Fig. 6. Simulated steady-state results of various combinations of liming and tolerance strategies on relative yields of species with varying dependence on subsurface soil water and nutrient supply.

liming at less than the acid addition rate. We are therefore faced with a situation in which rational short-term economic decisions are likely to result in continuing acidification of the subsoil and consequently to a decline in the long-term productivity and diversity of Australian pastures.

INTERACTING CLIMATIC AND EDAPHIC STRESSES

Although we have treated climatic and edaphic constraints as separate factors they often act together to affect legume persistence in pastures. In this section we stress the importance of these interactions with a small number of examples. Some nutrients, notably P are often localized in the surface soil and in fertilizer bands, remaining deficient in the rest of the profile. These situations are susceptible to surface drying which reduces P diffusion and uptake (Scott, 1973; Simpson & Lipsett, 1973; Cornish & Myers, 1977; Cornish, 1987), so plant response to the surface P supply can be critically dependent on soil moisture. Furthermore in surface-dry conditions, the lack of subsurface phosphate may restrict the response of plants to the elimination of subsoil problems such as Al toxicity (Pinkerton & Simpson, 1986a,b).

Soil temperature affects competition for S in mixtures of subterranean clover and annual ryegrass grown with an adequate supply of N. Two factors are relevant here: At low temperatures S-deficiency was not nearly as severe as at high temperatures and the higher relative yields of ryegrass at 7 & 13°C appear to be related to poor N_2 fixation by the legume-rhizobium symbiosis (Gilbert & Robson, 1984b).

Russell (1976) hypothesized that in addition to salt

tolerance, rooting patterns of legumes make them more vulnerable in periods of extreme moisture stress in the salt-free surface layers. This idea was also expounded by Keating *et al.* (1986) who observed that relatively sensitive indigenous legume species which persist on the cracking clay soils (verisols) of the subhumid regions of north-eastern Australia have shallow roots and an ephemeral character of rapid growth and extensive seed production when water is available.

CONCLUSION

Legume persistence in Australian pastures is becoming more difficult to manage. This is largely due to changing climatic and edaphic conditions. Climatic changes which are anticipated over the next fifty yr are likely to be sufficient to require a re-assignment of cultivars and species within the pastoral and cropping zones. Even the boundaries of these zones are likely to change. Models for predicting climatic change are being developed for various planning uses (flood control, coastal management, domestic energy demand, etc.). Our concern as agricultural scientists is to minimize the loss of production in the adjustment period. This task cannot be achieved with current genetic evaluation methodology. To hasten the evaluation process we need to develop reliable models of how plants will be affected by climatic changes and of the way in which we can match plant characteristics to environmental constraints.

Reversing soil degradation requires increased inputs by producers. In the short-term, soil acidification can be reduced if we can persuade graziers to use lime. This requires a significant research effort to improve our ability to diagnose lime responsive soil-pasture systems. In the longer term, our concern is that current cost-benefit ratios and soil amelioration options make a significant degree of continued acidification inevitable. To overcome this problem we need research to develop soluble soil amelioration materials, to investigate less-acidifying production systems and to select and breed acid tolerant species to cope with acid subsoils.

REFERENCES

Adams, F., and B. L. Moore. 1983. Chemical factors afecting root growth in subsoil horizons of coastal plain soils. Soil Sci. Soc. Am. J. 47:99-102.

Adams, F., R. W. Pearson, and B. D. Doss. 1967. Relative effects of acid subsoils on cotton yields in field experiments and on cotton roots in growth-chamber experiments. Agron. J. 59:453-456.

Aitken, Y. 1955. Flower initiation in *Medicago tribuloides* Desr. and other annual medics. Aust. J. Agric. Res. 6:258-264.

Archer, K. A., E. C. Wolfe, and B. R. Cullis. 1987. Flowering time of cultivars of subterranean clover in New South Wales. Aust. J. Exp. Agric. 27:791-797.

Arnold, G. W., and G. W. Anderson. 1987. The influence of nitrogen level, rainfall, seed pods, and pasture biomass on the botanical composition of annual pastures. Aust. J. Agric. Res. 38:339-354.

Arnold, G. W., P. G. Ozanne, K. A. Galbraith, and F. Dandridge. 1985. The capeweed content of pastures in south-west Western - Australia. Aust. J. Exp. Agric. Anim. Husb. 25:117-123.

Balasubramanian, V., and S. K. Sinha. 1976. Nodulation and nitrogen fixation in chickpea (*Cicer arietinum* L.) under salt stress. J. Agric. Sci. 87:465-466.

Barrow, N. J. 1975a. The response to phosphate of two annual pasture species. I. Effect of the soil's ability to adsorb phosphate on comparative phosphate requirment. Aust. J. Agric. Res. 26:137-143.

Barrow, N. J. 1975b. The response to phosphate of two annual pasture species. II. The specific rate of uptake of phosphate, its distribution, and use for growth. Aust. J. Agric. Res. 26:145-156.

Blaikie, S. J., and F. M. Martin. 1987. Limits to the productivity of irrigated pastures in south-east Australia. p. 119-122. In J. L. Wheeler *et al.* (ed.) Temperate pastures their production, use and management. AWC/CSIRO, Australia.

Bolan, N. S., A. D. Robson, and N. J. Barrow. 1987. Effects of phosphorus application and mycorrhizal inoculation on root characteristics of subterranean clover and ryegrass in relation to phosphorus uptake. Plant Soil 104:294-298.

Bolland, M. D. A. 1985a. Responses of serradella and subterranean clover to phosphorus from superphosphate and Duchess rock phosphate. Aust. J. Exp. Agric. 25:902-912.

Bolland, M. D. A. 1985b. Effects of phosphorus on seed yields of subterranean clover, serradella and annual medics. Aust. J. Exp. Agric. 25:595-602.

Bolland, M. D. A. 1986. Efficiency with which yellow serradella and subterranean clover use superphosphate on a deep sandy soil near Esperance, Western Australia. Aust. J. Exp. Agric. 26:675-679.

Bolland, M. D. A., and W. J. Collins. 1987. Persistence of seed produced from single seed crops of subterranean clover on sandy soils near Esperance, Western Australia. Aust. J. Exp. Agric. 27:81-85.

Bolton, J. L., and F. E. McKenzie. 1946. The effect of early spring flooding on certain forage crops. Sci. Agric. 26:99-105.

Broom, D. M., and G. W. Arnold. 1986. Selection by grazing sheep of pasture plants at low herbage availability and responses of the plants to grazing. Aust. J. Agric. Res. 37:527-538.

Carter, E. D. 1981. Seed and seedling dynamics of annual medic pastures in South Australia. p. 447-450. In Proc. XIV Int. Grassl. Congr., Lexington, KY.

Clarkson, N. M., and J. S. Russell. 1975. Flowering responses to vernalisation and photoperiod in annual medics (*Medicago* spp.). Aust. J. Agric. Res. 26:831-838.

Collins, W. J. 1981. The effects of length of growing season, with and without defoliation, on seed yield and hard-seededness in swards of subterranean clover. Aust. J. Agric. Res. 32:783-792.

Collins, W. J., and B. J. Quinlivan. 1980. The effects of a continued water supply during and beyond seed development on seed production and losses in subterranean clover swards. Aust. J. Agric. Res. 31:287-295.

Collins, W. J., R. C. Rossiter, and A. Ramos Monreal. 1978. The influence of shading on seed yield in subterranean clover. Aust. J. Agric. Res. 29:1167-1175.

Cornish, P. S. 1987. Root growth and function in temperate pastures. p. 79-98. In J. L. Wheeler *et al.* (ed.) Temperate pastures their production, use and management. AWC/CSIRO, Australia.

Cornish, P. S., and L. F. Myers. 1977. Low pasture productivity of a sedimentary soil in relation to phosphate and water supply. Aust. J. Exp. Agric. Anim. Husb. 17:776-780.

Cregan, P. D., and K. R. Helyar. 1986. Non-acidifying farming systems. p. 49-62. In Acid soils revisited. 15th Riverina Outlook Conference. Aust. Inst. Agric. Sci./Agric. Technol. Aust., Wagga, Australia.

Dahmane, A. B. K., and R. D. Graham. 1981. Effect of phosphate supply and competition from grasses on growth and nitrogen fixation of *Medicago trancatulata*. Aust. J. Agric. Res. 31:761-772.

Dear, B. S., and B. Loveland. 1985. A survey of seed reserves of subterranean clovr pastures on Southern Tablelands of New South Wales. Proc. 3rd Aust. Agron. Conf., Hobart, Tasmania. p. 214.

Donald, C. M. 1963. Competition among crop and pasture plants. Adv. Agron. 15:1-118.

Drew, C. M. 1983. Plant injury and adaptation to oxygen deficiency in the root environment: A review. Plant Soil 75:179-199.

Evans, L. T. 1959. Flower initiation in *Trifolium subterraneum* L. I. Analysis of the partial processes involved. Aust. J. Agric. Res. 10:1-16.

Francis, C. M., and M. L. Pool. 1973. Effect of waterlogging on the growth of annual *Medicago* species. Aust. J. Exp. Agric. Anim. Husb. 13:711-713.

Finn, B. J., S. J. Bourget, K. F. Nielson, and B. K. Dow. 1961. Effects of different soil moisture tensions on grass and legume species. Can. J. Soil Sci. 41:16-23.

Fischer, R. A., and N. C. Turner 1978. Plant productivity in the arid and semiarid zones. Annu. Rev. Plant Physiol. 29:277-317.

Gilbert, M. A., and K. A. Shaw. 1981. Residual effects of sulfur fertilizers on cut swards of a *Stylosanthes guianensis* and native grass pasture on a euchrozem soil in north Queensland. Aust. J. Exp. Agric. Anim. Husb. 21:334-342.

Gilbert, M. A., and A. D. Robson. 1984a. Studies on competition for sulfur between subteranean clover and annual ryegrass. I. Effect of nitrogen and sulfur supply. Aust. J. Agric. Res. 35:53-64.

Gilbert, M. A., and A. D. Robson. 1984b. Sulfur nutrition of temperate pasture species. I. Effects of nitrogen supply on external and internal sulfur requirments of subterranean clover and ryegrass. Aust. J. Agric. Res. 35:379-388.

Gilbert, M. A., and A. D. Robson. 1984c. Sulfur nutrition of temperate pasture species. II. A comparison of subterranean clover, medics and grasses. Aust. J. Agric. Res. 35:389-398.

Gilbert, M. A., and A. D. Robson. 1984d. Effect of sulfur supply on the root characteristics of subterranean clover and annual ryegrass. Plant Soil 77:377-380.

Godyn, D., P. D. Cregan, B. J. Scott, K. R. Helyar, and Z. Hochman. 1987. The cost of soil acidification in southern New South Wales. 31st. Annu. Conf. Agric. Econ. Soc., Adelaide.

Gonzalez-Erico, E., E. J. Kamprath, G. C. Naderman, and W. V. Soares. 1979. Effect of depth of lime incorporation on the growth of corn on an oxisol of central Brazil. Soil Sci. Soc. Am. J. 43:1155-1158.

Hall, R. L. 1974. Analysis of the nature of interference between plants of different species. II. Nutrient relations in a Nandi *Setaria* and Greenleaf *Desmodium* association with particular reference to potassium. Aust. J. Agric. Res. 25:749-756.

Hannam, R. J., and D. J. Reuter. 1987. Trace element nutrition of pastures. p. 175-190. In J. L. Wheeler *et al.* (ed.) Temperate pastures their production, use and management. AWC/CSIRO, Australia.

Hamblin, A. P., and J. Hamblin. 1985. Root characteristics of some temperate legume species and varieties on deep, free-drainig entisols. Aust. J. Agric. Res. 36:63-72.

Helyar, K. R., Z. Hochman, and J. P. Brennan. 1988. The problem of acidity in temperate area soils and its management. In Natl. Soil Conf. Review papers 1988 (ed. J. Loveday) Aust. Soc. Soil Sci, Inc., Univ. W.A., Perth (In press).

Helyar, K. R., and W. M. Porter. 1988. Acidification in soils. In A. D. Robson (ed.) Soil acidity and plant growth. (In press). Academic Press, Sydney, Australia.

Hochman, Z. 1984. "Lime-It" - a computer based model of lime use in long term subterranean clover pastures. p. 71. In Proc. Natl. Soil Acidity Workshop, AWC, Perth, Australia.

Hochman, Z. 1987. Quantifying vernalization and temperature promotion effects on time of flowering of three cultivars of *Medicago truncatula* Gaertn. Aust. J. Agric. Res. 38:279-86.

Hochman, Z., D. L. Godyn, and B. J. Scott. 1988. The integration of data on lime use by modelling. In A. D. Robson (ed.) Soil acidity and plant growth. (In press). Academic Press, Sydney, Australia.

Holford, I. C. R. 1987. Sulfur nutrition of temperate pastures. p. 155-158. In J. L. Wheeler *et al.* (ed.) Temperate pastures their production, use and management. AWC/CSIRO, Australia.

Jackman, R. H., and M. C. H. Mouat. 1972. Competition between grass and clover for phosphate. II. Effect of root activity, efficiency of response to phosphate, and soil moisture. N. Z. J. Agric. Res. 15:667-675.

Jones, P. D., T. M. L. Wigley, and P. B. Wright. 1986. Global temperature variations between 1861 and 1984. Nature 322:430-434.

Kaiser, A. G., and M. L. Curll. 1987. Improving the efficiency of forage conservation from pastures. p 397-411. In J. L. Wheeler *et al.* (ed.) Temperate pastures their production, use and management. AWC/CSIRO, Australia.

Keating, B. A., and J. J. Mott. 1987. Growth and regeneration of summer-growing pasture legumes on a heavy clay soil in south-eastern Queensland. Aust. J. Exp. Agric. 27:633-641.

Keating, B. A., R. W. Strickland and M. J. Fisher. 1986. Salt tolerance of some tropical pasture legumes with potential adaptation to cracking clay soils. Aust. J. Exp. Agric. 26:181-186.

Likens, G. E., R. F. Wright, J. N. Gallaway, and T. J. Butter. 1979. Acid rain. Sci Am. 241:39-47.

Ludlow, M. M. 1978. Light relations of pasture plants. p. 35-49. In J. R. Wilson (ed.) Plant relations in pastures. CSIRO, Melbourne.

Maas, E. V., and G. J. Hoffman. 1977. Crop salt tolerance-current assessment. J. Irrig. Drainage Div. ASCE. 103:115-134.

Mahoney, G. P., Jones, H. R., and J. M. Hunter. 1981. Effect of lime on lucerne in relation to soil acidity factors. p. 299-302 In Proc. XIV Int. Grassl. Congr., Lexington, KY, USA.

McCaskill, M. R. 1987. Modelling S, P and N cycling in grazed pastures. Ph. D. thesis, University of New England.

McIvor, J. G. 1984. Phosphorus requirements and responses of tropical pasture species: native and introduced grasses, and introduced legumes. Aust. J. Exp. Agric. Anim. Husb. 24:370-378.

McLachlan, K. D. 1975. Sulphur in Australasian agriculture. Sydney University Press, Sydney.

McWilliam, J. R. 1978. Response of pasture plants to temperature. p. 17-34. In J. R. Wilson (ed.) Plant relations in pastures. CSIRO, Melbourne.

Mehanni, A. H. 1987. Reclamation of a saline/sodic soil by aquifer pumping, application of tillage and gypsum and reuse of saline groundwater. Aust. J. Exp. Agric. 27:381-387.

Meissner, A. P., and A. L. Clark. 1977. Response of mown pasture to potassium fertilizer in south-eastern South Australia. Aust. J. Exp. Agric. Anim. Husb. 17:765-775.

Morely, F. H. W., and L. T. Evans. 1959. Flowering time in *Trifolium subterraneum* L. II. Limitations by vernalisation, low temperatures and photoperiod, in the field in Canberra. Aust. J. Agric. Res. 10:17-26.

Mount, J. H., and D. L. Schuppan. 1978. The effects of saline irrigation water and gypsum on perennial pasture grown on a sodic, cley soil at Kereng, Victoria. Aust. J. Exp. Agric. Anim. Husb. 18:533-538.

Nicholas, D. J. D., and A. R. Egan. 1975. Trace elements in soil-plant-animal systems. Academic Press, New York.

Nicholls, A. O., J. D. Williams, and R. M. Moore. 1987. Competition between *Chondrilla juncea* and *Trifolium subterraneum*: the influence of canopy areas and heights. Aust. J. Agric. Res. 38:329-337.

Noble, C. L., and D. W. West. 1987. Maximising clover growth in the presence of salinity. p. 123-125. In J. L. Wheeler *et al.* (ed.) Temperate pastures their production, use and management. AWC/CSIRO, Australia.

Nye, P. H., and P. B. Tinker. 1977. Solute movement in the soil-root system. Blackwell Scientific, Oxford.

Ozanne, P. G., J. Keay, and E. F. Biddiscombe. 1969. The comparative applied phosphate requirements of eight annual pasture species. Aust. J. Agric. Res. 20:809-818.

Ozanne, P. G., K. M. W. Howes, and Ann Petch. 1976. The comparative phosphate requirements of four annual pastures and two crops. Aust. J. Agric. Res. 27:479-488.

Peak, D. C. I., G. D. Stirk, and E. F. Henzell. 1975. Leaf water potentials of pasture plants in a semi-arid subtropical environment. Aust. J. Exp. Agric. Anim. Husb. 15:645-654.

Pearson C. J., and B. C. Jacobs. 1985. Root distribution in space and time in *Trifolium subterraneum* Aust. J. Agric. Res. 36:601-614.

Pinkerton, A., and J. R. Simpson. 1986a. Responses of some crop plants to correction of subsoil acidity. Aust. J. Exp. Agric. 26:107-113.

Pinkerton, A., and J. R. Simpson. 1986b. Interactions of surface drying and subsurface nutrients affecting plant growth on acidic soil profiles from an old pasture. Aust. J. Exp. Agric. 26:681-689.

Pittock, A. B. 1987. The greenhouse effect. Engineers Australia. (February) 6:40-43.

Pittock, A. B., and M. J. Salinger. 1982. Towards regional scenarios for a CO_2-warmed earth. Climatic Change 4:23-40.

Probert, M. E., and R. K. Jones. 1977. The use of soil analysis for predicting the response to sulphur of pasure legumes in the Australian tropics. Aust. J. Soil Res. 15: 137-146.

Reeves, T. G., P. J. Haines, and D. R. Coventry. 1984. Growth of wheat and subterranean clover on soil artificially compacted at various depths. Plant Soil 80:135-138.

Robson, A. D. 1969. Soil factors affecting the distribution of annual *Medicago* species. J. Aust. Inst. Agric. Sci. 35:154-167.

Robson, A. D., and J. F. Loneragan. 1978. Responses of pasture plants to soil chemical factors other than nitrogen and phosphorus, with particular emphasis on the legume symbiosis. p. 128-142. In J. R. Wilson (ed.) Plant relations in pastures. CSIRO, Melbourne.

Rossiter, R. C. 1977. What determines the success of subterranean clover strains in south-western Australia. Proc. Ecol. Soc. Aust. 10: 76-88.

Russell, J. S. 1976. Comparative salt tolerance of some tropical and temperate legumes and tropical grasses. Aust. J. Exp. Agric. Anim. Husb. 16:103-108.

Russell, J. S. 1981. Geographic variation in seasonal rainfall in Australia - an analysis of the 80-year period 1895-1974. J. Aust. Inst. Agric. Sci. 47:59-66.

Scott. B. J. 1973. The response of barrel medic pasture to top-dressed and placed superphosphate in central - western New South Wales. Aust. J. Exp. Agric. Anim. Husb. 13:705-710

Shaw, N. H., and C. S. Andrew. 1979. Superphosphate and stocking rate effects on a native pasture oversown with *Stylosanthes humilis* in central coastal Queensland. Aust. J. Exp. Agric. Anim. Husb. 19:426-436.

Simpson, J. R. 1987. Nitrogen nutrition of pastures. p. 143-154. In J. L. Wheeler *et al.* (ed.) Temperate pastures their production, use and management. AWC/CSIRO, Australia.

Simpson, J. R., and J. Lipsett. 1973. Effects of surface moisture supply on the subsoil nutritional requirements of lucerne (*Medicago sativa* L.). Aust J. Agric. Res. 24:199-209.

Spencer, K., and A. G. Govaars. 1982. The potassium status of pastures in the Moss Vale district of New South Wales. Div. Plant Industry Tech. Paper No. 38, CSIRO, Australia.

Tanner, C. B., and T. R. Sinclair. 1983. Efficient water use in crop production: Research or re-search? p. 1-27. In H. M. Taylor *et al.* (ed.) Limitations to efficient water use in crop production. ASA, CSSA, and SSSA, Madison, WI.

Taylor, G. B. 1979. The inhibitory effect of light on seed and burr development in several species of *Trifolium* Aust. J. Agric. Res. 30:895-907.

Taylor, G. B., R. C. Rossiter, and M. J. Palmer. 1984. Long term patterns of seed softening and seedling establishment from single seed crops of subterranean clover. Aust. J. Exp. Agric. Anim. Husb. 24:200-212.

Taylor, P. A., and F. C. Greenhalgh. 1987. Significance, causes and control of root rots of subterranean clover. p. 249-251. In J. L. Wheeler *et al.* (ed.) Temperate pastures their production, use and management. AWC/CSIRO, Australia.

Turkington, R., and J. L. Harper. 1979. The growth, distribution and neighbour relationships of *Trifolium repens* in a permanent pasture. J. Ecol. 67:201-218.

Wilson, J. R. 1970. Response to salinity in *Glycine*. VI. Some effects of a range of short-term salt stresses on the growth, nodulation and nitrogen fixation of *Glycine wightii* (formerly *javanica*) Aust. J. Agric. Res. 21:571-582.

Wilson, J. R. 1985. Comparative response to salinity of the growth and nodulation of *Macroptilium atropurpureum* cv. Siratro and *Neonotonia wightii* cv. Cooper seedlings. Aust. J. Agric. Res. 36:589-599.

White, P. J., M. J. Whitehouse, L. A. Warrell, and P. R. Berrill 1981. Field calibration of a soil sulfate test on sward lucerne on the eastern Darling Downes, Queensland. Aust. J. Exp. Agric. Anim. Husb. 21:303-310.

Wit, C. T. de. 1958. Transpiration and crop yields. Versl. Landbouwk. Onderz. 64.6.

Wit, C. T. de. 1960. On competition. Versl. Landbouwk. Onderz. 66.8.

Wit, C. T. de. 1978. Summative address. p. 405-410. In J. R. Wilson (ed.) Plant relations in pastures. CSIRO, Melbourne.

Witschi, P. A., and D. A. Michalk. 1979. The effect of sheep treading and grazing on pasture and soil characteristics of irrigated annual pastures. Aust. J. Agric. Res. 30:741-750.

Wong, C. C., and J. R. Wilson. 1980. Effects of shading on the growth and nitrogen content of Green Panic and Siratro in pure and mixed swards defoliated at two frequencies. Aust. J. Agric. Res. 31:269-285.

Zillman, J. W. 1986. The Villach conference - An assesment of the role of carbon dioxide and of other greenhouse gases in climate variations and associated impacts. Search 17:183-184.

DISCUSSION

El-Swaify: How long did it take your acid soil profiles to develop?

Hochman: In an extensive production system, problem acid soils have developed 30 to 60 yr after legume-based pastures were first introduced. We predict, with the SPAM model, that it will take a similar period to develop steady-state profiles. The key variables are: net acid input, soil buffering capacity, rainfall, and the distribution of roots and N sources in the soil profile.

Hoveland: You note that subsoil acidity is of lesser importance in pasture or hay production. Research by Malcolm Sumner and associates at the Univ. of Georgia shows conclusively that subsoil acidity (pH 4.5-4.9) limits productivity of lucerne and birdsfoot trefoil. Deep incorporation of lime (uneconomic) and surface application of gypsum have increased yields 40 to 60%. Comment on this, please, as to the difference between U.S. and Australia.

Hochman: The references to my statement include U.S. authors. The Australian reference applies to a high rainfall environment. In a semiarid environment, with deep acid soils in western Australia, Bill Porter has shown that while surface liming does not increase yield, large responses are obtained with deep incorporation of lime. However, the model we propose in the paper would hold for these situations as they are simply cases in which the B parameter value is 1.0.

In Australia, gypsum has been shown to have a minimal, or even negative, effect on reducing Al toxicity. We believe that the major benefit of gypsum (in addition to the minor beneficial effects of high Ca on Al toxicity) is likely to be from Fluro complexing of monomeric Al and subsequent leaching of this complex. This benefit would be restricted to gypsum produced as a byproduct of high P fertiliser manufactured from Fluorapatite rock phosphate. [In Helyar et al. (1988) we discuss Al complexing and movement to depth by various agents including F and organic anion complexing and movement].

Matches: What is the physiological basis for selecting acid-tolerant plants, including legumes?

Hochman: In wheat, I believe tolerance is related to whether plant uptake of N is preferentially via ammonia or nitrate, and the way this modifies the pH of the rhizosphere (the zone around the roots). In legumes, internal tolerance mechanisms seem more likely. Some plants detoxify Al by complexing it with organic acids, proteins or other liquids. Legume tolerance is complicated by the effect on N2-fixation of interactions among host genotype, rhizobial genotype and environmental factors. This subject was recently reviewed by Foy (Foy, C.D. 1988. Plant adaptation to acid, aluminium-toxic soils. Commun. in Soil Sci.Plant Anal., 19 (7-12):959-987) and Munns (Munns, D.N. 1986. Acid soil tolerance in legumes and rhizobia. Adv. Plant Nutr. 2:63-91).

Fisher: I would like to comment on the acid soil problem from a tropical American perspective. The oxisols and ultisols of the Latin American savannas have pH's 3.5 to 4.2 and high levels (ca. 80%) of Al saturation. The Tropical Pastures Program of the Centro Internacional de Agricultura Tropical (CIAT) is adopting a germplasm approach to seeking pasture species tolerant of these conditions and has identified a number of grasses and legumes displaying good adaptation. It is of interest that some of them, notably the grass Andropogon gayanus, did not originate from acid soils with high Al saturation.

Hochman: The CIAT project is dealing with soils that have already reached steady state. In this case, liming is uneconomic until tolerant species are used. In the Australian context, we are more concerned with arresting acidification. To do this, the best strategy is to use tolerance and lime together.

Sheath: What is the current state of research on acid soils in Australia?

Hochman: Keith Helyar, coordinator of acid soils research in Australia, publishes a newsletter that covers the research activities. Key projects are in progress at the Univ. of Queensland, Wagga Wagga (NSW), Rutherglen (Victoria), and the Univ. of Western Australia.

ENVIRONMENTAL SELECTION OF LEGUMES

D. Scott, J.H. Hoglund, J.R. Crush, and J.M. Keoghan

SUMMARY

A perspective on environmental limitations on herbaceous legumes in New Zealand should include the caveat that this is from a country which had only one, rare, herbaceous legume in its native flora. Also, the soils were massively deficient in P, and S. The discussion is therefore of introduced pasture legumes, in association with historically cheap phosphatic fertilizers. Also legumes are being considered primarily in relation to their N_2-fixing abilities, and only secondarily as fodders. A consequence is that rhizobial relationships are as important as the feeding characteristic of the host species. The paper takes an ecological approach by considering the environmental factors and vegetation processes which determine legume success, with as much attention to choosing the species for the environment as in modifying an environment to suit a species.

SPECIES SUITABILITY IN RELATION TO ENVIRONMENTAL GRADIENT

The concept of environmental gradients, and the confinement of individual species to part of a continuum is well established for natural vegetations. Legume evaluations on a range of different sites and management systems suggest the same concepts apply to agriculture. Scott et al. (1985) suggested that the suitability of legumes, and the productivity of pasture, is related to the four environmental gradients of moisture, temperature, soil fertility, and the management of grazing animals.

Soil moisture is generally recognized as a dominant gradient, usually expressed as annual rainfall, but modified by seasonal distribution, soil depth and local topography - in most cases integrated in soil types.

Temperature is a function of latitude, altitude, slope, aspect etc., and many studies use these indirect measures. While temperature and moisture have fundamental influence on species distribution, they are seldom capable of management manipulation, except, to a limited extent by irrigation and drainage.

Soil fertility is a concept relating observed plant growth to that measured under nutrient non-limiting conditions determined from nutrient amendment, soil analysis, or tissue analysis. For legumes in New Zealand this relates principally to P and S nutrition. Soil fertility is the most important factor in agriculture in the sense that it is the easiest to manipulate.

Given estimates of moisture, temperature and fertility, it is possible to make reasonable estimates of the potential production

(Fig. 1). The features are:- the logarithmic decreases in potential productivity with either decreasing moisture or temperature, and the five-fold increase with fertilizer changing soils of low natural fertility to high fertility status.

Legume species have similarities and differences in achieving these potentials. In New Zealand most species can be grown on a wide range of sites. All will be most productive on warm, wet, high fertility sites, and probably achieve more than half the potential productivity at all sites. However, for any site with a particular combination of moisture, temperature and fertility, there are probably only one or two legume species capable of expressing the full site potential productivity. Figure 1, places 16 legumes in environments where they most closely realize the site potential.

It should be emphasized that Fig. 1 does not necessarily indicate where each species grows best, but rather where each species would achieve dominance in competition with other legumes. Cultivars or lines within species would appear as satellites around the modal position of each species (e.g. the New Zealand red clover cv. Pawera will tolerate colder, wetter conditions, 'Turoa' drier, warmer conditions, and 'Hamua' moister warm conditions).

The fourth environmental factor is management of the frequency, intensity, and duration of grazing in relation to the position and growth rate of adventitious growing points. Legumes differ in their tolerance of both frequency and closeness of defoliation

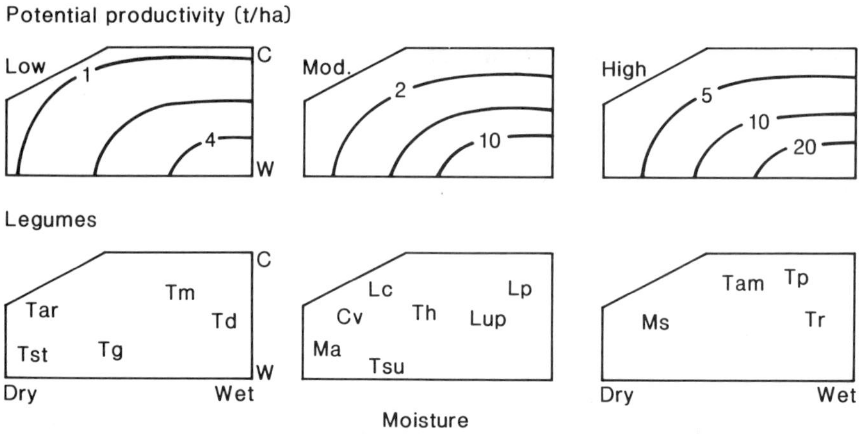

Fig. 1. Potential productivity (tonne DM/ha per yr) [upper] and legume suitability [lower] in relation to gradients of moisture (abscissa - dry to wet), temperature (ordinate - warm to cold), and for low, moderate and high soil fertility. Cv = Coronilla varia L., Lc = Lotus corniculatus L., Lp = L. pedunculatus Cav., Lup = Lupinus polyphyllus Lindl., Ma = Melilotus alba Medic., Ms = Medicago sativa L., Tam = Trifolium ambiguum Bieb., Tar = T. arvense L., Td = T. dubium Sibth., Tg = T. glomeratum L., Th = T. hybridum L., Tm = T. medium L., Tp = T. pratense L., Tr = T. repens L., Tst = T. striatum L. and Tsu = T. subterraneum L. (Based on Scott et al., 1985).

(Hoglund and White, 1985). To take two extremes, Medicago sativa is tolerant of close defoliation but intolerant of frequent defoliation, whereas Trifolium subterraneum is tolerant of frequent defoliation but intolerant of close defoliation. Trifolium repens tends to be tolerant of both regimes, and T. pratense intolerant. The palatability of different legume species also interacts strongly with management.

The base information for such concepts came from experiments of different designs and the next two sections illustrate contrasting approaches.

GENOTYPE/ENVIRONMENTAL ANALYSIS

In determining the suitability of legume species or cultivars for a wide range of environments, there is the problem of defining the environments. Finlay and Wilkinson (1963) made the conceptual breakthrough of defining an environment by the growth of plants in that environment i.e., an environment or treatment combination, in which the mean growth of all lines under test was high, is by definition a "good" environment. The concept was further expanded by Knight (1970), Hill (1975), and Scott (1985). After scaling all values relative to the grand mean, and using the mean yield of all species in a particular treatment as an environmental scale (Fig. 2 left), the performance of an individual species can be assessed by regression from their mean intercept (= yield advantage) and slope (= adaptabilility).

The presentation of these pairs of values on a common diagram (Fig. 2 right) enables comparison of their mean yield under the average environment (horizontal deviation from vertical axis), or good environment (as horizontal deviation from dotted line, for a one standard deviation better than average environment) as it relates to the adaptability coefficient. While further statistical analysis is possible where the 'environments' are some balanced combination of treatments, the diagram enables comparison when the environments are ill defined, or where mortality causes some treatment combinations to disappear.

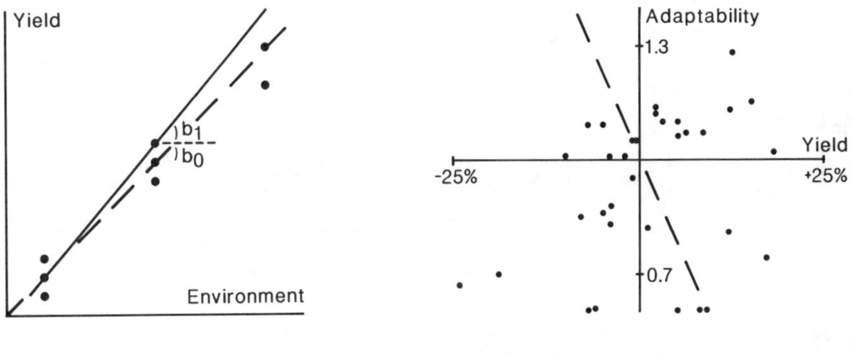

Fig. 2 Genotype/environmental analysis, from 32 Lotus lines in 3 sites x 2 fertilizer x 3 seasons x 3 yr environments (Scott, 1985).

The method has advantages in determining the source of legumes for new areas. For example, further screening of Lotus corniculatus lines for upland New Zealang areas indicated that material from Portugal, Yugoslavia, Italy, and France was best suited to dry intermontane basins, while material from Holland, Sweden, Canada, and Russia was better suited to cold upland infertile soils (Widdup et al., 1987).

ENVIRONMENTAL SELECTION USING SPECIES MIXTURES

The concept of species niche outlined in the first section was based on observations of many few-species trials in different environments. The concept is being tested in two long term experiments in tussock grassland at Lake Tekapo, New Zealand on a low fertility, dry, cool site (Fig. 1). Eleven legume species and 14 grasses and herbs were sown in a mixture, and a number of contrasting environments imposed, in the expectation that different legumes will achieve dominance in different environments.

The variables in the first experiment were 27 combinations of P and S fertilizers at annual rates of 0, 5, 10, 20, 50, and 100 kg element/ha, as triple phosphate and elemental S. Four combinations also had micro-nutrient additions. The second experiment had combinations of five fertilizer levels (0, 4P + 14S, 8P + 19S, 20P + 50S and 45P + 55S kg/ha using superphosphate with a decreasing proportion of S, plus irrigation at the highest level), three stocking rates (lax, optimal, and hard grazed), and two stocking methods (mob and sustained) (Scott and Covacevich, 1987). Both trials were established for over a year before sheep grazing commenced. Pasture composition is measured by ranking species according to their contribution to the sward after a common ungrazed period in the spring.

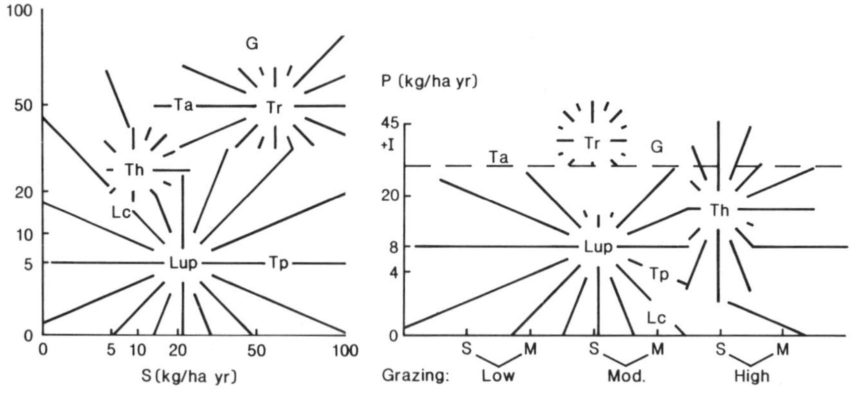

Fig. 3. Dominance of legumes in multiple species sowing followed by imposition of multiple environments of P and S levels (left); and P, grazing intensity and method (right). Most suitable environment for species indicated by position of symbol; and regions of dominance relative to other legumes by shading.

The results are given for the 6th yr when the principal legumes in both trials were perennial lupin (Lupinus polyphyllus), alsike clover (Trifolium hybridum), red clover (T. pratense), white clover (T. repens), birdsfoot trefoil (Lotus corniculatus), and caucasian clover (T. ambiguum). There were occasional plants of the other sown legumes: lotus (L. pedunculatus), hybrid lotus (L. cornic. x ped.), zig-zag clover (T. medium), lucerne (Medicago sativa), but no crownvetch (Coronilla varia).

In both experiments, there was sorting of species according to environment, with Lupinus polyphyllus being abundant in most environmental combinations, and dominant at all the lower phosphate levels. Trifolium hybridum became abundant and dominant at the higher phosphate levels, with T. repens becoming important at the highest levels. Lotus corniculatus persisted as a minor species at low fertilizer inputs, as did T. pratense and T. ambiguum. There was secondary sorting out due to grazing intensity in the second experiment, with a small effect due to stocking method.

NUTRIENT LIMITATIONS

Phosphorous and S are two elements generally deficient for legumes in the New Zealand environment. There are also Mo deficiencies, subsoil Al toxicity and occasional K deficiency. Phosphorous and S have been deliberately linked because of their similar relationship to soil characteristics and their often common supplementation through the use of superphosphate (Metson & Blakemore, 1978). The relative deficiencies of the two is related to the leaching and weathering sequence of soils (Fig. 4).

Phosphorous deficiency increases with rainfall, and to lesser extent with the age of the soil. Typically yellow brown or podzolized yellow brown soils require initial fertilizer rates of 300-800 kg/ha superphosphate and annual maintenance rates of 100-200 kg/ha. By contrast in the brown grey earths of the lowest rainfall areas, S is the most deficient mineral in the topsoil, though it is present in the subsoil. Initial fertilizer can be elemental S, or elemental S fortified superphosphate at ca.

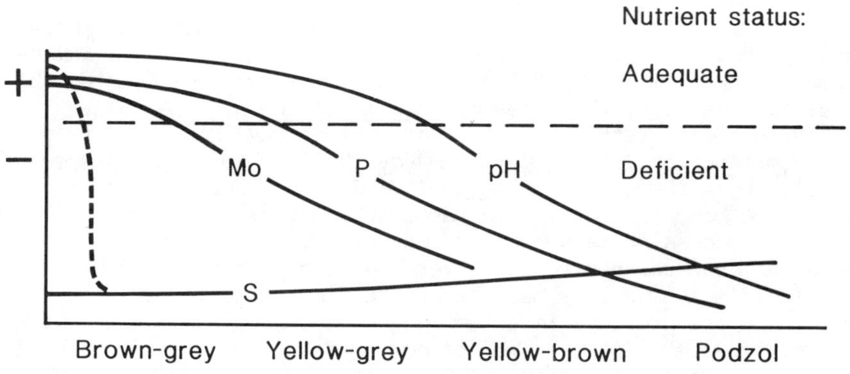

Fig. 4. Nutrient status of soils as related to rainfall and/or leaching (Sinclair and Floate, 1984).

100 kg/ha and annual maintenance rates of 20-50 kg/ha. Molybdenum also becomes deficient in the high rainfall leached soils with fertilizer amendments of 20 g Mo/ha required every 3-5 yr (Cornforth & Sinclair, 1982). Potassium has been shown to influence legumes' ability to compete with grass, and B deficiency has been seen in Medicago.

Lime use in New Zealand has decreased over recent decades. It is more efficient to direct coat lime onto legume seed (Lowther, 1974, 1975). Intensification of pastoral agriculture is inherently acidifying on soil primarily because of nitrate leaching, so pH adjustment will eventually be required on most sites.

Studies on the seeds of a range of pasture legumes have indicated there is little relationship between the nutrient content of the seed, their availability, and the nutrient requirements of the early seedling stage (Fenner & Lee, 1987, personal communication). Nitrogen and K are the most limiting nutrients, followed by Ca for legumes, then other nutrients, with S generally the least deficient.

There has been some development of direct nutrient coating of legume seed with the main constraint being to find forms of nutrients which are chemically inactive while on the seed and during the early germination stage. Thus even though the least deficient in seed reserves the greatest success has been with elemental S seed coatings, with mixed results from Mo coating and minor effects with P coats (Scott & Archie, 1978).

LEGUME SUCCESSION IN RELATION TO FERTILITY

In the development of pastures using legumes plus P and S fertilizer, the observed succession usually has an early legume dominant phase, where the coarse rooted but N_2-fixing legumes have competitive advantage assisted by mycorrhiza. This comparative advantage is surpassed by the greater nutrient scavenging abilities of fine-rooted grasses, as organic matter and mineralized N increases. However the time course of these legume to grass successions warrant examination in relationship to the suitability of species to different conditions, particularly fertility.

With high fertility development and species adapted to those conditions, there are plenty of documented cases of legume dominance by the likes of Trifolium repens for 1-3 yr, followed by the grass dominance by the likes of Lolium perenne L.

At moderate fertilities the same trends have been seen (e.g., the dominance by Trifolium medium for 10-15 yr followed in succession by Arrhenatherum elatius (L.) J & C. Presl. These changes appear to be on a logarithmic rather than linear time scale. The delayed succession is generally interpreted as both the lower rate of N build-up related to the lower legume growth rates, and continued leaching loss because of the time scale involved.

There are problems in extending these concepts and explanations to very low fertilities. We are not aware of documented cases of a legume to grass succession of the order of several decades. Very low fertility situation tend to remain grass or at least non-legume dominant over long periods. Legumes have higher threshold concentration for growth for nutrients like P and may not be able to compete with grasses in deficient environments. Nitrogen input from free-living bacteria, blue-green algae and rainfall can be adequate to sustain low fertility grasses.

Many young soils and natural plant successions start on freshly weathering mineral substrate, which may be relatively rich in non-organic sources of P, and S, but low N, which allow N_2-fixing plants to thrive. On mature soils under natural vegetation the N, P, and S and other elements are predominantly in organic matter, with low levels of inorganic P and S.

In soils impoverished through burning, overgrazing, windblow, etc., there can be direct loss of this organic matter, as well as increased mineralization and greater potential for nutrient loss. Many agricultural soils, without careful management could probably be classified as impoverished in that they do not have a fresh input of mineral material other than fertilizers, and do not support 'natural' equilibrium organic matter levels.

MICROBIOLOGICAL LIMITATIONS

The principal microbiological requirement of legumes for N_2 fixation is a suitable symbiont. It is assumed that this topic will be a theme of other papers and in the present context the only points that need be emphasized are: introducing legumes to new areas require the simultaneous introduction of rhizobia; rhizobia are often species specific; New Zealand and Australia have developed technology for direct coating of rhizobia onto the seed, protecting it from both seed coat and aerial effects; and that the initial root infection stage requires adequate pH either from soil liming or seed coating.

Mycorrhiza are the other microbiological requirement. This symbiotic relationship with fungi increases the root area of the generally more coarse rooted legumes, and enhances phosphate uptake on phosphate deficient soil. For example, Crush (1975) found that 70% or greater of roots were mycorrhizal in 14 legumes from a range of sites and varying fertilizer histories. Mycorrhiza are highly beneficial to legumes in low to moderate fertility soils, but are neutral or even slightly parasitic at high soil fertilities. Thus their role is likely to be more important in natural grasslands than in developed, legume based pastures. Mycorrhiza are less species specific than rhizobia.

ALLELOPATHY

The exudation or release by decay of substances from one species which influence its own or other species can also be considered an environmental factor. In temperate zones the allelochemicals are generally water soluble and reduce germination, but more particularly reduce root hair formation (Rice, 1974).

Some New Zealand evidence suggest that self allelopathy is one of the causes of failure of *Trifolium repens* sowings in marginal areas (Scott, 1974; Macfarlane et al., 1982). In laboratory germination tests T. repens was the most inhibitory of 20 shoot materials on the germination of 7 legumes and grasses including its own seed. In field trials on a range of sites, including added N, dried T. repens shoot material was spread at rates of 0, 0.5, 1, 2, and 4 t DM/ha, and seeds or plants of various species transplanted into it. On the most extreme sites there was a marked initial depression of germination and growth, presumably through allelopathy, to be followed late in the growing season by enhanced growth, presumably through mineralisation. On more favorable sites the initial depression was slight or absent presumably because of the greater growth rate of seedlings.

OTHER ENVIRONMENTAL CONSTRAINTS

The temperature optimum for growth of a range of New Zealand legumes is about 2° higher than that of comparable grasses (Mitchell, 1956; Mitchell & Lucanus, 1960). The temperature for root N uptake and N_2 fixation are about 2° lower than that for organic matter mineralization, which in turn is lower than that for legume nodulation (Hoglund & Brock, 1987). This means that if legumes can overwinter nodules they may often have a competitive advantage over grasses in early spring during the period prior to soil warming sufficient to stimulate mineralization of organic matter. Conversely if nodules do not overwinter, legumes may be at a competitive disadvantage until they re-nodulate. Seed of temperate zone legumes apparently do not have adverse reactions to chilling temperatures during inbibition as reported for subtropical legumes (Scott & Hanson, 1977).

In legumes there is a strong inverse interaction between N_2 fixation and N uptake. However in any particular situation N_2 fixation is proportional to legume growth for particular levels of soil available N (Hoglund & Brock, 1987). So in the search for legumes with maximum N_2 fixation in particular environment the legume species or cultivar with maximum growth rate in that environment is probably a sufficient measurement.

A final comment on the legume composition of vegetations or pastures is warranted. It is commonly assumed that the proportion of species in a vegetation is an integration of their individual responses to environment constraints including competition and other interaction between species. However, there is some evidence that there are some properties of a vegetation as a whole within which these other interactions may be constrained. Besides the environmental constraint to the upper potential yield referred to in the first section, there is the 3/2 thinning law constraining density/size relationship, and the near linear relationship between the logarithm of the proportion of different species and their rank from the most to least important in the vegetation.

RESUME

While this section of the workshop was devoted to the topic of edaphic and climate limitations to legumes we have taken a more neutral ecological stance in the belief that in practice as much attention is given to selecting a legume species suited to a particular environment, as in seeking to modify that environment.

The first section proposed that the four most important environmental factors in agriculture were moisture, temperature, soil fertility and grazing management. An attempt was made to place a range of legumes in their most suitable environment, based on qualitative assessment of many few-species trials. The second section on genotype-environment analysis defined an environment, and gave a specific quantitative method for defining the relative merits of a large group of like species or cultivars. The third section described an empirical method of sowing multi-species mixtures and direct environmental selection of legume species. There was a brief discussion of the suitability and succession of different legumes and grasses under different soil fertility conditions.

On most agricultural sites the environmental factor easiest to modify is soil fertility, which for legumes is generally the macro-nutrients P and S, and micro-nutrient Mo. It is shown how these requirements vary with leaching regime, which related to the rainfall gradient.

ACKNOWLEDGMENT

Miss E.L. Hellaby Indigenous Grasslands Trust, Mr J.S. Robertson and the typists of DSIR.

REFERENCES

Cornforth, I.S., and A.G. Sinclair. 1982. Fertiliser and lime recommendations for pastures and crops in New Zealand. MAF, Wellington.

Crush, J. 1975. Occurrence of endomycorrhiza in soils of the Mackenzie Basin, Canterbury, New Zealand. N.Z. J. Agric. Res. 18: 361-364.

Finlay, K.W., and G.M. Wilkinson. 1963. The analysis of adaption in a plant breeding programme. Aust. J. Agric. Res. 14: 742-754.

Hill, J. 1975. Genotype-environment interaction - a challenge for plant breeding. J. Agric. Sci. 85: 477-493.

Hoglund, J.H., and J.L. Brock. 1987. Nitrogen fixation in managed grasslands. In R.E. Snayden (ed) 'Managed Grasslands', B. Analytical Studies'. Elsevier Publishers, B.V., Amsterdam.

Hoglund, J.H., and J.G.H. White. 1985. Environmental and agronomic constraints in dryland pasture and choice of species. In R.E. Burgess, and Brock, J.L. (eds). Using herbage cultivars, N.Z. Grassl. Assoc., Palmerston North.

Knight, R. 1970. The measurement and interpretation of genotype environment interactions. Euphytica 19: 225-235.

Lowther, W.L. 1974. Interaction of lime and seed pelleting on the nodulation and growth of white clover. I. Glasshouse trials. N.Z. J. Agric. Res. 17: 317-325.

Lowther, W.L. 1975. Interaction of lime and seedling pelleting on the nodulation and growth of white clover. II. Oversown trials. N.Z. J. Agric. Res. 18: 357-360.

Macfarlane, M.J. et al. 1982. Allelopathic effects of white clover N.Z. J. Agric. Res. 25: 503-518.

Metson, A.J. and L.C. Blakemore. 1978. Sulphate retention by New Zealand soils in relation to the competitive effect of phosphate. N.Z. J. Agric. Res. 21: 243-253.

Mitchell, K.J. 1956. Growth of pasture species under controlled environment. I Growth at various levels of constant temperatures. N.Z. J. Sci. Technol. A38: 230-245.

Mitchell, K.J., and R. Lucanus. 1960: Growth of pasture species under controlled environment. II. Growth at low temperatures. N.Z. J. Agric. Res. 3: 643-655.

Rice, E.L. 1974. Allelopathy. Academic Press, New York.

Scott, D. 1974. Allelopathic interaction of resident tussock grassland species on germination of oversown seed. N.Z. J. Exp. Agric. 31: 135-141.

Scott, D. 1985. Plant introduction trials : genotype-environmental analysis of plant introductions for the high country. Ibid 13: 117-127.

Scott, D., and W.J. Archie. 1978. Sulphur, phosphate and molybdenum coating of legume seed. N.Z. J. Agric. Res. 21: 643-649.

Scott, D. and N. Covacevich. 1987. Effects of fertiliser and grazing on a pasture species mixture in high country. N.Z. Grassl. Assoc. 48: 93-98.

Scott, D., and M.A. Hanson. 1977. Effect of low temperature during initial germination of some New Zealand pasture species. N.Z. J. Exp. Agric. 5: 41-45.

Scott, D. et al. 1985. Limitations to pasture production and choice of species. p. 9-15. In R.E. Burgess, and J.L. Brock (ed.) Using herbage cultivars. N.Z. Grassl. Assoc., Palmerston North.

Sinclair, A.G., and M. Floate 1984. Nutrient deficiencies and fertiliser requirements in tussock grassland soils. MAF Wellington.

Widdup, K.H. et al. 1987. Breeding Lotus corniculatus for South Island tussock country. Proc. N.Z. Grassl. Assoc. 48: 119-124.

DISCUSSION

Brougham and Curll: How is grazing accommodated in your species niche concept?

Scott: Only examples of species adapted to grazing are listed in Fig. 1. However further subdivison of species according to their suitability to either lax or continuous grazing is given in the references.

Cramshaw: Your Fig. 1 shows increasing yields with increasing temperatures. How far would that extend?

Scott: The estimates were based on NZ experience which only extend to the warm temperate. I would expect the trend to continue into the tropics with any decrease being through associated moisture limitation rather than a temperature depression per se. Yields of 35-50 t DM/ha or more have been recorded from Hawaii for highly fertilised, irrigated, sugar cane.

Watson: Why is white clover such an important legume in NZ?

Scott: Most of NZ agriculture is pastures grown in the moderately warm moist areas which were temperate broadleaf rainforest, and with pastures now maintained by high fertiliser rates - the niche to which white clover is suited.

Hoveland, Curll, Sheath and Jones: A number of related questions on the use of multiple species mixtures.

Scott: The use of multiple species mixtures was a research technique to determine the fertiliser and grazing management matching at particular sites. It is not a practice to be advocated to farmers. Only those suited to the particular combination of conditions would be suggested. However on the general question of the relative desirability of a many or a few species mixture a general ecological perspective suggests there may be a trade off between stability or flexibility in a pasture as related to species diversity, and productivity as related to dominance by one or a few species. There is a strong empirical evidence from many thousands of years of agriculture from many millions of farmers than near mono-cultures are the more productive. The legume/grass combination may be the exception.

Watson: Pointed out that plot scale screening does not include the full suite of pests and diseases that may ultimately affect a species advocated.

Brougham and Fisher: Suggested that shrub legume/grass vegetations may be examples of low fertility legume to grass transitions.

MAJOR CLIMATIC AND EDAPHIC STRESSES IN THE UNITED STATES

Dwayne R. Buxton

SUMMARY

Plants grown under field conditions often experience stress. The major stresses on forage legumes in the USA result from cold and variable winter temperatures, hot summer temperatures, water saturated soils, water deficits during the growing season, low soil pH, shading, and injury from insects and diseases. Winter injury results from desiccation, flooding, ice encasement, soil heaving, and diseases in addition to the direct effects of freezing temperatures. Prolonged or intensive stress kills plants and reduces stands. In the northern USA, cold-tolerant forages undergo fall hardening by increasing sugar, protein, lipid, and dry matter concentrations, which increases their tolerance to freezing temperatures. Temperature fluctuations during the fall and winter can interfere with or undo the hardening process and cause stand loss. Most forage legumes are not tolerant to drought and survive by drought avoidance. In some species, drought is avoided or minimized by extracting water from deep in the soil profile. A critical need exists to better define physiological processes associated with stress resistance so that these plant characteristics can be altered by plant breeding and used as selection criteria for developing cultivars more resistant to plant stresses.

INTRODUCTION

Climatic and edaphic factors largely determine the geographical regions in which crops can be grown successfully. Most forage legumes are now grown extensively outside of the ecosystems in which they evolved. In these new environments, forage legumes often are confronted with edaphic and climatic stresses for which they may not be well adapted. Under field conditions, almost all plants undergo some environmental stress. Mild stress may be reflected only in yield reduction. Prolonged or intensive stress, however, kills plants and reduces stands. Legumes generally have a narrower range of adaptation and have less resilience to environmental stress and grazing than do grasses. As a result, legumes require better management than grasses to persist and remain productive.

Climate in the United States as well as on a global basis has shown cyclic variation. Thompson (1988) observed a general warming period in parts of the North Central Region of the USA from 1890 until the late 1930s, followed by cooling through the

1970s. The warming trend has reoccurred in the 1980s. In his study, summer rainfall was negatively correlated with summer temperature. Thompson concluded that in addition to these long-term climate changes, cyclical weather patterns exist with a period of approximately 20 yr. These may be caused by lunar cycles, whereas the longer-term changes may be related to changes in atmospheric transparency from volcanic dust and other sources. The present longer-term warming trend may be occurring at an accelerated rate because of the rise in CO_2 concentration of the atmosphere from intensive fossile fuel use (Hanson et al., 1981). As with the 1930s and 1940s, weather has been more variable since 1973 than occurred from 1956 to 1973 (Thompson, 1988).

It is the microclimate near plants that affects their growth and persistance. As noted by McCloud & Bula (1985), temperature and humidity on a clear day may be more variable within 1 m of the soil surface than the variation within several hundred square kilometers at the height of standard meteorological shelters.

OVERVIEW OF THE PROBLEM

Following his review of environmental limitations to legume persistence, Marten (1985) concluded that a better understanding is needed concerning the influence of climatic variables on basic physiological processes associated with stand depletion. He identified winter injury from cold and variable temperatures, injury from excessive wet conditions in poorly drained soils, and water deficit during the growing season as major stresses encountered by established forage legumes in the northern USA. Burns (1985), on the other hand, mentioned injury from insect and disease complexes as the major limitations to persistence of established forage legumes in the southern USA. But Burns noted that insect populations and disease outbreaks are highly dependent on climatic conditions and edaphic factors. He further observed that the impact of insects and diseases on legume persistence is aggravated when additional environmental stresses are placed upon forages.

Secondary cell-wall formation and associated lignification seems to be one of the most important resistance mechanisms evolved by plants against abiotic and biotic environmental factors capable of imposing stress. These allow plants to hold photosynthetic tissue in a position to compete for photosynthetically active radiation. After plant cells have expanded, secondary cell-wall formation occurs. As discussed by Buxton and Russell (1988), this process involves covalent bonding between lignin and hemicellulose to form a rigid matrix.

Not coincidentally, lignified tissue is relatively resistant to maceration and digestion by ruminants and insects, and attack by disease pathogens. Unfortunately, lignification of cell walls limits the quality of forages. Ruminants, even with their specialized digestive systems, are unable to fully utilize the energy of cellulose and hemicellulose once it is lignified (Buxton et al., 1987). Forage legumes typically have highly lignified stems with less lignified leaves. Indeed, the lignin and cell-wall portion of legume leaves often remains relatively constant during maturation (Buxton & Hornstein, 1986). Stems, on the other hand, increase in cell-wall concentration and lignin throughout most of the life of the herbage.

Alfalfa (Medicago sativa L.) is the most popular and one of the best adapted forage legumes in the USA (Knight, 1985), and often is the standard by which other forage legumes are compared. When properly managed, alfalfa has persisted as long as 34 yr on rangelands of the western USA (Marten 1985). Other important legumes are red clover (Trifolium pratense L.), white clover, (T. repens L.), and several other clovers; birdsfoot trefoil (Lotus corniculatus L.); vetches (Vicia spp.); annual and perennial lespedezas (Lespedeza spp.); sweetclovers (Melilotus spp.); and a few tropical species (Knight, 1985).

TEMPERATURE

Optimal Ranges

The range of optimal growth temperatures for forage legumes depends upon species, physiological stage, maturity, and plant part. Fick et al. (1988) concluded that growth rate of alfalfa is greatly reduced by temperatures outside the range 10 to 37°C. Further, that growth of new alfalfa seedlings is most rapid at 20 to 30°C, but that the optimum temperature declines to 15 to 25°C in established seedlings. They noted a similar shift in the optimal temperatures for regrowth shoots. The optimum is 30 to 33°C during the 1st week of regrowth, but drops to 10 to 27°C for older shoots. The upper temperature limit for growth of white clover is near 30°C (Kendall & Stringer, 1985). Maximum root production of white clover occurs at about 8°C below the optimum temperature for shoot growth (Richardson & Syers, 1985).

When harvested at the same harvest date, greatest yields are usually obtained when forages are grown at temperatures near the lower boundary of the optimal range (Fick et al., 1988). The longer period of development, as well as production of a greater number of shoots, at these temperatures compared with higher temperatures probably accounts for the high yield. The maximum number of leaves is produced at about 17°C in white clover (Kendall & Stringer, 1985). In this species, the threshold temperature for leaf appearance is near 3°C, and temperatures over 6°C are necessary for leaf expansion. Also, the rate of appearance of new alfalfa leaves increases with temperature to about 30°C (Fick et al., 1988).

Rate of plant development and maturation also increases with temperature. When grown under hot temperatures, temperate forage legumes tend to be shorter and to bloom earlier than when grown under cool temperatures. This response is part of the reason that forage yields of temperate legumes are lower during hot summers than during cool springs and falls.

Growth temperature also affects the morphology of legumes. Fick et al. (1988) concluded that the leaf-to-stem ratio of alfalfa is greatest at low growth temperatures when measured at a given age, but occurs at intermediate temperatures (15 to 30°C) when measured at a given growth stage. They also noted that high growth temperatures of alfalfa decrease the area of fully expanded leaflets, leaf thickness, and leaflet density. Furthermore, as temperature increases, leaflet shape tends to change from obovate to oblanceolate.

Moreover, the structure and composition of legume stems is modified by temperature. High growth temperatures decrease stem diameter and increase the rate of maturation and rate at which lignin is deposited (Fick et al., 1988; Marten et al., 1988). Because lignin limits digestibility of forages (Buxton & Russell, 1988), forage legumes grown at high temperatures are of lower quality at a given age than forages grown at lower temperatures.

When growth temperatures are outside the optimal range, stress is imposed on forages. Severity of the stress depends upon extent of the temperature deviation from the optimal range, length of the deviation, and maturity and physiological status of the plants. During the past several years, there has been much progress in the use of foliage temperature to quantify plant stress (Burke, 1988). Most of the concepts developed during this time use a comparison of foliage and air temperature to estimate plant stress. Nearly all biological processes are affected by temperature stress. Temperature directly affects the rate of respiration and the rate of cell division is closely related to temperature of the meristems. Temperatures below 15°C severely restrict nodulation of legumes (McKenzie et al., 1988). Stress imposed during an early stage of development may evoke physiological responses and precondition (harden) plants. Preconditioned plants are better able to withstand subsequent stress.

Cold Stress

Effects on Plants

Low-temperatures can cause chilling injury or, by preconditioning plants, can enhance survival during extended periods of subfreezing temperatures. Low-temperature stress may occur under two types of conditions. These are subfreezing temperatures during the growing season and freezing temperatures during the winter. In both situations, the stress effects may be cumulative over time or be catastrophic and result in plant or tissue death (Andrews, 1987). Forage legumes may be subject to low-temperature stress throughout their growth cycle, but more commonly during spring and late summer in the northern USA and during the winter in the southern USA.

Winter injury, as occurs in the northern USA, may result from several conditions beyond the direct effect of freezing temperatures. These are desiccation, flooding, ice encasement, soil heaving, and diseases (Andrews, 1987). Sublethal winter injury decreases plant vigor during the subsequent growing season and lethal winter injury reduces stand and limits the geographical distribution of forage legumes. Where winter temperatures are extreme, survival is dependent on a buffer of snow to insulate the legumes from natural air temperatures. Proper stubble management can increase the depth of snow and improve the likelihood of plant survival.

Plant Adaptations

In the fall, forages usually undergo physiological and morphological changes that increase their tolerance to freezing temperatures. These changes are closely associated with fall dormancy. Indeed, fall dormancy often is used as a measure of alfalfa cold tolerance in the USA (McKenzie et al., 1988). McKenzie et al. (1988) noted that short photoperiods with cool, fluctuating temperatures are essential for initiation of cold tolerance in cold-tolerant plants. Cold-sensitive legumes may lack the photoperiodic sensing mechanism to induce essential physiological changes associated with winter survival. They further state that cold hardening in the field begins in the fall when mean air temperatures are near 10°C and accelerate as temperatures approach 5°C. Freezing temperatures are necessary to attain maximum cold tolerance.

McKenzie et al. (1988) reported that the physiological status and environment of forages may limit or prevent development of cold tolerance. This can occur when nonstructural carbohydrates (TNC) and other reserves are reduced to critical levels, when soils become saturated with water, or when vigorous growth occurs in the fall. A soil moisture level of 40% of field capacity seems to be optimal for alfalfa survival (McKenzie, 1951). Indeed, drought conditions during hardening increase cold tolerance, and winter injury and plant death are often most severe on poorly drained soils.

Alfalfa seedlings are most susceptible to freezing injury until four or five leaves have formed (McKenzie et al., 1988). Cold tolerance usually is greatest for crowns, intermediate for roots, and least for leaves. The crown and many regenerative buds of alfalfa and other winterhardy forages are usually below the soil surface and protected against freezing by snow cover during the winter and by the low thermal conductivity of soil. Among alfalfa species, _M. falcata_ produces more underground buds than _M. sativa_. In birdsfoot trefoil, Empire, a late-maturing, semierect pasture-type cultivar, is more winter hardy than Viking, an erect hay-type cultivar (Grant and Marten, 1985).

Plant hardening to low temperature is related to increases in sugar, protein, lipid, and dry matter concentrations (McKenzie et al., 1988). During freezing, ice forms between the cells of hardened plants. In unhardened plants, water freezes within the cells causing cell rupture and death of the tissues. The process of solute accumulation, mostly sucrose, and fatty acid synthesis are both more active in cold-tolerant cultivars than in cultivars that are cold sensitive. The raised solute concentrations allow more water to be held against the concentration gradient to external cellular ice (Andrews, 1987). Proline concentrations are higher in cold-tolerant cultivars than in cold-sensitive cultivars (McKenzie et al., 1988).

Lipids that accumulate in hardened plants are primarily the polyunsaturated linoleic and linolenic acids, which have lower melting points than saturated fatty acids with an equal number of carbons (McKenzie et al., 1988). Genetic variation for physiological and biochemical functions during low temperatures has been positively correlated with unsaturated/saturated fatty acid ratios (Bartkowski et al., 1977; Clay et al., 1976, Clay et al., 1977). This seems to be a function of fatty acid composition of membranes. Membranes of plant tissues subject to chilling injury undergo a phase transition from liquid-crystalline to a solid gel structure. Crystallization at low temperatures can result in loss of membrane integrity and dysfunction of membrane-bound enzymes (McKenzie et al., 1988). The temperature at which the transition occurs is related to lipid composition. Most respiratory enzymes are associated with mitochondrial membranes, and are, therefore, subject to phase changes in the lipid membrane as temperatures decrease (Andrew, 1987). Selection of lines with high unsaturated/saturated fatty acid ratios may result in more cold-tolerant forage legumes.

Temperature fluctuations that interfere with or undo the hardening process during the fall and winter also cause stand loss (Marten, 1985). Dehardening of alfalfa may occur if temperatures rise above 10°C for a few days during the fall hardening period (McKenzie et al., 1988). Cold-sensitive cultivars deharden more rapidly and initiate growth faster during periods of warm weather than cold-tolerant cultivars.

Ice Encasement

Flooding of overwintering forage legumes may occur with the eventual formation of ice sheets. Furthermore, ice encasement can form over plants during rain or during mid-winter thaws followed by freezing temperatures. Injury to plants is caused by restricted gas exchange. These anaerobic conditions promote fermentation resulting in accumulation of ethanol and lactic acid, which are toxic to plants (Andrews, 1987).

The movement of upper soil layers resulting from the formation of ice layers in the soil is known as frost heave (Andrews, 1987; Van Keuren, 1988). Plant crowns and surface soil are forced upwards by the expanding ice. During thaws, the soil returns to its original level, frequently leaving plants with exposed crowns and broken roots. These plants usually desiccate and die before new root systems can be formed. Frost heave can be a major problem when snow cover is low and soils are of fine texture and not well drained. Van Keuren (1988) concluded that leaving plant foliage, including leaves, for winter protection is a dependable method for protecting alfalfa from frost heave. This may be most important in areas where adequate snow cover does not occur to insulate forages from the cold. Prostrate-type plants or those with a branched, creeping rooted habit survive heaving better than plants with a taproot system (McCloud & Bula, 1985). Grass grown in association with legumes often reduces the adverse effects of frost heave on legumes. Birdsfoot trefoil roots can produce new shoots, and segments taken below the crown will develop shoots and roots. This characteristic may aid plant survival during frost heave (Grant & Marten, 1985).

Heat Stress

Effects on Plants

As noted by McKenzie et al.(1988), direct heat injury occurs only when forage legumes are exposed to temperatures considerably greater than the maximal temperature for growth. For example, extremely high temperatures when soil is water saturated can cause "scalding" and plant death in the southwestern USA (Sheaffer et al., 1988). More common, however, is indirect heat injury, which occurs when temperatures are near maximum for growth for extended periods. This type of plant stress is common during hot summers in the southern USA. High-temperature stress frequently occurs concurrently with water stress and, thus, it is difficult to separate the two effects.

Plant metabolic dysfunctions at high temperatures are caused by changes in rate of enzyme catalyzed reactions (McKenzie et al., 1988). Net photosynthesis is one of the most heat-sensitive aspects of growth at high temperatures. In temperate plants, which lose CO_2 to photorespiration as well as dark respiration, plants may die of starvation when temperatures are above the temperature compensation point. Forage legumes grown under heat stress maintain only low levels of TNC (Feltner & Massengale, 1965) and have low rates of N_2 fixation (McKenzie et al., 1988). As with cold tolerance, there is evidence that plant membranes are sites of heat injury.

Plant Adaptations

High-temperature injury is less severe when temperatures are raised gradually over several days than when there is an abrupt change to hot weather (McKenzie et al., 1988). Plants adapt to high temperatures by producing heat-shock proteins at temperatures above those for normal growth. This hardening process is associated with very slow growth and low photosynthetic rates. Because of these slow rates, heat hardening is advantageous only during high-temperature stress. The hardening response is lost in plants after they return to optimal growth conditions. McKenzie et al. (1988) caution that heat tolerance of species or cultivars should be compared only when plants are in their hardened state.

As further noted by McKenzie et al. (1988), plants have several mechanisms for reducing energy loads and, thus, tissue temperatures. Morphological adaptations influence reflectance of radiant energy and variation in leaf size and angle of inclination. Plants also regulate their temperature through convection, radiation, and transpiration.

WATER RELATIONS

Excess water on poorly drained soils, particularly in the spring, and lack of water in the the summer are major constraints to forage-legume yield and persistence in many regions of the USA. Water relations in alfalfa are covered in detail by Sheaffer et al. (1988). They note that maximum daily water use is about 5 to 11 mm. Seasonal rates range from 400 mm in the northern USA to nearly 2000 mm in the arid southwestern USA. In many arid regions, especially in the western USA, water is supplied through irrigation. Proper water management is necessary for maximum production and stand persistence, because deficit irrigation causes reduced yield and excessive irrigation causes stand loss (Marten, 1985; Wright, 1988).

Excess Water

Poorly drained soils in high-rainfall or humid regions of the USA create an unfavorable environment for persistence of forage legumes. Short periods of excessive rainfall that result in temporary waterlogged conditions are common in these regions as well as in other areas. Cultivars vary in their tolerance of these conditions. Fick et al. (1988) reviewed research that showed that waterlogged conditions reduce stems per plant and decrease growth of forage legumes with the greatest reduction occurring at high temperatures. Flooding of the soil stopped root growth almost immediately, and prolonged flooding caused deterioration of the root system. Reduced growth occurs not only because of anaerobic conditions in the root zone, but also because waterlogged soil allows development of fungal diseases, particularly damping off (Pythium) and root rot (Phytophthora). A number of cultivars are resistant to Phytophthora.

Direct injury to plants from waterlogged soil occurs because of the anaerobic conditions as discussed for ice encrustment (Barta, 1987). Heinrichs (1970) reported that birdsfoot trefoil and white clover were the most tolerant forage legumes tested on waterlogged conditions, red clover and alfalfa were intermediate, and sweetclover and cicer milkvetch (Astragalus cicer L.) were the most sensitive.

Water Deficit

When transpiration exceeds water absorption, stress usually occurs, which adversely affects most physiological processess and growth of forages. Photosynthetic rates are usually affected less by drought than rate of respiration and growth, causing an increase in TNC levels (Kendall & Stringer, 1985). If the stress is severe or prolonged, plants desiccate and die. Young forage legumes are extremely vulnerable to dry conditions because of their limited root system. Growth rates decrease linearly as soil water is reduced below field capacity (Hattendorf et al., 1988). Water potentials required to limit growth of forage species can be variable because of osmotic adjustment as plants adapt to drought conditions (McCloud & Bula, 1985). Nevertheless, Brown & Tanner (1983) concluded that the critical water potential for leaf growth of alfalfa is -1.0 MPa.

Sheaffer et al. (1988) outlined several changes that occur as forage legumes adapt to arid conditions. Legumes grown under stress have less leaf area than nonstressed plants because the stressed plants produce smaller cells and cup their leaves to reduce energy load and water loss. Several morphological and anatomical traits are associated with low water use and high water-use efficiency in legumes. These include leaf pubescence and multiple layers of palisade cells in the leaf. Shoot removal reduces drought resistance of forage legumes (Kendall & Stringer, 1985).

Effects on Plants

Water deficit during early development reduces the number of basal buds, stems per plant, stem diameter, and internode number and length as well as leaf size (Fick et al., 1988; Sheaffer et al., 1988). Forage legumes grown under drought often are more digestible than those grown with adequate water and harvested at the same time. Much of the reason is because water deficit delays plant maturation, resulting in a slower reduction in the leaf-to-stem ratio with aging and a slower rate of secondary cell-wall formation of stems with accompanying lignification (Halim et al., 1989).

The effect of water deficit may differ if the stress is

applied late. Late-applied stress sometimes hastens maturation of perennial forages (Wilson & Ng, 1975). Also, droughts usually coincide with high temperatures, which speed maturation of forages (Vough & Marten, 1971).

Water deficits depress N_2 fixation. The principal reason may be because of lower photosynthetic rates during drought. Alfalfa cultivars vary in their ability to fix nitrogen during drought (Sheaffer et al., 1988). Effects of water deficit on crude-protein concentration of herbage have been inconsistent. This may be because of differential effects of water deficit on N_2 fixation and on growth of leaves and stems. Halim et al. (1989) found that water deficit caused an increase in crude protein concentration of alfalfa stems, whereas it caused a decrease in leaves. They speculated that the decline in leaves resulted from leaf senescence and translocation of amino acids to other plant parts.

Water deficit accelerates the rate of senescence of older leaves. During severe water deficits, leaves abscise, reducing both the leaf-to-stem ratio and forage quality. If the stress occurs only at the vegetative or bud stages, plants may recover and return to normal leaf-to-stem ratios with little loss in yield. If the stress occurs only at the flower stage, however, there is little time for recovery before harvest and forage quality can be lowered (Halim, 1986).

Resistance to Drought

Resistance of plants to drought occurs because of drought avoidance, drought tolerance, or both. Most perennial legumes can survive at least short periods of intense drought by becoming dormant. There is little direct evidence of drought tolerance in forage legumes (Kendall & Springer, 1985). In some species, drought is avoided or minimized by extracting water from deep in the soil profile. Alfalfa is particularly suited for this because of its potential for developing a deep root system (Sheaffer et al., 1988). As a result, alfalfa may continue to grow long after other legumes and most grasses have become dormant.

On the other hand, white clover is a shallow-rooted legume, especially those plants that are more than 1 to 2 yr old and that have lost their taproot through disease. These plants depend upon the shallower fibrous root system developed at stolon nodes (Metcalfe & Nelson, 1985). This characteristic makes white clover particularly susceptible to drought and often unproductive during the summer in much of the USA. Thus, its contribution to pastures is much greater during wet periods in the spring or winter, or under irrigation (Marten, 1985). Because it is of greater nutritive quality than most other forage legumes (Buxton et al., 1985; Buxton & Hornstein, 1986), improvement of its productivity and persistence under limited water environments could greatly improve pasture and livestock productivity in much of the USA.

Canopy Temperature

Water deficits cause stomatal closure, reduced transpiration, and elevated canopy temperature. Canopy temperature is a sensitive indicator of stresses placed on plants in response to soil-water stress, disease, or other environmental stresses (Temple & Benoit, 1988). If sufficient water deficits develop in transpiring plants such that stomates begin to close, the energy formerly used in transpiration heats the leaves and is partitioned into convective and radiated heat loss. Because canopy temperature - ambient air temperature differentials ($T_c - T_a$) are functions of transpiration rates, factors that reduce transpiration also reduce $T_c - T_a$. The increase in $T_c - T_a$ from decreased plant-water potential is linear to -3.0 MPa (Ehrler et al., 1978). Infrared thermometry is a practical method for determining canopy temperatures because it is nondestructive and rapid. Indeed, it may be practical to use remotely sensed data from satellites to estimate crop stresses (Maas, 1988).

One canopy-temperature-based stress indicator is the Crop Water Stress Index (CWSI). The CWSI is $T_c - T_a$ normalized for vapor pressure deficit. A CWSI of 0 indicates that the crop is not stressed and transpiring at its potential rate. A CWSI of 1.0 indicates that water stress is maximal and that transpiration has ceased. Hattendorf et al. (1988) found that alfalfa yields were exponential functions of CWSI. Even a mild stress (CWSI = 0.05) caused yield reductions of nearly 10%. Short-term stresses with average CWSIs of 0.3 caused leaf drop and lowered yields by about one-third (Halim, 1986).

IRRADIANCE

Photoperiod

Photoperiod varies with latitude and season. Minimum variation occurs in the southern USA and maximum variation occurs in the northern USA, particularly in Alaska where photoperiods approach 24 h during the summer. Under long photoperiods, leaf and stem growth tends to be erect, whereas under short photoperiods, growth tends to be prostrate and axillary, or adventitious bud activity increases (McCloud & Bula, 1985). Temperate perennial legumes, especially those adapted to the northern USA, usually are long-day plants and flower during the long photoperiods of summer (Marten, 1985).

Shade Response

As discussed by Trott et al. (1988), when water and nutrients are adequate, photosynthetic photon flux density becomes the major factor limiting plant growth. Short plants are shaded by tall plants within the canopy, and lower leaves are shaded by both upper leaves and upper main stems. Most forage legumes are relatively intolerant of shade. The rate of photosynthesis in individual leaves is saturated at somewhat less than maximum daylight. Nearly full sunlight is required for maximum relative growth rates (Kendall & Stringer, 1985). Common adaptations to reduced irradiance include reduced leaf thickness, specific-leaf weight, and root growth and increased shoot-to-root ratio, leaf-area ratio and stem length (Cooper, 1967; Cooper & Qualls, 1967; Dennis & Woledge, 1983). The increase in shoot-to-root ratio from shading can persist in the following production year after the shade stress is removed (Buxton & Wedin 1970a, b)

Adaptations to shading serve to maximize productivity and persistence within the constraints of the environment. Large, thin leaves possess a greater capacity for irradiance interception. Shading may induce stem elongation at the expense of root growth and stem girth so that leaves are lifted to higher elevations where they can be exposed to the sunlight. Also, as shoot length increases, shading of the lower leaves becomes an increasing limitation to alfalfa production because of leaf senescence (Marten, 1985).

Moderate shading can enhance stolon growth and result in greater persistence of white clover. This benefit of shading has been attributed to reduced temperatures at the soil surface rather than enhanced photosynthetic activity (Kendall & Stringer, 1985).

SOIL CHARACTERISTICS

Edaphic limitations to yield and persistence of forage legumes are associated with the capacity of the soil to supply water and nutrients. Ideal characteristics are deep and well-drained soils with high water-holding capacity, pH of 6.0 to 7.0, high nutrient supply, and no chemical or physical impediments to root penetration. Most soils do not have all of these characteristics, but many can be amended chemically before establishing forages (Burns, 1985).

In regions where dry periods frequently occur, high water-holding capacity is critical for forage legume growth and persistence during dry periods. The water-holding capacity of soil has a great influence on both forage yield and on the species and cultivars that can be grown. Species adapted to soils with low water holding capacity are usually early maturing and have low productivity (Helyar, 1985).

Soil pH

Large areas in the humid region of the USA have acid soils. These soils were acidified by leaching during rainfall with a mild acid, such as carbonic acid. This caused cations, mostly Ca, Mg, K, and Na, to be dissolved and replaced in the soil by Al and Mn. Large concentrations of soluble Al and deficiencies of soluble Ca often occur in these soils, which have deleterious effects on root growth (Ulrich & Sumner, 1988).

Acid subsoils form a chemical barrier to deep root penetration. This restriction is manifested in increased sensitivity to drought and reduced productivity and persistence of legumes. The best remedy is deep lime incorporation or the surface application of gypsum (Rechcigl et al., 1988; Sumner et al., 1986). The time required to modify soil pH depends on initial soil pH, liming material, and amount of mixing of the soil. In most situations, liming materials should be applied 26 wk or more before seeding (Lanyon & Griffith, 1988).

As Keeney (1985) pointed out, forage legumes grown in the USA generally are less adapted to acid soils than are grasses. A soil pH of 6.6 to 7.5 usually is recommended for alfalfa production (Lanyon & Griffith, 1988). Clovers and birdsfoot trefoil are slightly more tolerant to soil acidity, but grow best when the pH is 6.2 or higher (Grant & Marten, 1985).

Keeney (1985) noted that legume species that are more tolerant to acid soils have more efficient P uptake and transport mechanisms, are able to decrease the acidity of the rhizosphere, and are more tolerant to high Mn, low Ca, and low Mo availability than are acid-intolerant species. Genetic variation for tolerance to acid soils has been identified in alfalfa (Buss et al., 1975; Bouton & Sumner, 1983), but only limited effort has been expended to develop cultivars of legumes capable of growing and persisting in acid soils.

In arid regions of the USA, soils are often alkaline and occasionally high in salinity. High salinity can markedly limit legume growth, although variation in tolerance among and within legume species exists. Alfalfa is more tolerant of salinity than are most other forage legumes.

Soil Nutrients

Legumes have less-fibrous root systems than grasses and are less efficient in competing for nutrients located in the upper soil profile. As a result, legumes require greater soil concentrations of nonmobile nutrients than grasses. The K requirement of forage legumes is greater than that of other nutrients. Stand loss and invasion by weeds and grasses often occur with inadequate soil K (Lanyon & Griffith, 1988). Likewise, the low mobility of P in soils gives a competitive advantage to grasses over legumes for this nutrient. Deficiencies of S occur for legumes in much of the USA, but are most common in sandy soils and where manure has not been applied recently (Keeney, 1985).

RESEARCH NEEDS

Burton (1986) emphasized that persistence is probably the best test of forage dependabilty. Forages that fail because of stress are costly to users as stand failure may cause farmers to sell animals. He further stated that lack of persistence is the greatest weakness of most of the legumes that have been tested in the USA. Increasing the tolerance of legumes to environmental stresses will improve their dependability. The potential benefits to agriculture in the USA are enormous because of the importance of forages to the economy.

At the last trilateral workshop, Burns (1985) concluded that the effort in legume germplasm introduction and enhancement was inadequate and called for more effort to develop pest resistance in perennial legumes adapted to hot summers. Marten (1985) called for more research on the influence of climatic factors on basic physiological processes of forages. These needs still exist as little additional effort has been devoted to them. With the prospect of climatic stress increasing in the future, increased effort is critical.

Development of cultivars that are resistant to temperature and water stress and tolerant to acid soils is clearly needed. More tolerance to these stresses will allow forage legumes to be grown in marginal environments where they are not now adapted. New cultivars capable of withstanding adverse winters have already allowed forages to be grown in more severe winter climates. Little progress has been made, however, toward increasing frost heave or ice sheet tolerance of forage cultivars. Less critical seems to be the need for improvement in tolerance to shading and nutrient availability.

Progress has been slow thus far, emphasizing the need to better understand the basic physiological and morphological responses to stress. Additional effort is needed to define specific factors associated with stress resistance so that these plant characteristics can be altered as needed and used as selection criteria where possible.

REFERENCES

Andrews, C.J. 1987. Low-temperature stress in field and forage crop production - an overview. Can. J. Plant Sci. 67:1121-1133.

Barta, A.L. 1987. Supply and partitioning of assimilates to roots of Medicago sativa L. and Lotus corniculatus L. under anoxia. Plant, Cell Environ. 10:151-156.

Bartkowski, E.J., D.R. Buxton, F.R.H. Katterman, and H.W. Kircher. 1977. Dry seed fatty acid composition and seedling emergence of pima cotton at low soil temperatures. Agron. J. 69:37-40.

Bouton, J.H., and M.E. Sumner. 1983. Alfalfa (Medicago sativa L.) in highly weathered acid soils. I. Field performance of alfalfa selected for acid tolerance. Plant Soil 74:430-436.

Brown, P.W., and C.B Tanner. 1983. Alfalfa stem and leaf growth during water stress. Agron. J. 75:799-804.

Burke, J.J., J.R. Maham, and J.L. Hatfield. 1988. Crop-specific thermal kinetic windows in relation to wheat and cotton biomass production. Agron. J. 80:553-556.

Burns, J.C. 1985. Environmental and management limitations of legume-based forage systems in the southern United States. p.129-137. In R.F Barnes et al. (ed.) Forage legumes for energy-efficient animal production. Proc. Trilateral Workshop, Palmerston North, NZ. 30 Apr.-4 May 1984. USDA-ARS Natl. Tech. Info. Ser., Springfield, VA.

Burton, G.W. 1986. Forages for the future. p.1-6. In Proc. Forage Grassl. Conf., Athens, GA. 15-17 Apr. 1986. Am. Forage Grassl. Counc., Lexington, KY.

Buss, G.A., J.A. Lutz, Jr., and G.W. Hawkings. 1975. Effect of soil pH and plant genotype on elemental concentration and uptake by alfalfa. Crop Sci. 15:614-617.

Buxton, D.R., and J.S. Hornstein. 1986. Cell-wall concentration and components in stratified canopies of alfalfa, birdsfoot trefoil, and red clover. Crop Sci. 26:180-184.

Buxton, D.R., J.S. Hornstein, W.F. Wedin, and G.C. Marten. 1985. Forage quality in stratified canopies of alfalfa, birdsfoot trefoil, and red clover. Crop Sci. 25:273-279.

Buxton, D.R., and J.R. Russell. 1988. Lignin constituents and cell-wall digestibility of grass and legume stems. Crop Sci. 28:553-558.

Buxton, D.R., J.R. Russell, and W.F. Wedin. 1987. Structural neutral sugars in legume and grass stems in relation to digestibility. Crop Sci. 27:1279-1285.

Buxton, D.R., and W.F. Wedin. 1970a. Establishment of perennial forages. I. Subsequent yields. Agron. J. 62:93-97.

Buxton, D.R., and W.F. Wedin. 1970b. Establishment of perennial forages. II. Subsequent root development. Agron. J. 62:97-100.

Clay, W.F., E.J.Bartkowski, and F.R.H. Katterman. 1976. Nuclear deoxyribonucleic acid metabolism and membrane fatty acid content related to chilling resistance in germination cotton (Gossypium barbadense). Physiol. Plant. 38:171-175.

Clay, W.F., D.R. Buxton, and F.R.H. Katterman. 1977. Cottonseed germination related to DNA synthesis following chilling stress. Crop Sci. 17:342-344.

Cooper, C.S. 1967. Relative growth of alfalfa and birdsfoot trefoil seedlings under low light intensity. Crop Sci. 7:176-178.

Cooper, C.S., and M. Qualls. 1967. Morphology and chlorophyll content of shade and sun leaves of two legumes. Crop Sci. 7:66-72.

Dennis, W.D., and J. Woledge. 1983. The effect of shade during leaf expansion on photosynthesis by white clover leaves. Ann. Bot. 51:111-118.

Ehrler, W.L., S.B. Idso, R.D. Jackson, and R.J. Reginato. 1978. Wheat canopy temperatures: relation to plant water potential. Agron. J. 70:251-256.

Feltner, K.C., and M.A. Massengale. 1965. Influence of temperature and harvest management on growth, level of carbohydrates in the roots, and survival of alfalfa (Medicago sativa L.). Crop Sci. 5:585-588.

Fick, G.W., D.A. Holt, and D.G. Lugg. 1988. Environmental physiology and crop growth. In A.A. Hanson, D.K. Barnes, and R.R. Hill, Jr. (ed.) Alfalfa and alfalfa improvement. Agronomy 29:163-194.

Grant, W.F., and G.C. Marten. 1985. Birdsfoot trefoil. p.98-108. In M.E. Heath et al. (ed.) Forages, the science of grassland agriculture. Iowa State University Press, Ames.

Hattendorf, M.J., R.E. Carlson, R.A. Halim, and D.R. Buxton. 1988. Crop water stress index and yield of water-deficit-stressed alfalfa. Agron. J. 80:871-875.

Halim, R.A. 1986. Water-stress effects on forage quality of alfalfa. Ph.D. diss. Iowa State Univ., Ames.

Halim, R.A., D.R. Buxton, M.J. Hattendorf, and R.E. Carlson. 1989. Water-stress effects on alfalfa forage quality after adjustment for maturity differences. Agron. J. 81:189-194.

Hanson, J., D. Johnson, A. Lacis, S. Lebedeff, P. Lee. D. Rind, and G. Russell. 1981. Climate impact of increasing carbon dioxide. Science 213:957-966.

Heinrichs, D.H. 1970. Flooding tolerance of legumes. Can. J. Plant Sci. 50:435-438.

Helyar, K.R. 1985. Edaphic limitations and soil nutrient requirements of legume-based forage systems in tropical areas. p.82-88. In R.F Barnes et al. (ed.) Forage legumes for energy-efficient animal production. Proc. Trilateral Workshop, Palmerston North, NZ. 30 Apr.-4 May 1984. USDA-ARS Natl. Tech. Info. Ser., Springfield, VA.

Keeney, D. 1985. Edaphic limitations and soil nutrient requirements of legume-based forage systems in the temperate United States. p.95-100. In R.F Barnes et al. (ed.) Forage legumes for energy-efficient animal production. Proc. Trilateral Workshop, Palmerston North, New Zealand. 30 Apr.-4 May 1984. USDA-ARS Natl. Tech. Info. Serv., Springfield, VA.

Kendall, W.A., and W.C. Stringer. 1985. Physiological aspects of clover. In N.L. Taylor (ed.) Clover science and technology. Agronomy 25:111-159.

Knight, W.E. 1985. The distribution and use of forage legumes in the United States. p.34-39. In R.F Barnes et al. (ed.) Forage legumes for energy-efficient animal production. Proc.

Trilateral Workshop, Palmerston North, NZ. 30 Apr.-4 May 1984. USDA-ARS Natl. Tech. Info. Serv., Springfield, VA.
Lanyon, L.E., and W.K. Griffith. 1988. Nutrition and fertilizer use. In A.A. Hansen et al. (ed.) Alfalfa and alfalfa improvement. Agronomy 29:333-372.
Maas, S.J. 1988. Using satellite data to improve model estimates of crop yield. Agron. J. 80:655-662.
Marten, G.C. 1985. Environmental and management limitations of legume-based forage systems in the northern United States. p.116-128. In R.F Barnes (ed.) Forage legumes for energy-efficient animal production. Proc. Trilateral Workshop, Palmerston North, NZ. 30 Apr.-4 May 1984. USDA-ARS Natl. Tech. Info. Serv., Springfield, VA.
Marten, G.C., D.R. Buxton, and R. F Barnes. 1988. Feeding value (forage quality). In A.A. Hansen et al. (ed.) Alfalfa and alfalfa improvement. Agronomy 29:463-491.
McCloud, D.E., and R.J. Bula. 1985. Climatic factors in forage production. p.33-42. In M.E. Heath et al. (ed.) Forages, the science of grassland agriculture. Iowa State University Press, Ames.
McKenzie, R.E. 1951. The ability of forage plants to survive early spring flooding. Sci. Agric. 31:358-367.
McKenzie, J.S., Roger Paquin, and S.H. Duke. 1988. Cold and heat tolerance. In A.A. Hanson et al. (ed.) Alfalfa and alfalfa improvement. Agronomy 29:259-302.
Metcalfe, D.S., and C.J. Nelson. 1985. The botany of grasses and legumes. p.52-63. In M.E. Heath et al. (ed.) Forages, the science of grassland agriculture. Iowa State University Press, Ames.
Rechcigl, J.E., K.L. Edmisten, D.D. Wolf, and R.B. Reneau, Jr. 1988. Response of alfalfa on acid soil to different chemical amendments. Agron. J. 80:515-518.
Richardson, A.C., and J.K. Syers. 1985. Edaphic limitations and soil nutrient requirements of legume-based forage systems in temperate regions of New Zealand. p.89-94. In R.F Barnes et al. (ed.) Forage legumes for energy-efficient animal production. Proc. Trilateral Workshop, Palmerston North, NZ. 30 Apr.-4 May 1984. USDA-ARS Natl. Tech. Info. Serv., Springfield, VA.
Sheaffer, C.C., C.B. Tanner, and M.B. Kirkham. 1988. Alfalfa water relations and irrigation. In A.A. Hanson et al. (ed.) Alfalfa and alfalfa improvement. Agronomy 29:373-409.
Sumner, M.E., H. Shahanden, J. Bouton, and J. Hammel. 1986. Amelioration of an acid soil profile through deep liming and surface application of gypsum. Soil Sci. Soc. Am. J. 50:1254-1258.
Temple, P.J., and L.F. Benoit. 1988. Effects of ozone and water stress on canopy temperature, water use, and water use efficiency of alfalfa. Agron. J. 80:439-447.
Thompson, L.M. 1988. Effects of changes in climate and weather variability on the yields of corn and soybeans. J. Prod. Agric. 1:20-27.
Trott, J.O., K.J. Moore, V.L. Lechtenberg, and K.D. Johnson. 1988. Light penetration through tall fescue in relation to canopy biomass. J. Prod. Agric. 1:137-140.
Ulrich, B., and M.E. Sumner. 1988. Soil acidity. Springer-Verlag, New York.
Van Keuren, R.W. 1988. Frost heave of alfalfa as affected by

harvest schedule. Agron. J. 80:626-631.
Vough, L.R., and G.C. Marten. 1971. Influence of soil moisture and ambient temperature on yield and quality of alfalfa forage. Agron. J. 63:40-42.
Wilson, J.R., and T.T. Ng. 1975. Influence of water stress on parameters associated with herbage quality of Panicum maxium var tricholume. Aust. J. Agric. Res. 26:127-136.
Wright, J.L. 1988. Daily and seasonal evapotranspiration and yield of irrigated alfalfa in southern Idaho. Agron. J. 80:662-669.

DISCUSSION

R. Smith. In your presentation you had a slide showing death of alfalfa over a strip associated with fall flooding. Was this actual flooding or ice sheet damage? It would appear to be typical ice damage in the North Central USA.

Buxton. The flooding occurred in the early fall on a well-drained soil and was not associated with ice sheet formation and damage. There is evidence that high soil water is associated with low-temperature injury of forage legumes.

Irwin. Possible changes in climate, such as a warming trend, will have a profound influence on the severity of pest and diseases that occur on any plant species. For example, increased stress from leafhoppers has occurred on alfalfa this year. Also, resistance levels would change depending upon temperature, etc.

Woomer. To what extent do you expect that the warming trend can be compensated for by the use of cultivars from slightly warmer areas such as southern adapted cultivars moved to the northern USA?

Buxton. It is possible that this will help to compensate and may happen slowly over time, but the extent to which it can help is largely unknown.

Hoveland. With respect to climatic changes and forage adaptation, we have experienced a 3-yr drought with higher than normal temperatures in the southeastern USA. Pastures in the southern USA often contain both warm- and cool-season species. During the past 3 yr, there has been a shift from the cool- to warm-season species on the more stressful soils. Weak legumes such as white clover have disappeared. The effects of climatic changes will be more noticeable in the transition zones.

Matches. What do you see as being done in plant breeding to improve legume drought tolerance or drought resistance?

Buxton. Drought tolerance rather than drought resistance may be more important for improving forage legumes. Deep rooting ability for drought avoidance is important and the potential for improving this plant characteristic needs more attentation.

Sheaffer. Pubescent alfalfa developed in Kansas has demonstrated improved water use efficiency.

RHIZOBIAL ECOLOGY IN TROPICAL PASTURE SYSTEMS

Paul Woomer and B. Ben Bohlool

SUMMARY

We have studied the ecology of rhizobia in the diverse climates and vegetation of Maui, Hawaiian Islands. Many of the sites represent predominant tropical pasture systems including short and tall grass savannas, and upland and lowland grassy pastures. The population size of the rhizobia in the soils and rhizospheres of these pastures is correlated with environmental factors. The importance of the legume component on rhizobial populations is established. The frequency of observing both associated legumes and rhizobia at a site is .94. The density of legumes correlates significantly with the density of rhizobia in soils. Conceptual and regression models are presented which predict the population size of rhizobia in tropical pasture systems and which describe the environmental influences of native and introduced rhizobia in the soils supporting host and non-host swards. Also presented is an update of the new rhizobial taxonomic system relevant to the interest of pasture scientists.

INTRODUCTION

An important benefit derived from the presence of legumes in pasture systems is the N contributed through biological N_2 fixation (BNF). This occurs in association with the rhizobial microsymbiont which infects the root systems of legumes and results in the formation of root nodules, the site of BNF. The success of the root nodule symbiosis in many pasture systems is largely dependent on the ability of rhizobia to persist in fallow soils and within the rhizospheres of non-host plants during periods of adverse soil conditions. Such factors as moisture stress, high temperatures (Marshall, 1964), and soil chemical factors such as low pH and the accompanying toxic levels of Al, Fe or Mn oxides (see Lowendorf, 1980)all contribute to the fate of rhizobia in soils. A greater understanding of the ecology of rhizobia in soil will contribute to better utilization of the legume symbiosis in pasture improvement efforts.

CHANGES IN RHIZOBIAL TAXONOMY

A discussion of recent changes in rhizobial taxonomy is useful when reviewing literature concerning the legume symbiosis and when ordering cultures from germplasm collections. Recent revisions in the taxonomy of the *Rhizobiaceae* have split the old genus *Rhizobium* into two genera, *Rhizobium* and *Bradyrhizobium* (Jordan, 1984). *Rhizobium* spp. are the fast-growing, acid-producing root nodule bacteria associated with many temperate pasture legumes (e.g., *Trifolium*, *Medicago*, *Lupinus*, and *Vicia* spp.). *Bradyrhizobium* spp. are slower growing, alkali-producing, bacteria associated with soybean (*Bradyrhizobium japonicum*) and many tropical legumes such as siratro, centrosema, and stylosanthes (*Bradyrhizobium* sp.). Rhizobia is the plural of both *Rhizobium* and *Bradyrhizobium*. Rhizobia that nodulate *Trifolium* spp., *Phaseolus vulgaris* and

Vicia/Pisum/Lathyrus spp. has been consolidated into one species, R. leguminosarum, and assigned biovar status, bvs. trifolii, phaseoli, viceae, respectively. Rhizobium meliloti is maintained as the group of rhizobia that nodulate the legume genera Melilotus, Medicago, and Trigonella.

Two new Rhizobium spp. have been identified in this decade, R. loti and R. fredii. Previously, two groups of rhizobia were recognized as associating with Lotus spp. and other genera (e.g., Lupinus, Cicer, Astragalus, and Ornithopus). The fast-growing, acid-producing strains associated with these legumes are now classified as R. loti (Jarvis et al., 1982). The slow-growing isolates associated with these species are grouped into Bradyrhizobium sp. A recently identified species from China is Rhizobium fredii (Keyser et al., 1982; Scholla & Elkan, 1984) which in nature is associated with the primitive ancestor of soybean, Glycine soya (formerly G. usuriensis). This species also nodulates many Chinese and Western cultivars of commercial soybean. Leucaena, Sesbania, and Sophora spp. (to mention a few) are legumes which are associated with Rhizobium sp. It is customary when using either Bradyrhizobium sp. or Rhizobium sp. to identify in parenthesis the host from which the bacterium has been isolated [e.g., Bradyrhizobium sp. (Macroptilium atropurpureum) or Rhizobium sp. (Leucaena leucocephala)].

MAJOR EVENTS IN RHIZOBIAL ECOLOGY

The ability of rhizobia to persist in the absence of its legume host is crucial to successful establishment and persistence of legumes in pasture systems. A conceptualized account of the critical stages of rhizobial ecology in an annual pasture system is presented in Fig. 1.

In the "saprophytic" phase of the cycle, rhizobia may persist in the absence of the legume hosts either in the bulk soil, the decaying root nodules from previous symbiosis, or in the rhizospheres of non-host plants (Bohlool et al., 1983; Parker et

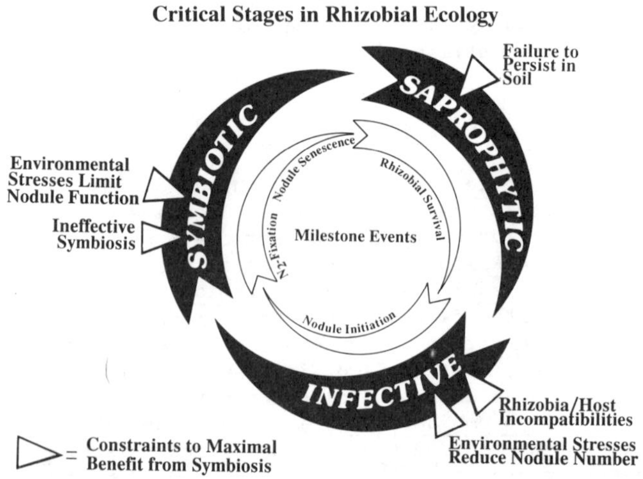

Fig. 1. The critical stages of rhizobial ecology in pastures.

al., 1977). The ability of rhizobia to persist in pastures in the absence of host legumes is also crucial to the BNF potential of pasture systems in succeeding years, because it is neither cost-effective nor practical to reinoculate pastures with rhizobia every year.

The "infective" phase in the life cycle of rhizobia represent a series of events involving both partners. The infection process is perhaps the most sensitive to stresses of the environment among the stages of the life cycle based on the results of Munns et al. (1977) with soil acidity and Singleton and Bohlool (1983) with salinity. Further, as highlighted by the subterranean clover experience in Australia (Brockwell et al., 1968), the compatibility of the rhizobia and the legume will have a profound influence of N input fron BNF into a pasture system. The "symbiotic" phase of the cycle refers to the functioning of the rhizobia inside the nodule. After infection of the cortical cells, rhizobia and the plant cells proliferate rapidly which, within a few days, result in the appearance of a visible nodule structure. The nodule will continue to increase in size and at some point, which varies with different legumes, will begin expressing the nitrogen-fixing property. The "symbiotic" phase is also subject to environmental factors that affect the root and the shoot of the host plant. Interstrain competition is an important aspect of rhizobial ecology (see Bohlool et al., 1986). It becomes of practical importance when production sites contain rhizobia that are inferior in nitrogen-fixing ability to selected strains in the inoculum. Often, the native rhizobia present a competitive barrier to the successful establishment of introduced organisms.

ABUNDANCE OF NATIVE RHIZOBIA IN TROPICAL PASTURE SYSTEMS

The effects of specific soil stresses on the survival of rhizobia in soil has been reviewed in detail (see Lowendorf, 1980). These and similar studies have been useful in identifying which components of the environment regulate the establishment of rhizobia in nature and which rhizobial species and strains are tolerant of specific stress conditions. The environmental influences on rhizobial diversity and abundance in grassland soils was examined by Woomer et al. (1988a). The density of host legumes and soil moisture was found to play significant roles in determining the number and species of rhizobia in soils. The species of rhizobia which are present correspond with their host legumes 94% of the time. A rhizobial species was seldom absent when its host legume was a component of the pasture, and vice versa.

The predominant pasture systems on Maui are presented in Table 1. From a west to east direction, as the mean annual rainfall increases from 350 mm/yr to 1870 mm/yr, the pH of these soils decreases from 7.9 to 4.6. The total rhizobia in the soil increases from 1.1 ($\log_{10}$) g^{-1} soil in semi-arid Torroxic Haplustolls to 4.6 ($\log_{10}$) g^{-1} soil in Homixic Tropohumults. These measurements of rhizobial density were conducted using plant infection counts (Woomer et al., 1988a) on six legume species from different cross-inoculation groups with non-rhizosphere soil from diverse sites. The data presented illustrate the importance of soil moisture as a regulator of rhizobial density and the

Table 1. Rhizobial densities in the soils of different pastures on Maui, Hawaiian Islands.

Pasture	Precipitation	pH	Total rhizobia/g soil
	mm/yr		($\log_{10}$)
Cenchrus ciliaris with isolated deciduous leguminous trees	322-380	6.8-7.9	1.1-2.8
Panicum maximum with leguminous forbs, shrubs, and trees	565	6.8	3.4
Pennisetum clandestinum with leguminous forbs	846-1800	5.2-6.8	3.8-4.8
Digitaria decumbens with leguminous forbs and isolated trees	1875	4.6	4.5

acclimatization of native rhizobial populations to prevailing soil acidity conditions. The data from these and other sites were used to generate regression models predicting the abundance of native rhizobia (Table 2). The models illustrate that the density of legumes in the plant community account for 75% of the variation observed between the sites. The model which includes a legume density component (% legume cover), an indirect measurement of soil moisture status (MAR), and a measurement of soil fertility (total extractable bases) accounts for 95% of the observed variation in the number of native rhizobia. The total extractable bases of the soils covaried with other soil physical and chemical properties such as organic carbon content, cation exchange capacity, soil water holding capacity, and soil pH. As sodic or alkaline soils were not included in this study, it is likely that this model may have to be modified for such conditions.

Other researchers have examined the abundance of indigenous rhizobia in soils. Lawson et al. (1987) developed a regression model which describes the densities of *Rhizobium leguminosarum* bv. *trifolii* as a function of clover height and solar radiation. Other investigators have related the number of rhizobia in soil with rainfall (Singleton & Tavares, 1986) and cropping history of the site (Weaver et al., 1972). Yousef et al. (1987) related the number of peanut rhizobia in Iraq to several soil parameters including soil pH, lime content, and cation exchange capacity.

The influence of nonhost rhizospheres on rhizobial densities is well documented (see Parker et al., 1977). The rhizosphere of

Table 2. Regression models describing native rhizobial populations on the island of Maui, Hawaiian Islands.

		- R -
Total rhizobia/g soil ($\log_{10}$)	= 2.07 + 0.056 (% legumes)	0.75**
	= 1.45 + 0.002 (MAR)[1]	0.82***
	= 1.33 + 0.030 (% legumes) + 0.0013 (MAR)	0.89***
	= 0.48 + 0.033 (% legumes) + 0.0016 (MAR) + 0.028 (total extractable bases)[2]	0.95***

** P = 0.01; *** - 0.001.

[1]MAR = Mean Annual Rainfall (mm); [2]Extractable bases in meq/100g soil.

Table 3. Stimulation of *Bradyrhizobium* sp. in the rhizosphere of tropical grasses.[1]

Sample	Bradyrhizobia/g soil
Non-rhizosphere soil	170
Rhizosphere soil	
Cenchrus ciliaris	1429
Panicum maximum	9739
Pennisetum clandestinum	636
Rhynchelytrum repens	3373

[1]Torroxic Haplustoll, mean annual rainfall 565 mm/yr.

selected pasture grasses, cereal crops, broad-leaved crop plants and weedy species have been shown to influence rhizobia. The stimulation of rhizobia in the rhizosphere of different tropical grasses is pronounced when compared to the densities observed in non-rhizosphere soil (Table 3). These measurements were conducted at a single site, a Torroxic Haplustoll, at 640 m elevation receiving 560 mm of rainfall annually. The rhizosphere of different tropical grasses are enriched for rhizobia from 3.7- and 57-fold compared to non-rhizosphere soil. The relatively low enrichment observed for *Pennisetum clandestinum* (kikuyugrass) may be due to its being sampled in the hot, dry extreme of that plant's ecological amplitude (Whitney et al., 1939). This is consistent with the view that the rhizosphere effect on rhizobia is in part related to the increased carbon availability (Parker et al., 1977), although many other soil properties are altered in the rhizosphere (Munns, 1977).

The increase in the number of rhizobia in soil was more pronounced when the host legume was present in the stand (Table 4). The kikuyugrass pasture in this study occurred on an upland Humoxic Tropohumult, a moderately acid soil (640 m; MAR 1800 mm; pH 5.3). The rhizobial density of the nonrhizosphere soil was not significantly different in a kikuyugrass pasture devoid of legumes from that in fallow soil 1.5 to 1.8 ($\log_{10}$) g^{-1} soil. The abundance of *Rhizobium leguminosarum* bv. *trifolii* and the legume host *Trifolium repens* are very well correlated, resulting in high populations of rhizobia (5.0 $\log_{10} g^{-1}$ soil).

Table 4. Abundance of clover rhizobia in the bulk soil of a kikuyugrass pasture.

	Total shoot mass	Clover	Rhizobia
	kg/ha	%	($\log_{10}$)/g soil
Fallow (12 months)	0	0	1.5
Kikuyugrass only[1]	10 316	0	1.8
Kikuyugrass and clover[2]	3 818	9	3.5
Kikuyugrass and clover	3 980	26	5.0

[1]*Pennisetum clandestinum* [2]*Trifolium repens*.

The survival of exotic rhizobia introduced directly into soils in the absence of host legumes and other vegetation presents an extreme condition under which to evaluate the saprophytic competence of a strain as described by Chatel et al. (1968). In a long-term study currently in progress on Maui, NifTAL scientists are evaluating the persistence of 18 elite strains of Rhizobium spp. and Bradyrhizobium spp. in diverse tropical soils. Rhizobial species are being enumerated using MPN plant infection procedures (Woomer et al., 1988b). The individual strains are monitored by serotyping the nodules which result in the MPN replicates. The data from this survival study demonstrate the pattern of rapid decline followed by survival equilibrium as has been reported by other researchers for Bradyrhizobium japonicum (Corman et al., 1987; Crozat et al., 1982). In Fig. 2, we present a number of hypothetical scenarios for the fate of exotic rhizobia released into the environment. Rhizobia are released into a soil at a given population size (A) against a background of native rhizobial population (B). Survival failure results when the organisms are no longer recoverable after a period of time (C). This was observed with R. leguminosarum bvs. trifolii and viceae and R. meliloti in Tropohumults (acid soils), consistent with the results of other investigators (Brockwell et al., 1968; Vincent, 1954). Similarly, B. japonicum failed to persist in Torroxic Haplustolls (hot, dry soils), supporting the observations of Jenkins et al. (1987) that Rhizobium sp. are more tolerant of dessication and extreme temperature than are Bradyrhizobium sp. Other strains are able to persist in very low levels (D) or become established at levels near that of native rhizobia (E). This was observed with Rhizobium spp. released in upland volcanic soils high in organic carbon (Andepts). Some introduced rhizobia are highly successful in colonizing certain soils at equilibrium populations exceeding that of the indigenous population (F). In our studies, this was the case with a strain of Rhizobium spp. (Leucaena) in the Torroxic Haplustolls tested.

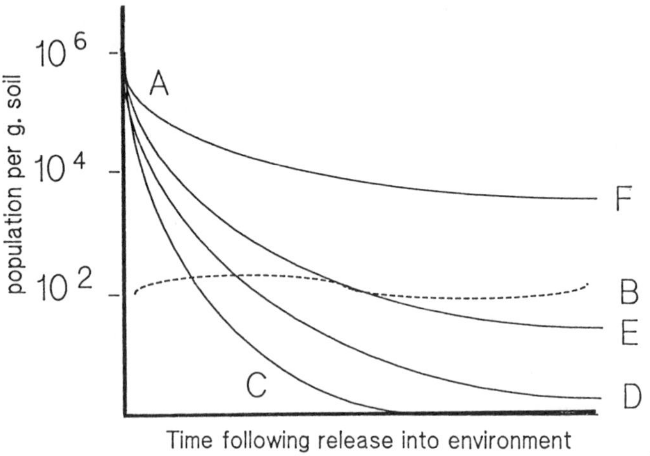

Fig. 2. The fate of exotic rhizobia released into the soil.

Our results suggest that introduced rhizobia respond to different sets of soil factors than those that affect abundance of native rhizobia. Soil organic carbon, water holding capacity and cation exchange capacity are factors which significantly influence population kinetics of introduced rhizobia. This suggests that indigenous rhizobia are adapted to the extremes in their native soil environment, something not guaranteed in introduced rhizobia.

RHIZOBIUM ECOLOGY IN PASTURES - A CONCEPTUAL MODEL

The conceptual model in Fig. 3 is designed after Odum and Odum (1976, p. 16-23 and 269-270) to describe the role of biotic and abiotic components of the soil in regulating the population size of native and inoculant rhizobia in a cropping system containing host and nonhost plants. The introduced rhizobia proliferate in the rhizosphere of host legumes. Infection of roots results in root nodules which act as another reservoir of rhizobia in the soil. The size of the introduced rhizobial population is determined by the growth (G) and death (D) rates which are regulated by other components of the environment. Influences on G and D include the quality (Q) and volume of the soil solution, the amount of available nutrients, e.g., organic carbon (OC), host and nonhost rhizosphere effects, the stimulation of the population due to the senescence of root nodules, and beneficial interactions with other soil microorganisms.

The death rate can also be regulated through microbial antibiosis, allelopathic rhizosphere effects, bacterial predators and parasites. While it is difficult to measure the actual growth and death rates of microorganisms in soils (Bohlool & Schmidt, 1972), the net outcome can be measured in terms of changes in soil populations over time.

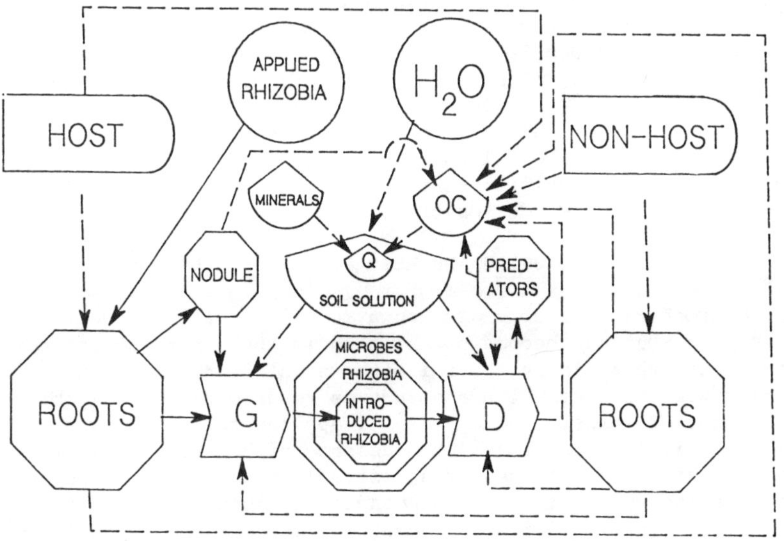

Fig. 3. Environmental regulation of rhizobia in the soil.

NUTRIENT EFFECTS ON THE SYMBIOTIC STATE

Following infection of the legume root by rhizobia, the processes of nodulation and N_2 fixation are sensitive to the mineral nutrition of the host legume. This topic has been reviewed by Munns (1977). Since then, additional studies have elaborated on the role of calcium nutrition in nodulation and N_2-fixation by many tropical and temperate legumes. Munns et al. (1977) found three responses of BNF due to liming of an N-deficient oxisol (pH 4.7 to 7.1). Some legumes showed little response to liming (*Arachis hypogaea*, *Vigna sinensis*). Some legumes (*Stylosanthes* spp.) demonstrated reduced growth and BNF in the limed soil. Many legumes responded to lime through increased growth and BNF. Soil acidity factors reduced the numbers of nodules and the efficiency of the nodules.

To some extent, tolerant rhizobial strains are able to overcome the effects of soil acidity. Comparing the yield potential of subterranean clover with different rhizobia in a limed and unlimed soil, Thorton and Davey (1983) found some acid resistant strains of *R. leguminosarum* bv. *trifolii* resulted in 90% of the yield potential in the low pH (pH 4.4) soil. Other strains produced less than half as much dry matter in the acid environment. Several rhizobia have been identified that are superior in BNF ability and tolerant of soil acidity and related stress factors (Keyser & Munns, 1979; Zaroug & Munns, 1980). For example, strain TAL 169 is considered a choice inoculum strain for several grain and pasture legumes (Somasegaran & Hoben, 1985).

Individual strains also differ in their abilities to nodulate subterranean clover in phosphate-depleted growth conditions (Leung & Bottomley, 1987). The effects of salinity on rhizobia and the legume/*Rhizobium* symbiosis indicate the tolerance of free-living rhizobia to saline conditions, yet the sensitivity of the symbiosis to salt stress. In culture, rhizobia isolated from coral sand soils were able to survive near sea water salinity (Singleton et al., 1982). When nodulated soybeans are exposed to 12mM NaCl 40 days after planting, plant yield and N_2 fixation are greatly reduced (Singleton & Bohlool, 1983).

The ability of rhizobia to withstand environmental extremes exceeds that of the host legume. This is due in part to the acclimation of rhizobia to salt, temperature and acidity-related stress conditions (Mendez-Castro & Alexander, 1976).

CONCLUSION AND RESEARCH PRIORITIES

Saprophytic rhizobia are known to proliferate in the rhizosphere but can become established in the non-rhizosphere soil. Some of the rhizobia in the rhizosphere infect the host root, forming root nodules which also act as reservoirs for rhizobial enrichment of soils. The ability of applied rhizobia to colonize pasture soils and persist in the absence of host roots, while maintaining the attributes of a desirable microsymbiont, is an important selection criteria of candidate inoculant strains. Research objectives which can advance the N_2-fixing potential of legumes through rhizobial performance in soils include: i) field testing numerous strains which have been identified as tolerant of specific soil stresses in rapid screening procedures;

ii) understanding and preventing the loss of symbiotic ability of some superior rhizobia that appears to be induced by specific soil stresses; iii) developing cost efficient methods of inoculant production, application, and establishment of highly effective rhizobia in competition with indigenous populations; and iv) advancing basic understandings of microbial ecology in the soil, allowing for improved environmental risk assessment required before the release of genetically altered rhizobia into crop and pasture systems.

ACKNOWLEDGMENT

This research was supported by National Science Foundation grant BSR 8516822 and U.S. Agency for International Development cooperative agreement DAN-4177-A-00-6035-00 (NifTAL Project).

We appreciatively acknowledge the use of the Maui Soil, Climate, and Land Use Network developed by H. Ikawa of the Dep. of Agronomy and Soil Science, Univ. of Hawaii; cooperation from Haleakala Ranch and Ulupalakua Ranch; technical assistance by Wendy Asano; and manuscript preparation by Susan Hiraoka and Patricia Joaquin.

REFERENCES

Bohlool, B.B., R.M. Kosslak, and R. Woolfenden. 1984. The ecology of Rhizobium in the rhizosphere: survival, growth and competition. p. 287-293. In C. Veeger and W.E. Newton (ed.) Advances in nitrogen fixation research. Martinus Nijhoff - Dr. W. Junk Publ., The Hague, Netherlands.

Bohlool, B.B., P.Nakao, and P.W. Singleton. 1986. Ecological determinants of interstrain competition in Rhizobium/legume symbiosis. p. 145-148. In W. Wallace and S.E. Smith (ed.) Proc. of the 8th Nitrogen Fixation Conf., Adelaide, Australia Aust. Inst. of Agricult. Sci. Parkville, Victoria.

Bohlool, B.B., and E.L. Schmidt. 1972. Fluorescent antibodies for determination of growth rates of bacteria in soil. p. 336-338. In T. Rosswall (ed.) Modern methods for the study of microbial ecology Vol. 17. Bull. Ecol. Res. Comm. (Stockholm).

Brockwell, J.H., W.F. Dudman, A.H. Gibson, F.W. Hely, and A.C. Robinson. 1968. An integrated programme for the improvement of legume inoculant strains. Trans. Int. Congr. Soil Sci., 9th. 2:103-114.

Chatel, D.L., R.M. Greenwood, and C.A. Parker. 1968. Saprophytic competence as an important character in selection of Rhizobium for inoculation. Trans. Int. Congr. Soil Sci., 9th. 2:65-73.

Corman, A., Y. Crozat, and J.C. Cleyet-Marel. 1987. Modelling of survival kinetics of some Bradyrhizobium japonicum strains in soils. Biol. Fert. Soils 4:79-84.

Crozat, Y., J.C. Cleyet-Marel, J.J. Giraud, and M. Obaton. 1982. Survival rates of Rhizobium japonicum populations introduced into different soils. Soil Biol. Biochem. 14:401-405.

Jarvis, B.D.W., E.C. Pankhurst, and J.J. Patel. 1982. Rhizobium loti, a new species of legume root nodule bacteria. Int. J. Syst. Bacteriol. 32:378-380.

Jenkins, M.B., R.A. Virginia, and W.M. Jarrell. 1987. Rhizobial ecology of the woody legume mesquite (Prosopis glandulosa) in the Sonoran Desert. Appl. Environ. Microbio. 53:36-40.

Jordan, D.C. 1984. Family III Rhizobiaceae CONN. 1938. p. 234-256. In N.R. Krieg (ed.) Bergey's manual of systematic bacteriology, Vol. 1. Williams and Wilkins, Baltimore, MD.

Keyser, H.H., B.B. Bohlool, T.S. Hu, and D.F. Weber. 1982. Fast-growing rhizobia isolated from root nodules of soybeans. Science 215:1631-1632.

Keyser, H.H., and D.N. Munns. 1979. Tolerance of rhizobia to acidity, aluminum and phosphate. J. Soil Sci. Soc. Am. 43:519-523.

Lawson, K.A., Y.M. Barnet, and C.A. McGilchrist. 1987. Environmental factors influencing numbers of Rhizobium leguminosarum bv. trifolii and its bacteriophages in field soils. Appl. Environ. Microbiol. 53:1125-1131.

Leung, K., and P.J. Bottomley. 1987. Influence of phosphate on growth and nodulation characteristics of Rhizobium trifolii. Appl. Environ. Microbiol. 53:2098-2105.

Lowendorf, H.S. 1980. Factors affecting survival of Rhizobium in soils. p. 87-123. In M. Alexander (ed.) Advances in microbial ecology. Plenum Press, New York.

Marshall, K.C. 1964. Survival of root nodule bacteria in dry soils exposed to high temperatures. Aust. J. Agric. Res. 15:273-281.

Mendez-Castro, F.A., and M. Alexander. 1976. Acclimation of Rhizobium to salts, increasing temperature and acidity. Rev. Lat-Am Microbiol. 18:155-158.

Munns, D.N. 1977. Mineral nutrition and the legume symbiosis. p. 353-391. In R.W.F. Hardy and A.H. Gibson (ed.) A treatise on dinitrogen fixation, Sect. IV. Agronomy and Ecology. John Wiley and Sons, New York.

Munns, D.N., R.L. Fox, and B.L. Koch. 1977. Influence of lime on nitrogen fixation by tropical and temperate legumes. Plant Soil 46:591-601.

Odum, H.T., and E.C. Odum. 1976. Energy basis for man and nature. McGraw-Hill Book Co., New York.

Parker, C.A., M.J. Trinick, and D.L. Chatel. 1977. Rhizobia as soil and rhizosphere inhabitants. p. 311-352. In R.W.F. Hardy and A.H. Gibson (ed.) A treatise on dinitrogen fixation, Sect. IV. Agronomy and Ecology. John Wiley and Sons, New York.

Scholla, M.H., and G.H. Elkan. 1984. Rhizobium fredii sp. nov., a fast-growing species that effectively nodulates soybeans. Int. J. Syst. Bacteriol. 34:484-486.

Singleton, P.W., and B.B. Bohlool. 1983. Effect of salinity on the functional components of the soybean-Rhizobium japonicum symbiosis. Crop Sci. 73:815-818.

Singleton, P.W., S.A. El-Swaify, and B.B. Bohlool. 1982. Effect of salinity on Rhizobium growth and survival. Appl. Environ. Microbiol. 44:884-890.

Singleton, P.W., and J.W. Tavares. 1986. Inoculation response of legumes in relation to the number and effectiveness of indigenous Rhizobium populations. Appl. Environ. Microbiol. 51:1013-1018.

Somasegaran, P., and H. Hoben. 1985. Methods in legume Rhizobium technology Univ. of Hawaii NifTAL Project, Paia, Hawaii.

Thorton, F.C., and C.B. Davey. 1983. Response of the clover-*Rhizobium* symbiosis to soil acidity and *Rhizobium* strain. Agron. J. 75:557-560.

Vincent, J.M. 1954. The root nodule-bacteria of pasture legumes. Proc. Linn. Soc. N.S.W. 79:1-32.

Weaver, R.W., L.R. Frederick, and C.C. Dunmenil. 1972. Effect of soybean cropping and soil properties on numbers of *Rhizobium japonicum* in Iowa soils. Soil Sci. 114:137-141.

Whitney, L.D., E.Y. Hosaka, and J.C. Ripperton. 1939. Grasses of the Hawaiian ranges. Hawaii Agric. Exp. Stn. Bull. 82. Univ. of Hawaii, Honolulu.

Woomer, P., P.W. Singleton, and B.B. Bohlool. 1988a. Ecological indicators of native rhizobia in tropical soils. Appl. Environ. Microbiol. 54:1112-1116.

Woomer, P.,P.W. Singleton, and B.B. Bohlool. 1988b. Reliability of the most probable number technique for enumerating rhizobia in tropical soils. Appl. Environ. Microbiol. 54:1494-1497.

Yousef, A.N., A.S. Al-Nassiri, S.K. Al-Azawi, and N. Abdul-Hussain. 1987. Abundance of peanut rhizobia as affected by environmental conditions in Iraq. Soil Biol. Biochem. 19:391-396.

Zaroug, M.G., and D.N. Munns. 1980. Screening strains of *Rhizobium* for the tropical legumes *Clitoria ternatea* and *Vigna trilobata* in soils of different pH. Trop. Grass. 14(7):28-33.

DISCUSSION

Clements. You mentioned that the NifTAL Project involved collaboration among scientists in many countries. This workshop is about collaboration, and the NifTAL Project seems to be a good example of what can be achieved. Could you tell us some more about it?

Bohlool. The NifTAL Project's , International Legume Inoculation Trials including 50 scientists, 18 legumes, and 250 experiments were designed to demonstrate the frequency of yield responses to inoculation of legumes with rhizobia in the tropics. Yield responses were obtained in 50% of all trials. Subsequent work is concentrating on predicting the response to inoculation.

Sheaffer. What delivery system was used to supply rhizobia. Did this affect successful nodulation?

Bohlool. Inoculation is applied through seed pelleting. Even at high inoculation rates, introduced rhizobia frequently cannot compete with native rhizobia.

Sheath. What is the source of competition between introduced and indigenous rhizobia before or after infection?

Bohlool. Competition occurs primarily during the infective stage where it is a plant controlled phenomena.

Brougham. What were the differences in soil factors that influenced persistence and survival of rhizobia in your work?

Woomer. Soil factors influencing rhizobial populations included acidity, desiccation, and high temperature stress. The organic carbon content of the soil appears to promote survival of introduced rhizobia in many cases.

Hockman. An example of the application in practice of the principles outlined in your paper can be found in the work done in Western Australia. They have been able to match acid-tolerant rhizobia to a medic host (*Medicago minima* L.). This has enabled medics to grow in an area where subclovers are not climatically adapted.

Woomer. The ecological amplitude of rhizobia is generally wider than that of their host legumes. Exceptions to this situation often result in problems. For example *R. meliloti* is normally sensitive to low pH soil conditions in Australia yet superior strains have been identified. At NifTAL, stress-tolerant strains of other *Rhizobium* species have been characterized.

Martin. Is the genetics of the host plant or that of the *Rhizobium* species most often operational in deciding the symbiotic relationship that develops.

Bohlool. It is the interaction of both the host and rhizobia. Examples can be given where the genotype of each has been most important in a specific environment.

Caradus. It may be, as you say, that rhizobia can survive in acid and aluminum toxic soils. As I understand it, it is the infection process that is inhibited by acid soil complexes. So whether the rhizobia survives or not is irrelevant if it will not nodulate in acid soils.

Bohlool. Yes. The infection process is the most sensitive step in nodulation of temperate legumes under low pH conditions.

Leath. Do the root-colonizing bacteria, that is *Pseudomonads*, interact significantly with *Rhizobium* species effectiveness or efficiency?

Bohlool. Yes. Dramatic enhancement has been demonstrated by the presence of *P. tabaci*. This was reported in a recent meeting of the Molecular Genetics of Plant and Microbe Interactions held in Acapulco, Mexico. Also several groups are trying to genetically engineer the *Pseudomonads* attribute into rhizobia.

Matches. What degree of improvment in nodulation or N_2-fixation do you forsee in the future by the identification of more effective rhizobia?

Bohlool. Response to applied N fertilizer is really the window for determining the potential improvement that is possible.

Matches. What percentage of new planting failures are due to poor nodulation or fixation?

Bohlool. As pictures of the plots showed in my presentation, failure of infection of rhizobia may result in no economic yield. Competition among rhizobia is extremely important. Competition from 50 to 100 rhizobia in a gram of soil can inhibit the establishment of an introduced rhizobia.

GENERAL DISCUSSION OF MAJOR EDAPHIC AND CLIMATIC STRESSES

A definition of legume persistence was proposed and discussed. This seemed necessary if identification of stable legume species/communities is to be conducted in a more efficient, co-ordinated manner in so far as edaphic and climatic stresses are concerned.

Comments:

Brougham: On a world-wide basis (e.g., Sub-Saharan Africa, Himalayan Foothills, Extensive Pastoral systems of China), maintenance and stability of vegetation cover is all important and then animal production follows.

Fisher: The Finlay and Wilkinson approach is relevant to identifying plastic plants that are suitable for a broad range of environments. The International Network of Evaluation of Tropical Pastures of CIAT tests over a range of environments and data can be used to identify pasture germplasm that is plastic or not to environment.

El-Swaify: Definition of persistence should include not only environmental controls but also animal pressures. In the long term, interactions which lead to soil/land degradation (not plant based) may need to be considered.

Jones: In seeking more adapted legumes, we must make better use of our accumulated knowledge (e.g., A.E. Kretschmer's Table 3, see chapter in this publication entitled "Tropical Forage Legume Development, Diversity, and Methodology for Determining Persistence").

Smith: Persistence could be best described as the survival of plant material against specific stresses unique to the existing environment.

Clements: Persistence is about plant numbers, not yield. Above a certain minimal level, yield is independent of numbers. Below that level, yield and numbers are positively correlated. We are concerned about persistence because it influences yield and stability. We can consider persistence as the maintenance of a certain minimum plant population. Incorporation of a yield concept is possible, but plant numbers must be the focal point. That is why a demographic approach to persistence is so useful.

Matches: If the desired level of legume is defined by a production component, it may be more relevant to consider needs of feed deficit periods rather than total production.

Kretschmer: Productivity is a relative term, i.e., 50% plant population of a poor-growing legume may be more productive than 10% of the most productive one. In this case, persistence is more important.

Rotar: In production agriculture, desired persistence will be defined by the desired production need, e.g., provide a source of high-quality feed for animals. In subsistence agriculture, the ability to survive severe stress and provide low production may be sought. Land stability may mean persistence without any production whatever. There is a large gradient across these expectations and requirements for persistence differ markedly.

Forde: If "productive output" sets the desired level, is it the forage value of the legume or the legume plus grass that is important? In other words, is a persistent but low-producing legume still worthwhile for its N_2 fixation. Persistency equates with stability at the desired level for the farm system and/or net profit.

Brougham: In considering persistence, we should first view it within the constraints of our agricultural systems and then to wider environments so as to contribute to Third World Agriculture. We attend to the wider environments poorly.

Sheath: Legume Persistence Definition: Where legume populations are at a stable density that achieves the expectations of the specific ecosystem (e.g., economic productivity, environment/cultural stability).

CULTURAL PRACTICES INFLUENCING LEGUME ESTABLISHMENT AND PERSISTENCE IN AUSTRALIA

D. Gramshaw and M.A. Gilbert

SUMMARY

Establishment, fertilizer, irrigation, mowing and pasture renovation practices that influence the persistence of important pasture and hay legumes in Australia are reviewed. Establishment requirements in high rainfall or irrigated environments are well known and, when adopted, give adequate establishment. Establishment and regeneration of annual temperate legumes in ley pastures are often constrained by cropping practices and by herbicides required for weed control. Low-cost legume establishment either by surface-sowing or by undersowing a crop can be unreliable. Fertilizer practices required to establish and maintain legumes in pastures are broadly known, although improved efficiency of fertilizer use is needed. Legume persistence in surface irrigated pastures is mainly limited by poor drainage and salinity although there are management practices that can increase the legume content of these pastures. Untimely mowing of annuals when they are seeding or too frequent mowing of lucerne decreases persistence, but mowing practices that minimize these effects are known. Mechanical renovation to encourage legumes is rarely practiced.

INTRODUCTION

Legume persistence in pastures or hay crops is often modified by cultural practices that influence establishment, plant survival and, in the case of annuals, self-regeneration. This paper overviews important cultural practices currently used by Australian farmers and graziers and notes situations where these may limit the productive persistence of legumes. Emphasis is given to practices associated with pasture establishment and regeneration, fertilizer use, irrigation, mowing, and pasture renovation.

LEGUME ESTABLISHMENT AND REGENERATION

The factors influencing pasture establishment in Australia have been recently reviewed (Campbell et al., 1987; Carter, 1987; Cook et al., 1987). We focus on the main problems associated with current commercial practices within the following pasture situations: intensive, cereal belt, and extensive pastures.

Intensive Pastures

Most of the temperate and tropical perennial legumes and many of the annual legumes that are important in Australia (Gramshaw et al., these Proceedings) feature in intensive pastures in the higher rainfall environments or under irrigation. Moisture is relatively favorable or can be manipulated for establishment and close attention is usually given to thorough seedbed preparation, weed control and fertilizer requirements. Under these conditions few establishment problems are experienced, provided the known requirements for seed treatment and sowing for each species are met (e.g., Jones & Rees, 1973; Teitzel & Middleton, 1980). Reduction of hardseededness, by mechanical scarification or heat treatment, is useful with many of the tropical legumes. Rhizobium specificity is important with some temperate and tropical legumes (Gramshaw & Cameron, 1988). Annual weeds can dominate newly sown perennial pastures, especially on fertile soils, although they decrease in importance in subsequent years (Jones, 1975).

Direct drilling of temperate legumes, mostly perennials, into perennial native or sown grasses is developing as an alternative to conventional cultivation and sowing, mainly in higher rainfall coastal and tableland areas or under irrigation in south eastern Australia (McDonald & Duncan, 1983). The availability of more refined drilling machinery and herbicides to reduce competition from resident plants is increasing the reliability of the technique. White clover (Trifolium repens L.) and subterranean clover (Trifolium subterraneum L.) have been successfully established in native pastures without herbicides. With more vigorous grasslands, and with perennial legumes other than white clover, herbicides are essential to reduce competition (Campbell et al., 1987). Difficulties with direct drilling are as follows: the refinement in the techniques and management required by farmers to ensure success; slow growth of seedlings and their increased susceptibility to insect pests and slugs compared with conventional sowing; adequacy of herbicides to control resident vegetation for sufficient time; the need for careful grazing management post-sowing to avoid uprooting and trampling of seedlings, but at the same time controlling competition; and lenient grazing management with perennials in the first 2 yr. Commercial methods for direct drilling tropical legumes are developing (Cook, 1982; Walsh & Cook, 1988).

For irrigated pastures and hay crops, high seeding rates are necessary for early productivity and persistence; dense seedling stands maximize the early season production of subterranean clover (Mason et al., 1987) and prolong the productive, weed-free life of lucerne (Medicago sativa L.) stands (Gramshaw, 1978). Higher than recommended sowing rates of tropical perennial legumes increases their contribution within the establishment year, but the effect decreases subsequently (Jones, 1975).

Cereal Belt Pastures

Pastures in the Cereal Belt are ley or semi-permanent pastures based on annual temperate legumes or lucerne. Cropping and pasture practices are interdependent. Pasture establishment and persistence are variable because rainfall amount and reliability can be lower than in intensive pastures situations. Most pasture and crop sowings now occur on old, often degenerated, pasture land where grass and broad-leaved weeds need to be controlled and diseases and pests are more prevalent (Gillespie et al., 1983; Carter, 1987). A number of pre-sowing broad-spectrum and post-sowing selective herbicides are used (Pool, 1987); it is estimated that more than 50% of wheat crops and 4% of pastures in Australia are now treated with herbicides (ABS, 1986).

Cultivation or herbicide use to control weeds and adoption of intensive cropping sequences can reduce legume seed production and deplete soil seed reserves in pasture leys. Initially deep cultivation and burial of legume seeds diminishes their contribution to regeneration with 'in-out' cropping but may help conserve seed for longer with continuous cropping (Taylor, 1985; Carter & Challis, 1987; Taylor & Ewing, 1988; Jones & Carter, these Proceedings). Desiccants or other 'knock-down' herbicides applied in spring to interfere with weed seed set before cropping in autumn may simultaneously decrease legume seed set (Gillespie et al., 1983; Thorn & Perry, 1983). Leaving mechanical or chemical weed control until after the first germination of seeds in autumn may destroy regenerating legume seedlings as well as delay pasture sowing (Gillespie et al., 1983), both potentially leading to decreased legume seed production in the establishment year (Quinlivan et al., 1973). Presently available post-emergence herbicides are insufficiently selective to discriminate in favor of volunteer legume seedlings, although new herbicides under evaluation show promise for total suppression of annual grass species (Thorn & Perry, 1983; Pool, 1987). More frequent 'in-out' cropping or continuous cropping using herbicides accelerates the depletion of seed reserves, with subterranean clover most at risk due to its lower hardseededness compared with other important annuals (Bolland, 1985, 1986a).

Pasture sowing techniques include low-cost surface-sowing into either crop stubble, dry pasture residues or shallow cultivation (oversowing); drilling into cultivation or direct drilling; and sowing beneath a crop (undersowing) (Carter, 1987; Cook et al., 1987). Crops are either conventionally or direct drilled. Broad-area drills, such as combine seeders and air seeders, are commonly used but provide limited opportunity for either accurate depth of sowing of pasture seed or the separate placement of seed and fertilizer (Cook et al., 1987).

Oversowing before the start of the growing season, although convenient and often used, has an associated risk of failed establishment when inadequate early season rains occur (Gillespie et al., 1983).

With direct drilling, high soil strength and crop residues can reduce legume establishment, although the slower growth of direct drilled crops may be less competitive with the legume seedlings (Cornish, 1985). High soil strength impedes subterranean clover burr burial, making seed more accessible to stock in summer, and restricts radicle entry and root growth of legumes (Carter et al., 1982). Crop residues may favor legume seedling diseases (Barbetti, 1986) and suppress subterranean clover and annual medic (Medicago spp.) seed germination (Gillespie et al., 1983; Quigley & Carter, 1985). Higher sowing rates are important where new subterranean clovers are direct-drilled to replace old, less productive ones (Schroder, 1987).

Undersowing a crop with a legume may result in inadequate legume establishment or depressed crop yields (Poole & Gartrell, 1970; McGowan & Williams, 1973; Scott & Brownlee, 1974; Brownlee & Scott, 1974). Undersowing has limitations with subterranean clover in Western Australia where poor establishment or failures are common (Gillespie et al., 1983; Bolland, 1987a). However, hard legume seed sown on the soil surface when a crop is planted can establish in the year following cropping (Bolland, 1986b). In dry environments furrow sowing, crop row arrangement and fertilizer banding techniques may have potential to increase the reliability of legume establishment by undersowing (Anonymous, 1986; Scott, 1985).

Insect pests, especially red-legged earth mite (Halytodeus destructor), may devastate establishing legume seedlings. Low-cost, seed applied systemic insecticides can provide temporary protection, although treatment precludes effective Rhizobium inoculation (Gillespie et al., 1983).

Early grazing after sowing damages juvenile legume seedlings (Carter, 1987). However, delaying grazing too long can reduce subsequent seed production with annuals (Collins, 1978; Gillespie et al., 1983).

Extensive Pastures

Surface-sowing (oversowing), with or without prior surface treatment such as rough cultivation, aerially applied herbicides, heavy grazing or burning, is used for extensive and low-cost pasture development. Legumes sown are stylos (Stylosanthes spp.) and siratro (Macroptilium atropurpureum (DC.) Urb) in northern Australia, and white clover, subterranean clover and, to a lesser extent, lucerne in south-eastern Australia. Aerial oversowing and fertilizing is often the only option for partially cleared land or with difficult terrain. A sequence of aerial herbicide application followed by aerial sowing is practiced in south eastern Australia on high rainfall hill country where there is also a need to control serious weeds (Campbell, 1974a, b). Most of the pasture seed and superphosphate applied aerially in Australia occurs in New South Wales (60-70%) or Victoria and Tasmania (15-20%) (ABS, 1983-84).

Oversowing can result in poor establishment, especially with perennials. Campbell (1974b) reported 1 to 27% establishment with temperate species, and Gramshaw & McKeon (1986, unpublished data) often recorded less than 5% establishment (range 0 to 74%) with oversown shrubby stylo (Stylosanthes scabra Vog.)

Factors limiting establishment from oversowing have been highlighted in a number of recent reviews (Gardener, 1984; McKeon & Mott, 1984; Campbell et al., 1987; Cook et al., 1987). Salient factors are a variable moisture supply for seeds and seedlings and substantial competition for moisture and nutrients from existing pasture or volunteer plants. Litter and vegetative cover may or may not improve establishment depending on rainfall sequences and, in some circumstances, may separate seed from the soil. Ants and termites can harvest seed.

Seed coatings of lime, reverted superphosphate or Terra sorb have not been successful in experimental sowings. Insecticidal seed dressings of either bendiocarb or permethrin decrease ant theft (Dowling, 1978; Campbell et al., 1987). The hard seed percentage of stylo seed is usually reduced either by mechanical scarification or heat treatment prior to surface-sowing (Cameron, 1985, 1988). However, the relative benefits of the alternative treatments is determined by the nature of the rainfall events that occur before and during establishment (McKeon & Brook, 1983; McKeon & Mott, 1984). Consistently superior establishment from heat-treated seed compared with untreated or scarified seed was recorded in multiple sowings in the inland subtropics (Gramshaw & McKeon, 1985).

Much of the knowledge on oversowing in Australia is site- and time-specific, either because there has been limited sampling of weather sequences or because studies have been confined to only components of the establishment process. A better definition of seed treatment, sowing time and sowing rate practices for different environments needs to account for weather variability using computer simulation techniques (e.g., Dowling & Smith, 1976; Leslie, 1982).

FERTILIZER

Many Australian soils in the virgin state are deficient in N and P. Sulfur, K and trace element deficiencies also occur in some soils (Williams & Raupach, 1983; Hochman & Helyar, these Proceedings).

Fertilizer Use

About 1.2m (million) t of fertilizer, 83% of which is superphosphate and 3% nitrogenous, is applied annually to sown pastures (ABS, 1986). Nitrogen and, to a lesser extent, K are only applied to intensive pastures in high rainfall and irrigation areas. Sulfur, as a component of single superphosphate, is applied extensively to legume-based pastures. Cu, Zn and Mo are sometimes applied as additions to phosphatic fertilizers.

The amount of phosphatic fertilizer used in Australia declined progressively between 1974 and 1986 in response to steep fertilizer price rises and changes in the relative returns from crops and livestock (Gramshaw et al., these Proceedings). The rate and the frequency of phosphate application on permanent pasture decreased, and ley pastures commonly subsisted on fertilizer residues remaining after cropping. Since 1980, increased freight costs also led to an increased use of high analysis phosphatic fertilizers on both crops and pastures; these fertilizers contain little S. These changes have important repercussions for the productivity and persistence of legumes in pastures. However, there are indications that P fertilizer use on pastures is now increasing in response to improved beef and wool prices and reduced grain prices.

Fertilizer Rates

The rate of fertilizer needed on legume-based pastures varies widely in relation to soil fertility, pasture productivity and the intensity of animal production. For intensive enterprises such as dairying and beef cattle fattening, P rates of 25-50 kg P/ha are initially applied, with annual maintenance rates of 15-30 kg P/ha. In contrast, under extensive conditions 10-20 kg P/ha at sowing is followed by 0-10 kg P/ha annually (Miller et al., 1989). Where S deficiency occurs, rates of 30 kg S/ha at sowing followed by 5-10 kg S/ha annually are common. Potassium deficiency in high rainfall areas is normally rectified by 30-120 kg K/ha every 2-4 yr.

Methods of Application

Fertilizer is broadcast onto pastures from ground machinery or aircraft. This results in poor placement of phosphatic fertilizers relative to plant roots, particularly in drier areas. However, in soils with high phosphate sorption capacity, incorporation of phosphate into the soil by cultivation may diminish availability (Bolland, 1987b). Pelleting fertilizers such as lime, S (Gilbert & Shaw, 1979) and Mo (Kerridge et al., 1973) onto seeds has been used to successfully establish legumes, although it is doubtful if this benefits annual legume persistence in the longer term.

Nodulation problems at establishment and during regeneration of annuals, particularly in old pastures or acid soils, are becoming more evident (Jones & Curnow, 1986). Cultivation that mixes the soil surface with the more acidic underlying layers may increase this problem (Coventry et al., 1985; Richardson et al., 1985). Lime pelleting is a low-cost practice for improving nodulation on acid soils, such as with new annual medics being introduced into mildly acid situations (Howieson et al., 1987).

Fertilizers and Pasture Composition

Sowing of legumes into grass pastures and the application of phosphatic fertilizer initially leads to legume dominance in both northern (Winks et al., 1974) and southern Australian pastures (Rossiter, 1964; Wolfe & Lazenby, 1973a). With time, a sustained P supply leads to varying degrees of grass dominance, depending on legume species, and is associated with increased N availability (Wolfe & Lazenby, 1973b; Cocks, 1980; Coates et al., 1989). Withholding P at the legume-dominant stage does not cause an immediate legume decline, but eventually pastures revert to native species (Cook et al., 1978; Jones et al., 1984). Similar effects result from the use of S (Hilder & Spencer, 1954; Gilbert & Shaw, 1981), K (Fitzpatrick & Dunne, 1956; Jones, 1966) and other essential elements. There is consequently a need to strategically supply fertilizers to pastures to maintain the legume component.

Nitrogen fertilizers are sometimes applied to irrigated legume-based pastures to boost production in seasons when pasture productivity is low and this leads to a temporary decline in the legume (e.g., Quinlan et al., 1981). However, high levels of N (>300 kg N/ha/yr) will eliminate legumes from a pasture (Jones, 1970).

Long-term use of fertilizer and pasture legumes in southern Australia has caused soil acidification over large areas resulting in poor establishment and persistence of sensitive legumes (Hochman & Helyar, these Proceedings). The use of lime to overcome the problem is hampered by high cost (Helyar, 1987).

Fertilizer Use Efficiency

Chemical analyses of soil and pasture, as well as test applications of fertilizer, are often used to predict fertilizer requirements, sometimes with considerable imprecision (Gartrell & Bolland, 1987). Mathematical models have been developed to define efficient fertilizer strategies for long-term maintenance of pasture productivity (e.g., Bowden & Bennett, 1976; Probert & Williams, 1985). Remote sensing of the phosphate status of pasture foliage is being explored and may prove a possible diagnostic technique (Vickery, 1983).

IRRIGATION

Pasture and hay legumes are grown on nearly 0.9m (million) ha or 55% of the irrigated land in Australia. Lucerne, used mainly for hay and occupying just under 0.1m ha (65% spray irrigated), is grown mainly on the eastern mainland from South Australia to the subtropics of Queensland. Annual pastures based on subterranean clover (0.5m ha) irrigated in winter, and perennial pastures based on white clover (0.3m ha) irrigated in summer, comprise most of the remaining irrigated pasture land; these are primarily (88%) located in the Murray and Murrumbidgee irrigation schemes in southern New South Wales and northern Victoria and 80% are surface irrigated (ABS, 1984).

Production from surface irrigated pastures is below potential for many reasons, including limitations of water supply, adverse soil physical properties, waterlogging, salinity and the frequent dominance of perennial grasses in white clover pastures (Smith et al., 1983; Grieve et al., 1986; Mason et al., 1987). Soil amelioration with high rates of surface-applied gypsum, or profile modification by deep ripping, gypsum injection or sub surface drainage, have substantially improved legume performance (Sedgley, 1962; Taylor & Olsson, 1987).

Lucerne is intolerant of waterlogging, particularly at high soil temperatures. Therefore, control of water application and drainage is essential for high yields and persistence. With spray irrigation there is less risk of temporary waterlogging, although not always so in the subtropics where heavy rain may unpredictably follow irrigation. Resistance to Phytophthora root rot (*Phytophthora megasperma* f. sp. *medicaginis*) normally confers a production and persistence advantage in waterlogged situations (Rogers et al., 1978), although there is no critical evidence that commercially used lucerne cultivars differ in tolerance to physiological waterlogging injury. Defoliated lucerne is highly susceptible to waterlogging (Cameron, 1973) and there is evidence that least damage occurs when labile carbohydrate content in the tap root is depleted (Gramshaw, 1981) in the mid-regrowth period.

Extending the irrigation season of subterranean clover increases early vegetative production and seed yield required for regeneration (Kelly & Mason, 1987a, b). Maintaining continuity of moisture supply after roots of white clover have been damaged by waterlogging may help its retention in perennial pastures (Grieve et al., 1986). The use of water-logging tolerant *T. subterraneum* ssp. *yanninicum* cultivars of subterranean clover (Francis & Devitt, 1969) reduces the effects of waterlogging in annual pastures.

Some grasses tolerate high salt levels in the soil compared with legumes (Russell, 1976). Grass dominance is a feature of white clover-perennial grass pastures growing on saline soils. Use of the salt-tolerant lucerne and strawberry clover (*Trifolium fragiferum* L.) may help maintain a legume component (Mehanni & Repsys, 1980, 1986).

MOWING

Less than 10% of legume-based pastures are mown for hay in Australia. Normally one harvest is taken and this seldom interferes with the persistence of perennial legumes in pastures. However, mowing of annual legumes during flowering reduces seed production with obvious implications for regeneration (Collins, 1978; Gillespie et al., 1983; Carter et al., 1988).

Repeated mowing in Australia is confined to irrigated lucerne in specialist hay enterprises in which mowing management involves a compromise among hay yield, hay quality and stand persistence. Lodge (1986) and Gramshaw (1983, unpublished data) found that commercial cultivars from different winter dormancy categories persisted similarly in response to cutting frequency, with all cultivars being less persistent when cut 4-weekly or more frequently. Height of cutting appears inconsequential for lucerne persistence in the subtropics (Lowe et al., 1985).

RENOVATION

Apart from cultivation and the use of herbicides prior to cropping, there are no other pasture renovation practices widely used in Australia that specifically affect legumes. Mechanical renovation of old siratro-based pastures can restore the contribution of this legume (Bishop et al., 1981), but this practice is not widespread.

CONCLUSION

Many cultural practices can be manipulated to enhance establishment and persistence of annual and perennial legumes in Australia. Reliability of these practices is least in less-favorable rainfall areas where extensive pasture sowing or ley farming is practiced. More reliable, low-cost cultural techniques need to be developed. Fertilizer practices are generally unsophisticated and techniques to increase the efficiency of fertilizer use, such as by modelling the fertilizer requirements of pastures, are of paramount importance. Surface irrigated pastures in Australia require considerable changes in soil and legume management if the potential pasture productivity of the irrigated land is to be achieved. Mowing may substantially decrease persistence of legumes, but the practices to minimize its effects are already known. Renovation of legume-based pastures to specifically increase the legume component is not widely practiced, although the use of selective herbicides is increasing and mechanical renovation may have a role.

REFERENCES

ABS. various dates. Australian Bureau of Statistics, Canberra.

Barbetti, M.J. 1986. Effect of season, trash and fungicides on fungi associated with subterranean clover. Aust. J. Exp. Agric. 26: 431-435.

Bishop, H.G., B. Walker, and M.T. Rutherford. 1981. Renovation of tropical legume-grass pastures in northern Australia. p. 555-558. Proc. XIV Int. Grassl. Congr., Kentucky, U.S.A.

Bolland, M.D.A. 1985. Serradella (Ornithopus sp.): maturity range and hard seed studies of some strains of five species. Aust. J. Exp. Agric. 25: 580-587.

Bolland, M.D.A. 1986a. A laboratory assessment of seed softening patterns for hard seeds of Trifolium subterranean sub spp. subterraneum and brachycalycinum, and of annual medics. J. Aust. Inst. Agric. Sci. 52: 91-94.

Bolland, M.D.A. 1986b. Establishment of serradella by sowing either pod segments or scarified seed under a wheat crop. Aust. J. Exp. Agric. 26: 441-444.

Bolland, M.D.A. 1987a. Sowing subterranean clover in wheat and oat crops on sandy soil near Esperance, Western Australia. J. Aust. Inst. Agric. Sci. 53: 205-206.

Bolland, M.D.A. 1987b. Effectiveness of top-dressed and incorporated superphosphate and Duches rock phosphate for subterranean clover on sandy soils near Esperance, Western Australia. Aust. J. Exp. Agric. 27: 87-92.

Bowden, J.W., and D. Bennett. 1976. The 'Decide' model for predicting superphosphate requirements. Proc. Phos. in Agric. Symp., 1974. Aust. Inst. Agric. Sci., Victorian Branch.

Brownlee, H., and B.J. Scott. 1974. Effects of pasture and cereal sowing rates on production of undersown barrel medic and wheat cover crop in western New South Wales. Aust. J. Exp. Agric. Anim. Husb. 14: 224-230.

Cameron, D.G. 1973. Lucerne in wet soils - the effect of stage of regrowth, cultivar, air temperature and root temperature. Aust. J. Agric. Res. 24: 851-861.

Cameron, D.G. 1985. Tropical and subtropical pasture legumes. 9. Caribbean stylo (Stylosanthes hamata cv. Verano): A complementary/replacement legume for Townsville stylo. Queensl. Agric. J. 111: 299-303.

Cameron, D.G. 1988. Tropical and subtropical pasture legumes. 16. Shrubby stylo (Stylosanthes scabra): The dry tropics perennial browse. Queensl. Agric. J. 114: 105-109.

Campbell, M.H. 1974a. Efficiency of aerial techniques for long-term control of serrated tussock (Nassella trichotoma). Aust. J. Exp. Agric. Anim. Husb. 14: 405-411.

Campbell, M.H. 1974b. Establishment, persistence and production of lucerne-perennial grass pastures surface-sown on hill country. Aust. J. Exp. Agric. Anim. Husb. 14: 507-514.

Campbell, M.H., W.J. Hoskins, D.A. Nichols, E.D. Higgs, and J.W. Read. 1987. Establishment of perennial pastures. p. 59-74. In J.L. Wheeler et al. (ed.) Temperate pastures. Aust. Wool Corp. Tech. Publ. AWC, CSIRO, Melbourne.

Carter, E.D. 1987. Establishment and natural regeneration of annual pastures. p. 35-51. In J.L. Wheeler et al. (ed.) Temperate pastures. Aust. Wool Corp. Tech. Publ. AWC, CSIRO, Melbourne.
Carter, E.D., M. Armstrong, and K.J. Sommer. 1988. Effects of tillage and haymaking practices on survival and productivity of annual pasture legumes. Waite Agric. Res. Inst. Bienn. Rep. 1986-87: N4-N5. Univ. of Adelaide.
Carter, E.D., and S. Challis. 1987. Effects of depth of sowing medic seeds on emergence of seedlings. p. 192. In Proc. 4th Aust. Agron. Conf., Melbourne.
Carter, E.D., E.C. Wolfe, and C.M. Francis. 1982. Problems of maintaining pastures in the cereal-livestock areas of southern Australia. p. 68-82. In Proc. Aust. Agron. Conf., Wagga Wagga, N.S.W.
Coates, D.B., P.C. Kerridge, C.P. Miller, and W.H. Winter. 1989. The effects of phosphorus on the composition, yield, stability and quality of legume-based pasture and their relationships to animal production. Trop. Grassl. 23: in press.
Cocks, P.S. 1980. Limitations imposed by nitrogen deficiency on the productivity of subterranean clover-based annual pasture in southern Australia. Aust. J. Agric. Res. 31: 95-107.
Collins, W.J. 1978. The effect of defoliation on inflorescence production, seed yield and hardseededness in swards of subterranean clover. Aust. J. Agric. Res. 29: 789-801.
Cook, S.J. 1982. Herbicide banding to aid establishment of sod-sown pastures in southern Queensland. p. 189. In Proc. Aust. Agron. Conf., Wagga Wagga, N.S.W.
Cook, S.J., G.J. Blair, and A. Lazenby. 1978. Pasture degeneration. II. The importance of superphosphate, nitrogen and grazing management. Aust. J. Agric. Res. 29: 19-29.
Cook, S.J., P.D. Cregan, M.H. Campbell, and J.W. Read. 1987. Tillage - its role in establishment and management of pastures. p. 94-125. In P.S. Cornish and J.E. Pratley (ed.) Tillage, new directions in Australian agriculture. Inkata Press, Melbourne.
Cornish, P.S. 1985. Direct drilling and early growth of wheat. p. 30. In Proc. 3rd Aust. Agron. Conf., Hobart.
Coventry, D.R., J.R. Hirth, T.G. Reeves, and H.R. Jones. 1985. Changes in soil populations of Rhizobium trifolii following crop establishment in a clover-ley rotation in south east Australia. p. 541-542. In Proc. XVth Int. Grassl. Congr., Kyoto.
Dowling, P.M. 1978. Effect of seed coatings on the germination, establishment, and survival of oversown pasture species at Glen Innes, New South Wales. N.Z.J. Exp. Agric. 6: 161-166.
Dowling, P.M., and R.C.G. Smith. 1976. Use of a soil moisture model and risk analysis to predict the optimum time for the aerial sowing of pastures on the Northern Tablelands of New South Wales. Aust. J. Exp. Agric. Anim. Husb. 16: 871-874.
Fitzpatrick, E.N., and T.C. Dunne. 1956. Potassium for subterranean clover. J. West. Aust. Dep. Agric. 5: 321-326.
Francis, C.M., and A.C. Devitt. 1969. The effect of waterlogging on the growth and isoflavone concentration of Trifolium subterraneum L. Aust. J. Agric. Res. 20: 819-825.

Gardener, C.J. 1984. The dynamics of Stylosanthes pastures. p. 333-357. In H.M. Stace and L.A. Edye (ed.) The biology and agronomy of Stylosanthes. Academic Press, Australia.

Gartrell, J.W., and M.D.A. Bolland. 1987. Phosphorus nutrition of pastures. p. 127-136. In J.L. Wheeler et al. (ed.) Aust. Wool Corp. Tech. Publ. AWC, CSIRO, Melbourne.

Gilbert, M.A., and K.A. Shaw. 1979. A comparison of sulfur fertilizers and sulfur seed pellets on Stylosanthes guianensis pasture on a euchrozem in north Queensland. Aust. J. Exp. Agric. Anim. Husb. 19: 241-246.

Gilbert, M.A., and K.A. Shaw. 1981. Residual effects of sulfur fertilizers on cut swards of a Stylosanthes guianensis and native grass pasture on a euchrozem soil in north Queensland. Aust. J. Exp. Agric. Anim. Husb. 21: 334-342.

Gillespie, D.J., M.A. Ewing, and D.A. Nicholas. 1983. Subterranean clover establishment techniques. J. Agric. West. Aust. 1: 16-20.

Gramshaw, D. 1978. A review of research on establishment, persistence and productivity of lucerne (Medicago sativa L.) at Biloela from 1950 to 1975. Queensl. Dep. Prim. Indust. Agric. Br. Tech. Rep. No. 20.

Gramshaw, D. 1981. A relationship between physiological waterlogging injury and root nonstructural carbohydrate in lucerne. Proc. XIVth Int. Grassl. Congr., Kentucky. Summaries of papers: p. 1987.

Gramshaw, D., and D.G. Cameron. 1988. Quick establishment - the key to early profits from sown pastures and forages. Queensl. Agric. J. 114: 83-88.

Gramshaw, D., and G.M. McKeon. 1985. Establishment of surface sown shrubby stylos. p. 202. In G.J. Murtagh and R.M. Jones (ed.) Proc. 3rd Aust. Conf. Trop. Past., Occas. Publ. 3., Trop. Grassl. Soc. Aust. Watson Ferguson, Brisbane.

Grieve, A.M., E. Dunford, D. Marston, R.E. Martin, and P. Slavich. 1986. Effects of waterlogging and soil salinity on irrigated agriculture in the Murray Valley: a review. Aust. J. Exp. Agric. 26: 761-777.

Helyar, K.R. 1987. Nutrition of plants on acid soil. p. 159-171. In J.L. Wheeler et al. (ed.) Temperate pastures. Aust. Wool Corp. Tech. Publ. AWC, CSIRO, Melbourne.

Hilder, E.J., and K. Spencer. 1954. Influence of sulfur on a natural Medicago pasture. J. Aust. Inst. Agric. Sci. 20: 171-176.

Howieson, J.G., M.A. Ewing, and C.W. Thorn. 1987. Inoculation and lime pelleting of medic seed. West. Aust. Dep. Agric. Farmnote No. 5/87.

Jones, H.R., and B.C. Curnow. 1986. Nodulation of subterranean clover growing in permanent pastures on acid soils in north central Victoria. Aust. J. Exp. Agric. 26: 31-36.

Jones, R.J. 1966. Nutrient requirements of improved pasture on podzolic soils developed in phyllite at North Deep Creek. Trop. Grassl. 6: 23-27.

Jones, R.J. 1970. The effect of nitrogen fertilizer applied in the spring and autumn on the production and botanical composition of two subtropical and grass legume mixtures. Trop. Grassl. 4: 97-109.

Jones, R.M. 1975. Effect of soil fertility, weed competition, defoliation and legume seeding rate on establishment of tropical pasture species in south east Queensland. Aust. J. Exp. Agric. Anim. Husb. 15: 54-63.

Jones, R.M., and M.C. Rees. 1973. Farmer assessment of pasture establishment reliability in the Gympie district, south east Queensland. Trop. Grassl. 7: 219-222.

Jones, R.M., C. Johansen, and D.A. Little. 1984. Effect of omission of annual superphosphate applications on Desmodium-pangola grass pastures. Trop. Grassl. 18: 205-215.

Kelly, K.B., and W.K. Mason. 1987a. Effects of irrigation timing on seedling establishment and productivity of subterranean clover pastures. Aust. J. Exp. Agric. 27: 545-549.

Kelly, K.B., and W.K. Mason. 1987b. Effects of irrigation timing in autumn and spring on seed production of subterranean clover, and the change in permeability and rate of germination of seed. Aust. J. Exp. Agric. 27: 799-805.

Kerridge, P.C., B.G. Cook, and M. Everett. 1973. Application of molybdenum trioxide in the seed pellet for subtropical pasture legumes. Trop. Grassl. 7: 229-232.

Leslie, J.K. 1982. WATPROF - A water balance model for simulation of grass establishment. p. 27-40. In R.F. Brown (ed.) Soil water balance modelling in agriculture: Components and applications. Queensl. Dep. Prim. Indust. Conf. and Workshop Series QC 84011.

Lodge, G.M. 1986. Yield and persistence of irrigated lucernes cut at different frequencies at Tamworth, New South Wales. Aust. J. Exp. Agric. 26: 165-172.

Lowe, K.F., T.M. Bowdler, and G.N. Schrodter. 1985. Effect of cutting height on lucerne (Medicago sativa) cultivars. Trop. Grassl. 19: 24-28.

McDonald, W., and M. Duncan. 1983. Direct drilling pastures. Dep. Agric. N.S.W. Agfact p. 2.2.5.

McGowan, A.A., and W.A. Williams. 1973. Factors affecting competition between subterranean clover and a barley cover crop. Aust. J. Exp. Agric. Anim. Husb. 13: 56-72.

McKeon, G.M., and K. Brook. 1983. Establishment of Stylosanthes species: Changes in hardseededness and potential speed of germination at Katherine, N.T. Aust. J. Agric. Res. 34: 491-504.

McKeon, G.M., and J.J. Mott. 1984. Seed biology of Stylosanthes. p. 311-332. In H.M. Stace and L.A. Edye (ed.) The biology and agronomy of Stylosanthes. Academic Press, Australia.

Mason, W.K., S.J. Blaikie, and C.R. Stockdale. 1987. New directions for irrigated pastures. p. 100-117. In Proc. 4th Aust. Agron. Conf., Melbourne.

Mehanni, A.H., and A.P. Repsys. 1980. Effect of saline irrigation water on pasture productivity and soil characteristics in the Goulburn Valley of Victoria. p. 199-210. In A.J. Rixon and R.J. Smith (ed.) Salinity and water quality. Proc. Symp., Darling Downs Inst. Tech., Toowoomba, Queensl.

Mehanni, A.H., and A.P. Repsys. 1986. Perennial pasture production after irrigation with saline groundwater in the Goulburn Valley, Victoria. Aust. J. Exp. Agric. 26: 319-324.

Miller, C.P., W.H. Winter, D.B. Coates, and P.C. Kerridge. 1989. Strategies for phosphorus use in north Australia beef production. Trop. Grassl. 23: in press.

Poole, M.L. 1987. Tillage practices for crop production in winter rainfall areas. p. 24-47. In P.S. Cornish and J.E. Pratley (ed.) Tillage, new directions in Australian agriculture. Inkata Press, Melbourne.

Poole, M.L., and J.W. Gartrell. 1970. Undersowing wheat with annual legumes - Effects on wheat yield and legume seed yield in the south eastern wheatbelt of Western Australia. Aust. J. Exp. Agric. Anim. Husb. 10: 84-88.

Probert, M.E., and J. Williams. 1985. The residual effectiveness of phosphorus for Stylosanthes pastures on red and yellow earths in the semi-arid tropics. Aust. J. Soil Res. 23: 211-222.

Quigley, P.E., and E.D. Carter. 1985. The effects of cereal straw on productivity of annual medic pastures. p. 205. In Proc. 3rd Aust. Agron. Conf., Hobart.

Quinlan, T.J., K.A. Shaw, and L.A. Worrell. 1981. The effect of nitrogen on growth and chemical composition of irrigated Panicum maximum var trichoglume cv. Petrie - Neonotonia wightii cv. Tinaroo on the Atherton Tableland, North Queensland. Trop. Grassl. 15: 26-31.

Quinlivan, B.J., A.C. Devitt, and C.M. Francis. 1973. Seeding rate, time of sowing and fertilizers for subterranean clover seed production. Aust. J. Exp. Agric. Anim. Husb. 13: 681-684.

Richardson, A.E., G.S. James, and R.J. Simpson. 1985. Enumeration and distribution of Rhizobium trifolii in an acid soil and implications for nodulation of subterranean clover. p. 51. In Proc. 3rd Aust. Agron. Conf., Hobart.

Rogers, V.E., J.A.G. Irwin, and G. Stovold. 1978. The development of lucerne with resistance to root-rot in poorly aerated soils. Aust. J. Exp. Agric. Anim. Husb. 18: 434-441.

Rossiter, R.C. 1964. The effect of phosphate supply on the growth and botanical composition of annual type pasture. Aust. J. Agric. Res. 15: 61-76.

Russell, J.S. 1976. Comparative salt tolerance of some tropical and temperate legumes and tropical grasses. Aust. J. Exp. Agric. Anim. Husb. 16: 103-109.

Schroder, P.M. 1987. Adoption of higher sowing rates of subterranean clover in south west Victoria. p. 207. In Proc. 4th Aust. Agron. Conf., Melbourne.

Scott, B.J. 1985. Effects of furrow sowing and row arrangement on the establishment of undersown barrel medic in the low rainfall wheatbelt of central western New South Wales. p. 43-47. In Z. Hochman (ed.) The ecology and agronomy of annual medics. Dep. Agric. N.S.W. Tech. Bull. No. 32.

Scott, B.J., and H. Brownlee. 1974. Establishment of barrel medic under wheat, oats, barley and linseed in central western New South Wales. Aust. J. Exp. Agric. Anim. Husb. 14: 785-789.

Sedgley, R.H. 1962. Physical effects of gypsum on some Riverine soils. Paper 55. In 3rd Aust. Conf. Soil Sci., Canberra.

Smith, R.C.G., W.K. Mason, W.S. Meyer, and H.D. Barrs. 1983. Irrigation in Australia: developmental prospects. p. 99-153. In D. Hillel (ed.) Advances in irrigation. Vol. 2. Academic Press, New York.

Taylor, A.J., and K.A. Olsson. 1987. Effect of gypsum and deep ripping on lucerne (Medicago sativa L.) yields on a red-brown earth under flood and spray irrigation. Aust. J. Exp. Agric. 27: 841-849.

Taylor, G.B. 1985. Effect of tillage practices on the fate of hard seeds of subterranean clover in a ley farming system. Aust. J. Exp. Agric. 25: 568-573.

Taylor, G.B., and M.A. Ewing. 1988. Effect of depth of burial on the longevity of hardseeds of subterranean clover and annual medics. Aust. J. Exp. Agric. 28: 77-81.

Teitzel, J.K., and C.H. Middleton. 1980. Pasture research by the South Johnstone Research Station. Queensl. Dep. Prim. Indust. Agric. Br. Tech. Rep. No. 22.

Thorn, C.W., and M.W. Perry. 1983. Regulating pasture composition with herbicides. J. Agric. West. Aust. 1: 21-26.

Vickery, P.J. 1983. Assessment of the superphosphate needs of improved pastures from Landsat data. p. 139-143. In Phosphorus for pastures. Aust. Wool Corp. Spec. Workshop, Adelaide.

Walsh, P.A., and S.J. Cook. 1988. Development of a three-row planter for band-seeding of pasture legumes. Proc. Conf. on Agric. Eng., Sydney, Australia (in press).

Williams, C.M., and M. Raupach. 1983. Plant nutrients in Australian soils. p. 777-793. In Division of Soils, CSIRO. Soils: An Australian viewpoint. CSIRO. Academic Press, Melbourne.

Winks, L., F.C. Lambert, K.W. Moir, and P.M. Pepper. 1974. Effect of stocking rate and fertilizer on the performance of steers grazing Townsville stylo-based pasture in north Queensland. Aust. J. Exp. Agric. Anim. Husb. 14: 146-154.

Wolfe, E.C., and A. Lazenby. 1973a. Grass-white clover relationships during pasture development. 1. Effect of superphosphate. Aust. J. Exp. Agric. Anim. Husb. 13: 567-574.

Wolfe, E.C., and A. Lazenby. 1973b. Grass-white clover relationships during pasture development. 2. Effect of nitrogen fertilizer with superphosphate. Aust. J. Exp. Agric. Anim. Husb. 13: 575-580.

DISCUSSION

Sheaffer: In aerial seeding, is there any seed coverage by soil? Does seed coating improve seedling survival?

Gramshaw: Partial seed coverage by soil may eventuate if soil and seed is washed into surface depression during rain, but much seed may remain either uncovered or under plant litter. Seed coatings with legumes generally have shown no conclusive benefits for seedling survival.

Curll: In aerial seeding, it is important that the seed falls onto a suitable land surface. If the surface is bare, there will be negligible establishment. With a rougher surface (e.g., tussocks, debris), establishment is better, but for most species is 2-3% of viable seed or less.

Matches: Frost seedings, which capitalize on soil movement caused by freezing and thawing, are commonly used in pasture renovation across the southern Corn-Belt of the USA. Red clover is often broadcast seeded in February or March, and freezing and thawing action covers the seed if soil moisture is adequate to permit heaving.

J. Hay: What work is being done in Australia on P-efficient genotypes of legumes?

Gramshaw: Appreciable differences have been found between, and within, legume species in their ability to use P. Marked variation has been detected in subterranean and white clovers in particular. The value of this in grazed pasture communities is unclear.

Reed: Most pastures in Australia are too low in P concentration for optimal animal nutrition. Phosphate-efficient legumes may slightly help persistence, but any substantial reduction in P will reduce productivity and reproductive performance of grazing animals.

Kretschmer: Has any attempt been made to utilize grazing animals to increase legume germinations in established grass pastures? If so, has it been successful?

Gramshaw: Hard seeds of legumes can be ingested, escape digestion, and be spread in animal feces. There are a number of examples where natural spread of legumes has been aided by livestock. The advantages of attempting establishment by deliberately feeding high-cost legume seed to animals compared to conventional establishment are unproven and are questionable on economic grounds.

Curll: Aerial seeding in Australia is cost-effective because seed is incorporated with the fertilizer and the cost of spreading seed and fertilizer is shared.

Hoveland: In the northern USA, sod-seeding with legumes is reasonably dependable but the probability of success declines sharply further south, even on cool-season grass sods. I conducted a series of sod-seeding experiments with white and red clovers on tall fescue in Alabama over a 5 yr period and the success rate was only 50%. What is the success rate in Australia?

Gramshaw: I am unable to give an authorative estimate of the success rate of sod-seeding on a national basis. However, I suspect experiences in Australia would be no better, and perhaps worse, than the Alabama experience. Success is strongly influenced by prevailing moisture conditions and the control of competition from resident plants post-sowing.

Sheaffer: Despite research in the USA for the past 60 yr, sod-seeding has not been widely accepted by producers. It is very risky and often it is more economical to use N fertilizer on permanent pasture.

Gramshaw: Use of N fertilizer is not always an economic proposition in pasture-based animal production systems in Australia. There is a greater reliance on the successful establishment of legumes in pastures.

Irwin: Does burning have a role for legume establishment and management in northern Australia?

Gramshaw: Legumes may be successfully established in native pastures after burning which temporarily reduces competition. However, burning confers the best advantage when newly cleared land is sown and a substantial ash-bed favors establishment. The deliberate use of fire to manipulate legume content of established pastures may be feasible with very hardseeded legumes, but valuable forage is simultaneously lost.

ASPECTS THAT LIMIT THE SURVIVAL OF LEGUME SEEDLINGS

W. L. Lowther, J.H. Hoglund, and M. J. Macfarlane

SUMMARY

Low rates of early seedling survival can be major limitations to the successful introduction of legumes, particularly following aerial oversowing of seed in hill and high country environments. Factors restricting seedling survival include competition from existing vegetation, nodulation failure and slow onset of symbiotic N_2 fixation.

In humid and subhumid environments, techniques have been developed to increase legume establishment by controlling the competition from existing vegetation. This involves hard grazing prior to oversowing, herbicide application to suppress or kill vegetation, use of stock trampling to disturb soil and vegetation, and managed grazing after oversowing. In soils devoid of rhizobia, nodulation failures can limit seedling establishment unless inoculation and pelleting techniques ensure high populations of rhizobia on the seed at sowing. However, low levels of available soil-N may limit seedling growth and survival in legume species with slow onset of symbiotic N_2 fixation.

Major limitations to seeding establishment still occur in semiarid environments and further research is required to develop reliable and cost-effective establishment techniques.

INTRODUCTION

The importance of inoculation, fertilizer and grazing management in establishing oversown legumes into existing swards on New Zealand hill and high country has been long known (Sewell, 1950; Suckling, 1954). However, establishment is often poor (Cullen & Ludecke, 1967; Charlton, 1977; Lowther, 1983) and in semiarid environments may be so low as to prevent the introduction of more productive legume species (Musgrave & Lowther, 1976). This paper discusses some recent New Zealand research into methods of improving the survival and persistence of establishing legume seedlings with emphasis on hill and high country environments where aerial oversowing of seed onto the soil surface is often the only practical and/or economical method of sowing.

EFFECT OF GRAZING MANAGEMENT AND HERBICIDES

On unplowable hill country, aerial oversowing is the only practical means of introducing new legume species or cultivars into the resident sward. Although herbicide application to reduce competition (Blackmore, 1964) and the treading of seed (Sewell, 1950) have been recognized as desirable management practices, little has been done to quantify the results of these treatments. A series of experiments in the North Island humid hill country have extended previous small-scale plot work (Sithamparanathan et al., 1986) into full-scale paddock oversowing situations on a practical farm basis to develop a reliable, cost effective development program (Macfarlane & Bonish, 1986).

In the absence of herbicide application, oversowing into the existing vegetation (2250 kg DM/ha) resulted in low establishment of white clover (Trifolium repens L.) seedlings (Table 1). Establishment was increased by grazing the existing sward down to 1100 kg DM/ha prior to oversowing, while grazing at a stocking rate of 200 sheep/ha for 48 h after oversowing, further increased establishment. Shepherding the stock around the paddock for 2 hours prior to their removal, disturbed more soil and vegetation, reduced the amount of the seed visible on the surface by 20%, and increased establishment.

Highest seedling establishment was obtained when grazing management treatments were coupled with herbicide (200 g a.i. paraquat/diquat 5:1) application. This light rate of desiccant was used to suppress the rate of regrowth rather than to kill the resident sward. Application of the desiccant in 200 L/ha of water was more effective in reducing competition and hence increasing establishment than application in 60 L. The overall effect of the lower rate was an establishment half way between using no herbicide and 200 L water/ha. However, the reduction in water rate saved 65-70% of application costs in both fixed wing and helicopter operations and a savings of 25-30% of the per ha development cost. Further research into the effect of water rate and active ingredient applied is required to find the most cost effective combination. If complete control of resident vegetation is required the use of a herbicide such as glyphosate is necessary.

Table 1. Effect of grazing management and herbicide on the percentage of white clover seed established.[1]

	Existing vegetation	Graze only	Graze + treading	Graze + treading + shepherding
Nil	14	30	42	46
Herbicide[2] in 60 L	29	55	51	66
Herbicide in 200 L	40	71	63	76

[1] Adapted from Macfarlane and Bonish (1986).
[2] 200 g a.i. paraquat/diquat (5:1) in 60 L or 200 L of water per ha.

Post grazing management was critical for seedling survival and plant establishment. When the competing pasture was allowed to accumulate beyond 1200 kg DM/ha before the first grazing, seedling establishment was reduced by nearly 50% in all treatments. This indicates the continued need for control of the sward to ensure competition is reduced to a minimum.

Practical guidelines for hill country pasture renovation, have been developed (Macfarlane & Bonish, 1986), and include the following criteria:

1. Hard graze before oversowing to reduce competition.
2. Spray to suppress, or control the existing vegetation.
3. Graze and shepherd stock after oversowing to further reduce competition and to increase soil disturbance.
4. Grazing management after oversowing must balance frequency and severity of defoliation to encourage legume seedling growth but restrict competition from the resident vegetation.

The tussock grasslands of the South Island are almost devoid of legumes, and development is dependent on successful legume establishment. Principles for maximizing legume establishment are similar to those for hill country but usually they cannot be implemented fully, due to climatic and management constraints (Allan et al., 1987).

In the subhumid and humid tussock grassland, dense vegetation can limit survival of legume seedlings and is normally cleared by using large mobs of stock or by burning a year prior to oversowing. Oversowing in late winter or early spring is recommended to allow establishment prior to the onset of dry summer conditions. The use of stock to trample seed into the soil is rarely practiced but soil disturbance and hence seed burial can occur with frost heave. Although seedling establishment is often less than 10% of seed sown in these environments (Cullen & Ludecke, 1967), the number of seedlings is usually sufficient, particularly with the ability of white clover to spread by stolons.

In semiarid tussock grassland (450 mm annual rainfall), establishment problems are accentuated by the severe environmental conditions. For example, Musgrave and Lowther (1976) reported that less than 2% of lucerne (Medicago sativa L.) seeds reached the unifoliate leaf stage following oversowing onto a range of denuded sites, unless very favorable weather conditions followed oversowing. Recent research (M. H. Douglas, 1988, unpublished data) has reiterated the importance of seed burial by stock trampling and control of competition. In the absence of trampling by sheep there was less than 1% establishment of lucerne at the end of the first growing season. On a denuded site, trampling resulted in 15% establishment with little further effect from herbicide treatment. In contrast, in the presence of existing cover (approximately 1500 kg DM/ha at sowing) trampling resulted in only 1% establishment which was increased to 11% with herbicide application. Further research is required to define minimum stocking rates to effect a satisfactory level of trampling.

EFFECT OF RHIZOBIUM ON SEEDLING SURVIVAL

Early seedling survival of pasture legumes is frequently dependent on nodulation by rhizobia and the onset of symbiotic N_2 fixation. Rhizobia of pasture legumes are not native to New Zealand but naturalized populations have become established since European settlement. With the exception of extensive areas of undeveloped tussock grasslands where rhizobia are absent (Lowther & Trainor, 1988), rhizobia infective on clovers are now relatively widespread in pastoral zones. However, distribution may be patchy and establishment responses to inoculation have been recorded (Macfarlane & Bonish, 1986). Populations of rhizobia vary in their N_2 fixing effectiveness but attempts to increase symbiotic N_2 fixation and hence growth of white clover through the introduction of highly effective strains into an existing soil population have been unsuccessful due to low nodule occupancy by the introduced strain (Pankhurst & Greenwood, 1983). Rhizobia for specific legumes (e.g. lucerne or lotus) are absent from large areas of New Zealand where rhizobia for clovers may be plentiful (Greenwood, 1965).

Although complete nodulation failures are easily recognizable, the influence of partial nodulation failures on legume establishment and yield is often overlooked. Low seedling establishment due to partial nodulation failure can lead to an underestimation of the yield potential of a legume under experimental conditions or to perceived poor productivity under farming conditions. Although the establishment phase is important in all legumes, the rapid spread of plants such as white clover by stolons and natural reseeding can "thicken-up" stands with time. However in the absence of vegetative spread [lucerne, birdsfoot trefoil (*Lotus corniculatus* L.)] or where vegetative spread is slow ('Grassland Maku' lotus (*L. pedunculatus* Cav.); Wedderburn & Lowther, 1985) establishment limitations have a long-term effect on the potential productivity of a legume sward.

There are well publicized methods for inoculating and pelleting legume seed, but experience in New Zealand has indicated that partial or complete nodulation failures still occur. Existing recommendations have often proved unsatisfactory under farm scale operations and/or where seed is sown into severe environmental or soil conditions.

There are three major areas where nodulation problems can limit legume establishment in hill and high country environments:

1. Nodulation failures due to low numbers of rhizobia on the seed at sowing.
2. Nodulation failures on acid soils due to ineffective pelleting.
3. Slow nodulation and initiation of symbiotic N_2-fixation in low fertility soils.

Peat based inoculants are recommended, to supply high numbers of rhizobia and to enhance survival of rhizobia on the seed after inoculation. However even with peat based inoculants, the number of viable rhizobia on the seed can decline very rapidly. For example, the number of rhizobia on white clover seed declined to 7.5% and then 2.9% of that applied after 1 and 6 h, respectively (Hale, 1977). Hale attributed the rapid decline to desiccation and toxic seed diffusates.

The very rapid decline in the number of viable rhizobia on the seed has important implications for legume establishment, particularly in relating experimental procedures to farming practice. Under farming conditions, delays are common between inoculation and sowing and values for number of rhizobia per seed at inoculation may bear little relationship to those at sowing. For example, in hill and high country environments, seed is commonly inoculated the day prior to aerial oversowing. For this reason, inoculation treatments have been evaluated with minimum storage intervals between inoculation and sowing of 20-24 h. Treatments have also been evaluated with the longer storage intervals typical of those that can occur when sowing is delayed (e.g., due to unfavorable weather) or where inoculation is carried out by a commercial firm.

Improvements in survival of rhizobia have been obtained by adding gum arabic to the inoculant slurry (Lowther & Littlejohn, 1984), by selection of better adapted strains of rhizobia (Lowther & Johnstone, 1978) and by seed pelleting (Lowther, 1975). The majority of research has been with white clover because of its importance in the pastoral development of the South Island tussock grasslands and if the recommendations of Lowther and Trainor (1988) are followed, satisfactory nodulation and hence survival of seedlings can be obtained using either commercially pelleted seed or seed inoculated and pelleted by individual farmers. However, limitations in seedling nodulation still occur with some other legumes.

Maku lotus is nodulated by slow growing, acid tolerant rhizobia (Norris, 1967) and New Zealand results have confirmed the suggestion of Norris to use low concentrations of adhesive in the inoculant slurry to improve survival of rhizobia on the seed (Lowther & Littlejohn, 1984). However, the highest nodulation of Maku lotus seedlings occurred when the inoculation level was increased five-fold from the rate stipulated by the manufacture. This result cannot be attributed to low quality inoculants as the number of rhizobia exceeded the New Zealand Inoculant and Coated Seed Testing Service standard of of 1.5×10^9 per g peat and supplied 6×10^3 rhizobia per seed at the normal rate. Similar responses to increasing the inoculation rate occur with birdsfoot trefoil (Table 2).

The present recommendation is to inoculate lotus seed at five times the stipulated rate, with the addition of 10% gum arabic to the slurry, and sow within 2 weeks. However recently, bulk supplies of gum arabic have become almost unobtainable in New Zealand resulting in the urgent need for an alternative material. Substituted celluloses have been recommended but in general, comparisons have shown them to be inferior to gum arabic in maintaining high populations of rhizobia on the seed and promoting nodulation (Lowther, 1975; Lowther & Littlejohn, 1984) and they are not recommended in New Zealand. Roughley (1970) lists a range of materials reputed to prolong the survival of rhizobia on the seed but further experimental work is required. Preliminary evaluations of milk powder and sucrose have been inconclusive (Lowther & Littlejohn, 1984) and further research is underway particularly into materials that are low cost and readily available.

Nodulation in Acid Soils

One of the earliest developments of seed pelleting was to replace the need for heavy dressings of lime on acid soils (Loneragan et al., 1955). Early trials in New Zealand failed to demonstrate any positive effects from lime pelleting white clover seed oversown onto acid tussock grassland soils, although marked responses occurred in glasshouse trials (Adams, 1965). This lack of response has since been attributed to low numbers of rhizobia on the pelleted seed due to low quality inoculants and unsuitable adhesives (Lowther, 1975).

A similar lack of response to seed pelleting is now occurring with birdsfoot trefoil (Table 2) apparently related to the use of a strain of rhizobia with poor survival characteristics. At the normal inoculation level (2.5×10^4 rhizobia/seed), lime pelleting had no apparent effect on the percentage of seedlings nodulated on an acid (pH 5.2) soil. At the higher inoculation level, lime pelleting increased the percentage of nodulated seedlings from 24 to 49 and increased the weight of nodulated seedlings by 70%. After storage for 2 weeks, few seedlings nodulated. The results clearly show the importance of seed pelleting in promoting establishment of birdsfoot trefoil in soils of marginal pH. However, further research is required into maintaining adequate populations of rhizobia on the seed, possibly by selection of a better adapted strain of rhizobia. In addition an alternative is required for gum arabic as an adhesive before pelleting can be carried out on a large scale.

Nodulation and Nitrogen Fixation in Low Fertility Soils

The application of fertilizer-N at sowing is not recommended for establishing clover based pastures on hill and high country soils even under low fertility conditions (Cullen, 1971). However, recent research with birdsfoot trefoil strongly suggests that N-stress can restrict early seedling growth and limit the survival of seedlings especially where there is an early onset of drought conditions. Birdsfoot trefoil was overdrilled into two sites in the semiarid and subhumid zones of Central Otago (mean annual rainfall 380 and

Table 2. Effect of inoculation level, pelleting, and soil pH on the percentage of birdsfoot trefoil seedlings nodulated.[1]

	Soil pH	
	5.2	5.8
No inoculation	0	0
Normal inoculation(I)	10	11
I + gum arabic	20	43
I + lime pellet	19	40
5-times inoculation(5I)	11	36
5I + gum arabic	24	60
5I + lime pellet	49	64
SED	7.0	

[1] W.L. Lowther, 1988, unpublished data.

600 mm, respectively) as part of an inoculation study. The semiarid site at had been regularly topdressed with P and S fertilizer and there was a vigorously growing sward dominated by annual clovers. In contrast, the subhumid site had not been topdressed and was severely S-deficient. Although there were some annual clovers present, plants were small and lacking in vigor. Both sites were topdressed with P and S fertilizer prior to overdrilling.

There was a rapid germination of birdsfoot trefoil seed on both sites but over the following 12 weeks visual observations indicated that seedling growth was more rapid on the semiarid site with seedlings averaging 10 cm in height and multi-stemmed. On the subhumid site few seedlings exceeded 5 cm in height. The difference in growth was reflected in plant foliage weights of 180 and 40 mg DM, respectively. Inoculation had no effect and there were no visual differences in growth between nodulated and non-nodulated seedlings. Rainfall figures for each of the 3 months were higher for the subhumid site indicating that the better growth at semiarid site was not attributable to more favorable soil moisture conditions.

The results strongly suggest that initial onset and rate of symbiotic N_2 fixation in birdsfoot trefoil is slow, limiting early seedling growth in soils low in N. This then exposes the small seedlings with restricted root systems to drought stress with the onset of dry summer conditions. On sites with higher soil-N, seedling growth is rapid and the larger, deeper rooted seedlings are more able to withstand the onset of drought.

Results from a glasshouse experiment support this conclusion and indicate marked species differences in seedling response to N. All legume species tested showed better seedling growth under controlled conditions when supplied with mineral N as compared to seedlings with complete dependence on N_2 fixation (J. H. Hoglund, 1988, unpublished data; Table 3). The magnitude of this mineral-N response varied from 2- to 15-fold with different species. Nitrogen response was greatest in species with a high nodule

Table 3. Response of legume seedlings to combined-N, and shoot/nodule weight ratio at N_0.[1]

Legume	Shoot yields (mg DM/pot) N_0	N_+	N-response index N_+/N_0	Shoot/ nodule ratio
M. sativa	190	430	2.3	26
T. pratense	330	840	2.5	36
T. hybridum	130	340	2.6	24
T. repens	150	400	2.7	22
Desmodium intortum	20	120	6.0	4
L. pedunculatus	20	260	13.0	2

[1] J.H.Hoglund, 1988, unpublished data.

mass relative to seedling size. Poor nodule efficiency in terms of N_2 fixation per unit nodule mass places a greater overhead on the seedling, not only in terms of the greater time required to develop such nodules to comparable N_2 fixation output, but also because of the greater N requirement of the developing nodule which competes directly with the seedling. *Lotus pedunculatus* was the most responsive species tested and Pankhurst and Jones (1979) increased nodulation and N_2 fixation in this species by alleviating N-stress through the application of mineral N.

The technology challenge in overcoming the limitation on seedling establishment, from slow initiation of symbiotic N_2 fixation, is how to provide a small amount of N to the legume seedling in the field without affecting seed germination or stimulating the competing vegetation at the same time.

CONCLUSION

In New Zealand, problems with early seedling survival are recognized as a major limitation to the establishment and persistence of legumes particularly in hill and high country environments. Attention to the correct management techniques and seed inoculation recommendations can ensure successful establishment of oversown seed in subhumid and humid environments. However, major limitations still occur in semiarid environments and further research is required to develop reliable and cost-effective establishment techniques.

REFERENCES

Adams, A. F. R. 1965. Observations of legume establishment and growth on acid soils. Proc. N.Z. Grassl. Assoc. 26:115-122.

Allan, B.E., W.L. Lowther, and P.J. Walton. 1987. Pastures. High Country. Planning, establishment, management of oversown grasses and clover. New Zealand Ministry of Agriculture & Fisheries, AgLink FPP 886.

Blackmore, L.W. 1964. Chemical renovation of pastures in Southern Hawke's Bay and Northern Wairarapa. N.Z. J. Agric. 108:122-135.

Charlton, J. F. L. 1977. Establishment of pasture legumes in North Island hill country II. Seedling establishment and plant survival. N.Z. J. Exp. Agric. 5:385-390.

Cullen, N. A. 1971. Establishment of pastures on yellow-brown loams near Te Anau IX. The effect of nitrogenous fertilisers on establishment of new pasture. N.Z. J. Agric. Res. 14:40-46.

Cullen, N. A., and T. E. Ludecke. 1967. The effects of inoculation pelleting, rate of lime and time of sowing on establishment of white clover. Proc. N.Z. Grassl. Assoc. 28:96-103.

Greenwood, R.M. 1965. Populations of rhizobia in New Zealand soils. Proc. N.Z. Grassl. Assoc. 26:95-101.

Hale, C.N. 1977. Some factors affecting the survival of Rhizobium trifolii on white clover. Proc. N.Z. Grassl. Assoc. 38 (1):182-186.

Loneragan, J. F., D. Meyer, R. G. Fawcett, and A. J. Anderson. 1955. Lime pelleted clover seeds for nodulation on acid soils. J. Aust. Inst. Agric. Sci. 21:264-265.

Lowther, W. L. 1975. Pelleting materials for oversown clover. N.Z. J. Exp. Agric. 3:121-125.

Lowther, W. L. 1983. Influence of site on response of 'Grasslands Maku' Lotus pedunculatus establishment to seed pelleting and broadcast lime. N.Z. J. Agric. Res. 26:423-426.

Lowther, W. L., and P. Johnstone. 1978. Effect of strains of Rhizobium trifolii on establishment of oversown white clover (Trifolium repens). Soil Biol. Biochem. 10:293-295.

Lowther, W. L., and R. P. Littlejohn. 1984. Effect of strain of rhizobia, inoculation level, and pelleting on the establishment of oversown Lotus pedunculatus 'Grasslands Maku'. N.Z. J. Exp. Agric. 12:287-294.

Lowther, W. L., and K.D. Trainor. 1988. Legumes. Seed inoculation and coating. Tussock grassland procedures. New Zealand Ministry of Agriculture & Fisheries, Aglink FPP 887.

Macfarlane, M. J., and P. M. Bonish. 1986. Oversowing white clover into cleared and unimproved North Island hill country - The role of management, fertiliser, inoculation, pelleting and resident rhizobia. Proc. N.Z. Grassl. Assoc. 47:43-51.

Musgrave, D. J., and W. L. Lowther. 1976. Effect of sowing date, inoculation level, and pelleting on the establishment of oversown lucerne. N.Z. J. Exp. Agric. 4:65-70.

Norris, D. O. 1967. The intelligent use of inoculants and lime pelleting for tropical legumes. Trop. Grassl. 1:107-121.

Pankhurst, C. E., and R. M. Greenwood. 1983. Establishment and persistence of Rhizobium trifolii in a developed pasture soil. N.Z. J. Exp. Agric. 11:165-169.

Pankhurst, C. E., and W. T. Jones. 1979. Effectiveness of Lotus root nodules III. Effect of combined nitrogen on nodule effectiveness and flavolan synthesis in plant roots. J. Exp. Bot. 30:1109-1118.
Roughley, R. J. 1970. The preparation and use of legume seed inoculants. Plant Soil 32:675-701.
Sewell, T. G. 1950. Improvement of tussock grassland. N.Z. J. Agric. 81:293-299.
Sithamparanathan, J. M. J. Macfarlane, and S. Richardson. 1986. Effect of treading, herbicides, season, and seed coating on oversown grass and legume establishment in easy North Island hill country. N.Z. J. Exp. Agric. 14:173-182.
Suckling, F. E. T. 1954. Pasture management trials on unploughable hill country at Te Awa I. Establishment of experimental area and results for 1949-51. N.Z. J. Sci. Technol. 36a:237-273.
Wedderburn, M. E., and W. L. Lowther. 1985. Factors affecting establishment and spread of "Grasslands Maku" lotus in tussock grasslands. Proc. N.Z. Grassl. Assoc. 46:97-101.

DISCUSSION

Sheaffer: Is seed coating with rhizobia effective in increasing establishment of alfalfa?

Lowther: It is effective on acid soils devoid of rhizobia.

Hoveland: Pelleting response have been obtained in southern USA where seed is sown in early autumn when soil temperatures are high and moisture is limiting.

Hochman: In Australia, lime pelleting is recommended when inoculated seed is mixed with fertilizer at sowing.

Reed: Sales representatives in Australia claim that nearly all legume seed aerially sown in New Zealand is coated with lime and other nutrients.

Lowther: Lime coating of inoculated clover seed is recommended for aerial oversowing. Sulfur and molybdenum may be added to the coat.

Jones: We should not forget the possibility of using seed dispersal in feeces to spread legumes into non-arable areas.

Lowther: This is an effective technique if rhizobia are present in the soil.

J. Hay: Does the lack of insect pollinators limit seed set in tussock grassland environment?

Lowther: There are enough pollinators to ensure sufficient seed set for natural reseeding.

Smith: Is there any research being carried out in New Zealand into using modern biotechnical techniques to encapsulate germinated seeds for aerial establishment in marginal areas?

Lowther: Not to my knowledge.

LEGUME ESTABLISHMENT AND HARVEST MANAGEMENT IN THE U.S.A.

C. C. Sheaffer

SUMMARY

Strategies for tilled and no-till seedbeds which maximize stand establishment and seedling persistence provide a favorable environment for seedling development by optimizing availability of essential resources. Reduction of competition for light and water by companion grasses and weeds is important for tilled and no-till seedbeds. In tilled seedbeds, legumes are frequently established using companion crops or herbicides for weed control. No-till methods include interseeding, sod-seeding, frost-seeding, and topseeding.

Selection of a harvest strategy is influenced by the relative importance of forage yield, forage quality, and stand persistence. Increasing harvest or grazing frequency results in higher quality forage but often reduces stand persistence. Legume stand persistence is best predicted by considering the interaction of harvest strategy with environmental and edaphic conditions and with plant factors such as disease resistance and stand age.

ESTABLISHMENT

The risk of forage legume establishment failure is greater than for other major crops. This risk is associated with small seed size and a lack of seedling vigor. Legume seedlings are vulnerable to competition from non-legumes. Limitation in water, light, and nutrients are major constraints to legume seedling persistence. Establishment strategies are aimed at optimizing seedling emergence and survival while minimizing soil and wind erosion.

Establishment Constraints

Water

Small seeded legumes are particularly vulnerable to moisture deficits since seeds are sown near the soil surface. After legume seeds have absorbed sufficient water for germination (typically greater than 100% of dry weight) and radicle emergence occurs, dehydration causes seedling mortality (Cardwell, 1984). Weeds, companion crops, and grass sods compete with legume seedlings for water, and therefore, control of competition is important.

Firm soil, soil-seed contact, controlled planting depth, and surface residues are important in minimizing moisture effects on germination. For a silt loam soil with no irrigation following seeding, emergence of alfalfa (Medicago sativa L.) seedlings improved progressively by increasing planting depth from 0 to 2.5 cm and by increasing soil compaction from 0 to 83 kPa (Triplett & Tesar, 1960). When 1.3 cm of irrigation was applied, a 0.6 to 1.3 cm planting depth with a compaction of 42 kPa resulted in the greatest emergence. Recommended dates for legume establishment vary with region but usually occur during periods of favorable rainfall and temperature (Van Keuren & Hoveland, 1985).

Light

Shading reduces both herbage and root growth of forage legumes, but shaded legumes partition relatively less dry matter into roots than stems and shoot/root mass ratio increases (Cooper, 1967). Gist and Mott (1957) reported that increasing light caused a curvilinear increase in herbage growth and a linear increase in root growth of alfalfa, red clover (Trifolium pratense L.), and birdsfoot trefoil (Lotus corniculatus L.). A barley (Hordeum vulgare L.) companion crop reduced alfalfa root penetration one-third and birdsfoot trefoil root penetration two-thirds (Cooper & Ferguson, 1964). Shading limits legume seedling response to soil moisture by reducing rooting. With no shading by competing companion crops or perennial grass sods, seedling growth rate was increased with increases in soil moisture, but with shading no response to increasing soil moisture was observed (Cooper & Ferguson, 1964; Groya & Sheaffer, 1981). In sod-seeding, high levels of soil moisture may promote competition for light by stimulating perennial grass growth.

Legumes are generally considered to be less tolerant of shading than grasses (Kendall & Stringer, 1985), but legumes differ in their tolerance of shading. Gist and Mott (1957) found that red clover was less affected by shading than alfalfa which was more tolerant of shading than birdsfoot trefoil. Hart (1976) reported that alfalfa and crownvetch (Coronilla varia L.) had similar tolerance to shade and that both species had greater shade tolerance than ladino clover (Trifolium repens L.).

Nutrients

The importance of supplying nutrients for successful legume establishment and long-term persistence is well recognized (Duell, 1974; Lanyon & Griffith, 1988). Phosphorus and K are generally considered the most important, but Ca, Mg, S, and B are also required for some soils. Calcium and Mg are often supplied during liming. Phosphorus is particularly important during seedling development because of its role in root formation (Lanyon & Griffith, 1988). Seeding legumes over a band of P fertilizer (band seeding) promotes seedling development on low P soils (Haynes & Thatcher, 1950). Tesar et al. (1954) found that alfalfa seeded directly over a band of P produced 66% more herbage than when placed 5.1 cm from the fertilizer. Duell (1964) also reported that adding K and N to P in the band promoted legume seedling development; however, legume emergence was reduced when seeds were sown in contact with banded N and K.

Soil pH influences the availability of toxic and essential elements, and biological dinitrogen fixation (Lanyon & Griffith, 1988). A soil pH of 6.5 to 7.0 is generally recommended for alfalfa, but red clover, white clover, birdsfoot trefoil, and subterranean clover (Trifolium subterraneum L.) are more tolerant of acid soils (Pearson & Hoveland, 1974). Lime should be applied 6 months before seeding because of the time required to change soil pH. Incorporation of lime into the root zone is also recommended (Barber, 1984). While Koch and Estes (1986) found that seeding year yields of sod-seeded alfalfa were greater with incorporated than unincorporated lime, yield in subsequent years was not affected by lime incorporation.

Allelopathy

Allelopathy, a chemically induced effect of one plant on another, has been implicated as a cause of legume stand establishment failure. Chemicals are excreted from growing plants or are leached from plant residues. Extracts of pasture weeds (Smith, 1986); tall fescue (Festuca arundinacea Schreb., Peters & Zam, 1968); alfalfa (Guenzi et al., 1964); and red clover (Tamura et al., 1969) have reduced legume seed germination and seedling development. Luu et al. (1982) indicated that allelopathic effects of tall fescue on birdsfoot trefoil germination and root growth were reduced by burning. Alfalfa autotoxicity in tilled seedbeds was reduced by delaying planting for 2 wk after plowing and in no-tillage by delaying planting for 3 wk after killing existing alfalfa with an herbicide (Tesar, 1986).

Establishment Strategies

Tilled Seedbeds

Tilled seedbeds are produced by varying amounts of primary and secondary tillage. Tillage provides the best opportunity to seed into a firm, weed-free seedbed, and for incorporation of fertilizer and lime into the plant root zone (Decker & Taylor, 1985). Moderate amounts of crop residues on the soil surface are beneficial in reducing soil erosion and crusting, and increasing moisture retention; however, the potential for soil and plant loss due to wind and water erosion is still greater under tilled than no-tilled systems.

Companion crops are frequently used for legume and legume-grass mixture establishment. Small grains are the most frequently used companion crops. Companion crops emerge faster than legumes and provide ground cover when legumes are in the initial stage of development. An effective companion crop suppresses weeds, reduces wind and water erosion, and ensures forage production in the seeding year while providing minimum competition to legume seedlings. Companion crops compete with forage seedlings for nutrients, water, and light. Competition can be reduced by selection of short stature, early maturing small grain species and varieties; reduction of small grain seeding rate; and removal of the small grain for forage before maturity. When small grains are harvested for grain, prompt removal of straw will prevent smothering of forage seedlings. Companion crops are not recommended for summer seedings because of their competition with legume seedlings for soil moisture.

In the northern USA, oat (Avena sativa L.) has traditionally been used as a companion crop in spring seedings although spring barley and spring wheat (Triticum aestivum L.) are also effective companion crops. Fall seeded small grains are not generally used as companion crops because of excessive competition (Klebesadel & Smith, 1959). Brink and Marten (1986) compared spring oat and barley companion crops for establishment of alfalfa. They found that although barley developed greater canopy area and subsequently provided greater light competition than oat following emergence, alfalfa stands were similar when oat and barley were harvested at vegetative or soft dough stages. Sheaffer et al. (1988) compared nonvernalized winter rye (Secale cereale L.), spring wheat, and spring oat as companion crops for spring alfalfa establishment and found no consistent difference in seeding year yields or stands when the different companion crops were harvested at vegetative or boot stages.

"Solo seeding" describes establishment of forages without companion crops. In solo seeding with herbicides ("clear seeding"), preplant and sometimes postemergence herbicides are used to eliminate competition from annual grasses and broadleaf weeds (Tesar & Marble, 1988). Clear seeding is effective on land without erosion potential and results in high-yielding legume monocultures in the seeding year. Preplant incorporated herbicides are not used for establishment of perennial legume-grass mixtures because of lack of herbicide selectivity between annual and perennial grasses.

Solo seeding without herbicides is a low-cost option to the use of companion crops and clear seeding for establishment. In this procedure, annual weeds are used as companion crops and grazed or harvested before maturity. Sheaffer et al. (1988) reported that alfalfa stands were similar and seeding year alfalfa yields were greater when weeds were used as companion crops than when spring seeded oat was used.

No-tillage

No-till or minimum tillage seeding includes those made by broadcasting into cereals or drilling into grass sods or stubble of small grains or row crops (Decker & Taylor, 1985). A no-till strategy reduces erosion, increases moisture retention, and uses less energy than conventional tillage systems. On rocky, irregular lands or highly erodible sites, there are no practical alternatives to no-tillage procedures. In contrast to tilled seedbeds, there is much greater risk of seedling mortality due to competition from weeds and established grass sods (White et al., 1985) and due to predation by insects and mollusks (Byers et al., 1985).

Interseeding, sod-seeding. In interseeding or sod-seeding, legume seeds are placed into established grass sods and crop residues using special drills (Decker & Taylor, 1985). Shallow seed placement is an important feature of effective drills (Decker & Taylor, 1985). Interseeding is an effective way to increase the productivity of permanent pastures while minimizing soil erosion. For example, Koch et al. (1987) reported that sod-seeding increased permanent pasture digestible DM and crude protein yield 48 and 75%, respectively, compared to N-fertilized controls. Other research has shown up to a 250% increase in pasture DM productivity due to legume interseeding (Bryan, 1975; Martin et al., 1983; Taylor & Allinson, 1983).

Considerable research on procedures for successful no-till legume establishment (Table 1) has resulted in the following management recommendations described by Decker & Taylor (1985) and White et al. (1985): (i) reduce competition for light and moisture by grass sods or weeds by grazing, mowing, burning, or herbicides; (ii) protect legume seedlings from insects, slugs, and nematodes using pesticides; (iii) provide adequate fertility; (iv) seed during periods of favorable temperature and moisture; and (v) precisely place seed in the soil. In sod-seeding, often the perennial grass sod is only partially suppressed and not completely killed. Partial sod suppression provides weed control

Table 1. Legume interseeding research which has evaluated alternative management practices.

Reference	Location	Legumes	Management evaluated
Decker et al. (1969)	MD	BFT, CV	Seeding date, Seeding equipment, Species comparison
Taylor et al. (1969)	KY	ALF, WHC	Seed placement-surface vs. drilled, Sod suppression-herbicide
Decker et al. (1985)	MD	RDC, BFT, CV, ALF	Herbicide banding, Species comparison
Olsen et al. (1978)	IL	WHC, RDC	Seeding equipment, No-till vs. conventional seeding
Kalmbacher et al. (1980)	FL	RDC, WHC, ALF	Sod suppression-herbicide vs. burning, Interseeding vs. overseeding, Species comparison
Welty et al. (1981)	WY	ALF, AKC	Sod suppression-seeding interval
Martin et al. (1983)	MN	ALF	Seeding date, Sod suppression-herbicide comparison
Taylor & Allison (1983)	CT	ALF, BFT RDC	Legume species comparison, Grass species comparison
Mueller & Chamblee (1984)	NC	ALF, WHC	Seed placement-surface vs. drilled, Seeding date, Sod suppression-herbicide
Rogers et al. (1985)	NC	ALF	Sod suppression-herbicide, Insect control
Koch & Estes (1986)	NH	ALF, BFT	Lime incorporation
Mueller-Warrant & Koch (1980)	NH	ALF	Seeding date, Sod suppression-herbicide comparison, No-till vs. conventional seeding
Tesar (1986)	MI	ALF	Herbicide spraying-seeding interval
Brink et al. (1988)	MS	SUB	Sod suppression-burning vs. haying Seed placement-surface vs. drilled

ALF = alfalfa, AKC = Alsike clover, BFT = birdsfoot trefoil, CV = crownvetch, RDC = red clover; SUB = subterranean clover, WHC = white clover.

and allows for higher seeding year grass yields than broadcast herbicide application (Decker et al., 1985).

Frost-seeding, topseeding, overseeding. The simplest form of no-till involves broadcasting legume seed on the soil surface during periods with adequate soil moisture. This technique can be used to establish legumes in winter cereals or to introduce legumes into grass sods (White et al., 1985). Freezing and thawing action, animal trampling, or light disking provides seed coverage and soil-seed contact. Successful seedings most often occur from late fall to early spring. In the southern USA, annual legumes including crimson (Trifolium incarnatum L.), arrowleaf (Trifolium vesiculosum Savi) and subterranean clovers, and vetch (Vicia sativa L.) have been successfully overseeded into dormant sods of warm season grasses in late fall (Dunavin, 1982; Evers, 1985). Taylor et al. (1972) found that with a doubling of the seeding rate, white clover and alfalfa could be surface-seeded into Kentucky bluegrass (Poa pratensis L.) sod in late winter and early spring. In Iowa, alfalfa, red clover, and birdsfoot trefoil were successfully frost-seeded into perennial grasses in March, but frost-seedings of cicer milkvetch (Astragalus cicer L.), white clover, and alsike clover (Trifolium hybridum L.) were not successful (George, 1984).

HARVEST MANAGEMENT

Local growing conditions and producer goals influence harvest management strategies. Producer goals are usually based on the relative value of forage yield, forage quality, and stand persistence. For example, in southern Minnesota, dairy producers who routinely harvest alfalfa at bud stage (four cuts per season) are willing to sacrifice DM or nutrient yield and stand persistence for excellent quality forage because of its value as feed. In contrast, if hay is produced for a market where maximum tonnage is more valued than forage quality, harvesting at first flower (three cuts per season) may be the most profitable.

Yield, Quality, and Persistence

The effect of a particular harvest system on seasonal legume forage DM yield and forage quality and on stand persistence is related to the morphological development of the crop at each harvest within the system. With maturation, DM yield usually increases while forage quality declines.

Forage DM accumulation rate of alfalfa, red clover, and birdsfoot trefoil is rapid until flowering and then usually decreases due to leaf loss from the basal portions of the stem (Buxton et al., 1985). McGraw and Marten (1986) reported that total forage DM accumulation of alfalfa, birdsfoot trefoil, and cicer milkvetch was nearly linear until early flowering (5 to 15% flowering) and was greatest at 25 to 50% flowering. Sainfoin (Onobrychis viciifolia Scop.) DM accumulation increased at a linear rate until 50% flowering and maximum DM accumulation occurred at 100% flowering. Changes in forage quality with maturation are related to changes in the leaf/stem proportion of the forage and to increases in structural components of the stem. Leaf mass of alfalfa, birdsfoot trefoil, sainfoin, cicer milkvetch, and red clover is typically equal to or greater than stem mass at vegetative to early bud stages, but by flowering stem proportion typically exceeds that of leaves (Buxton et al., 1985; McGraw & Marten, 1986). Although legume species and varieties vary in leaf retention, leaf loss usually exceeds production after 10 to 50% flowering.

Frequent harvest of forage legumes at immature stages has been related to reduced stand persistence due to depletion of nonstructural carbohydrates in storage organs (Kust & Smith, 1961; Chatterton et al., 1974). Storage and utilization of root reserves in alfalfa and red clover follow a cyclic pattern of decreasing during initiation of vegetative regrowth and then accumulating until plants reach full flower (Smith, 1962). Carbohydrate reserves of birdsfoot trefoil, cicer milkvetch and sainfoin are lower than in alfalfa and have less cyclic fluctuation in response to cutting during the summer (Nelson & Smith, 1968; Cooper & Watson, 1968; Gabrielsen et al., 1985). Regrowth from these legumes is more dependent on carbohydrates synthesized from leaf area remaining after harvest than from stored reserves.

Harvest Schedules

Legume harvests are scheduled either using fixed intervals, stage of growth, or a combination of the two criteria. Because of the relationship between the stage of morphological development and forage yield, forage quality, and stand persistence, it is apparent why growth stage is frequently used as a criterion of when to harvest. Cutting according to stage of development takes into account the effects of environments and variety maturity and is superior to cutting at fixed intervals in obtaining consistent forage yield and quality. Nevertheless, some producers harvest on fixed intervals because it facilitates scheduling of harvests with other activities such as irrigation or because legumes do not always flower due to environmental limitations.

Seasonal legume forage DM yield and forage quality due to variable harvest managements are shown in Table 2. Generally, as harvest frequency increases, seasonal forage DM yield decreases and forage quality increases. Long-term stand persistence of most legumes is usually reduced by harvesting at intervals of 30 days or less or by harvesting at preflowering stages; however, stand persistence of procumbent species such as cicer milkvetch, crownvetch, and birdsfoot trefoil is not always decreased by frequent harvesting if residual leaf area remains (Townsend et al., 1978; Gabrielson et al., 1985; Seim, 1968). Smith and Nelson (1967) found that leaving a leafy stubble by cutting at 7.6 or 15 cm was more important for birdsfoot trefoil persistence than for alfalfa persistence. Leaving greater than a 2.5 cm stubble was only beneficial to alfalfa when harvested five and six times per season.

In the northern USA, fall cutting can pose a risk to legume stand persistence by reducing levels of stored carbohydrates and by removing stubble which can catch and hold snowfall (Sheaffer et al., 1988). The traditional recommendation that alfalfa should not be cut during the fall critical period (from 4 to 6 wk before the first killing frost) was aimed at reducing this risk (Smith, 1972). Recent research has shown that with modern alfalfa cultivars and recommended fertility practices, the risk of stand loss due to fall cutting is minimized (Tesar & Yager, 1985). Furthermore, in the southern USA, with milder winter temperatures, fall cutting does not consistently affect persistence (Mays & Evans, 1973; Sholar et al., 1983).

Table 2. Effect of harvest management on seasonal yield and forage quality of legumes.

Legume	Variety	Location	Harvest mgt.	DM Yield	CP	IVDDM	Reference
				Mg/ha	-----%-----		
Cicer milkvetch	Lutana	CO	2-7 cuts/yr	10.5-9.2	16.8-30.0	64.8-70.4	Townsend et al. (1978)
Cicer milkvetch	Monarch	CO	3-4 cuts/yr	12.3-9.5	20.7-22.2		Gabrielsen et al. (1985)
Alfalfa	Baker			12.6-11.6	20.2-22.7		
Alfalfa	DK120	MN	2-4 cuts/yr	10.8-12.7	14.3-20.3	57.9-62.5	Sheaffer (1986, unpublished data, Univ. of MN)
White clover	Sacramento			4.3-5.1	19.9-23.4	65.8-67.5	
Alsike	Common			6.9-5.3	17.1-24.4	62.9-70.1	
Red clover	Arlington			9.5-10.3	10.3-15.4	60.1-64.9	
Crownvetch	Penngift			7.8-5.3	16.8-23.6	64.3-68.6	
Cicer milkvetch	Monarch			9.7-5.6	16.5-22.9	62.5-66.6	
Birdsfoot trefoil	Norcen			11.1-9.5	15.7-25.5	58.9-64.4	
Crownvetch	Emerald	PA	1-6 cuts/yr	5.4-3.0	11.7-24.0	51.4-70.0	Seim (1968)
Alfalfa	Vernal	MT	2-4 cuts/yr	11.8-9.1			Carleton et al. (1968)
Sainfoin	Eski			11.6-7.3			
Alfalfa	Vernal	WI	2-3 cuts/yr	9.4-9.9	15.6-20.7	63.4-65.2	Smith (1965)
Red clover	Lakeland			9.9-7.6	14.6-21.3	68.3-73.3	

Interacting Factors

Prediction of legume persistence is best related to overall plant stress. While aspects of harvest management (e.g., harvest frequency, cutting height) are important in determining persistence, it is also influenced by plant, edaphic, and environmental factors. These factors have been defined for alfalfa production in the northern USA (Sheaffer et al., 1988), but additional documentation is required for other regions and for other legumes. Plant factors which interact with alfalfa harvest management include variety winter hardiness and disease resistance, and stand age. Edaphic factors include soil fertility and soil moisture level. Winter temperature and temperature fluctuations and fall and winter precipitation are important environmental factors.

Increased varietal winterhardiness and disease resistance have reduced the effects of stressful cutting practices such as fall cutting on alfalfa persistence in the northern USA (Tesar & Yager, 1985). Alfalfa stands in the first 2 years of production are more tolerant of mismanagement than older stands because of the lower incidence of disease and less crown damage due to harvesting (McKenzie & McLean, 1980). High levels of soil fertility, particularly K, reduce the effect of stressful management and environments (Sheaffer et al., 1986). In Wisconsin, Smith (1975) found that fall cutting of unfertilized alfalfa (no K) reduced final stands by 50%, compared with no fall cutting; however, at annual K rates of 448 kg ha^{-1} or greater, stand reductions were <13%.

In the northern USA, precipitation in the form of snow protects plants from the harmful effects of low winter temperatures (Sharratt et al., 1987). In contrast, rainfall in the fall and winter may decrease persistence by saturating soils or by forming ice sheets. Soil saturation affects persistence by promoting heaving (Portz, 1967) and reducing winterhardiness (McKenzie & McLean, 1980).

There is a need to develop models and decision making aids which will facilitate prediction of stand loss and which determine the economic consequences of alternative management systems. A simple tool for decision making is the payoff matrix (Hesterman et al., 1986). The payoff matrix allows enumeration of the economic outcome of alternative management actions as a function of uncontrolled events of nature such as snowcover or rainfall.

REFERENCES

Barber, S.A. 1984. Liming materials and practices. In F. Adams (ed.) Soil acidity and liming. 2nd ed. Agronomy 12:171-209.

Brink, G.E., T.E. Fairbrother, and R.L. Ivy. 1988. No-till establishment of subterranean clover in warm-season sod. p. 207-210. In Proc. Forage Grassl. Conf., Baton Rouge, LA. 11-14 Apr. Am. Forage and Grassl. Counc., Belleville, PA.

Brink, G.E., and G.C. Marten. 1986. Barley vs. oat companion crops: II. Influence on alfalfa persistence and yield. Crop Sci. 26:1067-1071.

Bryan, W.B. 1975. Effect of sod-seeding legumes on hill land pasture productivity and composition. Agron. J. 77:901-905.

Buxton, D.R., J.S. Hornstein, W.R. Wedin, and G.C. Marten. 1985. Forage quality in stratified canopies of alfalfa, birdsfoot trefoil, and red clover. Crop Sci. 25:273-279.

Byers, R.A., W.C. Templeton, R.L. Mangan, D.L. Bierlein, W.F. Campbell, and H.J. Donley. 1985. Establishment of legumes in grass swards: effects of pesticides on slugs, insects, legume

seedling numbers and forage yield and quality. Grass Forage Sci. 40:41-48.
Carleton, A.E., C.S. Cooper, C.W. Roath, and J.L. Krall. 1968. Evaluation of sainfoin for irrigated hay in Montana. p. 44-48. In C.S. Cooper and A.E. Carleton (ed.) Sainfoin Symp., Bozeman, MT. 12-13 Dec. Montana Agric. Exp. Stn. Bull. 627.
Cardwell, V.B. 1984. Germination and crop production. p. 53-92. In M.B. Tesar (ed.) Physiological basis of crop growth and development. ASA, Madison, WI.
Chatterton, N.J., G.E. Carlson, R.H. Hart, and W.E. Hungerford. 1974. Tillering, nonstructural carbohydrates, and survival relationships in alfalfa. Crop Sci. 14:783-787.
Cooper, C.S. 1967. Relative growth of alfalfa and birdsfoot trefoil seedlings under low light intensity. Crop Sci. 7:176-178.
Cooper, C.S., and H. Ferguson. 1964. Influence of a barley companion crop upon root distribution of alfalfa, birdsfoot trefoil, and orchardgrass. Agron. J. 56:63-66.
Cooper, C.S., and C.A. Watson. 1968. Total available carbohydrates in roots of sainfoil (Onobrychis viciifolia Scop.) and alfalfa (Medicago sativa L.) when grown under several management regimes. Crop Sci. 8:83-85.
Decker, A.M., R.F. Dudley, L.R. Vough, M.I. Spicknall, and T.H. Miller. 1985. Band vs. broadcast paraquat and alternate row species seeding in no-till pasture renovation. p. 254-261. In Proc. Am. Forage Grassl. Conf., Hershey, PA. 3-6 Mar. Am. Forage and Grassl. Counc., Lexington, KY.
Decker, A.M., H.J. Retzer, M.L. Sarna, and H.D. Kerr. 1969. Permanent pastures improved with sod-seeding and fertilization. Agron. J. 61:243-247.
Decker, A.M., and T.H. Taylor. 1985. Establishment of new seedings and renovation of old sods. p. 288-297. In M.E. Heath et al. (ed.) 4th ed. Forages. Iowa State Univ. Press, Ames.
Duell, R.W. 1964. Fertilizer-seed placement with birdsfoot trefoil (Lotus corniculatus L.) and alfalfa (Medicago sativa L.). Agron. J. 56:503-505.
Duell, R.W. 1974. Fertilizing forage for establishment. p. 67-93. In D.A. Mays (ed.) Forage fertilization. ASA, Madison, WI.
Dunavin, L.S. 1982. Vetch and clover overseeded on a bahiagrass sod. Agron. J. 74:793-796.
Evers, G.W. 1985. Forage and nitrogen contributions of arrowleaf and subterranean clovers overseeded on bermudagrass and bahiagrass. Agron. J. 77:960-963.
Gabrielsen, B.C., D.H. Smith, and C.E. Townsend. 1985. Cicer milkvetch and alfalfa as influenced by two cutting schedules. Agron. J. 77:416-422.
George, J.R. 1984. Grass sward improvement by frost-seeding with legumes. p. 265-269. In Proc. Forage Grassl. Conf., Houston, TX, 23-26 Jan., Am. Forage Grassl. Counc., Lexington, KY.
Gist, G.R., and G.O. Mott. 1957. Some effects of light intensity, temperature, and soil moisture on the growth of alfalfa, red clover, and birdsfoot trefoil seedlings. Agron. J. 49:33-36.
Groya, F.L., and C.C. Sheaffer. 1981. Establishment of sod-seeded alfalfa at various levels of soil moisture and

grass competition. Agron. J. 73:560-565.
Guenzi, W.D., W.R. Kehr, and T.M. McCalla. 1964. Water-soluble phytotoxic substances in alfalfa forage: variation with variety, cutting, year, and stage of growth. Agron. J. 56:499-500.
Hart, R.H. 1976. Seedling growth of crownvetch, ladino clover, and alfalfa under shade. Agron. J. 68:683-685.
Haynes, J.L., and L.E. Thatcher. 1950. Band seeding methods for meadow crops. Ohio Farm Home Res. 262:3-5.
Hesterman, O.B., J.H. Hilker, and J.R. Black. 1986. A tool for agronomic decision making: the payoff matrix. Agric. Economics Staff Paper 86-1. Michigan State Univ., East Lansing, MI.
Jung, G.A., D.E. Brann, and G.W. Fissel. 1981. Environmental and plant growth stage effects on composition and digestibility of crownvetch stems and leaves. Agron. J. 73:122-128.
Kalmbacher, R.S., P. Mislevy, and F.G. Martin. 1980. Sod-seeding bahiagrass in winter with three temperate legumes. Agron. J. 72:114-118.
Kendall, W.A., and W.C. Stringer. 1985. Physiological aspects of clovers. In N.L. Taylor (ed.) Clover science and technology. Agronomy 25:111-159.
Klebesadel, L.J., and D. Smith. 1959. Light and soil moisture beneath several companion crops as related to the establishment of alfalfa and red clover. Bot. Gaz. 121:39-46.
Koch, D.W., and G.O. Estes. 1986. Liming rate and method in relation to forage establishment-crop and soil chemical responses. Agron. J. 78:567-571.
Koch, D.W., J.B. Holter, D.M. Coates, and J.R. Mitchell. 1987. Animal evaluation of forages following several methods of field renovation. Agron. J. 79:1044-1048.
Kust, C.A., and D. Smith. 1961. Influence of harvest management on level of carbohydrate reserves, longevity of stands, and yields of hay and protein from Vernal alfalfa. Crop Sci. 1:267-269.
Lanyon, L.E., and W.K. Griffith. 1988. Nutrition and fertilizer use. In A.A. Hanson et al. (ed.) Alfalfa and alfalfa improvement. Agronomy 29:333-372.
Luu, K.T., A.G. Matches, and E.J. Peters. 1982. Allelopathic effects of tall fescue on birdsfoot trefoil as influenced by N fertilization and seasonal changes. Agron. J. 74:805-808.
Martin, N.P., C.C. Sheaffer, D.L. Wyse, and D.A. Schriever. 1983. Herbicide and planting date influence establishment of sod-seeded alfalfa. Agron. J. 75:951-955.
Mays, D.A., and E.M. Evans. 1973. Autumn cutting effects on alfalfa yield and persistence in Alabama. Agron. J. 65:290-292.
McGraw, R.L., and G.C. Marten. 1986. Analysis of primary spring growth of four pasture legume species. Agron. J. 78:704-710.
McKenzie, J.S., and G.E. McLean. 1980. Some factors associated with injury to alfalfa during the 1977-79 winter at Beaverlodge, Alberta. Can. J. Plant Sci. 60:103-112.
Mueller, J.P., and D.S. Chamblee. 1984. Sod-seeding of ladino clover and alfalfa as influenced by seed placement, seeding date, and grass suppression. Agron. J. 76:284-289.
Mueller-Warrant, G.W., and D.W. Koch. 1980. Establishment of alfalfa by conventional and minimum-tillage seeding techniques in a quackgrass-dominant sward. Agron. J. 72:884-889.

Nelson, C.J., and D. Smith. 1968. Growth of birdsfoot trefoil and alfalfa. III. Changes in carbohydrate reserves and growth analysis under field conditions. Crop Sci. 8:25-28.
Olsen, F.J., J.H. Jones, and J.J. Faix. 1978. Forage establishment in wheat stubble. Agron. J. 70:969-972.
Pearson, R.W., and C.S. Hoveland. 1974. Lime needs of forage crops. p. 301-322. In D.A. Mays (ed). Forage fertilization. ASA, Madison, WI.
Peters, E.J., and A.H.B. Mohammed Zam. 1968. Allelopathic effects of tall fescue genotypes. Agron. J. 73:56-58.
Portz, H.L. 1967. Frost heaving of soils and plants. I. Incidence of frost heaving of forage plants and meteorological relationships. Agron. J. 59:341-344.
Rogers, D.D., D.S. Chamblee, J.P.Mueller, and W.V. Campbell. 1985. Fall no-till seeding of alfalfa into tall fescue as influenced by time of seeding and grass and insect suppression. Agron. J. 77:150-157.
Seim, A.L. 1968. Yield, composition and persistence of crownvetch as affected by cutting treatments. p. 124-128. In 2nd Crownvetch Symp. Agron Mimeo 6. Pennsylvania State Univ. University Park, PA.
Sharratt, B.S., D.G. Baker, and C.C. Sheaffer. 1987. Environmental guide to alfalfa growth, water use, and yield in Minnesota. Minnesota. Agric. Exp. Stn. Bull. 581.
Sheaffer, C.C., D.K. Barnes, and G.C. Marten. 1988. Companion crop vs. solo seeding: effect on alfalfa seeding year forage and N yields. J. Prod. Agric. 1:270-274.
Sheaffer, C.C., G.D. Lacefield, and V.L. Marble. 1988. Cutting schedules and stands. In A.A. Hanson et al. (ed.) Alfalfa and alfalfa improvement. Agronomy 29:411-437.
Sheaffer, C.C., M.P. Russelle, O.B. Hesterman, and R.E. Stucker. 1986. Alfalfa response to potassium, irrigation, and harvest management. Agron. J. 78:464-468.
Sholar, J.R., J.L. Caddel, J.F. Stritzke, and R.C. Berberet. 1983. Fall harvest management of alfalfa in the southern plains. Agron. J. 75:619-622.
Smith, A.E. 1986. Potential allelopathic influence of certain pasture weeds on interseeded forage species. p. 48. In R.R. Hill, Jr. et al. (ed.) Proc. Int. Symp. Estab. Forage Crops by Conserv.-tillage:Pest Mgt., State College, PA. 15-19 June. U.S. Regional Pasture Research Lab., University Park, PA.
Smith, D. 1962. Carbohydrate root reserves in alfalfa, red clover, and birdsfoot trefoil under several management schedules. Crop Sci. 2:75-78.
Smith, D. 1965. Forage production of red clover and alfalfa under differential cutting. Agron. J. 57:463-465.
Smith, D. 1972. Cutting schedules and maintaining pure stands. In C.H. Hanson (ed.) Alfalfa science and technology. Agronomy 15:481-496.
Smith, D. 1975. Effects of potassium topdressing a low fertility silt loam soil on alfalfa herbage yields and composition and on soil K values. Agron. J. 67:60-64.
Smith, D., and C.J. Nelson. 1967. Growth of birdsfoot trefoil and alfalfa. I. Responses to height and frequency of cutting. Crop Sci. 7:130-133.
Tamura, S., C. Shang, A. Suzuki, and S. Kumai. 1969. Chemical

studies on clover sickness. Part I. Isolation and structural elucidation of two new isoflavoids in red clover. Agric. Biol. Chem. 33:391-397.
Taylor, R.W., and D.W. Allinson. 1983. Legume establishment in grass sods using minimum-tillage seeding techniques without herbicide application: forage yield and quality. Agron. J. 75:167-172.
Taylor, T.H., J.S. Foote, J.H. Snyder, E.M. Smith, and W.C. Templeton, Jr. 1972. Legume seedling stands resulting from winter and spring sowings in Kentucky bluegrass (Poa pratensis L.) sod. Agron. J. 64:535-538.
Taylor, T.H., E.M. Smith, and W.C. Templeton, Jr. 1969. Use of minimum tillage and herbicide for establishing legumes in Kentucky bluegrass (Poa pratensis L.) swards. Agron. J. 61:761-765.
Tesar, M.B. 1986. Re-establishing alfalfa after alfalfa without autotoxicity. p. 51. In R.R. Hill, Jr. et al. (ed.) Proc. Int. Symp. Estab. Forage Crops by Conserv.-tillage:Pest Mgt. State College, PA. 15-19 June. U.S. Regional Pasture Res. Lab., University Park, PA.
Tesar, M.B., K. Lawton, and B. Kawin. 1954. Comparison of band seeding and other methods of seeding legumes. Agron. J. 46:189-194.
Tesar, M.B., and V.L. Marble. 1988. Alfalfa establishment. p. 303-332. In A.A. Hanson et al. (ed.) Alfalfa and alfalfa improvement. Agronomy 29:3O3-332.
Tesar, M.B., and J.L. Yager. 1985. Fall cutting of alfalfa in the North Central USA. Agron. J. 77:774-778.
Townsend, C.E., D.K. Christensen, and A.D. Dotzenko. 1978. Yield and quality of cicer milkvetch forage as influenced by cutting frequency. Agron. J. 70:1O9-113.
Triplett, G.B., Jr., and M.B. Tesar. 1960. Effects of compaction, depth of planting, and soil moisture tension on seedling emergence of alfalfa. Agron. J. 52:681-684.
Van Keuren, R.W., and C.S. Hoveland. 1985. Clover management and utilization. In N.L. Taylor (ed.) Clover science and technology. Agronomy 25:325-354.
Welty, L.E., R.L. Anderson, R.H. Delaney, and P.F. Hensleigh. 1981. Glyphosate timing effects on establishment of sod-seeded legumes and grasses. Agron. J. 73:813-817.
White, H.E., D.D. Wolf, and E.S. Hagood. 1985. Forage establishment innovations. p. 19-25. In Proc. Forage Grassl. Conf., Hershey, PA. 3-6 Mar. Am. Forage and Grassl. Counc., Lexington, KY.

DISCUSSION

Allen. Is the payoff matrix aimed at use by farmers or consultants, or both?

Sheaffer. The payoff matrix is a tool for aiding decision making by farmers and agricultural advisors. In the process of developing a payoff matrix, researchers will discover that information is often lacking for specific management action-event of nature combinations and initiate new experiments.

Allen. I agree with the principle and I am looking at a similar method for farmers to determine the need to control insect pests. However, it will probably be difficult to get farmers with low value annual legume pastures to apply such an approach.

Buxton. What guidelines do you have for assigning risk to management decisions? This seems to be a very subjective procedure.

Sheaffer. The researcher or the extension specialists involved should make these decisions. As researchers who evaluate alternative managements in diverse environments, we should be best able to decide the importance of various factors. I agree that some subjectivity will be involved. However, quantification of risk is important in making recommendations to producers.

Gramshaw. I would suggest that delivery of risk information can be made more consumable for producers through expert systems.

Clements. In Queensland, Australia, the "what-if" questions are what the farmer sees, but within computer software are probability functions that are based, for example, on 100-yr climatic databases including information from hundreds of weather stations. There is no subjectivity in these functions as they are based on real data obtained usually within a few miles of the farm.

Hockman. Regarding the need to deliver the information in an expert system, many of the problems raised about what factors are relevant and how probability may be used can be resolved by existing knowledge elicitation techniques. With respect to who will use decision support tools, in my experience this is an extension approach which can be handled in the normal methods.

Brougham. You indicated in your presentation that much research was predictable and had been repeated under many environmental conditions. An opportunity exists to collaborate with scientists in other countries representing different environmental conditions in which a simple model could be developed.

Watson. Are there measurable components in the plant which can be used as an indicator of its cold tolerance?

Sheaffer. Traditionally, total root nonstructural carbohydrate levels have been measured. However, carbohydrate levels are not always a good indicator of persistence. Specific components which increase during cold hardening include sugars, soluble proteins, amino acids, and lipids. In the northern USA, there is a strong correlation between a morphological feature, fall growth habit, and winter survival. Varieties with a greater dormancy reaction (reduced fall herbage production) are less likely to suffer winter injury.

GROWTH AND COMPETITION AS FACTORS IN THE PERSISTENCE OF LEGUMES IN PASTURES

M. J. Fisher and P. K. Thornton

SUMMARY

The literature of the influence of growth and competition on the relations between grasses and legumes in pastures is reviewed with particular reference to Australia. We conclude that the topic has not received much attention since it was last reviewed in 1976. Moreover, when competition between the components of a pasture has been discussed, evidence for it is inferred rather than demonstrated, and there is usually little evidence to determine which scarce resource in the environment is subject to competition.

A simple analytical model, in which growth is related to leaf area index and senescence is taken into account, suggests that growth rate is an important factor in the control of plant relations in pastures. In tropical pastures, where the C_4 pathway of photosynthesis often confers an advantage on the grasses compared with the C_3 legumes with which they are grown, grass dominance is inevitable unless the grass is at some competitive or demographic disadvantage, or is grazed preferentially.

Animals grazing at pasture frequently suffer deficiencies of dietary N. Often the deficiency is caused by the low N status of the whole pasture system, for which it is common practice to attempt to incorporate a legume component in order to increase the supply of N. In doing so, the N status of the forage on offer is increased, both directly from the legume and indirectly from N transferred from it to the associated grass.

The grazing industries of New Zealand and southern Australia are based on such a practice, and strenuous efforts have been made to transfer the same approach to the tropics, both in Australia and elsewhere (Davies & Eyles, 1965; Toledo, 1985). Obviously, the success of the system depends upon the maintenance of a satisfactory proportion of legume in the pasture under grazing. One of the key factors in the survival of legumes in pastures is their ability to withstand competition from the cohabiting species in the sward.

In this paper, we shall review the literature on competition in pastures for light, water and nutrients, emphasizing particularly published Australian work. However, the literature of competition between the components of pastures was reviewed extensively in association with a symposium on plant relations in pastures held in Brisbane in 1976 (Wilson, 1978). We shall therefore only summarize the main points of the appropriate reviews from that meeting, but we shall refer where necessary to key papers, and draw attention to any new information that has been published since that time. We shall then go on to analyze the consequences of differences in the growth rate of the two components of a pasture on the relations between them.

COMPETITION

Terminology

Some words have been given specific meanings by ecologists, which differ from their commonly accepted ones, notably competition, crowding as used in crowding coefficient, and interference. In order to avoid misunderstanding in what follows, we shall define them.

Competition is restricted to circumstances where the reduction of the growth of a plant by its neighbor is greater if they are of different species than if they are the same. Donald (1963, p.6) gave a graphic description of the factors within the environment for which plants compete: "Most of the factors for which there is competition are found as a pool of material from which competitors draw their supplies. If the pool is of limited volume, or if it is subject to intermittent depletion by the competing plants, then the most successful competitor is the plant which draws most rapidly from the pool or which can continue to withdraw from the pool when it is at a low ebb or when its contents can no longer be be tapped by other plants. If all plants in the community are nearly equal in competitive ability, as in a wheat crop, they will tend to share equally in the supply until it is exhausted, and then, simultaneously, to suffer the effect of depletion of the pool."

Crowding coefficient (de Wit, 1960) is the proportional change in the ratio of yields of two components before and after a period of growth. The term seems at odds with the commonly accepted notion of Clements (1907, p.252), who postulated that competition for space occurs only rarely between plants, in the sense that there is insufficient space for an individual plant to exist by virtue of "crowding ... when grown too closely". However crowding as used by de Wit (1960) implies crowding for any scarce resource, and not simply for physical space. Nevertheless, the possibility for confusion exists, and the term is now usually avoided by ecologists (Donald, 1963, p.6).

Interference is used to describe any effect of one plant upon the growth of another, whether detrimental (its usual connotation) or not, in an attempt to avoid the multiplicity of meanings of the word competition both in common language and in different fields of biology (Harper, 1961).

Methodology

If a range of proportions of the components of a binary mixture is obtained by successively replacing the individuals of a monoculture of the first component by individuals of the second until a monoculture of the second is achieved, the result is a replacement series (de Wit, 1960). The yield data of a replacement series may be analysed to calculate the rate at which one component is replaced by another (relative replacement rate, RRR), and the time course of RRR gives an insight into the behavior of the mixture (de Wit et al., 1966).

It is now accepted amongst ecologists that the replacement series is the only way unequivocally to ascribe actual competition between the components of a two species mixture (Hall, 1978; Trenbath, 1978), as distinct from other forms of interference. Thus, although various effects of interference between plants are frequently attributed to competition, it is very rare that the replacement series methodology is used allow this conclusion to be reached validly.

In discussing the replacement series, Trenbath (1978) considered that only five distinct relations were possible (Fig. 1). If we examine the ratio diagrams (Fig. 1, bottom row), the nature of stability of binary mixtures and its relation with the different forms of the replacement diagrams becomes quite clear. Obviously, type I has no point of equilibrium, and will inevitably go to dominance of the component on whose side of the 45 degree line it lies. (This is the common situation with tropical grasses and tropical legumes, see next section). Type II has a point of equilibrium, but it is inherently unstable. Type III is neutral, that is, it is stable over its whole range, and suggests that the components are not having any influence on each other, as in too wide spacing. Types IV and V always tend towards a stable equilibrium point, and as such represent the ideal sought by all pasture agronomists, but unfortunately are rare. It is plausible,

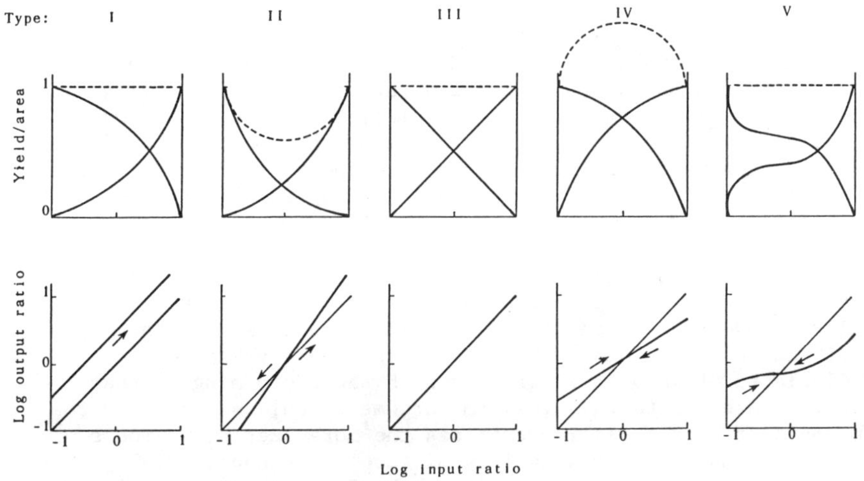

Fig. 1. The five types of results possible in a replacement series. Top row: relative yields of each component (solid lines), and relative yield totals (broken lines). Bottom row: corresponding diagrams relating the logarithms of the proportions of the components at the start and end of growth. The arrows indicate the direction of the change in proportions if the same relation holds in subsequent growth periods (after Trenbath, 1978).

of course, that any stable system should conform to either type IV or type V. Possibly they may be demonstrated in pastures growing on soils so low in N that growth of the grass is dependent on the satisfactory growth of the legume. In this case, the actual relation, and particularly the equilibrium point, may fluctuate quite widely with time, leading to a sort of dynamic stability between the components.

If we wish to know whether two species are likely to coexist in a mixture, we may grow the two of them together in a set of replacement series under conditions that cover those expected in the real world (Trenbath, 1978; Marshall & Jain, 1969). The difficulty is to take account of all the environmental variables that are likely to be encountered, especially the effects of herbivores (Harper, 1978), which are a fundamental part of the environment of a pasture. As we shall shortly show, herbivore grazing preference is one of the few factors available to pasture managers to manipulate. In fact, the pasture manager is unable to manipulate grazing preference to any substantial extent, because it is so imperfectly understood, and moreover, frequently he has scant opportunity to vary the disposition of his livestock.

We can only speculate on the reasons why the replacement series is used so infrequently in the field to elucidate competitive relations, but we believe that it is because of the unacceptably large amount of labor required to establish the experiments. For this reason, although competition is frequently invoked as the cause of interference between plants grown in the field, there is only indirect evidence to support the assertion.

Nutrients

Study of competition between the components of a pasture in Australia was stimulated by Donald (1951, 1958, 1961) and his colleagues, who sought to understand the dynamics of subterranean clover (*Trifolium subterraneum*) pastures, sown with or without a companion grass, and the cyclic variations that occur. He did several classical experiments, notably with the root systems of the clover and its associated grass intermingled or not. The shortcomings of this technique were outlined by Hall (1978), although it is tempting to many investigators to continue to use it, even in some modified form. The basis of Hall's criticism is that the plants growing in the separated soil volume have only half the absolute volume of soil available to them. If the legume is effectively nodulated, then the grass probably has more N available to it when it is growing with the roots mingled than when they are separated. The same argument could be made for the component of an association that has the more aggressive roots. If such a plant when growing in a mixture has a substantially larger root mass than when growing in half the volume of soil, then it is obviously at a disadvantage when it is so confined.

Hall (1974a,b) extended the use of the replacement series to study the effects of competitive interference for mineral nutrients. He reasoned that shortage of a limiting resource affects yield of that resource more than it affects yield of any other resource, or of dry matter. Therefore, if plants compete for a limited resource in the environment such as a soil nutrient, then the relative yield curves for that nutrient in a replacement series analysis should have a more exaggerated form (and higher RRR's) than those for dry matter. In a replacement series experiment between Desmodium intortum and Setaria anceps grown on a K deficient soil with and without added K, Hall (1974b) examined the relative yields of dry matter, N, P, and K. He clearly demonstrated that in the absence of added K, S. anceps competed strongly with D. intortum for K. In contrast, there was no competition for N, even when K was limiting. In this way, he emphasized the importance of competition for nutrients in controlling relations between plants in a pasture.

Hall's approach offers the opportunity for more efficient determination of whether or not there is competition for a particular nutrient, and perhaps to select other species or even cultivars that do not compete for a particular scarce resource. Unfortunately, the technique does not seem to have been taken up by other workers.

Water

There are few cases where the competition between grasses and legumes for water has been studied. Certainly, the replacement series technique has not been applied, and conclusions about the competition have therefore been made on an inferential basis. The C_4 grass Cenchrus ciliaris (buffel grass) and the C_3 legume Macroptilium atropurpureum (siratro) form successful associations in the subtropical environment of southeastern Queensland, where drought is an important feature of the climate. Yet these two species have quite contrasting physiological reactions to water deficits. The tissues of siratro are intolerant of desiccation and die if their water potential falls below about -2.4 MPa, while buffel grass is a desiccation tolerant species whose tissues are able to survive water potentials of less than -6 MPa. It is obvious, therefore, that the grass is able to maintain activity at levels of tissue hydration lethal to the legume. In order to resolve the apparent conflict, Sheriff and Ludlow (1984) investigated how the two species survive when grown together on a sandy soil in a subtropical environment by measuring the relation between their water relations, photosynthesis, stomatal conductance, and estimated transpiration, both when well-watered, and when exposed to drought.

The leaf water potential of the legume measured at dawn was found to be more highly correlated with the water potential of the soil at depth than in the shallower layers. Moreover, the roots of the legume were found to exploit deeper layers of the soil profile than those of the grass, and its stomata behaved in such a way as to remain open at those times of the day (early morning and late afternoon) when water loss was low due to lower temperatures, lower solar radiation and lower evaporative demand. Sheriff and Ludlow (1984) concluded that "Buffel grass was able to survive drought because of its inherent desiccation tolerance, whereas siratro was able to survive drought in a mixed sward because it did not compete with buffel grass for soil water." Hence, insofar as siratro is concerned, desiccation sensitivity and lack of osmotic adjustment were less indicative of performance than were patterns of rooting, and siratro is able to survive when growing in association with buffel grass despite their apparently incompatible physiological reactions to tissue desiccation.

The quote above, it must be remembered, is solely to do with survival of drought. Under more favorable conditions, that is when the water supply is adequate, these two species may well compete for water, even if only to a limited extent, and for other scarce resources in the environment, because the major part of the fibrous roots of siratro grow in close association with the superficial roots of buffel grass.

Light

There are two fundamental considerations with regard to the light relations of the component species in a pasture. Firstly, light is an instantaneous commodity, and thus any light that is not intercepted is lost for ever, and secondly, architecture of the canopy, that is the disposition of the individual leaves of the component species within the canopy, is important in determining the inter-relationships between them (Donald & Black, 1958). Furthermore, photosynthesis of C_3 legumes saturates (does not increase further) at about two-thirds of full sunlight, while the rate of photosynthesis of C_4 (tropical) grasses continues to increase (does not saturate) up to full sunlight, which gives the grasses an inherent advantage over the legumes in areas where light levels are high (Ludlow, 1978).

It is commonly argued that a high leaf area index, such that almost all the incident light is intercepted, confers on a pasture the capacity for higher growth rate because the incident light is spread more evenly over a larger leaf canopy, so that the majority of the leaves are at or below light saturation (Rhodes & Stern, 1978). However, we must also consider the influence of the different canopy architectures of grasses and legumes, and moreover how this interacts with the characteristic light responses of their photosynthesis. Grasses in general have erect leaves, so that light falls on them at an acute angle (reducing the level of irradiance) in contrast to the legumes, which frequently have leaves that are planofile or even have the capacity to track the sun (phototropism). Therefore, in a tropical pasture of a C_3 legume and a C_4 grass, a high LAI should confer a greater relative advantage on the legume compared with the grass, unless the grass is a taller-growing species, and by overtopping the legume, preempts its access to light. In contrast, in a temperate pasture in which both grasses and legumes are C_3 species, the outcome is more likely to be controlled by canopy architecture alone.

When radiation is reduced because of cloudy weather, as in the humid tropics, or in pastures grown as ground cover under the shade of plantation trees, legumes should also have a relative advantage over any cohabiting C_4 grasses. This expectation is confirmed by the general experience of the International Network for Evaluation of Tropical Pastures (RIEPT) of the Centro Internacional de Agriculture Tropical (CIAT). In the ecosystems of the humid tropics, the growth rate of legumes is often superior to those of the grasses during the wetter months of the year (Pizarro, 1985). Strictly, however, although this may be due differences in the light responses and canopy architecture between grasses and legumes, it may also be the consequences of competitive interference, or of superior adaptation of the legumes to other environmental conditions.

Studies Within the Life Cycle

It is of interest to consider competition at different stages of the growth cycle and under different circumstances, which is commonly not done, although the point was made cogently by Trenbath in his review (Trenbath, 1978).

In order for a pasture to maintain itself, it is necessary for new seedlings to establish in the face of (possibly) severe competition from the established plants. Few studies have been conducted to determine the nature of the competition, and its consequences in terms of plant survival. Establishment of plants from seed is superior in prepared seed beds, compared with when the seed is broadcast into an established grassland swards (Cook & Dolby, 1981), presumably because of competition from the existing plants. The interference may come from either or both of the components of the pasture, or from weedy species.

In general, germination is not affected by the presence of other plants, and indeed it is commonly enhanced by the rather more favorable microenvironment at the soil surface provided by the standing plant material (Cook & Dolby, 1981; McIvor & Gardner, 1981; Miller & Perry, 1968). However, survival of the new seedlings, leading to their establishment, is more affected by established plants as indicated by studies with roots of the germinating plants protected from competition by tubes driven into the soil, or inferentially by the killing of the established plants by herbicides or by reducing their ability to compete by defoliating the sward severely by mowing, grazing, or burning. None of these techniques allow the nature of the competition to be determined with certainty, although it is often possible to infer it. Valencia (1983) found that growth of *Stylosanthes capitata* seedlings was enhanced when competition from the roots of *Andropogon gayanus* was eliminated by the use of soil tubes, and concluded that the competition was principally for K.

Torssell and his colleagues (Torssell et al., 1975; Torssell & Nicholls, 1978), measured RRR at various stages during the growth of *Stylosanthes humilis* in competition with volunteer annual grasses, and with sown perennial grasses. They likened RRR to the behavior of filters in a separation process in engineering, and recognized a number of stages during the growth cycle in a dry monsoonal environment: germination, growth, seed production, and seed survival during the long hot dry season. They measured RRR for each of the various stages for a number of years and obtained mean values and a measure of their variances. The merit of their particular method lay in the identification as each of the stages as more or less independent, each with very different RRR's, and, by generating stochastic variation of each of them on the basis of the mean values and their variance, to model the behavior of different associations over a long run of years.

RRR, however, is simply descriptive of the changes that occur at particular stages in the growth of the association, but of itself does not indicate reasons for any changes, although it may be possible to speculate on them. Thus, if we wished to forecast what would be the outcome of competition between a grass and a legume grown in association, we could grow them together in a range of soils, at a number of levels of fertilizer and other management variables, and preferably a range of climates (years) and estimate the mean and variability of RRR. By means of applying a stochastic variation to this mean we could simulate the behavior of the association for whatever range of soils and managements or sequence of years that we wished. However, the validity of the forecast would be utterly dependent on the success with which we had sampled the potential range of environmental variation in our estimates of the various RRR's. Therefore, our ability to extrapolate would depend on how good was our sample of the environments. It is obvious, then, that we would not be in any sound position to extrapolate to other environments with which we had no experience, except at the risk of grave error.

The Role of Competition in Plant Demography

This topic is treated in depth by Jones and Carter (this book). Nevertheless, it is obvious that any factor that allows the proportion of one species of an association to increase at the expense of another will influence the populations of each of them. As a consequence, where such competition occurs, the demography of the population will inevitably be affected, if not immediately then certainly in the longer term. In contrast demographic studies may suggest that there is competition between the components of an association, and hence that there is a need to define or to investigate the exact competitive relations between them, possibly in order to indicate which management strategies are best to minimize or, if possible, to eliminate it.

GROWTH RATE AS A FACTOR IN COMPETITION IN TROPICAL PASTURES

Influence of Photosynthetic Characteristics

The higher crop growth rates of C_4 grasses compared with C_3 pasture plants is mainly due to higher rates of leaf net photosynthesis, and the absence of photorespiration associated with the C_4 pathway (Ludlow, 1985). Although there are also probably differences in the efficiency of light utilization in favor of the C_4 species, the difference appears to depend on temperature to a considerable extent, and it is possible that acclimation to higher temperatures may minimize the advantage.

Ludlow (1985) also ascribed the higher dry matter yield (as opposed to growth rates) of tropical C_4 grasses compared with C_3 legumes as "partly due to the longer growing seasons in the tropics where (they) grow best, and partly to their higher temperature requirements and tolerance". We do not support the notion that grasses have longer growing seasons than legumes in tropical areas, indeed observations in the neotropical savannas suggest that grasses and legumes cohabit there with no indication that their growing seasons are different (Pizarro, 1985). Whilst undoubtedly C_4 grasses do grow best in the tropics, many genera of legumes have their centres of origin in the tropics, and hence are necessarily well-adapted to tropical conditions.

It is important also to remember that temperatures in the tropics vary substantially, and that many tropical areas have temperature regimes that are frequently cooler than the summer in many parts of the sub-tropics. The difference is largely one of uniformity, in that in the tropics temperatures do not vary very much during the year. Certainly, in some parts of the tropics the productivity of grasses is superior to legumes, due largely to the differences in growth rates of C_3 and C_4 species. However, in environments with milder temperatures and low levels radiation, such as many areas of the humid tropics, there may be scant advantage in the total dry matter yield of C_4 compared with C_3 plants. Here the C_4 photosynthetic pathway does not confer the same advantages by virtue of the lower light levels (see above), and, moreover, under lower temperature conditions C_4 plants no longer have a comparative advantage (Charles-Edwards, 1978).

A Simple Model of a Grass/legume Association

The objective of the Tropical Pastures Program (TPP) of CIAT is to increase the production of beef and milk in Latin America by replacing or supplementing with legume-based pastures of higher nutritional status the low quality native pastures of the extensive areas of acid infertile soils of northern and central South America and of Central America. In addressing this objective, the TPP has chosen a germplasm approach, that is, the introduction and assessment of the adaptation of a wide range of grass and legumes from other areas with similar climate and soils. However, after selecting germplasm that shows gross adaptation to the soils, climate, pests, and diseases of the regions of interest, the problem is to determine which grass/legume associations will be compatible, and under what system of grazing management should further evaluation be carried out? Other questions to answer include: what is the capacity of the material to withstand mismanagement, and what is the optimum management for the material in commercial practice? Moreover, will the selected material survive under the give and take management of commercial practice? How can these questions be answered early in the introduction process to eliminate potential failures and, conversely, to identify potentially successful material? In order to answer some of these questions, an attempt was made to simulate the behavior of a grass and a legume growing in association, using simple growth analysis as a basis.

The conventional pattern of growth of a monospecific stand shows that dry matter yield commences at a low level, the actual level being determined by seed weight, and then goes through an exponential phase, then a linear phase, followed by a phase as yield approaches some maximum, and possibly a phase where yield declines, usually associated with reproductive development. However, in stands of plants that remain vegetative, there is usually only a trend towards an asymptotic maximum yield. The corresponding curves of growth rate show an increase from zero to a maximum rate at the stage just at the end of the exponential phase, followed by a broad peak corresponding the linear phase, and in turn followed by a gradual decline to zero coinciding with the attainment of the yield plateau.

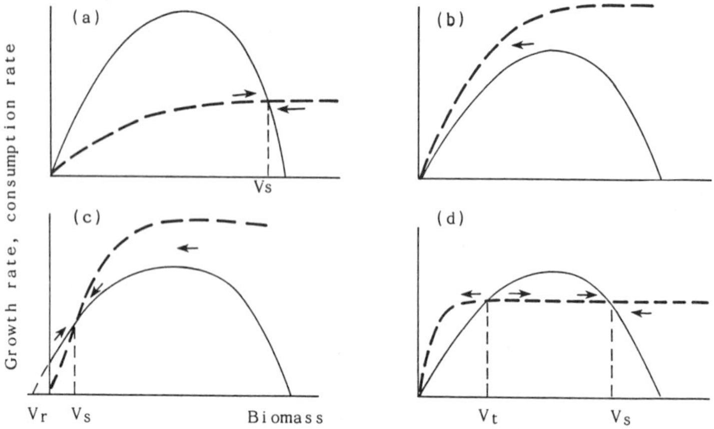

Fig. 2. The relation between the growth rate of a pasture and the consumption of grazing animals (after Noy-Meir, 1975). V_s = stable biomass, V_t = unstable turning point, V_r = ungrazable residue. —— growth rate, ---- rate of consumption.

It was on this pattern of growth rate that Noy-Meir (1975) based his model of stable grazing pressure (Fig. 2). A moment's consideration will show that this notion cannot be correct (J. Hodgson, 1985, personal communication), for under most circumstances, even in mature pasture stands, new growth is continually produced. For example, if we examine a grass plant in a pasture at the stage of maximum yield, there is clear evidence that new leaves are still being produced. Therefore, in order to account for the fall in net growth rate as yield approaches its maximum, the production of new growth must be balanced by the loss of old tissue, that is, the plants must be senescing at the same rate that they are growing. We believe, therefore, that the concept of tissue turnover (Hodgson, 1985) is of fundamental importance to the understanding of the behavior of the dynamic aspects of pasture growth.

We then considered the growth of two monospecific stands, a legume and a C_4 grass, which we wished to consider as growing in association. Photosynthesis, by which all autotrophic plants acquire the carbon necessary for growth, depends on the plants' ability to intercept incident solar radiation. Whilst the concept of leaf area index can be criticized on the basis of simplicity, it has the merit that for a given canopy architecture, growth rate is closely related to it. Moreover, again for a given genotype, leaf area index is generally closely related to biomass (Fisher et al., 1980). We therefore postulated three simple functional relationships of

- leaf area as a function of biomass,
- senescence as a function of biomass (Charles-Edwards & Fisher, 1980), and
- growth rate as a function of leaf area.

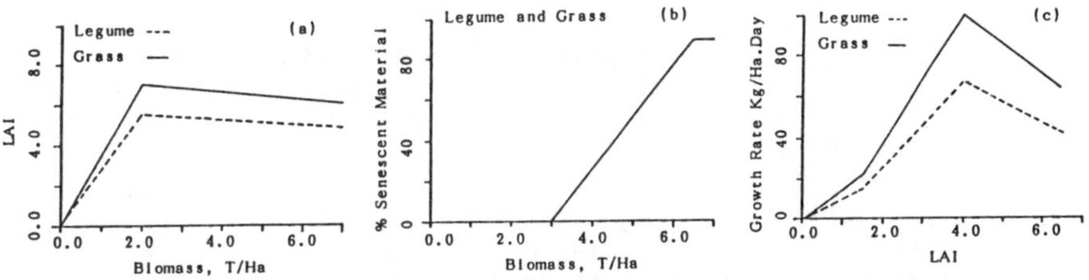

Fig. 3. Approximations of the functions controlling growth of a C_3 legume and a C_4 grass in association. a. Leaf area index (LAI) as a function of biomass. b. Senescence as a function of biomass. c. Growth rate as a function of LAI.

Based on general relationships for tropical C_3 legumes and C_4 grasses, we drew approximations in order to simplify coding of the simulation (Fig. 3a-c).

In general, tropical legumes have planofile leaves, which are often are phototropic (that is, have the capacity to orient themselves at right angles to the incident radiation) (Sheriff et al., 1986). In contrast, tropical grasses have erectofile leaves, and in consequence the relation between LAI and biomass of grasses and legumes is quite different (Fig. 3a and c), with the grasses in general having markedly higher LAI's per unit of biomass than the legumes. We assumed that the relation of senescence to biomass is the same for grasses and legumes (Fig. 3b).

These assumptions are not enough of themselves to allow us to simulate the growth of an association, because we must know how the plants within each species interact with each other. It is clear that their functional relationships as defined above will be different, and how do we take account of this difference? We invoked a conceptual treatment, derived from the definition of competition as we defined it above, that is, in the absence of competition each plant should react to its neighbor as if they were the same species, irrespective of whether they are the same or not. Therefore, to determine for each species the value of the dependent variable in each of the functional relations above, we used the total of both species for the independent variable. We then scaled the estimate of the dependent variable by the proportional contribution of the particular species to the total in the independent variable. This allowed us notionally to differentiate between the effects of like and unlike neighbors, and to separate the effects of growth and of competition.

Put in other words, we assume that for each functional response, each component affects the other as though each of them respond in exactly the same way, and that all other effects are interference. Expressed in mathematical notation, we write

$$Y_a = f(X_T) \, . \, X_a/X_T$$

where Y is the dependent variable of leaf area index, growth rate, or senescence rate,

X is the independent variable biomass, leaf area index, and biomass respectively, a is a subscript referring to whichever of the two components, and

X_a/X_T is the proportional contribution of component a to the mixture in the independent variable X.

We simulated the behavior of an association of a C_3 legume associated with a C_4 grass, using the response functions shown in Fig. 3a-c. We simulated the growth of a range of proportions of grass and legume, and monocultures of each of the components of the association for periods of 45 days, starting in each case with a yield of 1000 kg ha^{-1}. The proportions of grass and legume in the association after the simulation were plotted as a function of their proportions in the association at the start (Fig. 4). The result was a relation that is remarkable for its similarity to the idealized Fig. 1d, that is, showing competitive interference and mutual exclusion between the two species, in favor of the grass.

We conclude, therefore, simply on the basis of the different relations between leaf area and biomass and between growth rate and leaf area of C_3 legumes and C_4 grasses, an association of a tropical grass with a legume will move inexorably in composition towards the grass at the expense of the legume unless some external factor operates to counter the movement. Of course, if there is competitive interference, its effects will be in addition to the effects that we have described.

What are the practical consequences of these conclusions? It is inevitable that the majority of C_4 species have higher growth rates per unit of leaf area (that is higher net assimilation rates or unit leaf rates in classical growth analysis) than C_3 species by virtue of their superior photosynthetic characteristics.

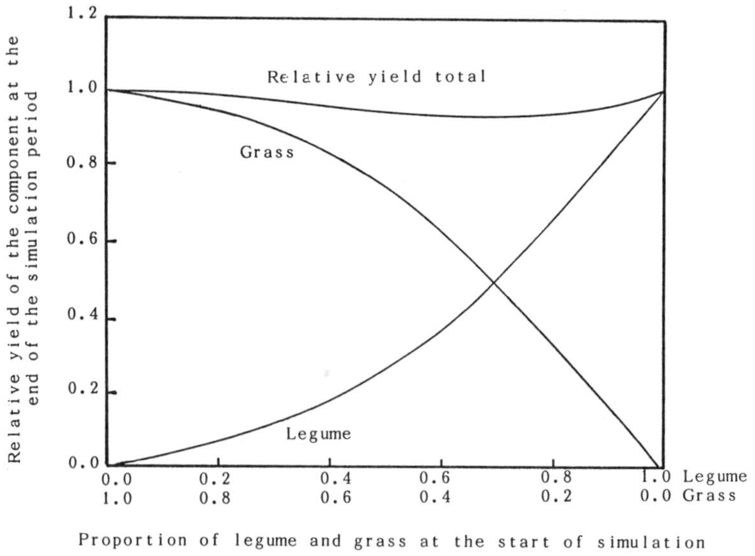

Fig. 4. Relative yields of two components after simulated periods of growth of 45 days, plotted as a function of the proportion of legume and grass at the start of the simulation. The functional relations in Fig. 3 were used in the simulation.

Moreover, grasses also have higher leaf area indices as we have discussed above by virtue of their different canopy architecture.

In order to maintain a legume component in a sward with a C_4 grass, it is necessary for one or more of the following to redress the growth rate advantage of the grass:

- The legume has a demographic superiority compared with the grass. Although it is clear that at times some legumes have superiority, the relative advantage is frequently dynamic, and influenced strongly by external factors, such as grazing pressure and climate in ways that are not always easily predictable (Torssell & Nicholls, 1978; Jones & Carter, this book).
- The legume has a superior competitive advantage for some scarce resource in the environment. At least as far as K is concerned (Valencia, 1983; Hall, 1971, 1974b) the legume is frequently at a competitive disadvantage compared with the grass.
- The species occupy different niches so that their growth rates are differently constrained, that is they do not strictly occupy the same space. We have already referred to the advantage conferred by the different rooting patterns of the two species in the association of *Cenchrus ciliaris* and *Macroptilium atropurpureum* in terms of survival during periods of water shortage (Sheriff & Ludlow, 1984). In soils low in N, the independence of the legume by virtue of the rhizobium symbiosis is a case in point, and undoubtedly is the reason why at least some of the grass-legume associations are successful in Australia, where the soils by world standards are notoriously infertile.
- The grass is preferentially consumed by the grazing animal. In many cases, the only way that the superior growth rates of the grass can be offset is if the animals graze the grass in preference to the legume. In many cases, the animals fortunately do show a marked preference for the grass, but in order to predict the outcome of the preference, it is necessary to understand the consequences of the defoliation on the subsequent growth rate. How does grazing pressure affect selectivity? Undoubtedly, the management of this type must be very sensitive, and relatively independent of grazing pressure.

Applicability

The extent to which the behavior may be forecast at a site other than the place in which the responses were measured depends on the extent to which the individual response functions are site independent. Here the environment of the neotropical savannas is particularly suitable, because of the low latitude. For example, at the Centro Nacional de Investigacion field station of Carimagua, on the eastern plains of Colombia at 4^0 N Lat, during the year the daylength varies only a little over 40 min, and the mean monthly maximum and minimum temperatures vary only a degree or so. During the rainy season there is abundant rain and dry periods are very rare (but the soils drain very freely, so that waterlogging does not appear to be a problem), and typically the onset and cessation of the short dry season is abrupt. Growth during the wet season appears to be controlled principally by edaphic factors and by the amount of radiation intercepted. Under these circumstances, we believe that we shall be able to extrapolate with confidence more widely than is customary.

REFERENCES

Clements, F. E. 1907. Plant physiology and ecology. Constable, London.

Charles-Edwards, D. A. 1978. An analysis of the photosynthesis and productivity of vegetative crops in the United Kingdom. Ann. Bot. (London) 42:717-731.

Charles-Edwards, D. A., and M. J. Fisher. 1980. A physiological approach to the analysis of crop growth data. I. Theoretical considerations. Ann. Bot. 46:413-423.

Cook, S. J., and G. R. Dolby. 1981. Establishment of buffel grass, green panic and siratro from seed broadcast into a speargrass pasture in southern Queensland. Aust. J. Agric. Res. 32:749-759.

Davies, J. G., and A. G. Eyles. 1965. Expansion of Australian pastoral production. J. Aust. Inst. Agric. Sci. 31:77-93.

Donald, C. M. 1951. Competition among pasture plants. I. Intra-specific competition among annual pasture plants. Aust. J. Agric. Res. 2:355-376.

Donald, C. M. 1958. The interaction of competition for light and for nutrients. Aust. J. Agric. Res. 9:421-435.

Donald, C. M. 1961. Competition for light in crops and pastures. Symp. Soc. Exp. Biol. 15:282-313.

Donald, C. M. 1963. Competition among crop and pasture plants. Adv. Agron. 15:1-118.

Donald, C. M., and J. N. Black. 1958. The significance of leaf area in pasture growth. Herb. Abstr. 28:1-6.

Fisher, M. J., D. A. Charles-Edwards, and N. A. Campbell. 1980. A physiological approach to the analysis of crop growth data. II. Growth of *Stylosanthes humilis*. Ann. Bot. 46:425-434.

Hall, R. L. 1971. The influence of potassium supply on competition between Nandi setaria and Greenleaf desmodium. Aust. J. Exp. Agric. Anim. Husb. 11:415-419.

Hall, R. L. 1974a. Analysis of the nature of interference between plants of different species. I. Concepts and an extension of the de Wit analysis to examine effects. Aust. J. Agric. Res. 25:739-47.

Hall, R. L. 1974b. Analysis of the nature of interference between plants of different species. II. Nutrient relations in a Nandi *Setaria* and Greenleaf *Desmodium* association with particular reference to potassium. Aust. J. Agric. Res. 25:749-756.

Hall, R. L. 1978. The analysis and significance of competitive and non-competitive interference between species. p.163-174. *In* J. R. Wilson (ed.) Plant relations in pastures. CSIRO, Melbourne.

Harper, J. L. 1961. Approaches to the study of plant competition. Symp. Soc. Exptl. Biol. 15:1-39.

Harper, J. L. 1978. Plant relations in pastures. p.3-14. *In* J. R. Wilson (ed.) Plant relations in pastures. CSIRO, Melbourne.

Hodgson, J. 1985. The significance of sward characteristics in the management of temperate sown pastures. p.63-66. In Proc. Int. Grassl. Congr., 15th, 1985, Osaka.
Ludlow, M. M. 1978. Light relations of pasture plants. p.35-49. In J. R. Wilson (ed.) Plant relations in pastures. CSIRO, Melbourne.
Ludlow, M. M. 1985. Photosynthesis and dry matter production in C_3 and C_4 pasture plants, with special emphasis on tropical C_3 legumes and C_4 grasses. Aust. J. Plant Physiol. 12:557-572.
Marshall, D. R., and S. K. Jain. 1969. Interference in pure and mixed populations of *Avena fatua* and *A. barbata*. J. Ecol. 57:251-269.
McIvor, J. G., and C. J. Gardner. 1981. Establishment of introduced grasses at different stages of pasture development: effects of seedbed. Aust. J. Exp. Agric. Anim. Husb. 21:417-423.
Miller, H. P., and R. A. Perry. 1968. Preliminary studies on the establishment of Townsville lucerne (*Stylosanthes humilis*) in uncleared native pasture at Katherine, N.T. Aust. J. Exp. Agric. Anim. Husb. 8:26-32.
Noy-Meir, I. 1975. Stability of grazing systems: an application of predator-prey graphs. J. Ecol. 63:459-480.
Pizarro, E. A. 1985. Red Internacional de Evaluation de Pastos Tropicales, resultados 1982-1985 (2 volumes). (In Spanish).
Rhodes, I., and W. R. Stern. 1978. Competition for light. p.175-189. In J. R. Wilson (ed.) Plant relations in pastures. CSIRO, Melbourne.
Sheriff, D. W., M. J. Fisher, G. Rusitzka, and C. W. Ford. 1986. Physiological reactions to an imposed drought by two twining pasture legumes: *Macroptilium atropurpureum* (desiccation sensitive) and *Galactia striata* (desiccation insensitive). Aust. J. Plant Physiol. 13:431-445.
Sheriff, D. W., and M. M. Ludlow. 1984. Physiological reactions to an imposed drought by *Macroptilium atropurpureum* and *Cenchrus ciliaris* in a mixed sward. Aust. J. Plant Physiol. 11:23-34.
Toledo, J. M. 1985. Pasture development for cattle production in the major ecosystems of the tropical American lowlands. p.74-78. In Proc. Int. Grassl. Congr., 15th, 1985, Osaka.
Torssell, B. W. R., and A. O. Nicholls. 1978. Population dynamics in species mixtures. p.217-232. In J. R. Wilson (ed.) Plant relations in pastures. CSIRO, Melbourne.
Torssell, B. W. R., C. W. Rose, and R. B. Cunningham. 1975. Population dynamics of an annual pasture in a dry monsoonal climate. Proc. Ecol. Soc. Aust. 9:157-171.
Trenbath, B. R. 1978. Models and the interpretation of mixture experiments. p.145-162. In J. R. Wilson (ed.) Plant relations in pastures. CSIRO, Melbourne.
Valencia, I. M. 1983. Root competition between *Andropogon gayanus* and *Stylosanthes capitata* in an oxisol in Colombia. Ph. D. thesis. Univ. of Florida, Gainesville (Diss. Abstr. 45/07:1976b).
Wilson, J. R. 1978. (ed.) Plant relations in pastures. CSIRO, Melbourne, Australia.

Wit, C. T. de. 1960. On competition. Versl. Landbouwk. Onderz. 66.8, 1-81.

Wit, C. T. de, P. G. Tow, and G. C. Ennik. 1966. Competition between legumes and grasses. Versl. Landbouwk. Onderz. 687:1-29.

DISCUSSION

Hodgson: What is the likely influence of row planting on both the selection and the competition function?

Fisher: Row planting is conventional to limit competition for scarce nutrient resources especially during the establishment phase, but only recently has investigation started of alternative planting strategies for these soils. Creeping legumes penetrate grass rows, but there is a tendency for animals to graze in the rows and, in tall-growing bunch grasses, to walk on the interrow, that is, on the legume.

Clements: You said that you have 8 months' data from your Carimagua experiment. Can you tell us in brief terms what has happened to the grass/legume balance in the four mixtures?

Fisher: Each mixture has proceeded towards a balance characteristic of that particular mixture. For example, under medium forage allowance the mixtures containing *Desmodium ovalifolium* presently contain about 30% legume, irrespective of the starting proportion. Those containing *Centrosema acutifolium* have lower legume contents. Added to this, there is an over-riding effect of forage allowance.

Kretschmer: Do you expect that results obtained at Carimagua would be non-site specific if similar percentages of the grass and legume components were present?

Fisher: The competition function is likely to be highly site specific. For this reason, we are seeking to understand what factors influence the competitive relations between plants when grown on acid soils of widely contrasting textures.

Scott: The equilibrium between species will be the outcome of their interaction with the whole environment.

Sheaffer: It is worthy of note that, because the environment is dynamic, we should not expect to find a true equilibrium point.

Buxton: What may be the long-term effects of the current US drought on grass/legume mixtures?

Fisher: It will depend on the relative ability of the component species to survive the effects of the drought (high temperatures and water deficits), and hence its effects on their demography. Where one or other of the species has superior survival, one could expect large short-term effects on composition, but the longer-term consequences will depend on the demographic factors such as recruitment, seed bank, and so on.

Jones: This raises the question of demography in your study. Obviously, as it has only been in progress for about a year, it is too early for substantial demographic changes to have occurred. However, these will become more important with time.

Fisher: Certainly, and I am making demographic measurements in the experiment to document the changes that occur in the populations in each association.

COMPETITION FROM ASSOCIATED SPECIES ON WHITE AND RED CLOVER IN GRAZED SWARDS

R.J.M. Hay and W.F. Hunt

ABSTRACT

Competition of white (*Trifolium repens* L.) and red clover (*T. pratense* L.) with associated species is reviewed. White clover is emphasized, as it is the most important legume in grassland farming in New Zealand. Grazing management strategies exist which increase both white and red clover content in swards, and genotype differences can confer considerable additional competitive advantages. Both clovers have superior light interception, different growth rhythms, and N_2-fixing capability compared with grasses. These features allow them to compete successfully in vigorous grass dominated swards. However, the guerilla habit of white clover, bestowed on it through its stoloniferous character, ensures that this species persists in a wide range of grazed pasture.

INTRODUCTION

This paper emphasizes white clover, which is easily the most important grassland legume in grazing systems in New Zealand agriculture (Langer, 1969). Thus, we are assuming that competition within swards of annual legumes, particularly subterranean clover (*T. subterraneum* L.) and the medics (*Medicago* spp.) will have been covered by the Australian author, and that alfalfa (*M. sativa* L.), and to a lesser extent red clover will be dealt with in detail by our American colleagues. It is also true that compared with white clover there is a paucity of information on the competitive abilities of other pasture legumes, particularly field data under grazing.

WHITE CLOVER

Monoculture/Mixture Comparisons

In temperate climates, annual production of white clover monocultures range from about 4000 to 14 000 kg DM/ha. At Palmerston North (latitude 40°S), Anderson (1973) obtained annual yields under cutting of 12 500 (90% clover content) from pure sowings of cv. Huia on a silt loam without irrigation. Williams and Barclay (1975) obtained an average of about 8500 kg DM/ha over 2 yr from pure sowings of New Zealand and Spanish white clover on a silt loam at Palmerston North. By comparison, a pure sowing of Ruanui perennial ryegrass yielded 17 500 kg DM/ha where fertilizer

N was applied (Harris & Hoglund, 1977). The annual production potential for adequately fertilized, irrigated ryegrass/white clover pastures at Palmerston North has been estimated as 24 000 kg DM/ha (Brougham 1959), and with unlimited fertilizer N 29 000 kg DM/ha for Nui ryegrass alone (Hunt & Mortimer, 1982).

Brown (1973) recorded mean yields under cutting of 6250 kg DM/ha from white clover cv. Huia at Gore, Southland (latitude 46°S), while yields of up to 19 000 kg DM/ha were obtained with perennial ryegrass cv. Ruanui also under cutting and with adequate fertilizer at the same location (Harris et al., 1973). However when these two cultivars were sown together annual yields of only 12 500-14 500 kg DM/ha were obtained under grazing without N applications. Huia content ranged from 1800 to 3000 kg DM/ha.

It is the reaction of white clover to both competition from associated species and to the stress of grazing that this paper will address.

White Clover Content in Swards

Standardized measurements of grazed pastures at 11 geographically dispersed sites (Radcliffe, 1974) provide an assessment of variation in white clover content in swards on farms in New Zealand. Although the sites span more than 10° of latitude, no clearly defined differences of clover yield related to latitude were found. Neither were there any clear relationships between clover yield and total yield.

An indication of the range of variation of clover content at a given level of production can be obtained by examining the median range of annual total yields (10 000 to 12 000 kg DM/ha). Within this range, clover content varied from 11.8 to 38.5%. Year-to-year variation at several sites was greater than site-to-site variation. Between-year variation was least for the high rainfall sites which suggests that much of the year-to-year variation of clover content in New Zealand may be principally associated with rainfall. Likewise a national series of trials in the UK (J. Morrison, IGAP, 1979, personal communication) found the same correlation.

(a) Frequency

Harris (1988) explained the varied response of white clover content to defoliation frequency obtained by Brougham (1959), Harris and Thomas (1973), and Harris (1974), by viewing frequent defoliation as especially prejudicial to erect grass species, whereas the stoloniferous habit of white clover means that a smaller proportion of its total biomass is removed in any one defoliation. Therefore, where clover is at a competitive disadvantage, frequent cutting may increase clover content. Infrequent defoliation could be expected to increase clover where clover is already at a competitive advantage, or at least actively growing and able to take advantage of the longer spells between defoliations, or where infrequent defoliation is prejudicial to the associated species growth.

(b) Intensity

Various workers have demonstrated that extremely hard cutting intensity can reduce the competitive ability of some cultivars of white clover. Brown (1939) found that the ground area occupied by white clover when cut to a height of 1.25 cm was reduced compared with 2.5 cm cutting height, and that the reduction was particularly marked with Ladino rather than Kent white clover.

At defoliation heights above 2 cm, increased intensity of defoliation of mixed grass/white clover swards has often, but not invariably, increased the proportion of clover in the sward (Robinson & Sprague, 1947; Kishi, 1973; Clark et al., 1974). However, New Zealand workers have generally found that the effect of defoliating to heights of 2 or 8 cm was negligible compared with the effect of cutting frequency on the competitive interaction between ryegrass and white clover (Harris & Thomas, 1973). In *Agrostis tenuis*/white clover mixtures (Harris, 1974), *A. tenuis* was more aggressive under frequent defoliation to a height of 8 cm compared with 2 cm. The development of a mat of *A. tenuis* under higher defoliation impaired the growth of white clover.

Generally defoliation heights which reduce competition, but do not affect stolon density, favor white clover growth.

(c) Timing

The effect of frequent and intensive defoliation on white clover content varies with season of the year. The series of experiments carried out by Jones (1933) and Brougham (1959, 1960, 1965) still represent the most definitive work addressing this aspect.

Generally, frequent intensive defoliations at the start of the clover growing season reduce grass competition, increase total annual sward clover content, and sometimes reduce total herbage production. Hay and Baxter (1984) confirmed this finding at Gore, demonstrating that set stocking during spring doubled the white clover content in summer, compared with the same old pasture sward which had been rotationally grazed at 3 or 4-weekly intervals during this spring period. There was no loss in total annual DM production.

Jones (1933) showed that spelling pasture in autumn was beneficial to white clover content, more so if the spelling was in late autumn. Hard and frequent grazing in winter at Palmerston North produced high clover yields and highest total annual yield (Brougham, 1960). However, available information suggests that in regions with low winter temperatures, lenient infrequent defoliation ensures more successful overwintering of white clover than the frequent close defoliation usually considered to maintain clover in swards (Harris et al., 1973; Harris et al., 1983).

Associate Species

With some exceptions, white clover yields are generally inversely related to yields of the associated species (Harris, 1977). Specific cases of antagonism between white clover and other species are starting to be extensively documented. There are indications that white clover can associate more successfully with tufted rather than with stoloniferous or rhizomatous grasses.

(a) Perennial Ryegrass

In a study of old permanent pasture in North Wales, Turkington and Harper (1979a) showed T. repens and L. perenne to be positively associated with each other and mostly negatively associated with all other abundant species in the pasture. There was no clear evidence that this association was determined by variation of edaphic factors in the area studied. Analysis of seasonal changes in the pasture (Turkington & Harper, 1979b) suggested that the positive association between T. repens and L. perenne was related to the complementarity of the growth cycles of the two species.

Jones (1933) observed that wild white clover seemed to be governed by the ryegrass population - the more ryegrass the less white clover. Chestnutt and Lowe (1970) concluded that there were no marked differences between ryegrass cultivars in their compatibility with white clover. Those ryegrass cultivars which suppressed white clover least were of low yield potential or low persistency (Green & Corrall, 1965). In a 4-yr experiment, Camlin (1981) concluded that the compatibility of the ryegrass cultivars with clover was inversely related to the grass component yield, which was determined largely by density.

Recent work by Rhodes (1988) suggests however, that the performance of different varieties of white clover was greatly influenced by the choice of perennial ryegrass companion with which they are grown. There is a case for the conjoint selection of grasses and legumes for their "ecological combining ability" (Harper, 1967). Work in New Zealand has demonstrated marked depressions in white clover content in swards of high endophyte perennial ryegrass, compared with those in mixtures with nil or a very low endophyte presence (Sutherland & Hoglund, 1988; D.R. Stevens, 1988, personal communication). In addition, work reported by Young (1987) in Australia has shown lower subclover contents in pastures with high endophyte Ellett compared with low endophyte Ellett, causing in this instance, lower sheep productivity.

Indeed, work by Harris and Hoglund (1977) measuring N_2-fixation using the acetylene reduction technique, showed a 65% reduction in N_2-fixation from the mean of Nui-based swards with Huia and Pitau white clover clovers and Pawera red clover, compared with the mean of Ruanui-based swards in combinations with the same clovers. The line of Nui seed used in this experiment was subsequently found to contain 92% endophyte, and the Ruanui 6%. At the time, Harris and Hoglund attributed the N_2-fixation difference to the greater aggression of Nui compared with Ruanui, which was reflected in higher ryegrass DM in the Nui plots.

Toxins produced by the ryegrass endophyte possibly restrict clover growth, and there is a phytotonic effect on ryegrass growth (Latch et al., 1985) which offers increased competition to companion species. However, this effect is complicated by the fact that low endophyte ryegrass swards generally have lower tiller numbers per unit area than high endophyte swards. This is due to widespread tiller death caused by Argentine stem weevil (present in Australia and New Zealand), and thus less competition is offered to companion sward species. Argentine stem weevil is not present in the UK, and very few of the UK commercial cultivars of perennial ryegrasses sampled recently contained any endophyte (G.C.M. Latch, 1988, personal communication).

(b) Cocksfoot

It has often been observed that the white clover content of swards is low when in mixture with cocksfoot (Hughes, 1951; Turkington et al., 1977). Unpublished data of J.A. Lancashire, 1988, personal communication) at Palmerston North showed white clover contents in a Wana cocksfoot/Pitau white clover sward under cutting to diminish to 1% of total annual DM. Adjacent swards of Ariki ryegrass/Pitau white clover yielded 22% clover of total DM.

Three explanations for this effect are possible. Firstly, Heddle and Herriot (1954) noticed that yields of unsown species were very high in cocksfoot plots, and using these data Harris (1988) showed the yield of white clover to be inversely related to the combined yield of sown grass plus unsown species.

Secondly, Harris (1988) postulates that there could be strong selective grazing of white clover in association with cocksfoot, cocksfoot being less acceptable to grazing animals than ryegrass. Thirdly, a greater interference between cocksfoot and white clover is caused by similarities in their seasonal growth pattern (Turkington & Harper, 1979b). However tall fescue (*Festuca arundinacea* Schreb.) has a similar seasonal growth pattern to cocksfoot and is known to be a highly compatible companion to white clover (Hay, 1987).

(c) Browntop

Brown (1939), comparing grass species sown with white clover under frequent close cutting, observed that the content of white clover was much less with the turf-forming grasses *A. tenuis* and *Poa pratensis* than the species with more open stands, e.g., *L. perenne*, *F. pratensis*, and *D. glomerata*.

In the succession towards *Agrostis* dominant pasture in New Zealand, the development of positive associations between ryegrass and white clover have been observed (Harris & Brougham, 1968; Harris, 1973). *Poa* spp., mainly *P. annua*, also showed positive associations with ryegrass and white clover, and negative associations with *Agrostis*.

Chestnutt and Lowe (1970) suggested that clover has a higher requirement for P than most grasses, but that when P is deficient, clover is better able than grasses to satisfy its requirement for P. However, Jackman and Mouat (1972b) qualified the latter conclusion and demonstrated that at suboptimal levels of available P, white clover was less efficient in its response than was *A. tenuis*.

Jackman and Mouat (1972a,b) encountered difficulty in separating the effects of competition for P between *A. tenuis* and white clover from the effects of competition for "other factors". These authors considered it likely that shading of clover stolons by *A. tenuis* was important at high P levels even though the results suggested that clover was dominant. As the swards developed, a mat of *A. tenuis* 1 cm thick was formed which covered the white clover stolons and, by shading them, probably inhibited clover bud development.

(d) Allelopathic Effects

Newman and Rovira (1975) and other authors feel that allelopathy is widespread among grassland species. Because of the difficulties of separating allelopathic factors from the effects of established plants on the micro-environment, most experiments investigating allelopathy have examined the effects of plant extracts on seed germination and seedling growth. Freeze-dried shoot and fresh root-soil material from several tussock grasslands species retarded white clover germination, while shoot material of white clover itself, severely inhibited germination (Scott, 1975). Inhibition by *Notodanthonia* spp. of rhizobia viability and nodulation was reported by Janson and White (1971).

White Clover Genotype

Differences between white clover genotypes in their ability to compete with associate grasses, and their response to the modifying effects of N fertilization and defoliation management, have been related largely to genotype differences of petiole length and leaf size (Harris, 1988). Genotypes which form canopies of large leaves on long petioles are able to avoid shading from associate species, and differ in strategy from those genotypes forming dense networks of well-rooted stolons bearing small leaves on short petioles which are better able to withstand frequent and close defoliation.

Table 1 shows results of work done at Palmerston North by J.L. Brock, 1987 (unpublished data) with the four Grasslands white clover cultivars of different leaf sizes, under two grazing managements. Under set stocking the smallest-leaved cv. Tahora outyielded the larger-leaved cultivars Pitau and Kopu. Under rotational grazing the result was reversed, Kopu contributing almost 30% to annual total DM production which Stewart (1984), Frame (1987) and others have decreed is a desideratum for grazed pasture.

Rhodes and Harris (1979) demonstrated that although the gross yield of clover herbage harvested was similar for a range of different leaf-sized cultivars, the harvestable part of the long petiole-large leaved cultivars was greater. Thus, assimilates appeared to be partitioned to form a tall canopy at the expense of development of the stolon system. This disadvantages these large-leaved cultivars in their ability to survive periods of stress, as they have less of the basic white clover survival unit - stolon growing points.

The inclusion of Mediterranean genes in temperate white clover cultivars can influence the seasonal clover content in pastures irrespective of leaf size x grazing management interactions. For example, the increased clover content in autumn and winter by sowing Pitau has been observed to reduce grass and total yield during this time (Barclay, 1969; Brock, 1971, 1974).

Table 1. Percentage white clover of total annual DM production at Palmerston North (From J.L. Brock, 1987, unpublished data).

	Tahora	Huia	Pitau	Kopu
Set stocked	19.9	13.6	7.2	8.1
Rotationally grazed	17.7	14.1	20.6	28.8

Studies of the root systems of a range of white clover cultivars emphasizes the difference between genotypes with root systems dominated by several deep thick roots and genotypes with predominantly fibrous root systems (Caradus, 1977, 1981). Thick, deep root development is correlated with the large-leaved habit, and predominantly fibrous root development with the prostrate small-leaved habit.

Varietal differences in pest or disease resistance can also influence the persistence and clover content of swards, as clearly illustrated by Aldrich (1970) who reported that *Sclerotinia trifoliorum* reduced the yield of the susceptible cv. S100, whereas the resistant cv. Pajbjerg Milka gave high yields.

Competitive Strategies

(a) Stoloniferous Guerilla Habit

Detailed studies of mineral uptake of white clover in competition with grasses by Jackman and Mouat (1972a,b), and more recent studies on the mechanism of P uptake by white clover (A.L. Hart, 1987, personal communication), indicate that white clover should not be able to survive in hill country pastures low in P. Its considerable success in these areas, and vital contribution to the N economy can be explained only by its guerilla growth habit (M.J.M. Hay, 1988, personal communication) of exploiting gaps or discontinuous areas of grass growth. This is an extension of the conclusion of Ennik (1970) that the relative competitive abilities of grass and clover suggest that clover growth is limited to the "space" left by the grass.

Turkington and Cavers (1978) conclude that the formation of stolons by white clover within 7 wk of germination permitted it to "reproduce" sooner than the annual *Medicago lupulina* could reproduce by seed. In this way, white clover behaves as a colonizer of frequently disturbed habitats, and through genotypic variation and phenotypic plasticity, provides individuals also able to compete effectively in stable environments with high populations of other species.

The stoloniferous character allows the species rapidly to attain an equilibrium in plant communities (Ennik, 1970; Harris & Thomas, 1973), and is the reason why variation in seed rate of white clover has only a short-term effect on the clover content of new-sown swards (Brown, 1973).

A more detailed treatise on stolon formation and function is given in the paper by Forde et al. in this publication.

(b) Light Interception

A feature of the white clover canopy which distinguishes it from that of grasses is the orientation of the laminae. The horizontal arrangement of the laminae which, with phototropic movement, orientates them at right angles to the sun, results in a low critical LAI for white clover. The long slender leaves of the grasses, erect for most of their length, allow deeper direct penetration of light into the canopy, whereas light not absorbed by the horizontal laminae of white clover is mostly reflected back into the atmosphere (Harris, 1988).

Although grass-legume competition studies indicate that white clover is at a disadvantage with grass in terms of light competition, probably most grasses and white clover are more compatible in their light relations than white clover is with other dicotyledonous species. Brougham (1965) compared the regrowth for 60 days after defoliation of a monoculture of white clover with that of white clover mixed with red clover. Thirty days after defoliation, the leaf area of white clover in the mixture was less than 25% of that in the pure stand.

(c) Seasonal Growth Periodicity

This strategy has been mentioned earlier in this paper and is regarded by Harris (1988) as being very important and strongly related to the higher temperature optima of this species compared with grasses. Hoglund et al. (1979) suggest that some of the complementarity of ryegrass and white clover is related to the high potential growth rates of grass in late winter and early spring. This depletes soil N at these times, thus providing a soil environment more suited to N_2-fixation during the late spring and summer when temperature and light favor clover growth. This situation emphasizes the subtle way in which white clover is adapted to moist temperate pastures.

(d) Nitrogen Fixation

A key factor in white clover's competitiveness is its ability to fix atmospheric N, particularly where soil N levels are low, as it is a poor competitor for soil N. High levels of soil N may also directly inhibit N_2-fixation (Hoglund, 1973). A trial investigating N_2-fixation in grazed ryegrass/white clover pasture at eight geographical sites throughout New Zealand by Hoglund et al. (1979) showed a mean yearly fixation of 185 kg N/ha.

A more thorough discussion of white clover N_2-fixation and interactions with rhizobium and mineral nutrients will be presented by other authors in this publication.

RED CLOVER

Background

Red clover is an important herbage legume in New Zealand agriculture. New Zealand Department of Statistic figures show red clover to be the second most important legume after white clover in terms of average yearly tonnage of seed sown in New Zealand - approximately 1250 tonnes white clover and 350 tonnes red clover. In the 1950's and 1960's red clover usage declined. This can be attributed to several factors including the trend towards simplification of seeds mixtures, particularly for long-term pastures, and the unpredictability of the crop owing to its sensitivity to grazing management, pests and diseases. On grazed swards there is the risk of bloat (Johns, 1963), and high oestrogenic activity in breeding ewes (Hay et al., 1978).

What prevented the complete disappearance of this species from New Zealand agriculture was its high palatability, animal intake, nutrition and growth periodicity advantages compared with ryegrass/white clover swards (Ulyatt et al., 1977; Harris & Hoglund, 1977).

Competitive Interactions

(a) Seasonal Growth Periodicity

Harris and Hoglund (1977) and Harris et al. (1980) have been interested in utilizing the asynchronous growth rhythm and aggressiveness of red clover to improve the seasonal distribution of available high quality pasture for grazing animals. This was achieved, particularly with the late flowering Turoa and Pawera cultivars, with greater summer productivity, an important consideration for dairy farmers and heavy weight lamb finishing units.

Work done by Hay et al. (1978) is presented in Table 2 and shows how much the seasonal distribution of annual pasture DM in this environment can be amended firstly by using red clover cv. Pawera on its own, and secondly by using a mixture of Pawera with an annual ryegrass cv. Tama, overdrilled in the autumn. Thus it is possible to manipulate feed supply to meet specific feed deficits by the use of different pasture species, including red clover. To maximize seasonal distribution even further Pineiro and Harris (1978a,b) compared the growth of three red clovers in combination with *Bromus wildendowii* cv. Matua which has considerable winter activity. Again this showed an improved seasonal growth curve, with a slightly higher total yield compared with standard mixtures in the 2nd yr.

In both these cases above, considerable damage was done to the crowns of the red clover plants when the cool-season active grass component of the sward was utilized. This winter damage allowed the invasion of pathogenic bacteria and fungi (Hay, 1985) which markedly affected the persistence of red clover in swards such as these. In addition depletion of carbohydrate root reserves by frequent late summer/autumn defoliations lowered subsequent spring DM yield and persistence (Hay, 1985).

(b) Grazing Management Influences

Yield and seasonal distribution of yield in red clover pasture

Table 2. Production of pastures of different composition in a cool temperate region of New Zealand (kg DM/ha) mean of 2 yr. (From Hay et al., 1978).

Pasture type	Spring	Summer	Autumn	Winter	Total
Perennial ryegrass and white clover	7000	4100	2500	1200	14 800
Red clover	4500	7700	1200	150	13 550
Red clover and annual ryegrass	5750	5000	1750	2000	14 500

mixtures is markedly influenced by grazing management (Pineiro & Harris, 1978b; Harris, et al., 1980). Infrequent grazing provided 40 % more yield with a more peaked distribution than where mixtures were frequently grazed.

Grazing frequency, particularly in summer, was found by Cosgrove and Brougham (1985) to be an important management variable for manipulating ryegrass and red clover balance in grazed pasture. Their 3-yr trial with mixtures of ryegrasses cv. Nui and cv. G4708 with red clover cv. Pawera and white clover cv. Pitau, showed that although summer production was slightly reduced by frequent grazing in summer, a large increase in spring ryegrass production offset the reduced production from red clover. Frequent grazing in winter, as opposed to summer, is less effective in achieving a more equitable ryegrass/red clover distribution.

(c) Associate Species Effects

Harris (1977) reported on a large trial in which he evaluated six legumes (three of which were red clovers) in combination with nine grasses. The interactions between competing species were illustrated through the application of regression analysis of DM yield as described by Breese (1969). A high endophyte line of Nui ryegrass was shown to be the most aggressive grass towards legumes, and Hamua the most aggressive legume. Applied N generally increased the aggressiveness of the grasses and decreased the aggressiveness of the legumes.

Unpublished work by R.J.M. Hay at Gore, 1980, investigated the competitive effects of six grasses with red clover cv. Pawera and compared component yields with that from monocultures of Pawera. Two grazing management regimes were used, short (15 to 2.5 cm) and lax (20 to 5 cm). Results from this 4-yr trial are presented in Fig. 1 as regressions of sown red clover yields against sown grass. Clearly there are large differences between Years 1 and 2 and Years 3 and 4, particularly for Pawera content.

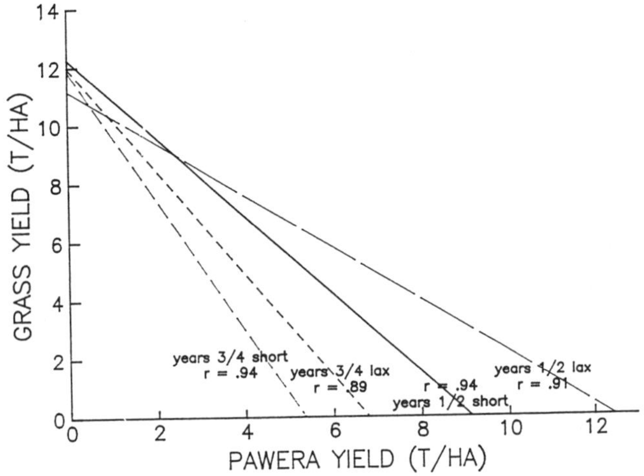

Fig. 1. Harvested yield of sown companion grass (Y) and Pawera red clover (X) in various mixtures, meaned for Years 1 and 2 and Years 3 and 4 under two defoliation regimes at Gore, New Zealand, 1980, (From Hay and Ryan, unpublished data).

Grazing management affected the amount of red clover more with some grasses than with others. Tall fescue cv. Roa was easily the most compatible grass with red clover and this can, in a large part, be attributed to its slow rate of establishment compared with the other grasses.

In general this work supports the finding of Harris (1977) that there were strong negative correlations between cultivar yield and suppressive effect on associates in mixture. Neither of these two trials reveal the desirable situation where a grass has a high yield and maintains a high legume content.

The persistence of red clover in Years 3 and 4 in grazed swards with other species is a major concern (Fig. 1), and this feature, along with its high phyto-oestrogenic content, have hindered the increase in use of this species in grassland farming in New Zealand (Hay et al., 1978).

SUMMARY

This paper reviews literature on aspects of grass/white and red clover competition. It concentrates on white clover as it is by far the most important pasture legume under grazing in New Zealand.

It is our view that deficiencies exist in production from grazed grass/clover swards obtained in practice compared with potential estimated figures. Strategies to significantly increase the white clover content in pasture through calculated hard grazings at specific times of the year are available but poorly understood by commercial farmers. Highly competitive high endophyte ryegrasses are of concern because they reduce the white clover content of pasture. Considerable differences in competitive ability are apparent between white clover genotypes due to root structure, leaf size, petiole length, pest and/or disease resistance, and expression of Mediterranean: temperate genetic background.

White clover has a distinct range of competitive strategies available to it, all of which can be employed in particular situations to great effect. Perhaps the most important strategy is the guerilla habit where it uses its stoloniferous character to rapidly establish in sward spaces. It then employs its superior light interception and different growth rhythm to successfully combat vigorous grass competition.

Late flowering red clover cultivars have an even more pronounced seasonal growth periodicity difference than ryegrasses. Because red clovers do not have a stoloniferous habit they are more sensitive to damage through inappropriate grazing management. Lack of persistence in grazed swards with associated grasses is the major drawback to the widespread use of red clovers.

REFERENCES

Aldrich, D.T.A. 1970. Clover rot (Sclerotinia trifoliorum) in white clover and its influence on varietal performance at different centres. p.143-146. Review in: J. Lowe. (ed.) White clover research. Occas. Symp. 6. Br. Grassl. Soc., Berkshire, UK.

Anderson, L.B. 1973. Relative performance of the late-flowering tetraploid red clover "Grasslands 4706", five diploid red clovers, and white clover. N.Z. J. Agric. Res. 14: 563-571.

Barclay, P.C. 1969. Some aspects of the development and performance of "Grasslands 4700" white clover. Proc. N.Z. Grassl. Assoc. 31: 127-134.

Breese, E.L. 1969. The measurement and significance of genotype-environment interactions in grasses. Heredity 24: 27-44.

Brock, J.L. 1971. A comparison of "Grasslands 4700" and "Grasslands Huia" white clovers in establishing ryegrass/clover pasture under grazing. N.Z. J. Agric. Res. 14: 368-378.

Brock, J.L. 1974. Effects of summer grazing management on the performance of "Grasslands Huia" and "Grasslands 4700" white clovers in pasture. N.Z. J. Exp. Agric. 2: 365-370.

Brougham, R.W. 1959. The effect of frequency and intensity of grazing on the productivity of a pasture of short-rotation ryegrass and red and white clover. N.Z. J. Agric. Res. 2: 1232-1248.

Brougham, R.W. 1960. The effect of frequent hard grazing at different times of the year on the productivity and species yields of a grass-clover pasture. N.Z. J. Agric. Res. 3: 125-136.

Brougham, R.W. 1965. The effect of red clover on the leaf growth of white clover under long spelling during the summer. N.Z. J. Agric. Res. 8: 859-864.

Brown, B.A. 1939. Some factors affecting the prevalence of white clover in grassland. J. Am. Soc. Agron. 31: 322-332.

Brown, K.R. 1973. Some effects of the time of sowing on the first twelve months growth of, "Grasslands Huia" white clover, sown at two seedling rates or associated with "Grasslands Manawa" ryegrass. N.Z. J. Exp. Agric. 1: 1-5.

Camlin, M.S. 1981. Competitive effects between ten cultivars of perennial ryegrass and three cultivars of white clover grown in association. Grass Forage Sci. 36: 169-178.

Caradus, J.R. 1977. Structural variation of white clover root systems. N.Z. J. Agric. Res. 20: 213-219.

Caradus, J.R. 1981. Root growth of white clover lines in glass-fronted containers. N.Z. J. Agric. Res. 24: 43-54.

Chestnutt, D.B.M., and J. Lowe. 1970. Agronomy of white clover-grass swards. p.35-40. Review in: J. Lowe. (ed.) White clover research. Occas. Symp. 6. Br. Grassl. Soc., Berkshire, UK.

Clark, J., C. Kot, and K. Santhirasegaram. 1974. The effects of changes in heights of cutting and growth on the digestible organic matter production and botanical composition of perennial pasture. J. Brit. Grassl. Soc. 29: 269-273.

Cosgrove, G.P., and R.W. Brougham. 1985. Grazing management influences on seasonality and performance of ryegrass and red clover in a mixture. Proc. N.Z. Grassl. Assoc. 46: 71-76.

Ennik, G.C. 1970. White clover/grass relationships: competition effects in field and laboratory. p.165-174. Review in: J. Lowe. (ed.) White clover research. Ocass. Symp. 6. Br. Grassl. Soc., Berkshire, UK.

Frame, J. 1987. The role of white clover in United Kingdom pastures. Outlook Agric. 16: 28-33.

Green, J.O., and A.J. Corrall. 1965. The testing of grass varieties in swards with clover: The effect of grass seed-rate on comparisons of grass yield. J. Brit. Grassl. Soc. 20: 207-211.

Harper, J.L. 1967. A Darwinian approach to plant ecology. J. Ecol. 55: 247-270.
Harris, A.J., K.R. Brown, J.D. Turner, J.M. Johnstone, D.L. Ryan, and M.J. Hickey. 1973. Some factors affecting pasture growth in Southland. N.Z. J. Exp. Agric. 1: 139-163.
Harris, W. 1973. Why browntop is bent on creeping. Proc. N.Z. Grassl. Assoc. 35: 101-109.
Harris, W. 1974. Competition among pasture plants. V. Effects of frequency and height of cutting between Agrostis tenuis and Trifolium repens. N.Z. J. Agric. Res. 17: 251-256.
Harris, W. 1977. An approach to evaluate a large number of mixtures under grazing. p. 401-409. In Proc. 13th Int. Grassl. Congr. Leipzig.
Harris, W. 1988. Population dynamics and competition. p.203-297. In White clover. M.J. Baker, W.M. Williams (Ed.).
Harris, W., and R.W. Brougham. 1968. Some factors affecting change in botanical composition in a ryegrass/white clover pasture under continuous grazing. N.Z. J. Agric. Res. 11: 15-35.
Harris, W., and J.H. Hoglund. 1977. Influences of seasonal growth periodicity and N-fixation on competitive combining abilities of grasses and legumes. p.239-243. In Proc. 13th Int. Grassl. Congr., Leipzig.
Harris, W., J. Pineiro, and J.D. Henderson. 1980. Performance of mixtures of ryegrass cultivars and prairie grass with red clover cultivars under two grazing frequencies. III. Herbage production and shoot numbers in the second year. N.Z. J. Agric. Res. 23: 339-348.
Harris, W., I. Rhodes, and S.S. Mee. 1983. Observations on environmental and genetic influences on the overwintering of white clover. J. Applied Ecol. 20: 609-624.
Harris, W., and V.J. Thomas. 1973. Competition among pasture plants. III. Effects of frequency and height of cutting on competition between white clover and two ryegrass cultivars. N.Z. J. Agric. Res. 16: 49-58.
Hay, R.J.M. 1985. Variety by management interactions with red clover (Trifolium pratense L.). Ph.D. Thesis, Lincoln College. pp.221. Abstr. In Proc. N.Z. Grassl. Assoc. 1987. 48: 219-220.
Hay, R.J.M. 1987. Understanding pasture growth. Proc. Massey Dairy Farmers Conf. 39: 55-63.
Hay, R.J.M., and G.S. Baxter. 1984. Spring management of pasture to increase white clover growth. Proc. Lincoln Farmers Conf. 34: 132-137.
Hay, R.J.M., R.W. Kelly, and D.L. Ryan. 1978. Some aspects of the performance of "Grasslands Pawera" red clover in Southland. Proc. N.Z. Grassl. Assoc. 38(2): 246-252.
Heddle, R.G., and J.B.D. Herriott. 1954. The establishment, growth and yield of ultra-simple grass seeds mixtures in the south-east of Scotland. J. Br. Grassl. Soc. 9: 99-110.
Hoglund, J.H. 1973. Bimodal response by nodulated legumes to combined nitrogen. Plant Soil 39: 533-545.
Hoglund, J.H., J.R. Crush, J.L. Brock, and R. Ball. 1979. Nitrogen fixation in pasture. XII. General discussion. N.Z. J. Exp. Agric. 7: 45-51.
Hughes, G.P. 1951. The seasonal output of pastures sown with ultra-simple seed mixtures. J. Agric. Sci. 41: 203-213.

Hunt, W.F., and B.J. Mortimer. 1982. A demographic analysis of growth differences between Nui and Ruanui ryegrass at high and low nitrogen input. Proc. N.Z. Grassl. Assoc. 43: 125-132.
Jackman, R.H., and M.C.H. Mouat. 1972a. Competition between grass and clover for phosphate. I. Effect of browntop on white clover growth and nitrogen fixation. N.Z. J. Agric. Res. 15: 653-666.
Jackman, R.H., and M.C.H. Mouat. 1972b. Competition between grass and clover for phosphate. II. Effect of root activity, efficiency of response to phosphate and soil moisture. N.Z. J. Agric. Res. 15: 667-675.
Janson, C.G., and J.G.H. White. 1971. Lucerne establishment studies on uncultivated country. II. A nodulation problem. N.Z. J. Agric. Res. 14: 587-596.
Johns, A.T. 1963. Bloat. p.23-30. In Dairyfarming Annual. Massey University.
Jones, M.G. 1933. Grassland management and its influence on the sward. J. R. Agric. Soc. 94: 21-41.
Kishi, H. 1973. Studies on competition between grass and legumes in a mixed sward. 1. The growth of two species in a cocksfoot/ladino clover sward. Proc. Crop Sci. Soc. Jpn. 42: 397-406.
Langer, R.H.M. 1969. Presidential address. Proc. N.Z. Grassl. Asson. 31: 5-8.
Latch, G.C.M., W.F. Hunt, and D.R. Musgrave. 1985. Endophytic fungi effects on growth of perennial ryegrass. N.Z. J. Agr. Res. 28: 165-168.
Newman, E.I., and A.D. Rovira. 1975. Allelopathy among some British grassland species. J. Ecol. 63: 727-737.
Pineiro, J., and W. Harris. 1978a. Performance of mixtures of ryegrass cultivars and prairie grass with red clover cultivars under two grazing frequencies. I. Herbage production in the establishment year. N.Z. J. Agric. Res. 21: 83-92.
Pineiro, J., and W. Harris. 1978b. Performance of mixtures of ryegrass cultivars and prairie grass with red clover cultivars under two grazing frequencies. II. Shoot populations and natural reseeding of prairie grass. N.Z. J. Agric. Res. 21: 665-673.
Radcliffe, J.E. 1974. Seasonal distribution of pasture production in New Zealand. I. Methods of measurement. N.Z. J. Exp. Agric. 2: 337-340.
Rhodes, I. 1988. Courting mixtures. Big Farm Weekly (Feb. 18) 1988. p.8.
Rhodes, I., and W. Harris. 1979. The nature and basis of differences in sward composition and yield in ryegrass/white clover mixtures. p.55-60. In A.H. Charles. (ed.) Changes in sward composition and productivity. Occas. Symp. 10. Br. Grassl. Soc., Berkshire, UK.
Robinson, R.R., and V.G. Sprague. 1947. The clover populations and yields of a Kentucky bluegrass sod as affected by nitrogen fertilisation, clipping treatments and irrigation. J. Am. Soc. Agron. 36: 107-116.
Scott, D. 1975. Allelopathic interactions of resident tussock grassland species on germination of oversown seed. N.Z. J. Agric. Res. 3: 135-141.
Stewart, T.A. 1984. Performance of pasture legumes in Scotland. p.93-103. In D.J. Thomson (ed.) Forage legumes. Ocass. Symp. 16. Br. Grassl. Soc., Berkshire, UK.

Sutherland, B.L., and J.H. Hoglund. 1988. Effect of the ryegrass endophyte (Acremonium lolii) on associated white clover and subsequent crops. Proc. N.Z. Grassl. Assoc. 50: (In press).

Turkington, R., P.B. Cavers, and L.W. Aarssen. 1977. Neighbour relationships in grass/legume communities. I. Interspecific contacts in four grassland communities near London, Ontario. Canadian J. Botany 55: 2701-2711.

Turkington, R., and P.B. Cavers. 1978. Reproduction strategies and growth patterns in four legumes. Canadian J. Botany 56: 413-416.

Turkington, R., and J.L. Harper. 1979a. The growth, distribution and neighbour relationships of Trifolium repens in a permanent pasture. I. Ordination, pattern and contact. J. Ecol. 67: 201-218.

Turkington, R., and J.L. Harper. 1979b. The growth, distribution and neighbour relations of Trifolium repens in a permanent pasture. II. Inter- and intra-specific contact. J. Ecol. 67: 219-230.

Ulyatt, M.J., J.A. Lancashire, and W.T. Jones. 1977. The nutritive value of legumes. Proc. N.Z. Grassl. Assoc. 38: 107-118.

Williams, W.M., and P.C. Barclay. 1975. Performance of a Spanish white clover population in New Zealand. N.Z. J. Agric. Res. 18: 45-49.

Young, P. 1987. A primary producers overview. Proc. 28th Conf. of Victorian Grassl. Soc. (In press).

DISCUSSION

Fisher: What is high endophyte/low endophyte ryegrass?

J. Hay: The endophyte is a saprophytic fungus, which is transmitted by seed, and is present in most New Zealand perennial ryegrass. If the seed is stored at normal temperatures, without humidity control, the endophyte dies and low endophyte material is obtained. It causes ryegrass staggers in sheep, but prevents damage by Argentine Stem Weevil.

Allen: You mentioned the conflict between Argentine Stem Weevil (ASW) control and reduced yields of white clover using high endophyte ryegrass. How do you see this conflict being resolved? Is the endophyte affecting clover growth the same as the endophyte causing ryegrass staggers or the endophyte conferring weevil resistance?

J. Hay: It is one endophyte species (Acromonium lolii) that produces peramine which deters ASW feeding, and lolitrem B which causes ryegrass staggers. We do not know what is causing the white clover suppression in high endophyte swards.

Watson: Concerning possible deleterious consequences of endophytic ryegrass on clover levels in pasture, my comment is that New Zealand perennial pastures all have high endophyte which enables them to persist.

J. Hay: Southland pastures are not subject to ASW pressure and are not high in endophyte. It is also true to say that old perennial ryegrass pastures (presumably high in endophyte) do not have high white clover contents e.g. hill country pastures often have 5-10% annual DM as white clover. This low clover production has been identified as a major limitation to productivity from these pastures.

Caradus: Is there a need to breed for an endophyte tolerant white clover?

J. Hay: This may be one solution to the problem of white clover suppression in high endophyte perennial ryegrass swards, the other is to identify the compound(s) responsible, and selecting ryegrass/ endophyte associations which don't produce them.

Sheath: In considering grass/clover balance, winter feed flow from grass is just as important as clover N and herbage to the total feed system. If endophyte in ryegrass depresses white clover then be careful in assessing the relative problem. It may be more important to retain ryegrass persistence and accept lower clover content.

Brougham: A counter view to Gavin Sheath's comment is that we should be thinking about not only persistence of legumes but also persistence of the associated grasses. By keeping persistence of both in perspective then we obtain highest production, regardless of environment.

Barnes: The potential is being explored for treatment of animals consuming endophyte-infected tall fescue to alleviate the effect of toxins. Preliminary results suggest a blockage of the metabolic effect of the toxin upon animal response. A report on the matter was presented at the 1988 American Society of Animal Science Annual Meeting [Lipham et al. 1988. Effects of meteoclopramide on steers grazing endophyte-infected fescue. J. Anim. Sci. 66 (Suppl. 1): 373]. Although antidotes may be available for tall fescue endophyte toxicity, the toxic principles and disease pathology in animals is different for tall fescue and ryegrass.

Reed: The effect of endophyte on persistence of perennial ryegrass at Hamilton, Vic., has varied with cultivar: Endophyte effect was not marked with Ellett but was most noticeable with Victorian.

Reed: How successful are the Hamua x Moroccan red clovers at improving persistence in marginal rainfall areas?

J. Hay: There has not been as full an evaluation to date, of either the diploid or tetraploid Hamua x Moroccan red clovers in marginal rainfall areas, as I would have liked. The best information on their performance in low rainfall areas is from Hamilton, Victoria, where they have outproduced all other red clovers after the second year. This is due to their superior persistence.

EFFECT OF COMPETITION ON LEGUME PERSISTENCE

C.C. Sheaffer

SUMMARY

Legume persistence is affected by competition with other species during establishment and in established stands. During establishment into tilled seedbeds, legumes compete with annual weeds and with small grain companion crops. In no-till establishment into perennial sods, legumes compete with established grasses. In established stands, the primary competitors are perennial grasses which are often established with legumes in mixtures.

Legumes differ in competitiveness. Important features include seedling vigor, growth habit, and rooting characteristics. Alfalfa and red clover are generally the most competitive legumes while birdsfoot trefoil is among the least competitive. Enhancement of traits affecting competitiveness through breeding would increase legume persistence.

Mixtures of species are seldom in equilibrium. Management practices influence the composition of mixtures by altering species competitiveness. Companion crop management during conventional establishment, sod suppression during no-till establishment, and defoliation frequency of established stands are examples of management practices influencing mixture composition by altering species competitiveness.

INTRODUCTION

Competition occurs when two or more plants seek the same essential resource and the supply of that resource is below the combined demand of the two plants (Donald, 1963). Water, light, and nutrients are the most consistently sought resources (Harper, 1977). In most instances, competition occurs between legumes and non-legumes, particularly with grasses, although legume-legume competition may also occur. Legumes are frequently seeded with annual grass companion crops and in mixtures with perennial grasses.

SOURCES OF COMPETITION

Establishment

Legume competition with non-legumes first occurs during establishment. In tilled seedbeds, small grain companion crops, which are planted to suppress weeds, compete with legume seedlings for light and moisture and may reduce legume persistence (Cooper & Ferguson, 1964; Tesar & Marble, 1988). Klebesadel and Smith (1960) reported that as an oat (<u>Avena</u> <u>sativa</u> L.) companion crop matured from vegetative through grain stages, soil moisture depletion increased and light penetration to alfalfa (Medicago sativa L.) seedlings decreased. However, alfalfa stands were only reduced by delaying oat harvest until maturity. Brink and Marten (1986) reported that although barley (Hordeum vulgare L.) and oat companion crops competed with alfalfa seedlings, alfalfa stand density was not affected unless the small grain was harvested for grain. The average reduction in alfalfa stands was only 5%.

Competition also occurs from rapidly growing summer annual weeds in spring seedings (Dawson & Rincker, 1982) and winter annuals in fall seedings (Peters & Linscott, 1988). Buxton and Wedin (1970) reported that weeds reduced incident solar radiation by 99% and alfalfa and birdsfoot trefoil (Lotus corniculatus L.) stands by 35 and 54%, respectively, compared to a weeded control. However, alfalfa stand loss due to weeds was less than that due to companion crop competition. Alfalfa stand establishment is not always reduced by weed competition. Brink and Marten (1986) and Sheaffer et al. (1988) reported that control of annual weeds did not increase stands of spring seeded alfalfa.

In sod-seeding, established perennial grasses provide the primary competition, but with no-till seeding into crop residues or killed sod, competition occurs from annual and perennial weeds (Decker & Taylor, 1985). Competition for light and moisture reduced ladino clover (Trifolium repens L.) establishment into orchardgrass (Dactylis glomerata L.) sod (Wilkinson & Gross, 1964). Competition reduced ladino clover seedling survival 34 and 51%, respectively. Groya and Sheaffer (1981) reported that light was the most limiting factor in establishment of alfalfa into Kentucky bluegrass (Poa pratensis L.) and smooth bromegrass (Bromus inermis Leyss.) sods. Only when shading by the grasses was eliminated did moisture become a limiting factor.

Established Stands

Perennial legume-grass mixtures are often established to reduce bloat, weed invasion, and erosion; and to increase hay drying rates. Mixtures provide insurance if legumes fail to persist due to hostile edaphic or environmental conditions. Theoretically, physiologically or morphologically different species in mixtures are able to utilize environmental resources more effectively than monocultures (Donald, 1963). However, Chamblee and Collins (1988) reviewed the literature on alfalfa-grass mixtures and reported that mixture yields often did not exceed alfalfa yield in monoculture. When mixture yields did exceed those of monocultures, maximum yield increases averaged 10 to 15%.

For compatible and stable legume-grass mixtures, species should have similar rates of development and palatability and be adapted to similar environmental conditions and to similar harvest managements (Decker & Taylor, 1985). However, these criteria are seldom met and usually the grass or legume component dominates the mixture.

In established stands in the northern USA, alfalfa is competitive with most perennial grasses if rotationally grazed or harvested at 30-d intervals. Miller et al. (1984) found that alfalfa became the dominant species in binary mixtures with smooth bromegrass and reed canarygrass (Phalaris arundinacea L.) by regrowing rapidly following harvesting and shading the grasses. Only orchardgrass, which has a rapid regrowth rate, persisted in binary mixture with alfalfa. Smith et al. (1973) evaluated binary mixtures of eight grasses with alfalfa and found that orchardgrass was an effective competitor with alfalfa but in contrast to the report of Miller et al. (1984) and Jones et al. (1988), they found that reed canarygrass persisted well. Chamblee and Lovvorn (1953) attributed poor alfalfa persistence when mixed with tall fescue (Festuca arundinacea Schreb.) to shading of alfalfa by the rapidly regrowing grass.

Tall growing grasses usually dominate mixtures with birdsfoot trefoil, crownvetch (Cornilla varia L.), and sainfoin (Onobrychis viciifolia Scop.). Sheaffer et al. (1984) evaluated the persistence of birdsfoot trefoil in binary mixtures with eight perennial grasses and found that in an environment which was favorable to perennial grass growth, birdsfoot trefoil only persisted in mixtures with Kentucky bluegrass and a short stature timothy (Phleum bertolonii DC. nodosum L.). With conditions less favorable for perennial grass growth, birdsfoot trefoil persisted best in mixtures with reed canarygrass, smooth bromegrass, and tall fescue. Mays and Evans (1972) reported that crownvetch did not persist when seeded in mixtures with alfalfa, tall fescue, or orchardgrass. Cooper (1972) evaluated sainfoin persistence in binary mixtures with Kentucky bluegrass, red fescue (Festuca rubra L.), birdsfoot trefoil, and ladino clover. He found that birdsfoot trefoil and Kentucky bluegrass were most compatible and that ladino clover was least compatible with sainfoin.

Weed invasion is a major concern in legume management (Peters & Linscott, 1988). In the northern USA, quackgrass (Elytrigia repens L. Nevski) and perennial broadleaf weeds including dandelion (Taraxacum officinale Weber) frequently invade alfalfa stands (Doll, 1986). While quackgrass invasion consistently reduced alfalfa yield and total forage quality (Dutt et al., 1979), all broadleaf weeds did not (Dutt et al., 1982; Sheaffer & Wyse, 1982). Marten et al. (1987) recommended that weed control decisions should be weed specific. While weeds compete with alfalfa and other legumes for essential resources and consequently may reduce yields; there is no conclusive evidence that reduced persistence is due to weed competition (Sheaffer et al., 1984; Cooper, 1972).

LEGUME FEATURES INFLUENCING COMPETITIVENESS

Seedling Vigor

Legumes differ in seedling vigor. These differences have been associated with differences in seed size (Kendall & Stringer, 1985), in biological N_2 fixation, and in CO_2 uptake (Rhykerd et al., 1959). Blaser et al. (1956) identified alfalfa and red clover (Trifolium pratense L.) as aggressive seedlings and ladino clover, white clover, and birdsfoot trefoil as non-aggressive seedlings. Cicer milkvetch (Astragalus cicer L.), crownvetch, and sainfoin also lack seedling vigor (Mays & Evans, 1972; Cooper, 1972; Townsend & McGinnies, 1972). Decker et al. (1985) evaluated legumes for use in a competitive environment caused by sod-seeding and found that red clover was relatively easy to establish and recommended its use over alfalfa, birdsfoot trefoil, and crownvetch for pasture renovation.

Growth Habit

Tall growing legumes with branched stems such as alfalfa and red clover intercept a large portion of the incident light and are effective competitors with grasses and weeds. In contrast, prostrate species such as birdsfoot trefoil and ladino clover are often shaded by tall growing grasses. Haynes (1980) reviewed the literature on foliage architecture and concluded that legumes are generally more prone to be shaded by competitors than grasses because legume leaves are generally horizontally inclined and absorb light from only a few layers while grass leaves are more upright and light is distributed more evenly throughout the canopy. In contrast to most legumes, the canopy structure of alfalfa allows light penetration to lower leaves. Rapid regrowth following defoliation also provides a competitive advantage for species such as alfalfa.

Defoliation interacts with growth habit in modifying competitiveness. Under frequent and close defoliation, prostrate legumes with leaf area near the soil surface are better able to persist than more erect legumes. Less frequent defoliation allows grasses to shade legumes, which favors upright legumes such as alfalfa. Dobson et al. (1976) evaluated the persistence of several legumes in binary mixture with tall fescue. They found that white clover had greater persistence than cicer milkvetch, crownvetch, birdsfoot trefoil, and red clover when frequently defoliated to a 5 cm height. Increasing cutting height to 10 cm decreased persistence of white clover, but increased persistence of the other legumes.

Ladino clover persistence in mixtures is reduced if shaded by tall growing grasses (Sprague & Garber, 1950). Blake et al. (1966) reported that ladino clover persistence was greater when a mixture with orchardgrass was frequently cut at 15 cm than when cut at 25 cm. They concluded that competition for light was more important than competition for moisture. Blaser et al. (1986) described the competition between Kentucky bluegrass and ladino clover under grazing. They reported that a 5-cm grazing height which allowed rapid regrowth of the apex leaves of bluegrass caused shading and slow development of clover leaves originating from stolons at the soil surface. In contrast, grazing to 1 cm depressed bluegrass growth because nearly all of its leaves were removed, but small leaves and stolons remained on ladino clover. Maximum survival of white clover in a bahiagrass (Paspalum notatum Flugge) pasture resulted from a combination of a short deferment period and moderate grazing pressure (Dzowela et al., 1986). A thick forage residue was conducive to root and stolon disease.

Rooting

Most legumes have prominent taproots which allow them to utilize soil water at a greater depth than grasses. Of the legumes, alfalfa is most noted for its taproot (Sheaffer et al., 1988). It extracts water to a greater depth than ladino clover and birdsfoot trefoil (Wolf, 1964). Bennett and Doss (1960) compared root characteristics of red clover, ladino clover, and alfalfa and found that ladino clover had the smallest and alfalfa the largest root mass. Ladino clover had a greater proportion of its root mass in the upper 15 cm than red clover or alfalfa. The taproot of ladino clover is frequently short-lived due to disease, and a fibrous root system arises from stolons (Frame & Newbould, 1986). The fibrous root system is less competitive with grasses for moisture, and white clover is susceptible to drought (Taylor et al., 1960).

Legumes are poor competitors with grasses for K. This has been associated with greater fibrous root mass and volume as well as a greater root cation exchange capacity of the grasses (Haynes, 1980). Hunt and Wagner (1963) reported that alfalfa dominated grasses when soil K was sufficient, but that grass dominated when K was limited.

Biological N_2 fixation allows legumes to independently meet their N needs (Heichel, 1987). Legume species and varieties vary in their potential to fix N_2, but it is unknown whether legume differences in N_2 fixation provide any competitive advantage since legumes also freely utilize soil N. It has been implied that poor establishment and seedling vigor of some legumes is due to failure to establish an effective symbiosis. Legumes benefit associated grasses in mixtures by transferring N. Ta and Faris (1987) reported that alfalfa contributed up to 13 kg N ha^{-1} yr^{-1} to timothy. Brophy et al. (1987) reported that reed canarygrass grown in association with alfalfa derived a maximum of 68% of its N from alfalfa and 79% from birdsfoot trefoil. Shading by competing grasses may reduce N_2 fixation by restricting supply of energy.

REFERENCES

Bennett, O.L., and B. D. Doss. 1960. Effect of soil moisture level on root distribution of cool-season forage species. Agron. J. 52:204-207.

Blake, C.T., D.S. Chamblee, and W.W. Woodhouse, Jr. 1966. Influence of some environmental and management factors on the persistence of ladino clover in association with orchardgrass. Agron. J. 58:487-489.

Blaser, R.E., R.C. Hammes, Jr., J.P. Fontenot, H.T. Bryant, C.E. Polan, D.D. Wolf, F.S. McClaugherty, R.G. Kline, and J.S. Moore. 1986. Forage-animal management systems. Virginia Agric. Exp. Stn. Bull. 86-7.

Blaser, R.E., T.H. Taylor, W. Griffeth, and W. Skrdla. 1956. Seedling competition in establishing forage plants. Agron. J. 48:1-6.

Brink, G.E., and G.C. Marten. 1986. Barley vs. oat companion crops. II. Influence on alfalfa persistence and yield. Crop. Sci. 26:1067-1071.

Brophy, L.S., G.H. Heichel, and M.P. Russelle. 1987. Nitrogen transfer from forage legumes to grass in a systematic planting design. Crop Sci. 27:753-758.

Buxton, D.R., and W.F. Wedin. 1970. Establishment of perennial forages: I. Subsequent yields. Agron. J. 62:93-97.

Chamblee, D.S., and M. Collins. 1988. Relationships with other species in a mixture. In A.A. Hanson et al. (ed.) Alfalfa and alfalfa improvement. Agronomy 29:439-461.

Chamblee, D.S., and R.L. Lovvorn. 1953. The effects of rate and method of seeding on the yield and botanical composition of alfalfa-orchardgrass and alfalfa-tall fescue. Agron. J. 45:192-196.

Cooper, C.S. 1972. Establishment, hay yield, and persistence of two sainfoin growth types seeded alone and with low-growing grasses and legumes. Agron. J. 64:379-381.

Cooper, C.S., and H. Ferguson. 1964. Influence of a barley companion crop upon root distribution of alfalfa, birdsfoot trefoil, and orchardgrass. Agron. J. 56:63-66.

Dawson, J.H., and C.M. Rincker. 1982. Weeds in new seeding of alfalfa (Medicago sativa L.) for seed production: Competition and control. Weed Sci. 30:20-25.

Decker, A.M., R.F. Dudley, L.R. Vough, M.I. Spicknall, and T.H. Miller. 1985. Band vs. broadcast paraquat and alternate row species seeding in no-till pasture renovation. p. 254-261. In Proc. Forage Grassl. Conf., Hershey, PA. 3-6 Mar. Am. Forage and Grassl. Counc., Lexington, KY.

Decker, A.M., and T.H. Taylor. 1985. Establishment of new seedings and renovation of old sods. p. 288-297. In M.E. Heath et al. (ed.) Forages. 4th ed. Iowa State Univ. Press, Ames.

Dobson, J.W., C.D. Fisher, and E.R. Beaty. 1976. Yield and persistence of several legumes growing in tall fescue. Agron. J. 68:123-125.

Doll, J.D. 1986. Do weeds affect forage quality? p. 161-170. In Proc. 16th Nat. Alfalfa Symp., Fort Wayne, IN. 5-6 Mar. Certified Alfalfa Seed Counc., Woodland, CA.

Donald, C.M. 1963. Competition among crop and pasture plants. Adv. Agron. 15:1-118.

Dutt, T.E., R.G. Harvey, and R.S. Fawcett. 1982. Feed quality of hay containing perennial broadleaf weeds. Agron. J. 74:673-676.

Dutt, T.E., R.G. Harvey, R.S. Fawcett, N.A. Jorgensen, H.J. Larsen, and D.A. Schlough. 1979. Forage quality and animal performance as influenced by quackgrass control in alfalfa with pronamide. Weed Sci. 27:127-132.

Dzowela, B.H., G.O. Mott, and W.R. Ocumpaugh. 1986. Summer grazing management of a white clover-bahiagrass pasture: legume survival. Exp. Agric. 22:363-371.

Frame, J., and P. Newbould. 1986. Agronomy of white clover. Adv. Agron. 40:1-88.

Groya, F.L., and C.C. Sheaffer. 1981. Establishment of sod-seeded alfalfa at various levels of soil moisture and grass competition. Agron. J. 73:560-565.

Haynes, R.J. 1980. Competitive aspects of the grass-legume association. Adv. Agron. 33:227-261.

Harper, J.L. 1977. Population biology of plants. Academic Press, New York.

Heichel, G.H. 1987. Legume nitrogen:symbiotic fixation and recovery by subsequent crops. p. 63-80. In Z.R. Helsel (ed.) Energy in plant nutrition and pest control. Elsevier Science Publishers B.V., Amsterdam, The Netherlands.

Hunt, O.J., and R.E. Wagner. 1963. Effects of phosphorus and potassium fertilizers on legume composition of seven grass-legume mixtures. Agron. J. 55:16-19.

Jones, T.A., I.T. Carlson, and D.R. Buxton. 1988. Reed canarygrass binary mixtures with alfalfa and birdsfoot trefoil in comparison to monocultures. Agron. J. 80:49-55.

Kendall, W.A., and W.C. Stringer. 1985. Physiological aspects of clover. In N.L. Taylor (ed.) Clover science and technology. Agronomy 25:111-159.

Klebesadel, L.J., and D. Smith. 1960. Effects of harvesting an oat companion crop at four stages of maturity on the yield of oats, on light near the soil surface, on soil moisture, and on the establishment of alfalfa. Agron. J. 52:627-630.

Marten, G.C., C.C. Sheaffer, and D.L. Wyse. 1987. Forage nutritive value and palatability of perennial weeds. Agron. J. 79:980-986.

Mays, D.A., and E.M. Evans. 1972. Effect of variety, seeding rate, companion species, and cutting schedule on crownvetch yield. Agron. J. 64:283-285.

Miller, D.W., C.C. Sheaffer, and G.C. Marten. 1984. Light distribution and canopy structure of alfalfa-grass mixtures. p. 132. In Agronomy Abstracts. ASA, Madison, WI.

Peters, E.J., and D.L. Linscott. 1988. Weeds and weed control. In A.A. Hanson et al. (ed.) Alfalfa and alfalfa improvement. Agronomy 29:705-735.

Rhykerd, C.L., R. Langston, and J.B. Peterson. 1959. Effect of light treatment on the relative uptake of labeled carbon dioxide by legume seedlings. Agron. J. 51:7-9.

Sheaffer, C.C., D.K. Barnes, and G.C. Marten. 1988. Companion crop vs. solo seeding: Effect on alfalfa seeding year forage and N yields. J. Prod. Agric. 1:270-274.

Sheaffer, C.C., G.C. Marten, and D.L. Rabas. 1984. Influence of grass species on composition, yield, and quality of birdsfoot trefoil mixtures. Agron. J. 76:627-632.

Sheaffer, C.C., C.B. Tanner, and M.B. Kirkham. 1988. Alfalfa water relations and irrigation. In A.A. Hanson et al. (ed.) Alfalfa and alfalfa improvement. Agronomy 29:373-409.

Sheaffer, C.C., and D.L. Wyse. 1982. Common dandelion (Taraxacum officinale) control in alfalfa (Medicago sativa). Weed Sci. 30:216-220.

Sprague, V.G., and R.J. Graber. 1950. Effect of time and height of cutting and nitrogen fertilization on the persistence of the legume and production of orchardgrass-ladino and bromegrass-ladino associations. Agron. J. 42:586-593.

Smith, D., A.V.A. Jacques, and J.A. Balasko. 1973. Persistence of several temperate grasses grown with alfalfa and harvested two, three, or four times annually at two stubble heights. Crop. Sci. 13:553-56.

Ta, T.C., and M.A. Faris. 1987. Effect of alfalfa proportions and clipping frequencies on timothy-alfalfa mixtures. II. Nitrogen fixation and transfer. Agron. J. 79:817-819.

Taylor, T.H., J.B. Washko, and R.E. Blaser. 1960. Dry matter yield and botanical composition of an orchardgrass-ladino white clover mixture under clipping and grazing conditions. Agron. J. 52:217-220.

Tesar, M.B., and V.L. Marble. 1988. Alfalfa establishment. In A.A. Hanson et al. (ed.) Alfalfa and alfalfa improvement. Agronomy 29:303-332.

Townsend, C.E., and W.J. McGinnies. 1972. Establishment of nine forage legumes in the central Great Plains. Agron. J. 64: 699-702.

Wilkinson, S.R., and C.F. Gross. 1964. Competition for light, soil moisture and nutrients during ladino clover establishment in orchardgrass sod. Agron. J. 56:389-392.

Wolf, D.D. 1964. Soil moisture extraction trends of several legume-grass mixtures as affected by cutting frequency and nitrogen fertilization. Agron. J. 56:467-469.

DISCUSSION

Fisher. You mentioned the difficulty of quantifying competition. The replacement series (de Wit, C.T. 1960. On competition. p. 1-82. In Versl. Landbouwkd. Onderz. No. 66.8. Inst. for Biol. and Chem. Res. on Field Crops and Herbage, Wageningen, Netherlands) allows adequate quantification of interference whether competitive or non-competitive.

Jones. What do you consider to be the value of the de Wit replacement series in analysis of legume/grass competition? Has it helped in understanding the result?

General audience response ranged from little value to a lot. In summary, the de Wit replacement series is useful in describing competitive relationships in mixtures. This research is time consuming and the results may be more efficiently derived by observation.

Fisher. In response to Jones' comments about the difficulty of using the replacement series to investigate competitive interference between grasses and clovers, the de Wit paper analyzes the relationship between white clover and perennial ryegrass. For white clover, the stolon length per unit area was used and in the grasses tiller number was used. The replacement diagrams, and hence the inferences were made on these observations.

Lowther. Does alternative row spacing reduce competition from companion grasses and is it used by farmers?

Sheaffer. It has some advantages in increasing light penetration into the canopy and reduces competition to the weaker component of the mixture. Although it is used in experimental conditions it is not widely practiced by producers.

GENERAL DISCUSSION OF
CULTURAL PRACTICES AND PLANT COMPETITION

Curll: Can we conclude from the three presentations on cultural practices that there are few knowledge gaps in this area? The exceptions seem to be restricted to some less favored areas in Australia and New Zealand. The economics of pasture establishment may affect the actual practices used. Similarly, economics may dictate certain cutting management practices despite known risks to persistence. My point is, the knowledge is there already.

Gramshaw: I challenge the concept that cultural practices are unimportant for legume persistence in Australia. In southern Australia, persistence of annual legumes through cropping sequences is said to be understood biologically, but application of management techniques is required, perhaps through expert systems. In my paper, I stressed that establishment techniques are not ideal for less favored areas in Australia. Profitability of legume pastures in northern Australia (particularly stylos) is markedly reduced where legume density is initially low and the legume component subsequently develops slowly.

Allen: The cultural practices needed to promote persistence of annual medic pastures in cereal/pasture rotations are generally understood. The problem lies with the lack of application of these practices by farmers. Generally, farmers have been more interested in cereal crops than pastures during the last 10 to 20 yr because of the better economic returns from cereals. More intensive cereal cropping in rotations, and the use of grain legumes, have reduced the areas of medic pasture on properties. This restricts the farmer's flexibility to apply correct livestock grazing management to ensure persistence of the medic pasture. For example, the farmer is less able to reduce grazing pressure on medic pastures during the spring in order to promote seeding.

Reed: Cultural practices are not seen as a major subject for research. Simple models for the medic ley situation have been developed, e.g., by Ted Carter. Much emphasis has been given to the maintenance of a minimal seed bank. During the 1970s and 1980s, seed banks have been eroded by the use of herbicides, increased stocking rates, greater crop areas, dry seasons, and insect problems.

Curll: Is equipment and technology available for pasture renovation work in the USA?

Sheaffer: Yes, technology is adequate. What is lacking is the management of the perennial grasses after legumes are interseeded. Also insect control is lacking. For successful pasture renovation, seed must be delivered in contact with soil, existing grasses suppressed, and volunteer grasses and weeds and insects controlled.

J. Hay: The development of the Massey drill has markedly improved the percentage success rate of direct-drilled pasture species in

New Zealand.

Hodgson: Collaborative work between biologists and engineers at Massey University has resulted in the development of effective minimal-cultivation technology, but there is still a need for intelligent engineering input to apply biological understanding of requirements for seed germination and establishment.

Jones: In the subtropics, a machine for over-sowing is being developed which uses a herbicide treated strip, 50-cm wide, with the seed being band-sown in the middle of this strip. Control of competition is important in the subtropical areas of Australia.

Matches: Producers in the USA often do not achieve good establishment because they do not apply proper canopy control. Seedlings die because of excessive shading. Also, rotational grazing is generally necessary to maintain stands of tall growing legumes such as red clover and alfalfa.

Curll: Work in the NSW Northern Tablelands on canopy control has shown the importance of control of competing vegetation in establishment of slot-seeded legumes.

Curll: Is harvest management (i.e., cutting) used to alter the composition of a pasture in each country?

Sheaffer: In the USA, harvesting to promote legume persistence is only used in the seeding year. Weeds and companion crops are harvested or clipped. In established stands, harvesting occurs for yield, not to alter persistence.

Gramshaw: In Australia, apart from lucerne stands managed for hay production, cutting of pastures for hay or silage would usually occur only once each year (if at all), during the early flowering stage of the legume. If this seriously reduces seed set in annuals it may be important in persistence, depending on the carryover seed reserves and other factors. There would be little effect on perennials.

Allen: Mechanically slashing grass seedheads prior to seed maturation in spring is used in subterranean clover-based pastures in southern Australia to shift the pasture towards clover dominance. Selective herbicide applied in spring is also used to reduce the seed set by grasses in these pastures.

Reed: Subterranean clover establishment can be weakened by under-utilized dry grass residues remaining at the end of the summer when autumn rains open the growing season. If this cover is made into hay and removed, the clover content of perennial ryegrass-subterranean clover pastures is improved considerably. Some of this benefit may be due to the removal of seed of weeds. However, cutting for hay production is often damaging to pastures consisting solely of annuals.

Curll: Turning to the papers on competition as a factor in persistence, can we see any innovative concepts that will help us to evaluate germplasm or formulate management strategies for persistence? In particular, what is the role of the gradient approach as outlined by David Scott, of theoretical models of competition to determine experimental treatments and research targets, and of expert systems?

Sheath: We need to define the optimal legume content for a given agricultural system. Ultimately, it is grazing management that modifies competition to determine the share of environmental resources (modified or not) obtained by each species, and whether target legume contents are achieved. Therefore in seeking more persistent or stable legumes, appropriate grazing pressures must be integrated into any assessment scheme.

Scott: Most investigators take the analytical approach of trying to understand species relationships by understanding all the mechanisms in detail. However, there is an alternative synthetic or "black box" approach, in that there may be predictable characteristics of the vegetation as a whole. Three of these characteristics are:

1. Potential productivity, as determined by moisture, temperature, fertilizer, and livestock.
2. Log-linear relationships between abundance of species and rank.
3. An inverse relationship between population density and the size of individuals in the species (the 3/2 thinning law).

Hodgson: The gradient approach and field evaluation of competition models are complementary rather than competitive approaches to understanding. It is important not to limit investigation procedures by currently acceptable practical conventions of management and control.

Marten: What we are leading to in this discussion is the moderator's third suggested way to study legume persistence problems, i.e., by the use of expert systems. We have emphasized expert systems recently in the USA because we recognize that there is a great amount of valuable knowledge gathered by senior researchers which does not lend itself to publication in scientific journals or is not quantifiable by statistical approaches. I expect the same is true in Australia and New Zealand. For example, experts must have decided, based on "fuzzy logic" or otherwise, that 30% white clover content is an ideal for typical ryegrass swards in New Zealand. This sort of information needs to be extended to practical use as well as to be documented, for the future well-being of science and humankind.

THE PLANT-ANIMAL INTERFACE AND LEGUME PERSISTENCE - AN AUSTRALIAN PERSPECTIVE

M. L. Curll and R. M. Jones

SUMMARY

The persistence of legumes in Australia's tropical, temperate, and mediterranean pastures can be substantially affected by the defoliation, treading, excretion and seed dispersal components of the grazing process.

The threshold beyond which defoliation becomes excessive and threatens legume survival is a function of the frequency and severity of defoliation, its timing in relation to plant development, and its selectivity. The threshold is different between and within legume species according to their growth habit, adaptability and whether survival is by seed or by vegetative means. The growing and reproduction points of stoloniferous, prostrate legumes may escape defoliation while those of erect, climbing, or trailing legumes are vulnerable to destruction unless they are able to favorably adapt their growth habit.

The effects of treading and excreta return on legume persistence is likely to be small relative to defoliation. However, under Australia's extensive grazing systems, the distribution of nutrients and seed returned in excreta, and the consequent effects on pasture heterogeneity and grazing behavior, have the potential to affect vegetative survival, seedling regeneration and colonization.

There is considerable scope for Australian farmers to improve legume persistence by paying more attention to grazing management. Management should ensure that established plants can recover from grazing and where appropriate produce seed and establish seedlings. In certain cases this may involve simple changes in seasonal stocking rate, while in other cases more complex procedures may be required. To develop those procedures, more needs to be known about the growth cycle of legumes under different seasonal conditions, their survival mechanisms and their reaction to grazing. Such understanding will enable the selection of more persistent legumes from breeding and evaluation programmes.

INTRODUCTION

Legume persistence in pastures is influenced by the response of the legume plant to grazing, by the effect of grazing on plant replenishment, and by the effect of grazing on the competitive relationships between the legume and its companion species. Species differences in persistence under grazing relate to differences in mode of reproduction and survival, and adaptability to grazing.

This paper discusses the role that interactions at the plant/animal interface have on legume persistence. It describes the effect of the grazing process - defoliation, treading, excreting and seed dispersal - on the persistence of pasture legumes from an Australian perspective. Defoliation frequency, severity, and selectivity are reviewed as are the effects of grazing on plant parts and the plant's reaction to grazing. The role that knowledge of the plant/animal interface can play in developing grazing management strategies to improve legume persistence is considered.

DEFOLIATION

The effects of defoliation by the grazing animal have rarely been disentangled from those of treading and excretion. Curll and Wilkins (1981) attempted to separate the effects of defoliation from those of treading and excreta return by using "graze through" cages (Smith et al., 1971). They showed that differences in defoliation between stocking rates had a far greater effect on the content and stolon density of Trifolium repens in a Lolium perenne/T. repens pasture than either treading or the return of excreta (Table 1).

Attempts have been made to simulate by cutting, the effects on botanical composition of defoliation by the grazing animal. But cutting is uniform and non-selective and so is not equivalent to grazing (Curll & Davidson, 1983). Individual plant reaction to cutting can however contributed to our understanding of the effect of defoliation by grazing.

Although pasture plants by definition are adapted to defoliation, defoliation can result in the death of plants. Death is more likely when plants weakened by defoliation are subjected to competition from companion species, water stress, attack by insects and pathogens, and other stress factors The threshold beyond which defoliation becomes excessive and threatens survival is a function of its frequency and intensity or severity, its timing in relation to the stage of plant development, and its selectivity. The threshold is different between and within legume species and varies according to the plant's mode of survival (seed or vegetative), growth habit and adaptability.

Table 1.Effect of defoliation, treading, and excreta return at high and low stocking rates on clover stolon length/unit area and % clover in a L. perenne/T. repens pasture (Curll & Wilkins, 1981).

Treatment	Stolon length	Clover
	m/m^2	%
50 sheep/ha		
Defoliation	43.0	31
Defoliation + treading	32.9	25
Defoliation + treading + excreta	25.4	18
25 sheep/ha		
Defoliation	173.4	75
Defoliation + treading	160.5	71
Defoliation + treading + excreta	94.8	45

Frequency and Severity

Decreasing frequency of grazing or long rest periods between grazings can reduce the content of relatively prostrate legumes such as Trifolium subterraneum in grass/clover pastures (Snaydon, 1981). In contrast, the survival of erect, non-stoloniferous species such as Medicago sativa and T. pratense is reduced by frequent grazing. Year-round grazing intervals of at least 5 wks are important for the survival of M. sativa in temperate-mediterranean regions of Australia (McKinney, 1974) and in the subtropics (Leach, 1979). The duration of grazing within the range 5 to 30 d appears less critical for M. sativa (McKinney, 1974) though shorter periods have been reported to favor plant survival (Smith, 1970; FitzGerald, 1974). In the subtropics, Leach (1979) reported that 16-d grazing periods for M. sativa regrowth were preferable to 4-d periods though survival was only marginally affected. The longer grazing periods allowed more complete grazing of sown or invading grasses which could shade M. sativa regrowth; excessively long periods of grazing would be expected to allow grazing of that regrowth. Twining legumes such as Macroptilum atropurpureum (siratro) are also vulnerable to frequent heavy grazing. Jones and Jones (1978) found that rest periods of 9 wks rather than 3 wks favored M. atropurpureum under high grazing pressure.

Stocking Rate

Studies on the relationship between stocking rate and the pattern of defoliation of individual plant units show that an increase in stocking rate results in an increase in the frequency and severity of defoliation. For example, Curll and Wilkins (1982) showed that a twofold increase in stocking rate on a L. perenne/T. repens pasture was coincident with an 80% increase in the frequency of defoliating individual clover plant units, and on a unit area basis 63% fewer leaves and 79% less stolon was defoliated each day at the low stocking rate than at the high stocking rate.

A number of Australian experiments have monitored the legume content of grazed pastures in relation to stocking rate. Hutchinson (1970), Langlands and Bennett (1973) and Robinson (1977) found that stocking rates most favorable to the survival of T. repens in L. perenne/T. repens pastures were of the order of 18-25 dry sheep equivalents per ha. At higher or lower stocking rates, clover content was reduced. Similar stocking rates favored T. subterraneum in P. aquatica/T. subterraneum pastures (Curll, 1977), though in drier temperate-mediterranean regions, the content of T. subterraneum in P. aquatica/T. subterraneum and L. perenne/T. subterraneum pastures was greatest at lower stocking rates (viz. 4-13 dry sheep equivalents: Cameron & Cannon, 1970; FitzGerald, 1979; Curll & Davidson, 1983). In some studies, plant density of M. sativa has been reduced by high stocking rates (Smith, 1970; Hall et al., 1985).

The sward content and plant density of tropical trailing or twining legumes such as Desmodium spp. and M. atropurpureum are usually reduced by increasing stocking rates (Jones, 1971; Bryan & Evans, 1973; Bisset & Marlow, 1974; Jones & Jones, 1978; Jones et al., 1982). Creeping legumes such as Vigna parkeri and Lotononis bainesii are less vulnerable to grazing at high stocking rates (Bryan & Evans, 1973; Jones & Clements, 1987). Survival and yield of the prostrate tropical species Stylosanthes humilis is also favoured by heavy grazing (Winks et al., 1974; Shaw, 1978; Gillard & Fisher 1978).

Thus, frequent, intense defoliation favors survival of temperate prostrate legumes such as T. subterraneum and T. repens and tropical legumes such as S. humilis that are prone to shading by companion grasses (Stern & Donald, 1962; Gillard & Fisher, 1978). On the other hand, the survival of twining or scrambling legumes that are able to avoid shading from vigorous grasses (Jones & Jones, 1978), is favored by less frequent and less intense defoliation. This suggests there is an optimum range of stocking rates that will enable an acceptable legume content to be maintained in the pasture.

Selective Defoliation

If the legume component of a pasture mixture is selectively defoliated by the grazing animal, its presence relative to its companion species may be reduced and its survival threatened. Australian studies have shown that sheep and cattle generally select leaf in preference to stem, and green (or young) herbage in preference to dead (or old) herbage (Arnold et al., 1966; Stobbs, 1973; Leigh & Holgate, 1978).

Selective defoliation of a plant species or plant parts is determined primarily by relative acceptability and accessibility. Differences in acceptability reflect differences in taste, odor and surface characteristics, modified by experience and satiation (Arnold, 1981; Black, 1987). Accessibility is a function of the relative proportions of the different species or plant parts and their position in the sward relative to other species or plant parts, and is modified by the quantity of pasture on offer, animal capability and the physical environment (Hodgson, 1981; Black, 1987; Curll et al., 1987).

Preference by grazing livestock for T. repens, T. subterraneum, and M. sativa in Australia has been related to a reduction in the legume content of grass/legume pastures (Curll, 1977; Leigh & Holgate, 1978; Leach & Clements, 1984). Other studies (Arnold et al., 1966; Hodge & Doyle, 1967) have reported no consistent preference for legume over grass though in those studies the total quantity of herbage on offer would have limited selection (Hamilton et al., 1973). A low preference by goats for clover in temperate pastures has been related to an increase in the clover content of those pastures (McGregor, 1984; Holst & Campell, 1987).

Between species differences in acceptability as well as seasonal differences within species, offers scope for controlling the legume content of grass/legume pastures. A low preference by cattle for Cassia rotundifolia and especially Calopogonium caeruleum is conducive to their survival under grazing (Middleton & Mellor, 1982; Clements, 1987; Valdes et al., 1987). Similarly, a reduction in preference for M. atropurpureum (Stobbs, 1977; Walker et al., 1982), S. hamata (Gardener, 1980) and S. scabra (McLean et al., 1981) during spring and early summer favours their survival under grazing. Preference for both temperate-mediterranean and tropical legumes by sheep and cattle may be increased by the application of phosphorus fertilizer to pastures (Ozanne & Howes, 1971; McLean et al., 1981; McLean & Kerridge, 1987). Withholding P fertilizer to reduce legume acceptability has been suggested as a method to assist legume survival (McLean et al., 1981). However, because temperate legumes generally have a higher phosphate requirement than grasses (Ozanne et al., 1976), their growth and persistence is likely to be reduced by withholding P fertilizer. Further, changing relative preference for a legume by making it less acceptable or its companion grass more acceptable, could reduce diet quality, particularly protein content.

The expression of preference depends on the quantity of herbage available and the heterogeneity of the pasture. However, harvesting takes time and if the most acceptable species or components are distributed too thinly for the animal's appetite to be readily satisfied, then a balance is struck between a lowered level of preference and a reduction in the amount eaten or defoliated. For example, with trailing tropical legumes, though leaf is greatly preferred, low leaf and sward density can prevent satisfactory intake necessitating consumption of less preferred species or plant parts by grazing cattle (Stobbs & Hutton, 1974). Preference can also be restricted in temperate swards despite a substantially higher sward density and lower sward height than in some tropical swards (Stobbs, 1973). For example, sheep were unable to graze T. subterraneum in stemmy swards of P. aquatica/T. subterraneum in late spring and summer (Arnold, 1964).

Though species selection is primarily determined by the characteristics of the pasture being grazed, inherent differences in animal capability and experience can effect selection. Because sheep have smaller mouths and teeth they can take smaller bites, they usually bite closer to the ground or deeper into the sward, and so are able to be more selective than cattle (Bedell, 1968; Dudzinski & Arnold, 1973; Wilson, 1976). When the quantity of pasture was not limiting, Bedell (1968) found that sheep selected three times as much T. subterraneum than cattle when grazing L. perenne/T. subterraneum pastures. Wilson (1976) reported that sheep ate more Medicago spp. than cattle grazing semi-arid grasslands in New South Wales. Differences also occur between individuals and breeds within animal species (Arnold & Hill, 1972; Arnold, 1964b). The influence of previous experience on selection was demonstrated by Arnold (1981) when he observed preference for M. sativa by sheep reared on M. sativa but not by sheep without previous experience of that pasture.

Effects on Plant Parts and Plant Reaction

Persistence of legumes under grazing can depend on the maintenance of sufficient growing points and leaf area for established plants and/or the recruitment of new plants from seed, stolons or rhizomes. Detailed information on the effects of grazing on plant parts and survival adaptations will help to favorably manipulate legume-based pastures. This information is limited for many species under Australian conditions.

Legumes vary in their growth habit and the location of their growing points. The growing points of stoloniferous prostrate legumes can escape defoliation while growing points of erect, climbing or trailing legumes are vulnerable to destruction. The effect of defoliating the growing points of legumes with contrasting growth habit has been studied by Clements (1986, 1987) and Clements and Jansen (1986). In grass/legume swards stocked by cattle at a range of stocking rates, 10-26% of runners of the twining legume M. atropurpureum were wholly or partly grazed every 3 wks compared with 21-50% of runners of the relatively prostrate T. repens. When a M. atropurpureum runner was defoliated, the terminal growing point was destroyed. However, only 1-5% of T. repens growing points were destroyed because most were below grazing height. Thus, though the frequency of grazing of T. repens runners was greater than that for M. atropurpureum, the destruction of M. atropurpureum growing points and associated reduction in the legume content of the pasture, contrasted with the survival and proliferation of T. repens growing points and the retention of the clover. Demographic studies of M. atropurpureum in M. atropurpureum/Setaria anceps pastures have shown that as defoliation frequency and severity increases with increasing stocking rates, the size of individual plants and the extent of stolon rooting and development is reduced. Root node density is important with stoloniferous species since perenniality or survival is dependent on new plants which are rooted stolons of the original plant (Hollowell, 1966).

The persistence of M. sativa depends on survival of the original plants. Defoliation must be sufficiently severe to remove apical dominance and to release basal buds from dormancy, but not severe enough to remove the dormant buds (Hodgkinson, 1973; Leach, 1978). Because of the legume's dependence on basal bud development, adaptations that would assist persistence under grazing would include an ability to develop adventitous shoots on roots, spreading crowns, crowns with basal buds partly buried for protection, and continual bud development (Hodgkinson & Williams, 1983; Leach & Clements, 1984; Waterhouse, 1988).

The survival of some legume species and cultivars under grazing is assisted by changes in growth habit in response to defoliation. For example, the tropical legume Vigna parkeri tolerates close grazing by growing close to the ground and rooting from stolons, but twines upward around associated grasses under light grazing (Jones & Clements, 1987). Cultivars of S. humilis are able to develop a well-branched flat crown under grazing with many bud sites inaccessible to grazing cattle (Humphreys, 1984). With T. subterraneum, the proportion of petioles that are prostrate will increase with frequent close grazing (Broom & Arnold, 1986). Under heavy grazing, the T. repens cv. Blanca can develop a prostrate growth habit with short petioles, a high leaf and root node density (per unit length of stolon) and many small leaves which helps it survive and/or escape defoliation. But under light grazing or when rested, the plant can take on a larger-leaved, longer-petioled habit with a lower leaf and node density (per unit lenth of stolon) which enables it to escape shading by its companion grass (Curll & Wikins, 1985).

In contrast to the petiole extension that occurs with T. repens under light grazing or resting (Curll & Wilkins, 1985), the stolons of T. semipilosum tend to ascend with the growing grass (Sproule et al., 1983). However, under heavy grazing, there is an increase in the density of T. semipilosum stolons, stolon roots and growing points close to the ground. These adaptive changes indicate the importance of phenotypic plasticity when seeking more persistent legumes.

Stoloniferous legumes tend to escape permanent damage from defoliation while those that are erect, twinning, or scrambling with a high proportion of shoots and auxillary buds accessible to the grazing animal, are more susceptible to permanent damage. The extensive nature of Australia's grazing industries, and a reluctance by many farmers to invest resources into grazing systems requiring numerous paddocks, would seem to indicate a need in legume improvement programs in Australia to select highly stoloniferous genotypes with a high density of root and leaf nodes. For legumes with other growth habits, selection for genotypes that have spreading crowns, staggered shoot regeneration, auxillary buds below ground level, and/or buds that are protected by inedible stem (e.g., Leucaena), may be desirable. The ultimate form of bud protection is shown in species with strongly developed underground rhizomes; for example, Arachis spp. in south-east Queensland can produce 6000 growing points or buds per square meter (Vos & Jones, 1986).

Natural re-seeding is essential for the persistence of annual legumes but is also important for the long term persistence of some perennial legumes (Jones & Carter, this Book). Grazing can effect flower formation, seed set, seed survival and seedling recruitment. Depending on the growth habit of the plant, grazing prior to early flowering can increase seed set (e.g., T. subterraneum ssp. subterraneum, M. polymorpha; Rossiter, 1978), or decrease seed set (e.g. T. subterraneum ssp. brachycalycinum; Bolland, 1987). Grazing during flowering and seed set can reduce seed yield (Koning & Carter, 1987). Grazing of mature pods or seeds can also reduce soil seed reserves, particularly where the seed pods can be easily prehended (e.g., Medicago spp.; Carter, 1983) and are grazed by sheep which excrete a lower proportion of ingested seed than cattle (Simao Neto et al., 1987). Grazing pressure can also affect seedling survival. For example, seedling survival of S. humilis (Gillard & Fisher, 1978) and M. atropurpeum (Jones & Bunch, 1988b) was reduced by competition from their lightly grazed companion grasses.

TREADING

The effects of treading on legume persistence has received little attention in Australia. This may be because studies elsewhere suggest that its impact relative to the other components of the grazing process is the least important (Curll & Wilkins, 1983), or because of the difficulty of realistically separating the effects of treading from those of defoliation and excreta return.

Treading can affect the plant directly by severing, bruising, or otherwise damaging growing points, leaves, stems and surface roots (Pott et al., 1983). Its indirect effects include restriction of gaseous diffusion by covering the plant with a film of mud or puddling and compacting the soil (Mullen et al., 1978), or reduction of soil moisture content, water infiltration rates, and root penetration by reducing porosity and increasing bulk density (Witschi & Michalk, 1979; Langlands & Bennet, 1973; Kelly, 1985). Treading may have beneficial effects by improving soil-seed contact, thereby assisting seed germination and emergence (Pott et al., 1983).

Generally, legumes are less tolerant of treading than grasses. For example, Curll (1980) recorded 20% more treading damage to marked plants of *T. repens* compared to marked *L. perenne* plants in a *L. perenne/T. repens* pasture; the difference between the legume and grass increased with stocking rate. Variation in species tolerance has been related to the quantity of sclerenchyma tissue in plant parts and to the extent that the more susceptible parts of the plant (i.e. growing points, leaf buds and surface roots) are exposed to damage by the hoof (Brown & Evans, 1973; Pott et al., 1983). Pott et al. (1983) related the treading resistance of *L. bainesii* plants to the development of a buried crown. Erect legumes such as *T. pratense* and *M. sativa* with basal buds aboveground would be particularly vulnerable to treading damage. With species that rely on natural reseeding for persistence, the vulnerability of seedlings to destruction by treading may be important (Carter & Sivalingam, 1977).

Stoloniferous legumes such as *T. repens* can maintain their plant populations under treading through their ability to establish new plants at each node and to increase the density of these nodes (per unit length of stolon) as stocking rate is increased (Pascoe, 1973; Curll, 1980). In a comparison of *L. perenne/T. repens* pastures that were defoliated and trodden with those that were defoliated only, Curll (1980) recorded that treading by sheep grazing at a rate of 25 yearlings per ha caused a 11% increase in node density (number per unit area) with a 7% reduction in stolon density (length per unit area) and 5% reduction in clover content. However, at twice that stocking rate, treading decreased node density by 6%, reduced stolon density by 23% and clover content by 20%. Depending on the structure of the sward, legumes in grass/legume mixtures should benefit by their companion grass species cushioning the impact of the grazing animals hooves (Edmond, 1964; Pott et al., 1983).

Overall, the impact of treading by grazing animals on legume persistence in Australia is likely to be relatively small. Its deleterious effects should assume greatest importance under intensive management, where high stocking rates are used over extended grazing seasons, and on irrigated pastures. However, care should be taken when drawing conclusions on the impact of treading during grazing from studies where treading is applied artificially by using large numbers of nongrazing animals to tread pastures for short periods . The effects on pasture and soil appear more severe with that technique (Edmond, 1964; Witschi & Michalk, 1979; Brown, 1968) than measured with grazing sheep (Curll & Wilkins, 1983) and cattle (Kelly, 1985). For example, with an equivalent stocking rate and pasture, Curll and Wilkins (1981) found that treading reduced *T. repens* yield by 6% and had no effect on *L. perenne*, while Brown (1968) using Edmond's (1958) technique to simulate treading, recorded a 46% reduction in *T. repens* yield and 25% reduction in *L. perenne* yield.

EXCRETION

Excreta can influence the legume content of a pasture by altering nutrient concentrations in the soil, by scorching or burning plants, by dispersing legume seed in feces, and by influencing grazing patterns. Most studies on the effect of excreta on the botanical composition of pasture have been with temperate pastures, primarily in New Zealand and the United Kingdom. Generally, urine and feces together, and urine alone, reduce the proportion of legume in grass/legume pastures (e.g., Sears et al., 1948; Curll & Wilkins, 1983). Feces may increase the proportion of legume in the short term (e.g., Weeda, 1967), but not in the long term (e.g., Weeda, 1977).

The reduction in the proportion of legume that follows from the return of both feces and urine to grass/legume pastures is due more to grass growth responding to the readily available source of N in the urine than to legume growth being suppressed. For example, at a stocking rate where sheep were excreting about 1 kg N per ha per day (70% in urine) and fertilizer inputs made P and K non-limiting, Curll and Wilkins (1983) found that the return of excreta reduced the proportion of T. repens in a L. perenne/T. repens pasture by 26%; this was associated with a twofold increase in the dry matter yield of L. perenne and a 13% reduction in the yield of T. repens. The density and size of L. perenne tillers was increased by excreta return. Node and leaf density per unit length of T. repens stolon was also increased but stolon density, stolon size, and leaf size decreased.

Legume suppression from excreta may not simply result from N increasing the competitive ability of its companion grass for light, moisture and nutrients (Donald, 1963). Clover has been reported to be more sensitive than grass to urine burn and to be less tolerant to a root growth inhibitor found in urine, as well as to the presence of free ammonia and the temporarily increased pH of urine-affected soils (Doak, 1954).

The spatial distribution of excreta is very irregular (Hilder, 1964, 1966; Taylor et al., 1987; Robinson et al., 1983) and by influencing grazing behaviour and sward heterogeneity, the area affected is much greater than the area covered by excreta (Boswell, 1971; Keogh, 1973; Jones & Ratcliff, 1983). Excreta can influence the legume content and its location within a pasture area (Hilder, 1964; Taylor et al., 1987). Sheep and cattle tend to reject herbage near feces (Jones & Ratcliff, 1983) and temporarily around urine patches (Boswell, 1971). The shifting patterns of grazing pressure that results has the potential to affect vegetative survival and seedling regeneration of legumes by overgrazing on preferred areas and undergrazing of rejected areas.

With sheep, there is a tendency for the concentration of excreta in a small portion of the paddock or field where the animals rest or camp at night (Hilder, 1964; Taylor, 1980). A zonation of plant species has been observed around these camp areas with legumes tending to dominate adjacent to the greatest concentration of excreta (Hilder, 1964; Whalley et al., 1978). The legumes benefit from an increased supply of soil nutrients from excreta, and increased grazing adjacent to the camp preventing shading from grass (Hilder, 1964; Whalley et al., 1978). Lower quality grasses tend to dominate further away where lower soil nutrient levels and shading by ungrazed grass may supress legumes.

The spatial distribution of excreta and its impact on botanical composition of a pasture area, varies with stocking rate. Hilder (1966), Donald and Leslie (1969) and Robinson et al., (1983) observed that increasing stocking rate generally gave a more even distribution of feces.

SEED DISPERSAL

Seed can be dispersed by animals through becoming attached to hooves, hides and wool, and by ingestion and subsequent excretion in feces. Seed dispersal via feces is the most important of these dispersal mechanisms for pasture legumes. Defecated seeds can germinate and establish seedlings. The role of animals in dispersing seeds of legumes in pastures where the legume content is already high, is probably of minor importance (Watkin & Clements, 1978). However, the ingestion of seeds by the grazing animal has the potential to impact on legume persistence where natural reseeding is important and if the quantity of seed consumed and digested significantly reduces soil seed reserves. Dispersal of defecated seeds can be important for colonization of previously unsown areas, and recolonization of marginal micro-environments where the legume has been temporarily lost.

The proportion of ingested legume seed that passes through the animal undigested has been related to animal species, seed size, roughage intake, digestibility of the associated diet, the amount of seed consumed, hard-seededness and whether the seed is contained in a pod or burr (Simao Neto et al., 1987). The percentage recovery of ingested seed can be as low as 2% (Carter, 1980) or over 50% (Simao Neto et al., 1987). Large numbers of legume seeds can be passed in the faeces of animals grazing legume based pastures. Averaging over the main seeding period, concentrations of 14 000 seeds per kg of dry feces have been recorded for *T. repens* (Jones, 1982), 8,000 seeds per kg of faeces for *S. scabra* (Kerridge et al., 1987), 50 seeds per kg of faeces for *M. atropurpureum* (Jones & Bunch, 1988b) and 15 seeds per kg of feces for *Desmodium intortum* (Jones, 1989). Low concentrations in the latter two species were attributed to plant growth habit and the pattern of seeding restricting the opportunity or need for the animals to ingest ripe seeds. The large numbers of seeds carried in the faeces indicates the potential for spreading legumes via the excreta of grazing animals. However, colonization by excreted seed will depend on factors such as competition from companion species. For example, only 1.5% of excreted seed of *Stylosanthes* may grow into established plants (C.J. Gardener, 1988 personal communication).

IMPROVING PERSISTENCE BY GRAZING MANAGEMENT

The impact that the grazing animal can have on legumes in pastures has been outlined. That impact indicates the potential importance of grazing management for legume persistence. Grazing management of legume-based pastures in Australia, with the possible exception of *M. sativa*, has generally been based on animal requirements, convenience, and maximization of short-term farm income. It has not been a priority to manage the grazing animal to ensure established legumes recover from grazing and where appropriate, to encourage persistence from natural reseeding. There has been a tendency to graze paddocks continually or at least for extended periods of time, with some obvious exceptions such as pastures on intensively managed dairy farms. This has been reflected in the fact that most grazing experiments in Australia use either year-round continuous set stocking or fixed rotational grazing. Perhaps it is surprising that our pasture legumes have persisted as long as they have under such simple management! However, increasingly there are reports of poor legume persistence (Gramshaw et al., this Book).

Poor persistence can simply reflect an inappropriate choice of cultivar. In the past, there have been instances where new legume cultivars, which subsequently failed to persist, were released without adequate testing under grazing. In more recent years, some failures may be due to premature recommendation of a cultivar in environments outside that in which it was originally tested. However, many failures with persistence can be clearly ascribed to inappropriate management. Frequently, the problem has been one of over-stocking and could have been largely overcome by using lower stocking rates (e.g., M. truncatulata, Carter, 1983; M. atropurpureum, Jones & Jones, 1978).

Nevertheless, even the most basic study of legume growth, development and demography, suggest that in many cases it should be possible to develop clearer guidelines to improve legume persistence with grazing management such as seasonal shifts in grazing pressure and grazing frequency, and strategic deferment. It should also be possible to more clearly define grazing management that is detrimental to legume persistence. The potential for more sensitive management has been known for years (Davies, 1946; Nunn & Suijendorp, 1954). However, there is a far greater probability of developing improved management practices from understanding the growth of legumes, their interaction with seasonal conditions and their reaction to grazing.

Matching grazing and rest intervals with the phenology of individual grass species has been used to manipulate botanical composition in natural grass pastures (Lodge & Whalley, 1985). However, patterns of plant development, i.e., yield, growth habit and quality changes, flowering and seeding, have been studied for only a few legumes under Australian field conditions. Even less attention has been given to determining the consequence of grazing on legume survival and regeneration in mixtures of species under different seasonal conditions. Greater quantitative definition of grazing activity and its impact in relation to the distribution of species and plant parts within the sward profile is required. The independent effects of the spatial arrangement of the legume and its companion species within the sward, the location of growing points, differences in herbage mass, sward height, and density need to be disentangled for many legume-based pastures in Australia.

There are examples in Australia where studies of this sort have established management guidelines for legume survival in grazed pastures, and appropriate strategies have been devised or suggested though not necessarily extensively tested. Long term persistence of M. atropurpureum in tropical and sub-tropical pastures depends on plant replacement. Plant replacement in turn depends on maintaining adequate reserves of seed in the soil and this can be aided by strategic resting of the pasture (Jones, 1988). Though considerable effort has been put into researching patterns of plant development of T. subterraneum and formulating management strategies appropriate to its persistence (Rossiter, 1978), those strategies are specific to particular environments. For example, the association between grazing, seed production, hard-seed content and seedling survival of T. subterraneum in native grass pastures in summer-rainfall environments is quite different from its usual annual pasture environment of moist winters and dry summers (FitzGerald, 1987). In summer-rainfall environments, grazing needs to ensure the production of large numbers of seeds and the provision of adequate cover to protect seedlings against dessication in summer, yet also allow adequate light penetration for their growth in late autumn and winter (FitzGerald & Archer, 1987). The specificity of grazing management to environment is also apparent with T. repens. Demographic studies have established that in the subtropics, close grazing of C4 grasses over summer favors T. repens persistence by enhancing seedling recruitment and stolon survival (Jones, 1987), whereas in temperate areas seedling establishment is favored by infrequent grazing (K.A. Archer, 1988, personal communication).

A greater understanding of the plant/animal interface is not necessarily an absolute prerequisite for better or more practical management strategies. Such strategies have on occassions been developed by enterprising farmers or agronomists with relatively little detailed knowledge of processes (e.g., Fleming, 1986). Nevertheless, there is ample evidence in this paper to conclude that a better understanding of the plant/animal interface and its effects on persistence will provide improved management guidelines for legumes persistence. Any strategies developed should be tested in practical farming systems before they are recommended. Compromise decisions must inevitably be made between grazing for legume persistence and the provision of a diet of adequate quality and quantity for the grazing animal, the need to minimize costs and to obtain adequate returns, and the need for new management strategies to fit in with the whole farming system.

REFERENCES

Arnold, G. W. 1964. Factors within plant associations affecting the behaviour and performance of grazing animals. p.133-154. In B. J. Crisp (ed.) Grazing in terrestial and marine environments. Blackwell, Oxford.

Arnold, G. W. 1964b. Grazing experience in early life and subsequent performance in sheep. Fld. Stat. Rec., CSIRO Div. Plant Ind. 3:13-20.

Arnold, G. W. 1981. Grazing Behaviour. p. 79-104. In F. H. W. Morley (ed.) World animal science B1: Grazing animals. Elsevier, Amsterdam.

Arnold, G. W., and J. L. Hill. 1972. Chemical factors affecting selection of food plants by ruminants. p. 71-101. In J. B. Harborne (ed.) Phytochemical ecololgy. Academic Press, London.

Arnold, G. W., J. Ball, W. R. McManus, and I. G. Bush. 1966. Studies on the diet of the grazing animal. 1. Seasonal changes in the diet of sheep grazing on pastures of different availability and composition. Aust. J. Agric. Res. 17:543-546.

Bedell, T. E. 1968. Seasonal forage preferences of grazing cattle and sheep in Western Oregon. J. Range Manage. 21:291-297.

Bisset, W. J., and G. W. C. Marlow. 1974. Productivity and dynamics of two Siratro based pastures in the Burnett coastal foothills of south-east Queensland. Trop. Grassl. 8:17-24.

Black, J. L. 1987. Nutritional criteria for forage nutritive value. p. 29-43. In K. J. Hutchinson (ed.) Improving the nutritive value of forage. Standing Committee of Agriculture Tech. Rep. No 20. CSIRO, Melbourne.
Bolland, M. D. A. 1987. Seed production of Trifolium subterraneum subsp. brachycalycinum as influenced by soil type and grazing. Aust. J. Exp. Agric. 27:539-544.
Boswell, C. C. 1971. Fouling of pastures by grazing cattle with special reference to herbage productivity and utilization. M.Phil. thesis, University of Reading.
Broom, D. M., and G. W. Arnold. 1986. Selection by grazing sheep of pasture plants at low herbage availability and responses of the plants to grazing. Aust. J. Agric. Res. 37:527-538.
Brown, K. R. 1968. The influence of herbage height at treading and treading intensity on the yields and botanical composition of a perennial ryegrass-white clover pasture. N.Z. J. Agric. Res. 11: 131-137.
Brown, K. R., and P. S. Evans. 1973. Animal treading - a review of the work of the late D. B. Edmond. N.Z. J. Exp. Agric. 1:217-226.
Bryan, W. W., and T. R. Evans. 1973. Effects of soils, fertilizers and stocking rates on pastures and beef production on the Wallum of south-eastern Queensland. 1. Botanical composition and chemical effects on plants and soils. Aust. J. Exp. Agric. Anim. Husb. 13: 516-529.
Cameron, I. H., and D. J. Cannon. 1970. Changes in the botanical composition of pasture in relation to rate of stocking with sheep and consequent effects on wool production. p. 640-643. In Proc. 11th Int. Grassl. Congr.
Carter, E. D. 1980. The survival of medic seeds folowing the ingestion of intact pods by sheep. Proc. Aust. Agron. Conf. 1:178
Carter, E. D. 1983. Seed and seedling dynamics of annual medic pastures in South Australia. p. 447-450. In Proc. 14th Int. Grassl. Congr.
Carter, E. D., and Sivalingam, T. 1977. Some effects of treading by sheep on pastures of the mediterranean climatic zone of South Australia. p. 307-311. In Proc. 13th Int. Grassl. Congr.
Clements, R. J. 1986. Rates of destruction of growing points of pasture legumes by grazing cattle. CSIRO Div. Trop. Crop Past., Ann. Rep. 1985-86. p. 73-74.
Clements, R. J. 1987. Rates of destruction of growing points of pasture legumes by grazing cattle. CSIRO Div. Trop. Crops and Past., Ann. Rep. 1986-87. p. 19-20.
Clements, R. J., and P. I. Jansen. 1986. Patterns of grazing of Siratro runners in Siratro/buffel grass pastures. CSIRO Div. Trop. Crop Past., Ann. Rep. 1985-86. p. 74-75.
Curll, M. L. 1977. Superphosphate on perennial pastures. 1. Effects of a pasture response on sheep production. Aust. J. Agric. Res. 28:991-1005.
Curll, M. L. 1980. The effect of grazing by set-stocked sheep on a perennial ryegrass/white clover pasture. Ph.D. thesis, University of Reading.
Curll, M. L., and J. L. Davidson. 1983. Defoliation and productivity of a Phalaris-subterranean clover sward, and the influence of grazing experience on sheep intake. Grass Forage Sci. 37:291-297.
Curll, M. L., G. E. Robards, and J. P. Langlands. 1987. Nutritive value improvement of existing forage systems by management. p. 60-

73. In K. J. Hutchinson (ed.) Improving the nutritive value of forage. Standing Committee of Agriculture Tech. Rep. No. 20. CSIRO, Melbourne.

Curll, M. L., and R. J. Wilkins. 1981. The effect of treading and the return of excreta on a perennial ryegrass/white clover sward defoliated by continously grazing sheep. p. 456-458. In Proc. 14th Int. Grassl. Congr.

Curll, M. L., and R. J. Wilkins. 1982. Frequency and severity of defoliation of grass and clover plants grazed by sheep at different stocking rates. Grass Forage Sci. 38:159-167.

Curll, M. L., and R. J. Wilkins. 1983. The comparative effects of defoliation treading and excreta on a Lolium perenne-Trifolium repens pasture grazed by sheep. J. Agric. Sci. 100:451-460.

Curll, M. L., and R. J. Wilkins. 1985. The effect of cutting for conservation on a grazed perennial ryegrass-white clover pasture. Grass Forage Sci. 40:19-30.

Davies, J. G. 1946. Grazing management. 3. A note on pasture management. CSIR Bull. No 201. p. 97-104.

Doak, B. W. 1954. The presence of root inhibiting substances in cow urine and the causes of urine burn. J. Agric. Sci. 44:133-139.

Donald, A. D., and R. T. Leslie. 1969. Population studies of the infective stage of some nematode parasites of sheep. 2. The distribution of faecal deposits on fields grazed by sheep. Parasitology 59:141-157.

Donald, C. M. 1963. Competition among crop and pasture plants. Adv. Agron. 15:1-114.

Dudzinski, M. L., and G. W. Arnold. 1973. Comparisons of diets of sheep and cattle grazing together on sown pastures in the southern tablelands of New South Wales by principal component analysis. Aust. J. Agric. Res. 24:899-912.

Edmond, D. B. 1958. The influence of treading on pasture. A preliminary study. N.Z. J. Agric. Res. 1:319-328.

Edmond, D. B. 1964. Some effects of sheep treading on the growth of 10 pasture species. N.Z. J. Agric. Res. 7:1-16.

FitzGerald, R. D. 1974. The effect of intensity of rotaional grazing on lucerne density and ewe performance at Wagga Wagga. p. 127-133. In Proc. 12th Int. Grassl. Congr.

FitzGerald, R. D. 1979. A comparison of four pasture types for the wheat belt of southern New South Wales. Aust. J. Exp. Agric. Anim Husb. 10:216-224.

FitzGerald, R. D. 1987. Subterranean clover for Northern New South Wales. 1. Special requirements for a summer dominant rainfall. p. 117-119. In Proceedings of the National Sub Clover Improvement Workshop, Wagga Wagga.

FitzGerald, R. D., and K. A. Archer. 1987. Improvement of natural pastures on the Northern Slopes of New South Wales. Final Report to the Australian Wool Corporation, Dan 10P.

Fleming, J. 1986. Profitable production from native pastures on the Northern Tablelands. p. 5-11. In Proc. Grassl. Soc. NSW.

Gardener, C. J. 1980. Diet selection and liveweight performance of steers on Stylosanthes hamata-native grass pastures. Aust. J. Agric. Res. 31:379-392.

Gillard, P., and M. J. Fisher. 1978. The ecology of Townsville stylo-based pastures. p. 340-351. In J. R. Wilson (ed.) Plant relations in pastures. CSIRO, Melbourne.

Hall, D. G., E. C. Wolfe, and B. R. Cullis. 1985. Performance of

breeding ewes on lucerne-subterranean clover pastures. Aust. J. Exp. Agric. 25:758-765.
Hamilton, B. A., K. J. Hutchinson, P. C. Annis, and J. B. Donnelly, 1973. Relationship between the diet selected by grazing sheep and the herbage on offer. Aust. J. Agric. Res. 24:271-277.
Hilder, E. J. 1964. The distribution of plant nutrients by sheep at pasture. Proc. Aust. Soc. Anim. Prod. 5:241-248.
Hilder, E. J. 1966. Distribution of excreta by sheep at pasture. p. 977-981. In Proc. 10th Int. Grassl. Congr.
Hodge, R. W., and J. J. Doyle. (1967). Diet selected by lambs and yearling sheep grazing on annual and perennial pastures in southern Victoria. Aust. J. Exp. Agric. Anim. Husb. 7:141-143.
Hodgson, J. 1981. Testing and improvement of pasture species. p. 309-318. In F. H. W. Morley (ed.) World animal science B1: Grazing animals. Elsevier, Amsterdam.
Hodgkinson, K. C. 1973. Establishment and growth of shoots following low and high cutting of lucerne in relation to the pattern of nutrient uptake. Aust. J. Agric. Res. 24:497-510.
Hodgkinson, K. C., and O. B. Williams. 1983. Adaptation to grazing in forage plants. p. 85-100. In J. G. McIvor and R. A. Bray (ed.) Genetic resources of forage plants. CSIRO, Melbourne.
Hollowell, E. A. 1966. White clover Triflium repens L., annual or perennial. p. 184-187. In Proc. 10th Int. Grassl. Congr.
Holst, P. J., and M. H. Campbell. 1987. The role of goats in the control of weeds of pastures. p. 262-263. In J. L. Wheeler et al., (ed.) Temperate pastures: their production, use and management. CSIRO, Melbourne.
Hutchinson, K. J. 1970. The persistence of perennial species under intensive grazing in a cool temperate environment. p. 611-614. In Proc. 11th Int. Grassl. Congr.
Humphreys, L. R. 1984. Grazing management and the persistence of yield in tropical pasture legumes. p. 1-11 In Asian pastures. FFTC Book Series 25.
Jones, R. J. 1971. Ph.D. thesis, University of New England Armidale
Jones, R. J., and R. M. Jones. 1978. The ecology of Siratro-based pastures. p. 363-367. In J. R. Wilson (ed.) Plant relations in pastures. CSIRO, Melbourne.
Jones, R. M. 1982. White clover (Trifolium repens) in subtropical south-east Queensland. 1. Some effects of site, season and management practices on the population dynamics of white clover. Trop. Grassl. 16:118-127.
Jones, R. M. 1987. Persistence of white clover under grazing. p. 6.6-6.10. In Proceedings of a Specialist Workshop on National White Clover Improvement, Armidale. McMahon, Glen Innes.
Jones, R. M. 1988. The effect of stocking rate on the population dynamics of Siratro in Siratro (Macroptilium atropurpeum)/setaria (Setaria sphacelata) pastures in south east Queensland. iii Effects of spelling on restoration of Siratro in overgrazed pastures. Trop. Grassl. 22:5-11
Jones, R. M. 1989. Productivity and population dynamics of Silverleaf desmodium (Desmodium uncinatum), Greenleaf desmodium (D. intortum) and two D. intortum x D. sanwicense hybrids in coastal south east Queensland. Trop. Grassl. 23:43-55.
Jones, R. M., and G. A. Bunch. 1988a. The effect of stocking rate on the population dynamics of Siratro (Macroptilium atropurpeum) - setaria (Setaria sphacelata) pastures in south east Queensland. i

Survival of plants and stolons. Aust. J. Agric. Res. 39:209-219.

Jones, R. M., and G. A. Bunch. 1988b. The effect of stocking rate on the population dynamics of Siratro (Macroptilium atropurpeum) - setaria (Setaria sphacelata) pastures in south east Queensland. ii Seed set, soil seed reserves, seedling recruitment and seedling survival. Aust. J. Agric. Res. 39:231-234.

Jones, R. M., and R. J. Clements. 1987. Persistence and productivity of Centrosema virginianum and Vigna parkerii cv. Shaw under grazing on the coastal lowlands of south-east Queensland. Trop. Grassl. 21:55-64.

Jones, R. M., and D. Ratcliff. 1983. Patchy grazing and its relation to deposition of cattle dung pats in pastures in coastal subtropical Queensland. J. Aust. Inst. Agric. Sci. 49:109-111.

Jones, R. M., J. C. Tothill, and L. 't Mannetje. 1982. CSIRO agronomic research on Siratro based pastures in south-east Queensland. p. 45-57. In R. F. Brown (ed.) Siratro in south-east Qeensland. Queensland Department of Primary Industries Conference and Workshop Series QC83002.

Kelly, K. B. 1985. Effects of soil modification and treading on pasture growth and physical properties of an irrigated red-brown earth. Aust. J. Agric. Res. 36:799-807.

Keogh, R. G. 1973. Pithomyces chartorum spore distribution and sheep grazing patterns in relation to urine-patch and inter-excreta sites within ryegrass-dominant pastures. N.Z. J. Agric. Res. 16:353-355.

Kerridge, P. C., R. W. McLean, and R. M. Jones. 1987. Seed of Stylosanthes scabra in faeces excreted by grazing cattle. CSIRO Div. Trop. Crop Past., Ann. Rep. 1986-87. p. 84-85.

Koning, C. T. de, and E. D. Carter. 1987. Impact of grazing by sheep on subterranean clover cultivars. Proc. Aust. Agron. Conf. 4:169.

Langlands, J. P., and J. L. Bennett. 1973. Stocking intensity and pasture production. 1. Changes in the soil and vegetation of a sown pasture grazed by sheep at different stocking rates. J. Agric. Sci. 81:193-204.

Leach, G. J. 1978. The ecology of lucerne pastures. p. 290-308. In J. R. Wilson (ed.) Plant relations in pastures. CSIRO, Melbourne.

Leach, G. J. 1979. Lucerne survival in south-east Queensland in relation to grazing management systems. Aust. J. Exp. Agric. Anim. Husb. 19:208-215.

Leach, G. J., and R. J. Clements. 1984. Ecology and grazing management of Alfalfa pastures in the subtropics. Adv. Agron. 37: 127-154.

Leigh, J. H., and M. D. Holgate. 1978. Effects of pasture availability on the composition and quality of the diet selected by sheep grazing mature, degenerate and improved pastures in the upper Shoalhaven Valley, NSW. Aust. J. Exp. Agric. Anim. Husb. 18: 381-390.

Lodge, G. M., and R. D. B. Whalley. 1985. The manipulation of species composition of natural pastures by grazing management on the Northern Slopes of NSW. Aust. Range. J. 7:6-16.

McGregor, B. A. 1984. Growth and fleece production of Angora wethers grazing annual pastures. Proc. Aust. Soc. Anim. Prod. 15:715.

McKinney, G. T. 1974. Management of lucerne for sheep grazing on the Southern Tablelands of New South Wales. Aust. J. Exp. Agric. Anim. Husb. 14:726-734.

McLean, R. W., and P. C. Kerridge. 1987. Effect of fertilizer and sulphur on the diet of cattle grazing buffelgrass/siratro pastures. p. 93-94. In M. Rose (ed.) Herbivour Nutrition Research. Occassional Publication, Australian Society of Animal Production.

McLean, R. W., W. H. Winter, J. J. Mott, and D. A. Little. 1981. The influence of superphosphate on the legume content of the diet selected by cattle grazing Stylosanthes-native grass pastures. J. Agric. Sci., 96:247-249.

Middleton, C. H., and W. Mellor. 1982. Grazing assessment of the tropical legume Calopogonium caeruleum. Trop. Grassl. 16:213-216.

Mullen, G. J., R. M. Jelley, and D. M. McAleese. 1978. Effects of cattle treading in winter and the level of nitrogen fertilizer on the total and seasonal pasture production. Ir. J. Agric. Res. 17: 141-148.

Nunn, R. M., and H. Suijdendorp. 1954. Station management - the value of deferred grazing. J. Agric. W. Aust. 3:585-587.

Ozanne, P. G., and Howes, K. M. W. 1971. Preference of grazing sheep for pasture of high phosphate content. Aust. J. Agric. Res. 22:941-950.

Ozanne, P. G., K. M. W. Howes, and A. Petch. 1976. The comparative phosphate requirements of four annual pastures and two crops. Aust. J. Agric. Res. 27:479-488.

Pascoe, W. B. 1973. Effects of grazing on the growth and development of white clover. M.Sc. thesis, Massey University.

Pott, A., L. R. Humphreys, and J. W. Hales. 1983. Persistence and growth of Lotononis bainesii-Digitaria decumbens pastures. 2. Sheep treading. J. Agric. Sci. 101:9-15

Robinson, G. G. 1977. Wool production from pastures - factors influencing response to superphosphate on the Northern Tablelands, N.S.W.. Wool Tech. Sheep Breed. 15:23-29.

Robinson, G. G., R. D. B. Whalley, and J. A. Taylor. 1983. The effect of prior history of superphosphate application and stocking rate on faecal and nutrient distribution on grazed natural pastures. Aust. Range. J. 5:79-82.

Rossiter, R. C. 1978. The ecology of subterranean clover-based pastures. p. 325-339. In J. R. Wilson (ed.) Plant relations in pastures. CSIRO, Melbourne.

Sears, P. D., V. C. Goodall, and R. P. Newbold. 1948. The effect of sheep droppings on yield, botanical composition and chemical composition of pasture. 2. Results of the years 1942-44 and final summary of the trial. N.Z. J. Sci. Tech. 30A:231-250.

Shaw, N. H. 1978. Superphosphate and stocking rate effects on a native pasture oversown with Stylosanthes humilis in central coastal Queensland. 1. Pasture production. Aust. J. Exp. Agric. Anim. Husb. 18:788-799.

Simao Neto, M., R. M. Jones, and D. Ratcliffe. 1987. Recovery of pasture seed injested by ruminants. 1. Seed of six tropical pasture species fed to cattle, sheep and goats. Aust. J. Exp. Agric. 27:239-246

Smith, A., R. A.Arnott, and J. M. Peacock. 1971. A comparison of the growth of a cut sward with that of a grazed sward, using a technique to eliminate fouling and treading. J. Br. Grassl. Soc. 26:157-162.

Smith, M. V. 1970. Effects of stocking rate and grazing management on the persistence and production of dryland lucerne on deep sands. p. 624-628. In Proc. 11th Int. Grassl. Congr.

Snaydon, R. W. 1981. The ecology of grazed pastures. p. 13-29. In

F. H. W. Morley (ed.) World animal science B1: Grazing animals. Elsevier, Amsterdam.

Sproule, R. J., H. M. Shelton, and R. M. Jones. 1983. Effects of summer and winter grazing on growth habit of Kenya white clover (Trifolium semipilosum) cv. Safari in a mixed sward. Trop. Grassl. 17:26-30.

Stern, W. R., and C. M. Donald. 1962. Light relationships in grass-clover swards. Aust. J. Agric. Res. 13:599-614.

Stobbs, T. H. 1973. The effect of plant structure on the intake of tropical pastures. 11. Differences in sward structure, nutritive value, and bite size of animals grazing Setaria anceps and Chloris gayana at various stages of growth. Aust. J. Agric. Res. 24:821-829.

Stobbs, T. H. 1977. Seasonal changes in the preference by cattle for Macroptilium atropurpureum, Siratro. Trop. Grassl. 11:87-91.

Stobbs, T. H., and E. M. Hutton. 1974. Variations in canopy structures of tropical pastures and their effects on the grazing behaviour of cattle. p.680-687 In Proc. 12th Int. Grassl. Congr.

Taylor, J. A. 1980. Merino sheep and pasture patterns. Ph.D. thesis. University of New England Armidale.

Taylor, J. A., G. G. Robinson, D. A. Hedges, and R. D. B. Whalley. 1987. Camping and faeces distribution by merino sheep. Appl. Anim. Behav. Sci. 17:273-288.

Valdes, L. R., R. J. Clements and R. M. Jones. 1987. Acceptability of Cassia rotundifolia to grazing cattle. CSIRO Div. Trop. Crop Past., Ann. Rep. 1986-1987. p. 20-21.

Vos, G., and R. M. Jones. 1986. The role of stolons and rhizomes in legume persistence. CSIRO Div. Trop. Crop Past., Ann. Rep. 1985-86. p. 70-71.

Walker, B., M. T. Rutherford, and P. C. Whiteman. 1982. Diet selection by cattle on tropical pastures in Northern Australia. p. 681-684. In Proc. 14th Int. Grassl. Congr.

Waterhouse, D. B. 1988. Development of improved lucernes for NSW. Final Report to the Australian Meat and Livestock Research and Development Corporation, Dan 30S.

Watkin, B. R., and R. J. Clements. 1978. The effects of grazing animals on pastures. p. 273-289. In J. R. Wilson (ed.) Plant relations in pastures. CSIRO, Melbourne.

Weeda, W. C. 1967. The effect of cattle dung patches on pasture growth, botanical composition and pasture utilization. N.Z. J. Agric. Res. 10:150-159.

Weeda, W. C. 1977. The effect of cattle dung patches on soil tests and botanical composition of herbage. N.Z. J. Agric. Res. 20:471-478.

Whalley, R. D. B., G. G. Robinson, and J. A. Taylor. 1978. General effects of management and grazing by domestic livestock on the rangelands of the Northern Tablelands of New South Wales. Aust. Range. J. 1:174-190.

Wilson, A. D. 1976. Comparison of sheep and cattle grazing on semi-arid grassland. Aust. J. Agric. Res. 27:155-162.

Winks, L., F. C. Lamberth, K. W. Moir, and P. W. Pepper. 1974. Effect of stocking rate and fertilizer on the performance of steers grazing Townsville stylo-based pastures in north Queensland. Aust. J. Exp. Agric. Anim. Husb. 14:146-154.

Witschi, P. A., and D. L. Michalk. 1979. The effects of sheep treading and grazing on pasture and soil characteristics of irrigated annual pastures. Aust. J. Agric. Res. 30:741-750.

DISCUSSION

Reed: Medium-leafed white clover persists better than large-leafed white clover (e.g., Haifa) when heavily stocked with sheep. However, Haifa persists well when heavily stocked with cattle in marginal white clover areas of Victoria. This may be due to its unpalatability in winter which in turn may be due to its hydrogen cyanide (HCN) content.

Curll: New Zealand cultivars also have relatively high HCN contents, but there does not seem to be any palatability problem with them, at least with sheep.

Caradus: An old paper by Lionel Corkill shows clearly no evidence of sheep being able to discriminate between high and zero cyanogenic plants of white clover. Similarly, goats are also unable to discriminate.

Watson: Could you comment on the effect of HCN content of white clover on selective grazing by slugs?

Jones: Slugs show selection for low HCN when offered a choice, but are not affected by feeding on cyanogenic clovers.

Berberet: Does there appear to be an association of increased disease incidence with higher stocking rates (more treading) in pastures?

Curll: Increased incidence of disease with higher stocking rares can be a problem in poorly drained soils but this is primarily due to the effects of defoliation rather than treading.

Irwin: Physical damage to crowns of lucerne plants caused by trampling would make it easier for crown pathogens to gain access.

Marten: Should legumes in Australian pastures be continuously or rotationally grazed? Is there variation between and within legume species?

Curll: As a generalization, prostrate temperate legumes may be continuously grazed while erect types require rotational grazing.

Jones: We know there are some special cases - for example, hedgerow leucaena should be rotationally grazed. For most legumes, stocking rate appears to be more important than grazing method.

Clements: There is some variation within species. For example, the lucerne cultivar Sheffield, relesed to commerce in 1980, has low, spreading crowns and is able to survive continuous heavy grazing that soon kills other lucerne cultivars.

Scheaffer: During your presentation, you mentioned the development of a lucerne which has staggered shoot regeneration. How does this plant grow and what is the advantage?

Curll: With staggered shoot regeneration, lucerne continually provides new shoots under grazing. These shoots come from a partially buried or prostrate crown.

Allen: You postulated that there is a need for seed pods of medics which are unacceptable to grazing sheep. What type of pod did you have in mind?

Curll: My comment was only a suggestion if it was necessary to reduce selective grazing of a legume - there are native legumes with spiny pods which could be used in breeding programs.

Allen: Spiny pods create a problem with contamination of wool, and breeding programmes have been directed at minimizing pod spininess.

Reed: Some annual clovers (e.g., T. globosum) with very hairy burrs build up bigger seed banks under grazing than those with non hairy burrs. Work on this was publishes recently by M.A. Bolland in the Journal of the Australian Institute of Agricultural Science.

M. Hay: Could you outline what work is being done to increase phenotypic plasticity in white clover in Australia?

Curll: There is no existing work, but selection for phenotypic plasticity is proposed in the National Improvement Program starting this year.

Forde: Can you in fact recognize strains of white clover which show a wider range of phenotypic plasticity than others?

Curll: Yes. The cv. Blanca shows a wide range of leaf size in response to grazing in the United Kingdom, though this does not occur in Australia. In Australia, the cultivar Haifa seems to have greater plasticity than local naturalized white clover.

Buxton: What is the physiological basis of the change in morphological form as a result of grazing?

Curll: It is not known, but the plant is probably responding to stress. This is an important phenomenon that would benefit from a research effort.

M. Hay: We have found that the system of grazing management (e.g., set stocked vs. rotationally grazed) influences the size of white clover plants but not the branching structure of plants. For branching characters, maybe plasticity is overrated.

R. Smith: In your final list of strategies to improve persistence you appeared to emphasize management of existing germplasm rather than improvement through selection. Do you really mean to ignore potential genotypic plasticity?

Curll: In the final summary I did emphasize management but I certainly don't wish to imply that what you suggest is not also important. In my presentation, I referred to the need for grazing tolerant legumes. I believe genetic improvement has to be in concert with management.

Kretschmer: What is the importance and impact of seed production on white clover persistence? Does higher seed production become more important as the environment becomes more subtropical or generally more stressful.

Curll: Seed production for white clover replacement is generally only important in the subtropics and perhaps drier marginal areas in Australia. Most cultivars used have more than adequate seed yields, and hard seed levels in the soil seed bank can be very high.

PLANT-ANIMAL FACTORS INFLUENCING LEGUME PERSISTENCE

G.W. Sheath and J. Hodgson

SUMMARY

In considering legume persistence, emphasis is placed on the importance of the vulnerability to grazing of reproductive and vegetative organs involved in plant propagation. Grazing effects that accentuate or reduce this vulnerability are highlighted. Grazing severity-timing interactions and animal species effects are most important. Many legumes play an opportunistic role in permanent pastures through either seed-set and seedling development, or vegetative extension. Where grazing creates spatial or temporal heterogeneity this opportunity increases. Selective grazing is discussed in detail. Selectivity is strongly influenced by pasture conditions (e.g., spatial leaf distribution, legume proportions) and animal species, but the operative factors governing discriminative behavior are not fully understood.

INTRODUCTION

Persistence is a loosely used concept that has different meanings depending on the legume species involved. It may describe the survival or death of discrete individual plants (e.g., red clover); the net balance between production and loss of propagated growth units (e.g., white clover); and the regenerative success or failure of annual legume populations (e.g., subterranean clover). Given this diversity, grazing can have both a direct and indirect influence on the many processes that determine persistence. Grazing of plant parts may eliminate a key process of persistence (e.g., reproductive organs of annual legumes); may accentuate biotic and environmental stress (e.g., reduced vigor or tolerance); and may act as a secondary agent in modifying plant aggressiveness in the face of competitors (e.g., regrowth potential).

In mixed species pastures, the major influence of the animal is likely to be observed in terms of a differential effect upon the survival or competitive success of different species growing in a mixed community, rather than a direct effect upon the survival of individual species. In the context of this paper, attention is concentrated on the legume component of mixed grass/legume pastures, with grasses as the main competitors.

REGROWTH ABILITY

Maintenance of plant vigor and competitiveness is reliant on the retention of sufficient photosynthetic tissue and growth sites to allow immediate leaf and stem production following grazing. The interaction between plant habit and grazing severity is the main determinant of regrowth ability in legumes and, to a certain extent, morphological adaptation occurs to accommodate grazing intensity. Large-leaved white clover cultivars reduce petiole and stolon internode length and so retain photosynthetic capacity under intense grazing (Korte & Parsons, 1985). However, morphological adaptation has its limits and long-term persistence and production under intensive, continuous stocking is clearly superior with small-leaved, prostrate white clover cultivars (Williams & Caradus, 1979; Brock, 1988).

Within diverse white clover communities that are intensively grazed, there is a genetic drift towards those genotypes that are prostrate and densely stoloniferous (Macfarlane & Sheath, 1984; Lambert *et al.*, 1986b). This morphological form adjusts the energy economy of the plant towards the retention of photosynthetic reserves and leaf primordia, and away from leaf tissue formation. It is therefore assumed to be better adapted for persistence than for herbage production. However, there is a need for a more critical assessment of the relative importance of current assimilation and energy storage, and of stolon extension and node establishment. Their role as determinants of grazing tolerance and explorative capability must be considered over the range of genetic variation in white clover. At the extreme, there can be an over investment in persistency strategies to the detriment of productive capability.

Legumes which elevate terminal growth sites into the grazing zone are most susceptible to intensive defoliation. For example, regrowth of *Lotus pedunculatus* is based on axillary shoots in residual stems and nodal shoots on rhizomes (Table 1). While the latter eventually dominate regrowth, they are slow in expressing leaf and stem production relative to axillary shoots. Lenient defoliation provides for better stubble shoot regrowth, but it results in low utilization of grown herbage. Further, not only does intense defoliation lead to slow regrowth and weak competitive ability in mixed pastures, but it also leads to a restriction of underground growth (Table 2). Rhizome expansion and plant fragmentation is an annual event which is the basis of spread and colonization in *Lotus pedunculatus* (Wedderburn & Lowther, 1985).

Table 1. Effect of defoliation frequency and severity on shoot and canopy production (t DM/ha) of *Lotus pedunculatus* (Source: Sheath, 1980a)

	Stubble shoots	Rhizome shoots	Sec. axil shoots	Total shoot	Net canopy
Defoliation					
Sec. axil* to 1.5 cm	0.3	12.1	0.9	13.3	11.0
6 wkly to 1.5 cm	1.6	9.2	0.6	11.4	8.5
Sec. axil. to 9.5 cm	5.2	8.2	0.8	14.2	9.0
6 wkly to 9.5 cm	4.2	8.5	1.0	13.7	9.6

*Defoliated when secondary axillary shoots growth recurred on dominant rhizome shoots.

Table 2. Effect of defoliation frequency and severity on rhizome dry weight (mg/plant), crown plus taproot dry weight (mg/plant) and plant density (/m2) of *Lotus pedunculatus* (Source: Sheath, 1980b)

	Rhizome		Crown & taproot		Density	
	Jan.	May	Jan.	May	Jan.	Sept
Defoliation						
3 wkly to 1.5 cm	342	1121	373	640	179	170
3 wkly to 9.5 cm	926	2688	987	1459	153	242
6 wkly to 1.5 cm	538	1562	566	1302	155	182
6 wkly to 9.5 cm	879	2160	886	1440	159	225

Consequently, more prostrate cultivars are better adapted to persist under intensive grazing (Lambert *et al.*, 1974).

There is a similar need for retention of shoots within the grazed residual of red clover. Frequent grazing, particularly in summer, leads to reduced persistence and lower production in mixed ryegrass-red clover pastures (Cosgrove & Brougham, 1985). Further, cultivars with a low propensity to initiate shoots require lax grazing to remain persistent (Pineiro & Harris, 1978). On the basis of persistence in mixed pastures, red clover seems more suited to cattle grazing than either continuous or rotational sheep grazing (Lambert *et al.*, 1986b).

Unlike *Lotus spp.* and red clover, basal flushes of new shoots are available for immediate regrowth in lucerne, provided there are adequate carbohydrate reserves, and recommended management takes this into account. Defoliation at 10% flowering was traditionally recommended as a compromise between maximum yield and quality (O'Connor, 1970), but in practice the management flexibility achieved by reducing spelling periods to 6 weeks still provides satisfactory vigour, health and persistence (Janson, 1982). Certainly, lucerne persistence is poor under frequent or continuous grazing, especially in spring when lucerne has no edaphic, competitive advantage.

REPRODUCTIVE REGENERATION

Maintaining the integrity of reproductive processes is vital to the persistence of annual legume populations. The direct removal through grazing of reproductive organs jeopardises reseeding processes. Subterranean clover cultivars that are open and erect in habit, and which are unable to retain runners below the grazing zone, reseed poorly (Sheath & Macfarlane, 1988). Similarly, grazing managements that encourage the elevation of flowers and burrs into the grazing zone reduce population regeneration. Pastures with a stable mass or height, as occur under continuous grazing, experience better regeneration (Table 3). The ability to morphologically adapt to stable grazing pressures even assists the reseeding of prostrate annuals that are terminal seeders such as *Lotus angustissimus* and *L. subbiflorus*. Poor regeneration can be expected where hard, rotational grazing occurs during reseeding (Table 4). Neither morphological adjustment, nor grazing strategies are sufficient to ensure the survival of erect terminal seeders such as *Trifolium vesiculosum* in intensive grazing systems (Sheath *et al.*, 1984).

Successful establishment of seedlings is also dependent on grazing time and pressure. Germination of subterranean clover seed within tall vegetation is unsatisfactory, not only because of competition, but also because seedlings become etiolated and are then vulnerable to grazing. Where pastures are spelled during summer and hard grazed following autumn rains, then reduced content of annual legumes can be expected (Table 4). The speed with which other species move to occupy vacant space also determines seedling establishment success. Where grazing management encourages white clover, subterranean clover levels are reduced (Bircham, 1977), because in mixed communities, the former appears to be the more aggressive in exploiting space.

VEGETATIVE REGENERATION : WHITE CLOVER

Where legume content is a balance between vegetative expansion and death of growth units, animal grazing can impact on the opportunity to expand and the expression of that potential. White clover is well suited to rapidly invade into non-competitive

Table 3. Effect of August-December sheep grazing on the winter frequency (% occurrence) of subterranean clover and annual *Lotus spp.* in mixed species hill pastures (Source: Sheath, 1988, unpublished data)

	1979	1981	1982	1983	1984
Subterranean clover					
continuous stocking	12	12	24	18	17
mob stocking (4 wkly)	13	9	15	8	7
SED	1.6	2.2	3.4*	1.7**	1.5**
Annual *Lotus spp.*					
continuous stocking	12	10	13	14	6
mob stocking (4 wkly)	9	6	4	5	3
SED	3.3	2.6	2.4**	2.4**	1.3*

Table 4. Effect of November-March mob grazing by sheep on the winter frequency (% occurence) of subterranean clover, annual *Lotus spp.* and white clover in mixed hill pastures (Source: Sheath & Boom, 1985)

	Lax[1]/ spell[2]	Lax/ hard	Hard/ spell	Hard/ hard	SED
Species					
Subclover	4	24	5	10	2.5
Annual *Lotus spp.*	7	2	8	1	1.4
White clover	30	25	35	39	3.7

1 grazing during November-December seeding.
2 grazing during January-March drought.

environments that are created by disturbance (Lambert *et al.*, 1986a; Chapman, 1986). Growth and survival is strongly dependent on stolon development and replacement (Chapman, 1983; Hay, 1983) and stolon length per unit area seems to be the strongest indicator of ultimate performance (Lambert *et al.*, 1986c). Maintenance of herbage production by the clover component of a sward then depends upon leaf development at nodes and the subsequent susceptibility of these leaves to defoliation (Chapman *et al.*, 1983).

Rotational sheep grazing allows a greater expression of potential exploration and growth compared with continuous sheep stocking, but it has been argued that it produces a less stable white clover population (Brock, 1988). While various combinations of sheep grazing management have little effect on the amount and the above and below ground distribution of stolon in the pasture (Hay *et al.*, 1983), year-round rotational grazing accentuates the natural spring fragmentation of white clover plants to even smaller units (Hay *et al.*, 1988). Such fragments appear less persistent if summer drought is subsequently encountered. As a consequence continuous stocking during spring is recommended (Brock, 1988). The reasons for this are not clear, but certainly where grass dominance is prevented during mid-late spring through either intensive continuous or hard rotational grazing, then white clover is better able to maintain its occupancy and invade unoccupied spaces (Hay & Baxter, 1984; Sheath & Boom, 1985 - Table 4). This expansion requires warm temperatures and adequate moisture that favour growth and stolon expansion and would therefore be absent during cold winters and dry summer conditions. The impact of different grazing managements in creating explorative opportunity and thereby modifying white clover content is dependent on seasonal growing conditions and the relative aggressiveness of competing species.

Frame and Newbould (1986) have emphasised the potentially deleterious effect of continuously stocked sheep upon the white clover content of mixed pastures. However, in New Zealand conditions the effects of contrasting grazing managements are generally small, particularly where clover populations have adapted to high grazing pressure (Clark *et al.*, 1984a; Lambert *et al.*, 1986a). Under continuous stocking, grazing frequency of white clover is similar to that of ryegrass (Curll & Wilkins, 1982; Clark *et al.*, 1984a), although this observation may be influenced by the relative sizes of the units of plant measurement involved. Grazing frequency may range from 19-88 days depending on the time of the year, accessibility being the major determinant. During winter when petioles are shortest relative to grass tillers, grazing frequency is least. However, where clover foliage is relatively concentrated in the surface strata of the canopy, the effect may still be a greater defoliation pressure on the clover than the grass component of the sward (Bircham & Hodgson, 1983). Under these circumstances, even minor differences in defoliation pressure may tip the balance in favour of the grass component, particularly if they are cumulative.

Differential grazing of white clover and grass in mixed swards is most likely to occur in summer. This will happen if white clover remains as a distinct green leaf component and/or becomes elevated into the grazing zone of a grass dominated canopy. Unpublished work of Bircham (Table 5) clearly indicates greater grazing severity of white clover during dry summer months. Certainly, white clover content is severely depleted if high mass pastures are hard grazed by concentrated mobs of sheep during summer (Table 4). However, stolon death is only enhanced by severe defoliation when there is simultaneous development of grass dominance, and it is not greatly affected by differences in sheep grazing management (Chapman *et al.*, 1984).

Table 5. Sheep grazing of white clover in a mixed species hill pasture under different managements throughout the year (Source: J.S. Bircham, 1988, unpublished data)

	Spring[1] cont.	Spring rotn.	Summer rotn.	Autumn-winter rotn.
Length of individual stolon removed (mm)	0.34	0.64	5.06	0.59
% total petiole nos. defoliated	23.4	34.5	58.5	37.6
% terminal petiole nos. defoliated	34.7	37.5	97.5	43.2

[1] Management Description:
spring cont. - continuous stocking
spring rotn. - 14 day grazing, 14 day spelling sequence
summer rotn. - 10 day grazing, 30 day spelling sequence
autumn-winter rotn.- 20 day grazing, 40 day spelling sequence

In contrast to the general view that grazing by sheep leads to a reduction in clover content in mixed swards compared to cutting management (Curll, 1982; Newton & Davies, 1987), cattle grazing has a relatively neutral effect (Frame & Newbould, 1986). It is inferred from this that sheep exert greater selective pressures on clover than cattle (Frame & Newbould, 1986; Newton & Davies, 1987), though critical evidence to support this contention is limited. Goat grazing, however, enhances clover content (Clark *et al.*, 1984b). There appear to be marked differences between sheep and goats in selection for clover (Clark *et al.*, 1982b; Nicol *et al.*, 1987), but it is not always clear whether effects on pasture composition reflect differences in grazing pressure and/or genuine differences in dietary preference.

Grazing animals also differ in the extent and distribution of the effects of treading and excretal returns. White clover is more susceptible than perennial ryegrass to treading damage, but it is suggested that the adverse effects of treading are only likely to become apparent at stocking rates where clover survival is already threatened by defoliation processes (Curll, 1982; Curll & Wilkins, 1983). White clover may have more opportunity for colonisation under cattle than under sheep grazing as a consequence of greater treading damage (Chapman, 1986) or of greater protection from grazing in the vicinity of dung pats. Excretal returns influence clover content and survival primarily by enhancing grass growth (Curll, 1982).

Table 6. Diet selection by sheep and cattle grazing mixed white clover-perennial ryegrass swards of contrasting clover distribution (Source: Clark *et al.*, 1986). All results based on % point contacts of lamina and petiole.

	Sward		
	1	2	SED
Clover proportions:			
in whole sward	13.6	31.5	3.9
in top half of sward	7.1	30.2	-
in diet of sheep	21.7	44.6	5.1
in diet of cattle	5.0	39.3	3.9

SELECTIVE GRAZING

Heterogeneity in space and time resulting from defoliation, treading and excreta may be of critical importance in influencing the survival and performance of an invasive species like white clover (Turkington & Harper, 1979). There are few reports on this aspect of the ecology of mixed grass-clover communities except under conditions of limited control, and the subject deserves more attention in future. It is apparent that the distribution of the clover and grass components in mixed swards has a marked influence upon the selective activities of grazing animals.

Where forage components are homogeneously distributed in actively growing swards, grazing may be largely indiscriminate within the surface strata (Milne *et al.*, 1982; L'Huillier *et al.*, 1984; Clark & Harris, 1985), though this is not always the case (Table 6). Discrimination becomes more apparent with increasing contrasts between the stages of maturity, or the growth activity of the sward components, and can be extreme in mature swards (Grant *et al.*, 1985) or dried-off pastures (Keogh, 1975). In these cases it is legitimate to question whether discrimination is for clover or for some general plant characteristics associated with immaturity or active growth. Milne et al (1982) showed that selection for clover was substantially greater in 3 wk regrowths than in 1 or 2 wk regrowths, and Laidlaw (1983) presented evidence of deliberate discrimination by sheep in favour of the clover component throughout the canopy of a red clover-perennial ryegrass sward. These effects may be explained by variations in sward structure (Milne *et al.*, 1982) or by developing differences with time in the maturity of grass and legume foliage. Where discrimination occurs, it is likely to be more extreme in sheep than in cattle (Grant *et al.*, 1985; Clark *et al.*, 1986; Table 6).

There is clearer evidence of discrimination where clover and grass components are clumped. The results of Clark and Harris (1985) demonstrate selection of strips of clover-dominant herbage in contrast to associated clover-free areas, and preference for high-clover patches within the clover-dominant areas. Clark and Hodgson (1986) showed that sheep were able to exercise fine discrimination between adjacent patches of vegetation differing in clover content. However, discrimination in favour of clover may decline as clover proportions increase towards 50% (Clark & Harris, 1985), and may even reverse at higher clover proportions (Milne *et*

Table 7. Correlation matrix (r^2) between the proportion of leaves removed from individual white clover plants (defoliation score), petiole length, leaf size, cyanogenic glucoside activity and leaf mark intensity (Source: Hodgson & Clark, 1987, unpublished data).

	Defoliation score	Petiole length	Median leafsize	Glucoside activity
Petiole length (mm)	0.142***	-		
Median leaf size (Williams scale, range 8-18)+	0.136**	0.366***	-	
Glucoside activity (Picrate test, scale 0-5)#	-0.045	0.022	0.065	-
Leaf mark intensity (Scale 0-5)	-0.002	0.035	0.043	0.084*

n=550; *, ** and *** significant at 5%, 1% and 0.1% levels of probability; other values not significant; + Williams *et al.*, (1964); # Corkill (1952)

al., 1982). The reasons for these behavioural changes are not clear.

There has been speculation about the characteristics which are likely to influence the selection of individual clover plants by grazing animals. Particular reference has been made to genetic polymorphism in leaf mark (Cahn & Harper, 1976) and cyanogenic glucoside activity (Corkill, 1952; Crowley, 1983). Recent studies by Hodgson and Clark (1988, unpublished data; Table 7) have involved grazing by sheep of dominantly perennial ryegrass plots which contained matrices of cloned white clover plants differing in leaf size, petiole length, leaf mark and glucoside activity. Results indicated that the latter two factors had little influence upon selective defoliation of the clover plant. In accord with earlier evidence, selectivity was largely related to leaf size and position within the sward canopy. There is no clear evidence from these or other studies that gross chemical characteristics (e.g., N concentration) are determining factors. In *Lotus pedunculatus*, however, a high tannin concentrations may help to protect the legume from grazing even when growing in association with mature grasses (Lowther & Barry, 1985).

REFERENCES

Bircham, J.S. 1977. Grazing management for the improvement of browntop pastures in hill country. Proc. N.Z. Grassl. Assoc. 38: 87-93.

Bircham, J.S., and J. Hodgson 1983. The influence of sward conditions on rates of herbage growth and senescence in mixed

swards under continuous stocking management. Grass Forage Sci. 38:323-331.
Brock, J.L. 1988. Evaluation of New Zealand bred white clover cultivars under rotational grazing and set-stocking with sheep. Proc. N.Z. Grassl. Assoc. 49: 203-206.
Cahn, M.G., and J.L. Harper 1976. The biology of the leaf mark polymorphism in *Trifolium repens L.* 2. Evidence for the selection of leaf marks by rumen fistulated sheep. Heredity 37: 309-325.
Chapman, D.F. 1983. Growth and demography of *Trifolium repens* stolons in grazed hill pastures. J. Appl. Ecol. 29: 597-608.
Chapman, D.F. 1986. Development, removal and death of white clover leaves under three grazing managements in hill country. N.Z. J. Agric. Res. 29: 39-47.
Chapman, D.F., D.A. Clark, and C.A. Land 1983. Leaf and tiller growth of *Lolium perenne* and *Agrostis spp.* and leaf appearance rates of *Trifolium repens* in set stocked and rotationally grazed hill pastures. N.Z. J. Agric. Res. 26: 159-168.
Chapman, D.F., D.A. Clark, and C.A. Land 1984. Leaf and tiller or stolon death of *Lolium perenne, Agrostis spp.* and *Trifolium repens* in set stocked and rotationally grazed hill pastures. N.Z. J. Agric. Res. 27: 303-312.
Clark, D.A., D.F. Chapman, and C.A. Land. 1984. Defoliation of *Lolium perenne* and *Agrostis spp.* tillers and *Trifolium repens* stolons in set stocked and rotationally grazed hill pastures. N.Z. J. Agric. Res. 27: 289-301.
Clark, D.A., and P.S. Harris 1985. Composition of the diet of sheep grazing swards of differing white clover content and spatial distribution. N.Z. J. Agric. Res. 28: 233-240.
Clark, D.A., and J. Hodgson 1986. Discrimination by sheep between swards of differing white clover content. Anim. Prod. 42: 456.
Clark, D.A., J. Hodgson, E. Robertson, and G.T. Barthram 1986. Diet selection by sheep and cattle from mixed grass/clover swards. Hill Farming Res. Org. Bienn. Rep. 1984-85. p. 28.
Clark, D.A., M.G. Lambert, and D.F. Chapman 1982a. Pasture management and hill country production. Proc. N.Z. Grassl. Assoc. 43: 205-214.
Clark, D.A., M.G. Lambert, and M.P. Rolston 1982b. Diet selection by goats and sheep on hill country. Proc. N.Z. Soc. Anim. Prod. 42: 155-157.
Clark, D.A., M.P. Rolston, and M.G. Lambert 1984. Pasture composition under mixed sheep and goat grazing on hill country. Proc. N.Z. Grassl. Assoc. 45: 160-166.
Corkill, L. 1952. Cyanogenesis in white clover (*Trifolium repens L.*). VI. Experiments with high-glucoside and glucoside-free strains. N.Z. J. Sci. Technol. 34A: 1-16.
Cosgrove, G.P., and R.W. Brougham 1985. Grazing management influences on seasonality and performance of ryegrass and red clover. Proc. N.Z. Grassl.. Assoc. 46: 71-76.
Crowley, M.J. 1983. Herbivory. The dynamics of animal plant interactions. Blackwell, Oxford.
Curll, M.L. 1982. The grass and clover content of pastures grazed by sheep. Herb. Abstr. 52: 403-411.

Curll, M.L., and R.J. Wilkins 1982. Frequency and severity of defoliation of grass and clover by sheep at different stocking rates. Grass Forage Sci. 37: 291-297.

Curll, M.L., and R.J. Wilkins 1983. The comparative effects of defoliation, treading and excreta on a *Lolium perenne* - *Trifolium repens* pasture grazed by sheep. J. Agric. Sci. 100: 451-460.

Frame, J., and P. Newbould 1986. Agronomy of white clover. Adv. Agron. 40: 1-88.

Grant, S.A., D. Suckling, H.K. Smith, L. Torvell, T.D.A. Forbes, and J. Hodgson. 1985. Comparative studies of diet selection by sheep and cattle: the hill grasslands. J. Ecol. 73: 987-1004.

Hay, M.J.M. 1983. Seasonal variation in the distribution of white clover (*Trifolium repens*) stolons among three horizontal strata in two grazed swards. N.Z. J. Agric. Res. 26: 29-34.

Hay, M.J.M., J.L. Brock, and R.H. Fletcher 1983. Effect of sheep grazing management on distribution of white clover stolons among three horizontal strata in ryegrass/white clover swards. N.Z. J. Exp. Agric. 11: 215-218.

Hay, M.J.M., J.L. Brock, and V.J. Thomas 1988. Seasonal and sheep grazing effects on branching structure and dry weight of white clover plants in mixed swards. Proc. N.Z. Grassl. Assoc. 49: 197-201.

Hay, R.J.M., and G.S. Baxter 1984. Spring management of pasture to increase summer white clover growth. Proc. Lincoln College Farmers Conf. 34: 132-137.

Janson, C.G. 1982. Lucerne grazing management research. p.85-90. In R.B. Wynn-Williams (ed.) Lucerne for the 80s. Spec. Publ. 1. Agron. Soc. N.Z.

Keogh, R.G. 1975. Grazing behaviour of sheep during summer and autumn in relation to facial eczema. Proc. N.Z. Soc. Anim. Prod. 35: 198-203.

Korte, C.J., and A.J. Parsons 1984. Persistence of a large leaved white clover variety under sheep grazing. Proc. N.Z. Grassl. Assoc. 45: 118-123.

L'Huillier, P.J., D.P. Poppi, and T.J. Fraser 1984. Influence of green leaf distribution on diet selection by sheep and the implications for animal performance. Proc. N.Z. Soc. Anim. Prod. 44: 105-107.

Laidlaw, A.S. 1983. Grazing by sheep and the distribution of species through the canopy of a red clover-perennial ryegrass sward. Grass Forage Sci. 38: 317-321.

Lambert, J.P., A.F. Boyd, and J.L. Brock 1974. An evaluation of five varieties of *Lotus pedunculatus* compared with "Grasslands Huia" white clover under grazing at Kaikohe. N.Z. J. Exp. Agric. 2: 359-363.

Lambert, M.G., D.A. Clark, and D.A. Grant 1986a. Influence of fertiliser and grazing management on North Island moist hill country. 2. Pasture botanical composition. N.Z. J. Agric. Res. 29: 1-10.

Lambert, M.G., D.A. Clark, and D.A. Grant 1986b. Influence of fertiliser and grazing management on North Island moist hill country. 3. Performance of introduced and resident legumes. N.Z. J. Agric. Res. 29:11-21.

Lambert, M.G., D.A. Clark, and D.A. Grant 1986c. Influence of fertiliser and grazing management on North Island moist hill country. 4. Pasture species abundance. N.Z. J. Agric. Res. 29:23-31.

Lowther, W.L., and T.M. Barry 1985. Nutritional value of "Grasslands Maku" lotus grown on low fertility soils. Proc. N.Z. Soc. Anim. Prod. 45: 125-127.

Macfarlane, M.J., and G.W. Sheath 1984. Clover - what types for dry hill country? Proc. N.Z. Grassl. Assoc. 45: 140-150.

Milne, J.A., J. Hodgson, R. Thompson, W.G. Souter, and G.T. Barthram. 1982. The diet ingested by sheep grazing swards differing in white clover and perennial ryegrass content. Grass Forage Sci. 37:209-218.

Newton, J.E., and D.A. Davies 1987. White clover and sheep production. p.79-87. In G.E. Pollott (ed.) Efficient sheep production from grass. Occas. Symp. 21. Br. Grassl. Soc., U.K. Berkshire.

Nicol, A.M., D.P. Poppi, M.R. Alam, and H.A. Collins. 1987. Dietary differences between sheep and goats. Proc. N.Z. Grassl. Assoc. 48: 199-205.

O'Connor, K.F. 1970. Influence of grazing management on herbage and animal production from lucerne. Proc. N.Z. Grassl. Assoc. 32: 108-116.

Pineiro, J., and W. Harris 1978. Performance of ryegrass cultivars and prairie grass with red clover cultivars under two grazing frequencies. N.Z. J. Agric. Res. 21: 665-673.

Sheath, G.W. 1980a. Production and regrowth characteristics of *Lotus pedunculatus* "Grasslands Maku". N.Z. J. Agric. Res. 23: 201-209.

Sheath, G.W. 1980b. Effects of season and defoliation on the growth habit of *Lotus pedunculatus Cav.* "Grasslands Maku". N.Z. J. Agric. Res. 23: 191-200.

Sheath, G.W., and R.C. Boom 1985. Effects of November-April grazing pressure on hill country pastures. 2. Pasture species composition. N.Z. J. Exp. Agric. 13: 329-340.

Sheath, G.W., M.J. Macfarlane, and P.M. Bonish 1984. Evaluation of arrowleaf clover (*Trifolium vesiculosum*) in North Island hill country. N.Z. J. Exp. Agric. 12: 209-217.

Sheath, G.W., and M.J. Macfarlane 1989. Evaluation of clovers in dry hill country. 2. Components of subterranean clover regeneration at Whatawhata. N.Z. J. Agric. Res.(in press).

Turkington, R., and J.L. Harper 1979. The growth, distribution and neighbour relationships of *Trifolium repens* in a permanent pasture. I. Ordination, pattern and contact. J. Ecol. 67: 201-218.

Wedderburn, M.E., and W.L. Lowther 1985. Factors affecting establishment and spread of "Grasslands Maku" lotus in tussock grasslands. Proc. N.Z. Grassl. Assoc. 46: 97-102.

Williams, W.M., and J.R. Caradus 1979. Performance of white clover lines on New Zealand hill country. Proc. N.Z. Grassl. Assoc. 40: 162-169.

Williams, R.F., L.T. Evans, and L.J. Ludwing 1964. Estimation of leaf area for clover and lucerne. Aust. J. Agric. Res. 15: 231-233.

DISCUSSION

Curll: The correlations between leaf size or petiole length and selection do not seem very high.

Hodgson: They may not be but bearing in mind the variability inherent in individual plant observations, they are sufficient to determine the relative importance of the factors under consideration.

Leath: Greater defoliation occurred when longer petioles and larger leaves were present. Since these usually occur together could you separate these two factors?

Hodgson: Yes. We did this by lowering large-leaved plants in the pasture canopy and concluded that leaf position i.e. long petioles, was more important than leaf size.

Matches: Do the relationships reported in the clover defoliation study hold true for both cattle and sheep?

Hodgson: We only have this type of information for sheep.

Brougham: Cattle can graze just as intensively as sheep. They use their tongues to pull stolons off the ground in grass/legume swards. This effect can be very pronounced and can markedly influence white clover persistency especially during the summer.

Hodgson: Agreed. In general cattle show less selection than sheep, but if plant contrasts are great enough they can be equally discriminating. Damage to the legume plant will be influenced by animal species differences in the grazing process.

Marten: We have just completed an experiment with a world collection of birdsfoot trefoil (*Lotus corniculatus L.*) in which we had similar results to yours with white clover in that grazing sheep preferred upright types over decumbent types. However, we found no association between a wide range of cyanogen concentration and palatability of genotypes. What was the range of cyanogen in your clover?

Hodgson: The cyanogen range was very large - from 0 to 4 or 5 in the picrate test. We selected and cloned plants within cultivars to achieve the widest range possible in cyanogen and leaf mark factors.

Hochman: I have some data on subclover cultivars of contrasting erectness which agrees with the white clover data you presented. After a period of grazing, large differences in height and dry matter were virtually eliminated.

Watson: High intensity grazing of clover by sheep in relation to clover density in the sward effectively establishes the upper level of clover in the sward.

Hodgson: Yes, but the indications are that selection moves progressively towards the minor component in a mixed sward - whether clover or grass. This implies instability, unless modified by the effects of changed sward distribution.

Watson: Selective grazing may account for the lower level of clover in ryegrass swards with endophyte compared with no endophyte.

Hodgson: I have no direct evidence on endophyte levels in either case, but indications are that in U.K. studies, endophyte would not be an issue.

Hodgson: These results fit the theories of foraging strategy which have been developing for herbivores grazing natural plant communities. They indicate that selection, where it is apparent, is directed to maintaining ease of defoliation and rate of intake.

LEGUME PERSISTENCE UNDER GRAZING IN STRESSFUL ENVIRONMENTS OF THE UNITED STATES

C. S. Hoveland

SUMMARY

Legumes used in pastures where frequent environmental stress occurs, such as the southeastern United States, are generally nonpersistent and undependable. Factors responsible for this unreliability are drought, subsoil acidity, nematodes, disease, insects, and competition from rhizomatous warm-season perennial grasses. These environmental effects are a greater problem than the grazing animal in legume persistence. Grazing animals influence legume persistence by selective grazing, treading, and excretion. Breeding efforts on pasture legume improvement in regions of stress have made progress. Birdsfoot trefoil (*Lotus corniculatus* L.) cultivars have been developed that are tolerant of subsoil acidity. Improved sericea lespedeza [*Lespedeza cuneata* (Dum.-Cours.) G. Don] cultivars have higher nutritive quality, greater palatability, and are persistent under good grazing management. Alfalfa (*Medicago sativa* L.) selection under continuous grazing has produced a productive and persistent cultivar that can be used either for hay or pasture. Future research needs to be intensified on screening of species and cultivar improvement for legume persistence in stressful environments.

INTRODUCTION

Legumes are generally considered to be a desirable component in pastures. They provide N, improve nutritive quality, and extend the productive season of warm season grass pastures. Burns and Standaert (1985), in an extensive review of grazing experiments in the United States, reported that steer daily gains from legume-grass were 0.14 kg/d more than from N-fertilized grass and calf daily gains were 0.15 kg/d higher on legume-grass in cow-calf experiments. Generally, 200 kg of N/ha was required for grass to produce the approximately 400 kg/ha gain expected from legume-grass mixtures. The higher animal performance obtained with legumes is attributed to higher crude protein, digestibility, and minerals, resulting in higher intake potential (Marten, 1985).

Perennial temperate legumes such as white clover (*Trifolium repens* L.), red clover (*T. pratense* L.), alfalfa (*Medicago sativa* L.), and birdsfoot trefoil (*Lotus corniculatus* L.) are commonly used in most of the United States. Winter annual clovers such as arrowleaf (*T. vesiculosum* Savi), crimson (*T. incarnatum* L.), subterranean (*T. subterraneum* L.), and rose (*T hirtum* L.) are grown in the Coastal Plains of the Gulf states and in winter rainfall areas of Pacific Coast states. Tropical legumes are restricted primarily to south central Florida.

LEGUMES AND STRESSFUL ENVIRONMENTS

Perennial legumes are an important pasture component in the humid eastern United States where temperature, rainfall, and soils are favorable. However, legumes become less dependable in the Great Plains where water is limiting. Likewise, perennial legumes are much less successful in the southern United States where various abiotic and biotic factors create especially stressful environments (Burns, 1985; Hoveland, 1986; Van Keuren & Hoveland, 1985). In the Coastal Plain region, perennial legumes are generally unsuccessful and annual legumes are grown but they also encounter many problems especially during establishment.

A large percentage of the beef cow-calf population is located in the southern United States (Hoveland, 1985). Yet, legumes are not generally an important component in pastures and most producers rely on N-fertilized grass. The instability and undependability of legumes is an important reason for the reliance upon grass-N pastures in this region, especially with warm season grasses (Burton, 1976). The lack of legume persistence is illustrated in small-plot studies over 3 yr at three Georgia locations (Table 1). Although all legumes had excellent stands initially, at the end of 3 yr the stands of ladino and red clovers were nearly eliminated. Alfalfa and birdsfoot trefoil persistence was better but stands were reduced even under a monthly-harvested regime which is less stressful than grazing.

The unreliability of ladino clover and birdsfoot trefoil with tall fescue (*Festuca arundinacea* Schreb.) as compared to grass-N under drought conditions is further illustrated by beef steer performance on continuously grazed pastures during spring in northwest Georgia (Table 2). The first year had adequate rainfall, the second year extreme winter and spring drought which resulted in little growth, followed by below average spring rainfall the next 2 yr. Average daily gains of steers on grass-legume systems were higher than on the tall fescue-N pastures. However, stocking rates on the legume systems were much lower, resulting in substantially lower gain/ha than on the grass-N. This is a result of a reduction in legume stands and reduced N_2 fixation. Thus, overall productivity of the grass-N pastures was more consistent than the grass-legume pastures over the 4-yr period.

Table 1. Legume stand at the end of 3 yr in mixtures with AU Triumph tall fescue when harvested monthly at three locations in Georgia, USA (Hoveland et al., 1986).

Legume	Mountains, North Georgia	Piedmont, North Georgia	Piedmont, Central Georgia
	% stand		
Apollo alfalfa	75	41	65
Arcadia ladino clover	20	5	16
Florie red clover	4	0	0
Fergus birdsfoot trefoil	89	90	22

Table 2. Beef steer performance during spring on pastures of endophyte-free AU Triumph tall fescue with N fertilizer or with legumes in northwest Georgia, USA, over 4 yr (C. S. Hoveland, 1988, unpublished data).

Animal performance	Year	Tall fescue + 134 kg N/ha	Tall fescue + Regal ladino clover	Tall fescue + Fergus birds-foot trefoil
Average daily gain, kg	1985	0.78	1.06	0.96
	1986	1.12	1.25	1.25
	1987	0.76	0.88	0.94
	1988	1.04	1.18	1.13
Gain, kg/ha	1985	512	608	488
	1986	134	146	128
	1987	437	358	346
	1988	372	264	277

Legume pastures received no N fertilizer.
Days grazed: 141 in 1985, 44 in 1986, 111 in 1987, and 84 in 1988.
Approximately 2 T/ha^{-1} of hay was harvested annually in summer from each system. In addition, pastures were grazed in autumn during 1 yr.

Causes of legume stand losses in pastures differ in the different stress zones. In the northern United States, legume stand losses are commonly a result of winterkilling, excess water in soils, and pathogens (Marten, 1985). Westward in the semiarid and arid zones, drought limits the persistence of many legume species. In contrast to the northern United States where forages are utilized mainly for stored feed, pastures are utilized over most of the year in the southern region. In this stressful environment, insects, diseases, nematodes, drought, and competition from warm season perennial grasses reduce establishment and stand persistence of most annual and perennial cool season legumes (Burns, 1985). Imposition of grazing animals over much of the year adds further stress on legumes already weakened by pests and drought. Thus, livestock producers in the southern United States generally choose N-fertilized grass pastures because of greater pasture dependability than provided by legumes.

ANIMAL EFFECTS ON LEGUMES IN PASTURE

Although other environmental factors are generally most important in legume persistence, grazing animals place additional stress on the legume component of pastures. Frequent defoliation is the most obvious effect of grazing animals on legumes in pastures but there are several other factors as well. Legumes more tolerant of heavy defoliation include those with prostrate growth form such as white clover and subterranean clover as compared to erect types such as red clover or alfalfa (Evans, 1982).

SELECTIVE GRAZING

In pastures, animals select some species rather than others and some plant parts in preference to others. Grazing animals tend to select cool season legumes over grasses, even when they are similar digestibility (Minson, 1982). However, the selectivity of species by grazing livestock is modified by plant maturity, sward height, density, and leafiness (Hodgson, 1982). Sheep tend to select more broad-leaf or legume herbage than do cattle. Antiquality constituents in certain legumes such as blue lupine (*Lupinus augustifolius* L.), sweet clover (*Melilotus alba* Desr.), sericea lespedeza (*Lespedeza cuneata* G. Don), and cicer milkvetch (*Astragalus cicer* L.) may cause them to be less palatable than grasses or weeds to livestock (Marten et al., 1987). Thus, legumes which are less palatable will be selected to a lesser extent and be more persistent in a mixed sward. Antiquality constituents may also contribute to pest resistance.

TREADING

Treading by livestock damages pastures, especially at higher stocking rates, fine-textured soils, and during wet weather (Edmond, 1966). White clover is damaged more than ryegrass (*Lolium* spp.). Red clover is much less tolerant of treading than red clover. The indirect effects of treading may be more important for legume persistence and productivity. On a clay soil in north Georgia, treading rotationally-grazed alfalfa pasture at nine steers ha^{-1} over 4 months increased soil bulk density and reduced water infiltration rate but had little effect on plant and shoot density (Tollner et al., 1983). The long-term effects of treading on persistence of pasture legumes in stressful environments of the southern United States has received little study so its importance is not known.

EXCRETION

Excretion of dung and urine causes herbage in pasture to be temporarily unpalatable. In the case of urine, the effect is short-lived but grass growth in urine patches is stimulated which may depress clover growth (Watkin & Clements, 1978). In continuously grazed pastures herbage near dung patches may be rejected by grazing animals for several months, resulting in the ungrazed grass becoming mature and unpalatable. Clover growth again may be adversely affected by shading. Since dung and urine patches cover only a small part of a pasture, the overall effect on legume persistence is probably small.

SOME IMPROVED PERENNIAL LEGUMES FOR STRESS ENVIRONMENTS

There is a need for improved perennial legumes in high-stress environments such as the southeastern United States. At the last Trilateral Workshop held in New Zealand, Burns (1985) stated that alfalfa, lespedeza, and birdsfoot trefoil were receiving little or no breeding effort in the southeastern United States. This was incorrect as there has been a full-time sericea lespedeza breeder in Alabama for over 30 yr from which a number of cultivars have been released with improved quality and persistence such as Serala, Serala 76, Interstate, Interstate 76, AU Lotan, and AU Donnelly (Hoveland & Donnelly, 1985). A part-time birdsfoot trefoil breeding program in Alabama has released the cv. AU Dewey (Pedersen et al., 1986) which is much more tolerant of subsoil acidity than other cultivars (Alison & Hoveland, 1986). Alfalfa breeding programs using recurrent selection in Florida have released a persistent hay-type cultivar (Horner & Ruelke, 1980) and in Georgia have developed a persistent grazing-type cultivar by selecting from nurseries subjected to continuous grazing over 5 yr (Smith et al., 1988). Breeding programs on white clover are active in Florida and Mississippi while annual clover improvement is in progress in Mississippi and east Texas.

The two perennial legumes with the greatest potential in the stressful environment of the southeastern United States are sericea lespedeza and alfalfa. Each one fits into a different ecological niche; sericea being tolerant of very acid soils (Joost & Hoveland, 1986) and alfalfa requiring higher soil pH and fertility. Both species have been mainly used for hay but can be utilized for pasture. Both species can be damaged by continuous hard grazing because of their erect growth habit so require special management.

Sun-cured high-tannin sericea hay was substantially improved in digestibility and intake as compared to fresh-frozen sericea in feeding trials with sheep (Terrill et al., 1983). When rotationally grazed (1 wk grazing, 2 wk rest) with beef steers from 22 April to 8 September over a 3-yr period, average daily gains were higher on low-tannin AU Lotan than on high-tannin Serala sericea (Table 3.)

Stand persistence of sericea was good in spite of the frequent rotation system used and less affected by grazing than alfalfa, probably a result of keeping 10 to 15 cm of stubble in the pastures. Sericea lespedeza offers potential as a low-cost, drought-tolerant, persistent perennial legume for either hay or pasture on acid, low-fertility soils. New low-tannin cultivars just released or soon to be released are more vigorous and higher yielding than the AU Lotan cultivar. Sericea can be grown alone or in association with a cool season annual or perennial grass (Hoveland & Donnelly, 1985).

Management recommendations for pasturing hay-type alfalfa cultivars are rotational grazing with a 7 to 10 d grazing period and a 30 to 40 d recovery period (Van Keuren & Matches, 1988). With a rotational system of 1 wk grazing and 2 wk rest, alfalfa stand persistence was fairly good and steer average daily gains of 1 kg were obtained from 30 March to 8 September over a 3-yr period (Table 3).

Under continuous grazing, hay-type alfalfa stands deteriorate rapidly, reducing alfalfa in a mixed sward (Table 4). However, the development in Georgia of a new grazing-type alfalfa offers more flexibility as yields are similar to hay types under hay management while under continuous grazing stands are much

Table 3. Beef steer performance on alfalfa and sericea lespedeza pastures in northwest Alabama, USA, 3-yr mean. (Schmidt et al., 1987).

	Cimarron alfalfa, rotational grazing	Serala sericea (high-tannin), rotational grazing	AU Lotan sericea (low-tannin)	
			Rotational grazing	Continuous grazing
Grazing season, days	163	139	139	139
Stocking rate, steers/ha	3.2	3.2	3.0	3.0
Steer average daily gain, kg	0.98	0.63	0.75	0.85
Legume stand loss at end of 3 yr, %	34	24	4	24

Table 4. Legume composition of AU Triumph tall fescue-Cimarron alfalfa pastures as affected by rotational grazing (grazed 1 wk and rested for different intervals) with beef steers in central Georgia, USA.

Weeks of rest between grazing periods	Alfalfa composition of sward		
	11 Feb.'87	21 Sept.'87	23 Feb.'88
	%		
0 (continuous grazing)	25	3	4
1	36	15	10
2	29	14	12
3	26	26	27

more persistent. When small plots of the experimental grazing type GA-GC were grazed continuously with beef steers from May to September over 2 yr, stands were more persistent and herbage yields higher than the hay type or other less-productive grazing-type cultivars (Table 5).

Improved pasture legume persistence in stressful environments depends on strong plant breeding programs in those regions. Unfortunately in the United States, commercial forage plant breeding programs are generally centered in favorable environmental regions so little improvement in tolerance to stress can be expected. Greater plant breeding efforts by university and U.S. Dep. of Agriculture scientists are needed in regions of environmental stress to produce legume cultivars that are more persistent in pastures.

Table 5. Plant counts, stand loss, and 2-yr dry herbage yield (from movable cages) of alfalfa cultivars continuously grazed by beef steers from May to September, Central Georgia, USA (Smith et al., 1987).

Cultivar	Plant stand			Dry herbage yield, 1987
	April 1986	October 1987	Loss	
	plants/m^2		%	kg/ha
Grazing types				
GA-GC	124	71	40	7830
Travois	155	67	52	6190
Spredor II	168	40	76	5850
Hay types				
Florida 77	132	15	89	3230
Apollo	97	6	95	3860

REFERENCES

Alison, M. W. Jr., and C. S. Hoveland. 1986. Root development of birdsfoot trefoil cultivars as affected by soil pH and aluminum. p. 106. In Agronomy abstracts. ASA, Madison, WI.

Burns, J. C. 1985. Environmental and management limitations of legume-based forage systems in the southern United States. p. 129-137. In R. F. Barnes et al. (ed.) Forage legumes for energy-efficient animal production. Proc. Trilateral Workshop, Palmerston North, NZ., 30 Apr.-4 May 1984. USDA, Washington, DC.

Burns, J. C., and J. E. Standaert. 1985. Productivity and economics of legume-based vs nitrogen-fertilized grass-based pasture in the United States. p. 56-71. In R. F Barnes et al. (ed.) Forage legumes for energy-efficient animal production. Proc. Trilateral Workshop, Palmerston North, NZ., 30 Apr.-4 May 1984. USDA, Washington, DC.

Burton, G. W. 1976. Legume nitrogen versus fertilizer nitrogen for warm-season grasses. p. 55-72. In C. S. Hoveland (ed.) Biological N fixation in forage-livestock systems. ASA, Madison, WI.

Edmond, D. B. 1966. The influence of animal treading on pasture growth. p. 453-458. In Proc. 10th Int. Grassl. Congr. Helsinki, Finland. 7-16 July. Finnish Grassland Association, Helsinki, Finland.

Evans, T. R. 1982. Overcoming nutritional limitations through pasture management. p. 343-361. In J. B. Hacker (ed.) Nutritional limits to animal production from pastures. Commonwealth Agricultural Bureaux, Farnham Royal, United Kingdom.

Hodgson, J. 1982. Influence of sward characteristics on diet selection and herbage intake by the grazing animal. p. 153-166. In J. B. Hacker (ed.) Nutritional limits to animal production from pastures. Commonwealth Agricultural Bureaux, Farnham Royal, United Kingdom.

Horner, E. S., and O. C. Ruelke. 1980. Florida 77 alfalfa and recommended management practices for its production. Florida Agric. Exp. Stn. Cir. S-271.

Hoveland, C. S. 1986. Beef-forage systems for the southeastern United States. J. Anim. Sci. 63:978-985.

Hoveland, C. S., M. W. Alison, Jr., G. V. Calvert, J. F. Newsome, J. W. Dobson, Jr., and C. S. Fisher. 1986. Perennial cool-season grass-legume mixtures in north Georgia. Georgia Agric. Exp. Stn. Res. Bull. 339.

Hoveland, C. S., and E. D. Donnelly. 1985. The lespedezas. p. 128-135. In M. E. Heath et al. (ed.) Forages, the science of grassland agriculture. Iowa State Univ. Press, Ames.

Joost, R. E., and C. S. Hoveland. 1986. Root development of sericea lespedeza and alfalfa in acid soils. Agron. J. 78:711-714.

Marten, G. C. 1985. Environmental and management limitations of legume-based forage systems in the northern United States. p. 116-128. In R. F. Barnes et al. (ed.) Forage legumes for energy-efficient animal production. Proc. Trilateral Workshop, Palmerston North, NZ., 30 Apr.-4 May 1984. USDA, Washington, DC.

Marten, G. C. 1985. Nutritional value of the legume in temperate pastures of the United States. p. 204-212. In R. F Barnes et al. (ed.) Forage legumes for energy-efficient animal production. Proc. Trilateral Workshop, Palmerston North, NZ., 30 Apr.-4 May 1984. USDA, Washington, DC.

Marten, G. C., F. R. Ehle, and E. A. Ristau. 1987. Performance and photosensitization of cattle related to forage quality of four legumes. Crop Sci. 27:138-145.

Minson, D. J. 1982. Effects of chemical and physical composition of herbage eaten upon intake. p. 167-182. In J. B. Hacker (ed.) Nutritional limits to animal production from pastures. Commonwealth Agrcultural Bureaux, Farnham Royal, United Kingdom.

Pedersen, J. F., R. L. Haaland, and C. S. Hoveland. 1986. Registration of AU Dewey birdsfoot trefoil. Crop Sci. 26:1081.

Schmidt, S. P., C. S. Hoveland, E. D. Donnelly, J. A. McGuire, and R. A. Moore. 1987. Beef steer performance on Cimarron alfalfa and Serala and AU Lotan sericea lespedeza pastures. Alabama Agric. Exp. Stn. Cir. 288.

Smith, S. R., Jr., J. H. Bouton, and C. S. Hoveland. 1987. Persistence and forage and seed yield of an alfalfa germplasm recurrently selected for survival under continuous grazing. p. 80. In Agronomy abstracts. ASA, Madison, WI.

Terrill, T. H., W. R. Windham, C. S. Hoveland, and H. E. Amos. 1987. Influence of forage preservation method on tannin concentration, intake, and digestibility of sericea lespedeza by sheep. p. 146. In Agronomy abstracts. ASA, Madison, WI.

Tollner, E. W., S. L. Fales, C. S. Hoveland, and G. V. Calvert. 1983. Determining the effects of natural and traffic-induced soil compaction. Proc. Am. Soc. Agric. Eng., St. Joseph, MI. Paper No. 83-1542

Van Keuren, R. W., and C. S. Hoveland. 1985. Clover management and utilization. In N. L. Taylor (ed.) Clover science and technology. Agronomy 25:325-354.

Van Keuren, R. W., and A. G. Matches. 1988. Pasture production and utilization. In A. A. Hanson et al. (ed.) Alfalfa and alfalfa improvement. Agronomy 29:515-538.

Watkin, B. R., and R. J. Clements. 1978. The effects of grazing animals on pastures. p. 273-290. In J. R. Wilson (ed.) Plant relations in pastures. CSIRO, Melbourne.

DISCUSSION

Buxton: What is the potential for developing alfalfa cultivars for acid soils?

Hoveland: The potential is somewhat limited; however, it should be possible to develop cultivars adapted to soil pH of 5.8.

Forde: I was interested to hear of increased palatability of low-tannin forms of sericea lespedeza. Many high-tannin forages like big trefoil (Lotus pendunculatus Cav.) and sulla (Hedysarum coronarium L.) are highly palatable.

Hoveland: The levels of tannin in high-tannin sericea lespedeza cultivars are many times higher than in lotus species and do reduce palatability. The low-tannin sericea lespedeza cultivars have levels of tannin as high as birdsfoot trefoil and are sufficient to prevent bloat.

Lowther:. I agree that L. pedunculatus Cav. (big trefoil) has low levels of condensed tannins (3-5%) under high fertility. However, under stress conditions of low soil water, low fertility, cold or acidity, plant growth may be reduced and tannin levels apportionately increased (10-15%) and may cause problems with palatability.

Kretschmer: Do you know what the new technique is for tannin analysis?

R. Smith: The newer techniques for measurement of tannin content are based on a protein precipitate bioassay using ethanol extracts. There is current evidence to suggest that tannins in the rumen help reduce bypass protein.

Hoveland: In our tannin research at the University of Georgia we have found that a number of changes should be made in the analysis to correctly measure tannin levels. Oven drying of sericea lespedeza herbage mincreases polymerization of tannins or complexation with other cellular constituents, cutting tannin levels to one-half or more. Freeze drying is the most efficient method of forage preservation for tannin analysis. Extracting with absolute methanol in the vanillin-HCl assay does not extract all the tannin. A 7:3 acetone:water solution has extracted the most tanin in our studies. Also, the use of catechin equivalents doubled the estimation of tannin content compared to the purified tannin assay. One paper on drying effects has been accepted for publication in Crop Science, another on the methods of analysis has been submitted to the same journal.

Rotar: Will commercial companies work in these areas when they may not be able to sell much seed?

Hoveland: One approach is for the cultivars from publicly funded experiment stations to be released exclusively for seed production and marketing by a single commercial company. This has been done with a number of forage cultivars such as AU Triumph tall fescue, Oasis phalaris, and Regal ladino clover.

Jones: You mentioned public institutions having to take the lead in plant breeding and also the success of an introduced line of birdsfoot trefoil from Yugoslavia. What do you see to be the relative role of plant introduction and plant breeding?

Hoveland: Most pasture legume cultivars in the USA, alfalfa and red clover excluded, are plant introductions that have not been modified or have been released after several cycles of recurrent selection. We would expect breeding to play a much more important role in the future.

Marten: What would be the effect of genetic incorporation of a rhizomatous growth habit on the persistence of birdsfoot trefoil in the southern USA?

Hoveland: The effect should be substantial and positive. However, incorporation of disease resistance is equally important in the southern USA.

Sheath: If you move to a rhizomatous plant, be careful that you are not being disadvantaged by slow regrowth. This can be a competitive disadvantage in mixed pasture communities.

A CASE STUDY OF WHITE CLOVER/RYEGRASS INTRODUCTIONS INTO KIKUYUGRASS ON A COMMERCIAL CATTLE RANCH IN HAWAII

Burt Smith

Kahua/Ponoholo Ranch is located in the Kohala Mountains on the Big Island, Hawaii. The ranch is composed of some 8100 ha and runs approximately 5000 head of cows (Bos taurus). The ranch raises its own replacement heifers and bulls through a combination of artificial insemination and embryo transfer programs. Stockers are retained until they reach 295 kg, at which point they are shipped to Oahu for fattening and eventual processing and marketing at a ranch-controlled facility. In addition, the ranch grows out, and breeds, Holstein heifers for an Oahu dairy. The yearly average cattle head count is around 7500.

The ranch also runs 1800 ewes (Ovis aries). The lambs are fattened on kikuyugrass/ryegrass/white clover (Pennisetum clandestinum Hochst. ex. Chiov./Lolium spp./Trifolium repens L.) pastures, with a daily supplement of 36 g of grain; average daily gain is 208 g. The finished lambs are sold to local markets, restaurants, and hotels. The wool is shipped to the mainland USA in high-density bales. There are about 70 head of horses (Equus caballus) maintained on the ranch which includes a breeding herd in addition to work horses. The ranch has diversified into wind farming (some 280 windmills), a quarter hectare of greenhouse carnations (Dianthus caryophyllus L.), a trucking company, and several tourist enterprises.

Until 1982, the grazing method employed by Kahua/Ponoholo, as well as most ranches of the state, was a form of "set stocking" practiced with large, up to 1600-ha pastures. Grazing management consisted of a combination of continuous and seasonal grazing with long rest periods coupled to long grazing periods (Smith et al., 1983). Performance of stockers ranged from less than 0.2 to 0.45 kg/day. "Weaner slump," the loss of 20 to 30 kg upon weaning was accepted as normal. The state average percent weaned calf crop, from 1977 through 1982, was 66.8% (Hawaii Dep. of Agric., 1983).

In 1982, the ranch adopted Intensive Grazing Management (IGM) and embarked on a program to bring all of its property under control; this was effectively accomplished in 1987. Presently the ranch is running more than twice the animals they were in 1982. In addition, the individual performance has significantly increased. A 1984 study on an IGM steer unit reported that average daily gains increased from 2.5 to 4.4 animals per hectare, this resulted in an annual per hectare production increase from 223 to 646 kg. This resulted in over a fourfold increase (based on value added) in net returns to capital and management over pre-IGM practices, which allowed the cost of development to be paid back in just 13 months (Leung & Smith, 1984). The management of Kahua/Ponoholo Ranch has stated that the overall ranch cost of production, per kilogram of liveweight calf, has dropped from $1.32 in 1982 to $.79 in 1987.

The Ranch operates in five distinct environments, or life zones (Holdridge Life Zone Classification scheme). These vary from tropical scrub thorn to pre-montane and montane wet, with elevations from sea level to 1310 m. The average annual rainfall varies from 125 to 1525 mm at sea level, to over 3050 mm in the mountains. The area of ryegrass/white clover introductions lies at an elevation between 915 to 1160 m, with annual rainfalls of 890 to 1145 mm, respectively. Temperatures at this elevation average 15 C in the winter and 21 C in the summer. Soils are derived from volcanic ash and are relatively fertile with pH's in the 6.0 to 6.2 range. Kikuyugrass makes up about 85% of the vegetative cover at this elevation.

Kikuyugrass is a rest climax dominate that rapidly builds up a thick mat (often in excess of 60 cm) under conditions of light to zero grazing pressure. As grazing pressure increases, the thickness of the mat decreases, reaching a level of around 15 cm at a yearlong pressure of one animal unit per hectare. Increasing the stocking rate further, under a continuous grazing regimen, has been found to reduce animal performance as well as per hectare production. However, trials have demonstrated that three to four applications of 1-d grazing, using mobs on the order of 197 mature animals per hectare, with a 21 to 28 d rest interval, will virtually eliminate the mat and result in the kikuyugrass assuming a more leafy growth pattern with a consequent increase in both quality and quantity. Further reduction of animal density has been found to increase the length of time for acceptable mat reduction to occur. Below 37 animals per hectare, there appears to be little long-term effect on mat residuals (B. Smith, 1986, unpublished data).

Kahua/Ponoholo Ranch has made several attempts in the past to introduce legumes, all of which experienced various degrees of failure. This was due either to inadequate establishment procedures, and/or lack of adequate animal control required for maintenance. Numerous attempts with mechanical preparation were tried, some with large off-set disk plows (0.9-m disks). However, all required many passes and left the area exceedingly lumpy, which posed a hazard to riders as well as vehicles. Poor seedling survival was also noted, but at the time it was believed due to the quality of the seedbed. However, in hindsight, slugs would seem to have been the culprits.

Commercial vegetable farmers in the Kamuela area have long used herbicides, coupled with mechanical preparation, to provide a suitable seedbed. However, the cost of clearing requires a high value crop in order to obtain a cash flow in the desired direction. A University of Hawaii researcher direct seeded legumes into kikuyugrass treated with different rates of Dalpon and found the process to be effective (S. Whitney, 1984, personal communication), but again the economics were questionable.

The economics of beef production in Hawaii are considerably more stringent than on the mainland USA. Virtually all material inputs are produced elsewhere and must be shipped in at a cost that is often 25 to 50% higher. In addition, the price paid to Hawaiian ranchers for their animals are dictated by the mainland price less the cost of getting them there. The end result are prices that are one-fourth to one-third less than what mainland ranchers receive for comparable animals (Garrod et al., 1987).

Two events were necessary to make legume introduction feasible on ranches in Hawaii. The first was to obtain the necessary control. Ranch management practices had to be upgraded so that paddocks could be grazed, at proper intensities, to the desired residual levels necessary to maintain an equable balance between grass and legume. Tight control is especially important when dealing with the rapid growth of C4 grasses. This level of control, prior to the introduction of IGM, was lacking.

Controlled grazing concepts were introduced to Hawaii in 1980 with clipping studies and mob grazing trials at Kahua/Ponoholo and elsewhere. These were followed up with workshops featuring Mr. Alan Savory, a private consultant from the mainland USA, Mr. Vaughn Jones, a representative of Gallagher Electronics of New Zealand, and Mr. Keith Milligan from the Ministry of Agriculture and Forestry, New Zealand. Numerous intensive grazing courses were taught throughout the state and in 1986 the first book in four decades on intensive grazing was made available to ranchers (Smith et al., 1986). The net effect of all this is that progressive ranchers in the state are now in a position to affect the pasture management required to maintain a legume/grass sward.

The second condition was the introduction of a suitable , low-cost method to introduce legume seeds into the heavy mat and sod of kikuyugrass. In theory, once control has been obtained, introduction of seeds can be accomplished through hard grazing (whipping it), followed by seed dispersal, than trampling in the seed through concentrated hoof action. A trial was set up using Haifa white clover and kikuyugrass; the results were disappointing.

Four major concerns emerged: The first was getting the kikuyugrass "whipped" down to the desired level. Cattle were the only animals available and while it had been demonstrated that several applications of non-lactating pregnant animals could be used to reduce the mat, without undue physical stress on the animal, it remained to be seen if forcing them to remove that "last centimeter" was economically feasible. After the first trial it was felt that it was not. The second concern stemmed from the first: While a grazing program could easily be designed to reduce the kikuyugrass to the desired level without over stressing the animals, such a program resulted in relatively few hectares being ready for winter seeding. The ranch was anticipating conversion of hundreds of hectares to the kikuyugrass/ryegrass/white clover mix. The third concern was getting the seed tramped into the ground. Since most of the forage had been removed, this meant herding the animals around the area, which in turn meant more animal stress and the consequent lowering of performance. The fourth concern, poor seedling survival, was at the time thought to be due to improper trampling action. It was later found to be due, at least in part, to slugs consuming the new sprouts.

The concerns raised by the trial prompted the ranch to purchase a large "Roto-Tiller" from a New Zealand firm in 1983. A trial was initiated that winter with 2 ha of kikuyugrass roto-tilled and planted with different combinations of white clovers ('Mt. Baker', 'Ladino', 'Haifa', 'Maku' trefoil, and 'New Zealand White') and annual ryegrass. A fertilization trial was superimposed on this planting (179 kg/ha of triple 16 (NPK), 4.5 t/ha of ground coral (lime), Mo, and a trace mineral mix). Of the clovers, Haifa appeared to be the best, with Ladino showing the poorest stand; Maku trefoil (Lotus corniculatus L.) failed to establish. Fertilization made little difference in either yield or stand.

In February and March 1984, a grazing unit of 13 ha, divided into 16 paddocks was planted with 11.3 kg of ryegrass (1/3 annual, 2/3 'Elliott' perennial) plus 2.7 kg of Haifa clover per hectare. The area was tilled three times prior to seeding. Seedling establishment of both clover and ryegrass was fair to poor. The suspected cause was the extreme fluffiness of the seedbed; later corrected by the incorporation of a cultipacker into the seeding process.

Grazing trials compared animal performance on this unit with one that was fertilized at 45 kg(N)/ha; both groups of animals were given molasses at the rate of 1.81 kg per head per day. The average daily gains were 1.12 kg/d for both groups. An "in-ranch" economic analysis indicated that the clover would have to last slightly over 2 yr to equal the returns from N fertilization. After 4 yr the ryegrass has almost disappeared from this area; however, there is still significant amounts of Haifa clover present.

To date, the ranch has converted 339 ha of kikuyugrass pastures into the ryegrass/white clover mix, at a cost of $442 per hectare, plus $52 for 68 kg/ha of 6-20-20. The methodology of establishment has undergone significant changes since 1983 and more are expected. Presently, the procedure is to first apply a hard grazing, followed by a rotary mower, then, in a single pass, the roto-tiller, with seeder and cultipack attached. Insect damage was suspected early and was addressed with the use of Diazinon; however this gave variable results. In early 1987, a visiting scientist from New Zealand, Dr. Deric Charlton, identified slugs as being the major pest. Since that time, a slug-bait has been used with excellent results. Certain problems have arisen with the above approach, the most note-worthy being poor survival of the ryegrass seedlings. It is surmised that the high levels of organic material being incorporated into the soil are causing N to be immobilized and unavailable to the seedlings. Two approaches are being considered: (i) Increase the amount of N fertilizer incorporated at time of planting. (ii) Till, then allow time for mineralization to occur, a month or so, depending on soil temperature, followed by another tilling or seeding.

Management of the kikuyugrass/ryegrass/white clover units has been to keep the area grazed close, with a fast cycle, 12 to 28 d, depending on the season. Dry cows are used once a year for pasture maintenance. This strategy has resulted in a gradual loss of the ryegrass but appears to be maintaining a significant clover component. At this time, it is unknown whether the ryegrass is disappearing due to the grazing treatment and/or environmental conditions. It is known, from both clipping and grazing trials conducted at this ranch and other locations, that white clover, in a kikuyugrass stand, can be increased through grazing management to where it makes over 30%, by weight, of available forage. This is accomplished by holding the level of the kikuyugrass mat at no higher than 2 to 3 cm. Techniques have also been demonstrated for increasing or decreasing other species of grasses and legumes. It is suspected that periodic applications of hard grazing (whipping it) is required to maintain a balance between the ryegrass, clover, and kikuyugrass. It is also suspected sheep, rather than cattle may be the animals of choice.

The 1987 annual production average of all kikuyugrass/ryegrass/white clover IGM units on the ranch was 2.04 kg/ha, with an average daily gain of 0.6 kg at a stocking rate of 4.3 animals per hectare. Production from similar areas, but with kikuyugrass only was 1.12 kg/ha. At today's beef prices ($1.10/kg) the ranch estimates the pay back, for the costs of seeding ryegrass and white clover into kikuyugrass, to be slightly over 10 months.

The maximum amount of energy that can be used to successfully pulse a system in an effort to obtain high magnification factors is dependent upon the magnitude and periodicity of the naturally occurring flows into that system (Smith et al., 1986). Thus, land that is naturally endowed with good growing conditions, high fertility, and adequate rainfall will always have a greater response to an "intelligent" application of outside energy than will marginal land. The use of "Heavy Metal" (large, high cost, pieces of heavy equipment that are capable of moving mountains, but usually spend most of their life just sitting around, depreciating) on a livestock operation is no more "bad" than is the use of fire, water, fertilization, etc. What distinguishes between "good" and "bad" is the how, when, where, and why of use.

Thus, their decision to go with Heavy Metal, rather than the slower, but initially cheaper, method using livestock to introduce legumes. With paybacks on the order of 1 yr, time becomes money. On other areas, where because of topography, climate, etc. the economic payback is measured in years rather than in months, the ranch has opted for the slower animal impact program to use such marginal areas and reclaim abused sites.

Kahua/Ponoholo has a long and rich tradition in Hawaii. The ranch is committed towards maintaining a viable livestock enterprise not only for itself, but also a healthy Statewide industry. As owner/operators they are quite cognizant of the long-term dangers involved with the "Harvard Business School" approach of maximizing for short-term profits. Conversely, being located on some of the most expensive agricultural land in the nation ($25,000/ha), they are forced to view profits, and what it takes to obtain them, a little more critically than many large, traditional cattle ranches are prone to do. As a result, management practices are always subject to review, and in fact have undergone significant changes since 1982.

REFERENCES

Garrod, P.L., L. Cox, C. Ingraham, S. Nakamoto, and J. Halloran. 1987. The Hawaii beef industry: Situation and outlook. Info. Text. Ser. 029 CTAHR, Univ. of Hawaii.

Hawaii Department of Agriculture. 1983. Statistics of Hawaii. Hawaii Dep. of Agric. Honolulu, HI.

Leung, P.S., and B. Smith. 1984. Economics of intensive grazing: A case in Hawaii. Res. Ext. Ser. 045, CTAHR, Univ. of Hawaii.

Smith, B., P.S. Leung, and G. Love. 1986. Intensive grazing management: Forage, animals, men, profits. The Graziers Hui, Box 1944, Kamuela, HI.

Smith, B., G. Love, and E. Spence. 1983. Livestock in the land of Aloha. Rangelands 5(3):99-103.

DISCUSSION

Kretschmer: Do you use fire to renovate kikuyugrass pastures?

B. Smith: No, the only time kikuyugrass is sufficiently dry to carry fire is during drought, at which time the danger of wildfire is very high.

Kretschmer: Are high inputs of fertilizer economical?

B. Smith: Yes, but only if the level of management is high. On two intensive grazing units, a fertilizer trial was made on kikuyugrass after using 22.3 kg/ha N in the form of ammonium sulfate. At the end of the 90-d trial we found not only an increase in average daily gain but in stocking rate as well. An economic analysis indicated a $2 return for every $1 invested in fertilizer.

Matches: Do they have a problem in maintaining clover at Kahua/Ponoholo Ranch?

B. Smith: We can maintain Haifa white clover by adjusting grazing pressure. However, problems occur in maintaining ryegrass in the stands. I tend to believe that the answer is increasing grazing pressure using sheep rather than cattle.

Brougham: The management effects described for kikuyugrass/ryegrass/white clover associations are very similar to those that occur in the North Island on New Zealand. However, under frequent rotational grazing systems, Rumball maintained excellent kikuyugrass/ryegrass/white clover pastures with very high production potential.

Hoveland: Is it possible that the poor growth of ryegrass in kikuyugrass and white clover pastures is the result of the extensive root system of the kikuyugrass being more competitive for the biologically fixed N from the clover?

B. Smith: It is possible that if you could put sufficient stress on kikuyugrass one might negate this advantage, if in fact it exists.

Woomer: On Maui, the ecological and productive amplitudes of kikuyugrass are considerably wider than that of white clover. Is this the case on the Island of Hawaii and do you feel that other legumes have potential in areas where white clover does not persist?

B. Smith: Yes, the same conditions occur on the Island of Hawaii, and no I am not aware of legumes that are of potential commercial value.

Jones: As a follow up to Burt Smith's response , presently there are no promising legumes other than Desmodium uncinatum DC (silverleaf desmodium), which is not a particularly desirable legume because of its high tannin content (10%).

Hochman: What other legumes can you grow with kikuyugrass?

Jones: Vigna parkeri Bak. and Kenya white clover could grow in some areas of your kikuyugrass country where white clover does not grow. However, they have their own temperature and rainfall requirements.

B. Smith: We tried Kenya white clover but did not have much success. However, this trial was done before we knew about the damage slugs could do to clover seedlings. We have not tried Vigna parkeri.

Jones: Without claiming that continuous grazing is the best way, experience in subtropical Australia shows that given adequate stock numbers it is possible to maintain good clover in continuously grazed kikuyugrass pastures.

B. Smith: We have not compared the systems side by side, but based on historical data of both animal and per acre performance, we believe out intensive grazing approach is the best for our situation. Certainly our package has improved animal efficiency.

Woomer: Do you feel that *Desmodium intortum* L. (greenleaf desmodium) can persist in the wetter, warmer extreme of kikuyugrass pastures?

B. Smith: It can persist only if the pasture is grazed for persistence of *Desmodium* and not for the maximum productivity of the pasture.

GENERAL DISCUSSION OF PLANT-ANIMAL INTERFACE

There was some initial discussion on the format of the published proceedings with pleas put by Dr. Brougham for full reference lists, Dr. Marten for an index as a preface, and Dr. Curll for a statement of purpose at the beginning of the proceedings.

J. Hay: Briefly summarized the papers presented in the session, emphasizing that this large topic of the Plant/Animal Interface is really the synthesis of all the separate aspects that have been discussed over the whole workshop. There was no argument from authors that animal grazing had a large impact on forage legume persistence. The following factors were identified as being the most important:

1. Treading (soil moisture interactions).
2. Selective grazing (animal species).
3. Nutrient return.
4. Seed dispersal from feces.
5. Frequency/intensity of grazing.
6. Response/vulnerability/opportunism of legume to above factors at critical stages in yearly growth cycle.

Leath: Can I open the discussion by commenting that it is still not known which does more damage to lucerne crowns, grazing or machinery from hay/silage management systems. Pathogens could be spread with seed in feces. How important is this in the field?

Irwin: Disease methodology on grazed pasture is not adequate.

Sheath: Grazing animals can influence legume persistence through their impact on removal/availability of growth sites. Outside this, grazing influenced legume content by determining opportunity to explore/grow rather than through persistence.

Curll: May I remind Dr. Sheath that although that may be true for New Zealand, he should recognize that there are important differences between New Zealand, Australia, and the USA in the critical importance of grazing management to legume persistence. Certainly for Australia this is the case. Legume persistence may be enhanced by selection and breeding, but appropriate grazing management strategies are critical.

Kretchmer: The morphology/ecology of the legumes has first to be known, to enable understanding of how its persistence is likely to be effected by grazing. For example, there were tropical and subtropical legumes which were able to persist under both lax and intense grazing pressure because of low crown or bud apices. Others could withstand lax but not intensive grazing because their bud apices are higher up the plant.

Sheaffer: It has been clearly demonstrated that frequent or continuous grazing of lucerne reduces persistence through depletion of carbohydrate reserves.

Watson: Biotic factors affecting legume persistence (grazing management, pests, diseases, etc.) are additive of abiotic factors (climate, soils, etc.). The more extreme the latter, especially drought, the more sensitive the system becomes to the farmer, within the adaptive range of any particular legume species. New Zealand has an equable climate relative to Australia and the USA in general.

B. Smith: Although you cannot transplant one system wholesale from one life zone to another, you can apply the basic principles, and design the management approach to fit the environment and the objectives of the operation. Hawaii has 27 different life zones described, and the University has been involved in placing successful intensive grazing units in 21 of them.

Sheaffer: Has any effort been made to model the persistence of forage legumes as affected by the grazing animal?

Hodgson: There is good contact between modelling groups worldwide, and effective management models are available. However, information on plant/animal relations is not good enough to construct models which are adequate at the biological level.

Curll: I agree. Various Australian Institutions are working on developing more sophisticated general models of plant/animal systems for different enterprises and environments, e.g., "Grazeplan" - CSIRO/DPI.

Clements: Researchers in the Australian tropics are rapidly developing the "Beefman" model and its component sub-models, as an aid to whole-farm development and animal production. There are significant knowledge gaps in legume ecology, but these are being researched very actively. Some examples are the work of Mike Hay and Dave Chapman in New Zealand, Ted Carter and his colleagues in South Australia, Reg Rossiter and Graham Taylor in Western Australia, and Dick Jones in Queensland.

Gramshaw: Grazing management modelling in Queensland is largely empirically based on water supply, pasture growth, and animal intake functions. The objective is to assess various combinations of forage options and their impact on animal production and economic returns. Details of selective grazing within a forage option (e.g., legume vs. grass) are not yet accommodated.

M. Hay: More ecological/demographic information was needed before accurate models could be described.

Curll: Seed set cycling and seed dynamics would have to be considered also.

Jones: If we are to make progress in mechanistic modelling, then we have to combine the forms of modelling as outlined by Dr. Fisher, with modelling demographic pathways. In general, there seems to be a trend in both the Australian and New Zealand papers on plant/animal interface to emphasize seasonal trends in grazing pressure, and there is much less conflict on the merits of rotational vs. continuous grazing.

Fisher: Dr. Thornton of CIAT has produced a pasture beef model which simulates the behaviour of beef herds both on savanna and exotic pastures. The relevant papers are referenced in Dr. Fisher's written paper.

J. Hay: Progress in forage legume persistence could be helped by selecting for greater phenotypic plasticity.

Caradus: "Plasticity" is simply a significant genotype x environment interaction. Genotype here is an individual genotype, not a collection of genotypes as you might find in a cultivar. A significant genotype x environment interaction is seen as a change in the phenotype. It is necessary to distinguish clearly between phenotypic plasticity and genetic variability.

R. Smith: We have to be careful in our use of the term plasticity. He believed that we were discussing phenotypic plasticity which can be associated with two forms - morphological change of given genotypes vs. genetic variability in the species. Therefore, we need to make sure how we are using the term plasticity.

Scott: Modelling and plasticity refer between species as well as within species. It is important to define the output system, e.g., sheep vs. cattle.

DISEASES OF PASTURE LEGUMES IN AUSTRALIA

J. A. G. Irwin

SUMMARY

Pasture legumes in Australia are subject to a range of acute and chronic plant diseases which limit persistence and/or productivity. Over the last decade, considerable progress has been made at developing cultivars of lucerne and subterranean clover with resistance to acute diseases caused by *Phytophthora* spp., *Colletotrichum trifolii* and *Kabatiella caulivora*. Work is in progress to develop cultivars of *Stylosanthes* with resistance to anthracnose (*Colletotrichum gloeosporioides*), which provides the major constraint to the full commercial utilization of the genus in northern Australia. Further work is needed to quantify disease losses in pasture legumes. This may require the development of new methodologies which will facilitate our understanding of the role which pathogen complexes, or interacting chronic diseases, play in reducing legume persistence. Emphasis should also be directed towards breeding schemes which provide durable forms of resistance, in particular in predominantly inbreeding species.

INTRODUCTION

Diseases have now been recognized as major constraints to pasture persistence and productivity throughout Australia. This review summarizes research on diseases of pasture legumes, and indicates where future research should be directed.

TEMPERATE PASTURE LEGUMES

Subterranean Clover

Subterranean clover has been used as a pasture legume in southern Australia for over 80 yr, and for most of this period it has been relatively free of serious diseases (Collins, 1987). Since 1970, there have been reports of devastating disease losses, particularly in the wetter regions. The most serious diseases are scorch (*Kabatiella caulivora*) and various fungal root rots. Both diseases have been recently reviewed (Barbetti & Sivasithamparam, 1986; Barbetti *et al.*, 1986).

Clover Scorch

Clover scorch is regarded as the most serious foliar disease of subterranean clover in Australia (Bokor & Chatel, 1973). The disease reduces hay production, yield of grazed pastures and seed production and is thought to contribute to deterioration of subterranean clover based pastures through death of seedlings. There is an interaction between *K. caulivora* and *Cercospora zebrina* on subterranean clover, with the presence of the latter pathogen delaying scorch development on petioles (Barbetti, 1985).

The epidemiology of the disease is relatively well understood, and epidemics are associated with cool moist conditions (Roberts, 1957). With Australian isolates of the fungus, disease development was optimal over the temperature range 8-20°C, and was highest under conditions of 100% RH and continuous leaf wetness for 4 d, and was further enhanced at very low light intensities (11.61 lx) (Helms, 1975a).

Both Hanson (1950) and Helms (1975b) have shown the existence of pathotypes of *K. caulivora* in North America and Australia, respectively. Marked differences in virulence (relative capacity to cause disease) have been reported on cv. Karridale for collections made in Western Australia (Barbetti & Gillespie, 1987).

Anderson *et al.* (1982) investigated the effects of the disease on pasture and sheep production, and the influence of sheep stocking rate on disease incidence over a 3 yr period in Western Australia. In certain treatments, control was achieved by benomyl application. Scorch had no consistent effect on pasture or animal production, except that pasture dry matter on offer in spring was reduced in unsprayed treatments at a low (6 sheep ha^{-1}) stocking rate. Higher stocking rates (10 and 12 sheep ha^{-1}) without benomyl application reduced the inoculum potential and thus reduced scorch incidence to a similar level to that observed in sprayed treatments. As a 0 sheep ha^{-1} stocking rate treatment was not used, it is not possible to obtain a measure of the full yield reducing potential of scorch, yet the results are directly applicable to the commercial situation.

Breeding for resistance to scorch has been practiced in Western Australia since 1972, and scorch resistance has also become a major objective in the Australian National Subterranean Clover Improvement Programme (ANSCIP) (Barbetti & Gillespie, 1987). In highly resistant lines of *T. subterraneum* ssp. *yanninicum*, resistance to scorch was conditioned by a single dominant gene (Beale & Thurling, 1980). Four genes appeared to be conditioning a moderately resistant response in two other lines tested. Genotype x environment interactions with respect to disease resistance were established (Beale & Goodchild, 1980).

The ANSCIP utilizes field screening of segregating lines in the F_3 generation and subsequently F_5 progenies of selected F_4 plants are also tested. Limited glasshouse screening (20-50 lines per year) with a wide range of isolates has recently been introduced for advanced lines under consideration for release (Collins, 1987). Currently, no cultivar is recommended unless it has been evaluated by this assessment procedure, and there has been considerable success with developing scorch resistant cultivars (Barbetti & Gillespie, 1987).

Reasonable levels of field control of scorch were obtained by spraying grazed plants with benomyl at the rate of 200 g ha^{-1} when the first visible symptoms of the disease appeared each season

(Anderson *et al.*, 1982), and in 1 yr, only one application was required to keep scorch incidence below 2%.

Future work should quantify losses due to scorch in a range of environments where subterranean clover is grown, thus providing more clearly defined goals for regional breeding and or selection programs. Given that pathogenic specialization has been observed in *K. caulivora*, it will be necessary to maintain diversity with respect to host resistance mechanisms if bred cultivars are to have durable resistance. Possible strategies that could be implemented for forages have been outlined by Irwin *et al.* (1984a). The establishment of a differential set of subterranean clover genotypes would assist in monitoring pathogen variability, and cooperation between Australian states and overseas countries would be essential before such a program could be successfully implemented.

Root Rots

Root rot diseases of subterranean clover seriously limit productivity and persistence of the legume in all temperate regions of Australia, and the situation has recently been reviewed (Barbetti *et al.*, 1986). Subterranean clover root rot is of complex etiology, and interactions between fungal pathogens are an important part of the syndrome. In the 1970s, fungal species commonly isolated from rotted roots and for which pathogenicity could be established included *Pythium irregulare* (Stovold, 1974), *Fusarium avenaceum* (Burgess *et al.*, 1973) and *Rhizoctonia* spp. (Ludbrock *et al.*, 1953). More recently, the newly described *Phytophthora clandestina* (Taylor *et al.*, 1985a) and *Aphanomyces euteiches* (Greenhalgh & Merriman, 1985) have been implicated as major pathogens of subterranean clover in Victoria, South Australia and New South Wales (Taylor, 1984; Greenhalgh & Flett, 1987) and Western Australia (Taylor *et al.*, 1985b).

The role of *P. irregulare* and *F. avenaceum* in causing subterranean clover root rot in Victoria has been questioned, since both pathogens caused different symptoms in controlled inoculations to those most commonly observed in the field, and they did not interact with *P. clandestina* or *A. euteiches* (Greenhalgh & Taylor, 1985). Losses in yield of subterranean clover attributed to root rot caused by *A. euteiches* and *P. clandestina* have been estimated in irrigated pastures by use of selective fungicides. Results from a single location in northern Victoria showed that root rot diseases reduced dry matter production by 71, 32, and 38% in the autumn, winter and spring of 1985, respectively (Greenhalgh & Flett, 1987). The data also demonstrated that tap root rot caused by *P. clandestina* was responsible for most of the lost production.

In contrast to the situation in Victoria, a range of fungi (*P. clandestina, P. irregulari, R. solani,* and *F. avenaceum*) are thought to have a role in subterranean clover root rot in Western Australia, and it is regarded as likely that *A. euteiches* may also be involved. No one fungus has been able to reproduce the wide range of different field disease symptoms observed in different locations in Western Australia (Barbetti, 1987).

In studies to determine the influence of temperature and soil water holding capacity (WHC) on root rots caused by *F.*

avenaceum, P. irregulare, R. solani, and *F. oxysporum*, rot was most severe where combinations of the fungi were used at temperatures of 10°C, less severe at 15 and 25°C, and least severe at 20°C (Wong *et al.*, 1984). Disease also occurred at WHCs of 45-65%, but was least severe under flooding, and there were often significant interactions between temperature and moisture. In contrast to the above, *P. clandestina* caused the most severe disease levels at 65-100% WHC and flooding, and at temperatures of 10°C (Wong *et al.*, 1986). Caution is needed in interpretation of interactions between pathogens unless natural levels of inoculum have been used. It is not possible to compare inoculum levels used above to those that might be occurring naturally. Further work is needed to quantify natural inoculum levels of the above-mentioned pathogens.

A glasshouse screening program to identify subterranean clover lines resistant to *P. clandestina* has been developed and used in Victoria (Greenhalgh & Flett, 1987). Where a pathogen complex is involved in causing root rot, as reported in Western Australia, control will be more difficult to implement than where a single pathogen is involved. In Western Australia, prior to 1984/85, screening for root rot resistance was conducted in the field only. A glasshouse procedure has now been developed for screening for resistance to *P. irregulare, P. clandestina, F. avenaceum*, and *R. solani* (Gillespie, 1987). Approximately 100-120 lines can be tested per year in this program. Cultivars varied widely in their susceptibility to different pathogens. Trikkala was little affected by *Pythium* and *Phytophthora*, but was highly susceptible to *Rhizoctonia*. Most of the commercial cultivars were susceptible to one or more of the pathogens (Gillespie, 1987).

Generally, fungicides have not been successful in controlling root rots of subterranean clover in Western Australia (Barbetti, 1987). The most success has been achieved with metalaxyl (specific to Oomycetes), but results have been inconsistent, reflecting the complex etiology of the disease in that region. In contrast, promising results have been obtained in Victoria for control of *P. clandestina* with foliar applications of phosphorous acid (partially neutralized formulation). In one experiment, in comparison to the untreated control, phosphorous acid at 0.313 kg a.i. ha^{-1} applied at either 8 or 14 d after first irrigation (when 7 and 52% of seedlings were infected by *P. clandestina*, respectively) reduced root rot severity and increased autumn yield by 103 and 85%, respectively. Winter/spring yield was increased by 36 and 29%, respectively (Greenhalgh, 1988, personal communication).

Virus Diseases

Prior to the mid 1950s, no virus diseases of subterranean clover of any consequence were recognized in Australia (Johnstone & Barbetti, 1987). After that time, subterranean clover stunt developed as an important problem, followed by soybean dwarf virus. These two viruses are now less important, due to the widespread occurrence of a new suite of viruses which are non-persistently transmitted by the recently introduced lucerne and pea aphids. This suite of viruses is also seed-borne and includes alfalfa mosaic virus, clover yellow vein virus, and subterranean clover mottle virus (Johnstone & Barbetti, 1987). Studies have commenced to systematically survey for the incidence of soybean dwarf, subterranean clover mottle, alfalfa mosaic and clover yellow vein viruses using the double sandwich enzyme-linked immuno- sorbent assay (ELISA) (Helms, 1987). When the relative importance of viruses has been determined in various regions, breeding programs can then be implemented, if warranted. There is tremendous variation in the subterranean clovers with respect to their reactions to viruses infecting them and to resistance to their aphid vectors (Johnstone, 1987).

Lucerne

Since the 1960s, the role of plant diseases, in particular root and crown rots, as contributing factors to poor lucerne persistence has been recognized (Purss, 1965; Irwin, 1977). Lucerne is both an outbreeder and an autotetraploid, both of which tend to preserve variation. Because of this, a lucerne cultivar consists of a heterogeneous mixture of plants, and this has important implications for the control of pathogens (Hanson, 1972).

Phytophthora Root Rot

Phytophthora root rot (PRR) is caused by *Phytophthora megasperma* f. sp. *medicaginis* (Pmm) and is regarded as one of the major factors limiting lucerne production on poorly drained soil types in Australia (Irwin, 1974; Rogers *et al.,* 1978).

Phytophthora root rot develops in lucerne when the soil remains excessively wet for a 7-10 d period (Frosheiser & Barnes, 1973). Root infection is initiated by zoospores of the fungus which swim in free water and contact roots. Detailed descriptions of the diseases caused by Pmm have been published by Erwin (1954) and Irwin and Dale (1982).

Disease resistant cultivars of lucerne have provided effective control of Pmm in Australia (Clements *et al.,* 1984). Most cultivars contain at least a small proportion of resistant plants, and rapid progress can be achieved through recurrent select on procedures (Bray & Irwin, 1978). The genetics of resistance to Pmm in lucerne have been studied by Lu *et al.* (1973), Irwin *et al.* (1981a & b), and Havey *et al.* (1987), and at least four different genetic systems have been identified. Although pathogenic races of Pmm have not been reported, individual isolates of Pmm are known to vary widely in virulence and Faris (1985) stressed the need to use a sample of Pmm chosen from within the area of adaptation, in order to screen lucerne genotypes for resistance to PRR.

Colletotrichum Crown Rot (Southern anthracnose)

Anthracnose, caused by *Colletotrichum trifolii* is a serious problem when lucerne is grown under warm humid conditions (Barnes *et al.,* 1969). It is the crown rot phase of the disease which is responsible for the premature thinning of stands in the humid areas of Eastern Australia (Irwin, 1977). Although many aspects of anthracnose have been extensively researched, not a lot of information is available concerning the epidemiology of the disease. In Queensland and New South Wales, general disease surveys and quantitative studies on diseases as factors contributing to poor lucerne persistence have demonstrated the importance of warm humid conditions for the development of stem anthracnose which subsequently leads to the development of crown rot.

During the 1970s, cultivars with resistance to *C. trifolii* were developed in North America (Devine *et al.*, 1971), and Australia (Irwin *et al.*, 1980). Work in Australia has shown that selection for resistance to stem anthracnose also confers crown rot resistance (Irwin *et al.*, 1980) probably because *C. trifolii* is a hemibiotroph. In 1978, a second pathogenic race of *C. trifolii*, with virulence on the cv. Arc and its derivatives, was discovered in North America (Ostazeski & Elgin, 1982).

Although there have been no reports of loss of anthracnose resistance in the Australian-based cultivars Trifecta and Sequel (Stovold & Francis, 1988), careful monitoring of the race spectra of *C. trifolii* in the region will be required. This material was evaluated for resistance to race 2 at Beltsville, and both cultivars contained resistant plants (J. Elgin, 1980, personal communication), thus there is scope for further recurrent selection within both cultivars for race 2 resistance should the need arise.

Acrocalymma Crown Rot

Acrocalymma medicaginis causes a root and crown rot disease of lucerne with symptoms indistinguishable from those caused by *Stagonospora meliloti* (Irwin, 1972, 1977; Alcorn & Irwin, 1987). The fungus has only been reported from Australia, and previously had been misidentified as *Stagonospora meliloti* in Queensland, New South Wales and South Australia (Alcorn & Irwin, 1987). *Stagonospora meliloti* has also been recorded from South Australia and New South Wales but is now thought to be of only minor significance as a root and crown pathogen of lucerne.

In pathogenicity tests under controlled environmental conditions, disease development is slow, with the rot extending only 1-2 cm on either side of a wound inoculation site in 4 months. Under the same experimental conditions and with the same lucerne clone, the rot caused by *S. meliloti* was somewhat less extensive (Alcorn & Irwin, 1987). *Acrocalymma medicaginis* has not been found associated with or isolated from leaf lesions, in contrast to *S. meliloti*, and is non-infective to leaves under conditions of artificial inoculation (Alcorn & Irwin, 1987).

Symptoms typical of those described above are present in the roots and crowns of all lucerne plants over 12 months of age sampled in Queensland and New South Wales (G.E. Stovold, 1988, personal communication). There is also a positive correlation between age of plants and the extent of crown rot. Observations suggest there is a synergistic interaction between the lucerne crown borer (*Zygrita diva*) and *A. medicaginis*, with the damage being more extensive where both agents are involved.

Further research work is warranted to determine: (i) the disease cycle (the teleomorph very closely resembles *Leptosphaeria pratensis*), (ii) the role of the pathogen in causing premature plant death and or loss of productivity, (iii) whether genetic resistance to the pathogen exists. Because of the slow growth of the fungus in lucerne crowns, it could be relatively difficult to achieve the latter two objectives.

Other Diseases

In Queensland, the following root and crown rot diseases have also been recorded: Violet root rot (*Rhizoctonia crocorum*), Fusarium wilt (*Fusarium oxysporum* f. sp. *medicaginis*), Rhizoctonia root and stem canker (*Rhizoctonia solani*) and Sclerotium crown rot (*Sclerotium rolfsii*) (Irwin, 1977). The latter pathogen is widespread, but its incidence within a field is usually very low. The other diseases do cause serious losses, but their occurrence is relatively localized. Because of this, no disease management strategies have been implemented.

Stemphylium vesicarium has recently become a serious foliar pathogen during the cooler weather in Australia (Irwin, 1984). Because of the widespread nature of the disease, two cycles of recurrent selection for *Stemphylium* resistance have been made within the cvs. Trifecta and Sequel. The level of *Stemphylium* resistance has been increased in both populations, while maintaining reasonable levels of resistance to PRR and anthracnose (Bray & Irwin, 1989).

In southern Australia, both bacterial wilt (*Corynebacterium insidiosum*) and stem nematode (*Ditylenchus dipsaci*) cause serious problems in certain regions (Kellock & Coster, 1968). Resistance to these pathogens is being developed in breeding programs in southern Australia. The cv. Lahontan, which has stem nematode and bacterial wilt resistance, has been used extensively in breeding programs in southern Australia (Rogers *et al.*, 1978). More minor fungal diseases of lucerne in temperate areas of Australia have been reviewed by Johnstone and Barbetti (1987).

Virus diseases, chiefly alfalfa mosaic, and witches' broom (mycoplasma) commonly occur, but are not generally considered of major importance (Johnstone & Barbetti, 1987).

Other Temperate Pasture Legumes

White Clover and Other Clovers

Little is known about the incidence or importance in Australia of diseases of *Trifolium* species other than subterranean clover (Johnstone & Barbetti, 1987). In these clovers, overall, diseases have generally been considered of minor economic importance, but sometimes capable of causing severe epidemics. Foliar pathogens of clovers recorded in Australia include *C. trifolii*, *Cercospora zebrina*, *Stagonospora meliloti*, *Cymadothea trifolii*, *Erysiphe polgonii*, *Leptosphaerulina trifolii*, *Peronospora trifoliorum*, *Phoma medicaginis*, *Stemphylium vesicarium*, *Stemphylium sarciniforme*, and *Uromyces trifolii* (Johnstone & Barbetti, 1987). Many of the above pathogens can also infect lucerne and/or subterranean clover.

Work in Queensland established that *Pythium middletonii* and possibly nematodes (*Meloidogyne incognita*, *Heterodera trifolii*, *Helicotylenchus dihystera*, *Pratylenchus* spp., and *Xiphinema radicicola*) were involved in a summer/autumn decline of white clover (Irwin & Jones, 1977). Many of the root pathogens reported for subterranean clover are probably also capable of attacking other *Trifolium* spp. A range of plant viruses have been reported on clovers in Australia (Johnstone & Barbetti, 1987), but again, no substantial data are available on their relative importance as factors contributing to poor productivity or persistance.

Annual Medics

Annual medics (*Medicago* spp.) are also susceptible to many of the pathogens of clovers and lucerne (Johnstone & Barbetti, 1987). The most important foliage disease of medics is black stem and leaf spot caused by *Phoma medicaginis*, which can cause complete defoliation and premature death, particularly in dense stands during wet weather (Barbetti, 1983). Root rots are regarded as a serious problem of medics in Victo ia, especially on clay soils (Bretag, 1985). Pathogenicity has been established for *F. acuminatum, F. avenaceum, P. irregulare, P. ultimum,* and *R. solani*. Bretag and Kollmorgen (1985) established that medic cultivars varied in their reactions to root rot, indicating the potential to develop root rot resistant cultivars. Medics are also susceptible to many of the viruses infecting clovers and lucerne, but there is little information on their occurrence in southern Australia (Johnstone & Barbetti, 1987).

TROPICAL PASTURE LEGUMES

Stylosanthes spp.

The early promise of sown tropical pastures based on *S. humilis* (Townsville Stylo) and *S. guianensis* has not materialized, due to disease susceptibility. The main species now used in commerce are *S. hamata* and *S. scabra*.

Anthracnoses (Colletotrichum gloeosporioides)

There are two anthracnose diseases of *Stylosanthes* spp. in Australia caused by different strains of *C. gloeosporioides* (Irwin & Cameron, 1978). The diseases have different symptomatology on the same host species. The Type A causal agent infects all *Stylosanthes* spp., producing severe disease on *S. scabra, S. humilis, S. viscosa,* and *S. fruticosa*. The disease is characterized by limited leaf and stem lesions with light centres and dark margins. This pathogen was first observed in Australia in 1973 and was responsible for the almost complete annihilation of 0.5 M ha of naturalized *S. humilis* pastures by 1976 (Staples, 1981; Vinijsanun *et al.*, 1987a). This fungus can also infect a wide range of other tropical pasture legumes under controlled inoculation conditions (Vinijsanun *et al.*, 1987b).

The other disease has been designated Type B anthracnose, also caused by *C. gloeosporioides*, and is confined to *S. guianensis* and *S. montevidensis*. This disease is characterized by a general necrosis resulting in a terminal shoot blight. It has now been responsible for the almost complete annihilation of *S. guianensis* based pastures in Australia, with the exception of Oxley fine stem stylo. The strain of *C. gloeosporioides* attacking *S. guianensis* has a very narrow host range, and it has thus been designated *C. gloeosporioides* f. sp. *guianensis* (Vinijsanun *et al.*, 1987b). Both Type A and Type B diseases show relatively high rates of seed transmission (Davis, 1987).

Morphological and biochemical differences between Type A and Type B disease-producing strains of *C. gloeosporioides* have been reported (Ogle *et al.*, 1986; Dale *et al.*, 1988).

Yield loss studies in *S. guianensis* cv. Graham, *S. scabra* cv. Fitzroy and *S. hamata* cv. Verano have been made in North Queensland (Davis *et al.*, 1986). Yield losses were severe in all three cultivars. Average dry matter yields of the unsprayed controls were only 22, 67, and 53%, respectively of the yields of the disease free plots of Fitzroy, Verano, and Graham. Corresponding seed yields were 16, 49, and 42%, respectively. No data is available for grazed swards, but dry matter losses of the magnitude reported here may be less in a lower density pasture.

Controlled environment studies on the influence of temperature, leaf wetness period after inoculation, and relative humidity on the development of the Type A and Type B diseases were reported by Irwin *et al.* (1984b). Severe disease of both types developed in plants incubated at temperatures of 20-30°C and given 24 h of leaf wetness after inoculation. Provided these conditions were imposed, high levels of disease developed irrespective of the relative humidity (40-50% or 95%). Neither disease developed at constant temperatures of 15 or 37°C, and only the Type A disease developed at 34.5°C.

Disease progress curves were generated for 17 accessions from six *Stylosanthes* spp. at two sites in North Queensland over a 3-yr period (Davis *et al.*, 1987). Disease (Type A and Type B) development occurred at comparable levels each year at both the Silkwood (SW) (AAR 3847 mm) and Southedge (SE) (AAR 1230 mm) sites. Rainfall appeared to be the main environmental difference between the sites over the three seasons. The averages for the three seasons were 1933 mm of rainfall and 59 wet days at SW, and 472 mm of rainfall and 42 wet days at SE. These data clearly demonstrate that anthracnose can contribute to severe losses in semiarid areas, and in some instances, (e.g., *S. scabra* cv. Fitzroy), final severities were higher at the drier site.

Extensive pathogenic specialization and variation for virulence have been identified within strains of *C. gloeosporioides* that cause anthracnose of *Stylosanthes* spp. (Irwin & Cameron, 1978; Davis *et al.*, 1984; Irwin *et al.*, 1986). In Australia, previously resistant cultivars of *S. guianensis*, *S. scabra,* and *S. viscosa* have been overcome by new pathotypes of the fungus, while in South America eight different isolate groups have been identified (Irwin *et al.*, 1984a). More recently, Davis *et al.* (1987) reported seven pathogenicity groups within both pathogen groups in Australia, and some of the isolates were relatively complex, showing specificity towards more host genotypes than previously observed. South American studies have shown that the most complex pathotypes of *S. capitata* existed in the native habitat of the species (Lenne & Calderon, 1984). As anthracnose of *Stylosanthes* has only been observed in Australia since 1973 (Pont & Irwin, 1976), it would be expected initially that the races would be relatively simple, increasing in complexity over time as more genetically diverse host genotypes are deployed.

The fungicide benomyl, when applied on a regular basis to seed production crops of *Stylosanthes*, affords adequate levels of disease control, even in highly susceptible cultivars such as Fitzroy (Davis *et al.*, 1986). However, the use of benomyl on grazed *Stylosanthes* pastures is both uneconomic and impractical. Davis (1987) has demonstrated that dusting of de-hulled seed with benomyl significantly reduced rates of seed transmission of anthracnose, and this practice is now recommended in commerce.

The rapidity with which *C. gloeosporioides* strains can overcome previously resistant *Stylosanthes* genotypes has posed a challenge to research workers involved in the development of cultivars with durable resistance (Irwin *et al.*, 1984a). Major gene resistances have been identified for both diseases (Cameron & Irwin, 1983; D.F. Cameron, 1988, personal communication), and damage to previously resistant cultivars has resulted from the apparent matching of single dominant genes in the host by virulence genes in the pathogen. Current breeding programs are directed towards providing host diversity with respect to genetic mechanisms conferring disease resistance.

A breeding program was commenced in *S. scabra* in 1983 utilizing five resistant accessions identified from screening programs and genetic studies (D.F. Cameron, 1988, personal communication). Resistance in each accession appears to be controlled by either one or two major genes. Pedigree selection for anthracnose resistance, early flowering and high dry matter and seed yields was applied to progenies derived from each of the five resistance sources. In the F5 generation seven composites differing in agronomic characters were formed by mixing selections derived from each resistance source. These composites are now undergoing evaluation at four sites in Queensland.

An alternative to the approach discussed above is to use partial resistance, which is also sometimes race-non-specific in its action. Several *S. scabra* lines have been identified from field and glasshouse studies which show lower disease severity scores than cv. Fitzroy (Chakraborty *et al.*, 1988). Work is now in progress to determine the mechanism of their partial resistance.

Currently anthracnose is being controlled in commercial situations by producers sowing mixtures of *S. hamata* cv. Verano and *S. scabra* cv. Seca, both of which possess moderate to high levels of field resistance.

Other Diseases of *Stylosanthes* spp.

Apart from the anthracnose diseases, few other disease problems have been encountered in Australia, although a relatively large number of diseases have been recorded (Simmonds, 1966; Irwin & Davis, 1985; J.L. Alcorn, 1988, personal communication). Under seed production conditions, *Botrytis cinerea* (grey mold) can cause a blight of terminal flowering shoots (O'Brien & Pont, 1977). Benomyl has provided good control of grey mold. Web blight (*Rhizoctonia solani*) has occasionally damaged *Stylosanthes* pastures during the wet season (O'Brien & Pont, 1977). Another anthracnose disease, caused by *Colletotrichum dematium*, has also been recorded in Australia, but it is thought to be of only minor importance (Shipton, 1979). Davis and Irwin (1982) recorded a leaf spot of *S. guianensis* caused by *Curvularia eragrostidis* in North Queensland. A review of diseases of *Stylosanthes* recorded worldwide has been prepared by Lenne and Calderon (1984), and a range of fungal, bacterial, mycoplasma, nematode, and viral pathogens has been observed. It is apparent that considerable additional input into disease management will be required in the domestication of *Stylosanthes* as pasture legumes.

Macroptilium atropurpureum (Siratro)

Macroptilium atropurpureum (DC.) Urban has been widely sown in tropical pastures throughout the world. It was probably the second most important tropical pasture legume grown in Australia, until a rust disease became established on it in the late 1970s. Rust substantially reduces both herbage yield and digestibility (Jones, 1982) and seed production is also adversely affected (English & Hopkinson, 1983).

Rust (*Uromyces appendiculatus*)

Rust on *M. atropurpureum* was first reported in 1910 in South America (Guyot, 1957). The disease was not observed in Australia until 1978, but since then it has become established everywhere the legume is grown, and is regarded as the most important disease of Siratro in Australia (Cameron, 1985). The fungus has previously been identified as *Uromyces appendiculatus* but it is currently thought that it is a new variety of *U. appendiculatus*, since it shows morphological and physiological differences to bean rust (J.A.G. Irwin, 1988, unpublished data). No pathogenic specialization has yet been reported for the fungus that causes *Macroptilium* rust.

The rusts from bean and Siratro showed significantly different responses to dew period and post dew period temperatures for infection (Code *et al.*, 1985). For *Macroptilium* rust, optimal conditions for maximum disease development during the pre-penetration stage of the disease cycle were 20° and 24 h of dew, and in the post-penetration stage, 20-26°C.

Bray (1988) reported on the genetics of resistance to rust in 16 accessions of *M. atropurpureum*. In seven accessions, resistance was dominant, regulated by a single locus in four of these, but by more than one locus in the three others. At least three of these loci were identified as non-allelic. In one accession, resistance was near-recessive and regulated at a single locus, while combinations of dominant and recessive alleles at different loci explained the segregation patterns in other accessions. Ogle *et al.* (1988) made quantitative histological studies on some of this material, and four histologically different resistant responses were observed. These studies may have identified sufficient genetic diversity for the development of a multi-line cultivar or mixture with effective rust resistance, especially if the genes are functionally different.

A major limitation to the Australian studies reported above is that only one collection of the rust was used. Stage III of this rust has never been observed in Australia, yet it regularly occurs in South America and Mexico (G. Cummins, 1988, personal communication). It would seem highly likely that pathotypes of *Macroptilium* rust do occur in those areas, and work should be initiated to study the reactions of the accessions studied by Bray (1988) to a wider range of rust collections.

Other Diseases of *Macroptilium atropurpureum*

In Australia, web-blight, caused by *R. solani*, can cause serious losses in Siratro seed crops, especially during prolonged showery conditions (J.M. Hopkinson, 1984, personal communication). Halo blight, caused by *Pseudomonas syringae* pv. *phaseolicola*, can also cause severe defoliation on *M. atropurpureum* following prolonged rainfall. A range of other fungal, viral and nematode diseases have been reported on *M. atropurpureum* in Queensland (Simmonds, 1966; J.L. Alcorn, 1988, personal communication), but they are not considered to be responsible for serious economic losses.

Other Tropical Pasture Legumes

In the sub-tropics, temperate legumes (white clover, lucerne, medics) have been more successful than tropical species, most of which have failed to persist in the paddock (Mears & Partridge, 1986). Presently, the role of disease in this relatively poor persistence of many of these legumes is unknown. Further intensive research is needed on t e etiology, epidemiology, and yield loss assessment for diseases of these legumes if their full commerical potential is to be realized.

Aeschynomene falcata cv. Bargoo (joint vetch) is known to be highly susceptible to *C. gloeosporioides* in the field. The strain of the fungus attacking *Aeschynomene* is physiologically different to the forms which attack *Stylosanthes* spp. (Vinijsanun *et al.*, 1987b). Other pasture legumes for which a range of pathogens have been recorded include *Desmodium intortum*, *Dolichos* spp., *Cassia* spp., *Glycine wightii*, *Lotononis bainesii*, and *Vigna* spp. These records are contained in Simmonds (1966), and unpublished records of the Plant Pathology Branch of the Queensland Department of Primary Industries (J.L. Alcorn, 1988, personal communication).

CONCLUSION

Diseases provide a considerable restraint on the productivity and persistence of pasture legumes in the temperate and tropical areas of Australia. Since the 1970s in particular, this has been recognized, and new cultivars of disease resistant temperate and tropical legumes have been developed or are in the process of being developed. Programs such as the National Subterranean Clover Improvement Program have done much to provide a rational basis for the identification of production contraints and subsequent cultivar improvement. Such an integrative approach allows scientific input from all state and federal bodies involved in agricultural research in Australia. White clover is Australia's second most important pasture legume (Clements, 1987), yet very little work has been done to identify key pathogens. Work in this area is essential, so that breeding objectives for the proposed National White Clover Improvement Program can be soundly based. The existing National Subterranean Clover Improvement Program provides a model system for the establishment of similar programs for other species such as white clover.

In pasture legume species which currently only occupy relatively small areas, continual monitoring of disease incidence and severity is desirable. If this is done, then significant pathogens can be taken into account during research and development projects. For example, *C. gloeosporioides* should be considered in any further selection or breeding work done on *Aeschynomene* spp. in sub-tropical and tropical regions of Australia. The potential of *C. gloeosporioides* as a highly significant pathogen of pasture legumes has already been demonstrated in Australia with *Stylosanthes* spp. This pathogen caused considerable economic losses in a less than 3-yr period through the devastation of *S. humilis*, which prior to that time was probably the most important pasture legume species in Northern Australia (Staples, 1981).

In recent years, new exotic pathogens and their vectors have become established in Australia apparently following introduction from overseas (Johnstone & Barbetti, 1987). Considerable emphasis should continue to be placed on measures to prevent introductions of further exotic pathogens and vectors, and for some legume species, there may be a strong case for making post-entry quarantine requirements even more stringent.

Breeding or selection for resistance offers the most feasible method of controlling almost all pasture diseases. This strategy has been particularly effective in species such as lucerne, where resistance has generally been durable over time to almost all pathogens (Irwin *et al.*, 1981). This is probably because of the population buffering provided by the heterogeneity of a lucerne cultivar which comprises a mixture of highly heterozygous outbreeding individuals (Hanson, 1972). For inbreeding species, breeding schemes which allow the deployment of a wide range of host resistance mechanisms should be used. Durability of resistance is essential in perennial species such as *S. scabra*, which are established over large tracts of land under conditions of minimal management. In such situations, composites or mixtures would appear to offer several advantages over pure line cultivars, both from agronomic and disease management aspects.

Up to now, virus and virus-like diseases have been largely ignored in pasture legumes in Australia (Johnstone & Barbetti, 1987). Recent advances in the rapid detection of viruses through recombinant DNA technology have now made it possible to effectively survey pasture legumes for the incidence of virus diseases in a region, and to assign correct priorities to breeding programs. Other advances being made in molecular biology, such as the transformation of plants with foreign DNA, as has been achieved for *S. humilis* (Manners, 1988), may soon have direct application to the practical control of plant diseases. Every effort should be taken to focus the advances made in molecular biology to the solving of the problem of poor persistence of many of our pasture legumes.

ACKNOWLEDGEMENT

Mr. G.E. Stovold, Mr. F.C. Greenhalgh, Dr. I. Kaehne and Mr. M.J. Barbetti provided information on pasture legume disease research in southern Australia. Mrs. P. Chattin typed the manuscript. All of this assistance is gratefully acknowledged.

REFERENCES

Alcorn, J. L., and J. A. G. Irwin. 1987. *Acrocalymma medicaginis* gen. et sp. nov. causing root and crown rot of *Medicago setiva* in Australia. Trans. Br. Mycol. Soc. 88:163-167.

Anderson, W. K., R. J. Parkin, and M. D. Dovey. 1982. Relations between stocking rate, environment and scorch disease on grazed subterranean clover pasture in Western Australia. Aust. J. Exp. Agric. Anim. Husb. 22:182-189.

Barbetti, M. J. 1983. Fungal foliage diseases of pasture legumes. J. Agric., W. Aust. 1:10-12.

Barbetti, M. J. 1985. I fection studies with *Cercospora zebrina* on pasture legumes in Western Australia. Aust. J. Exp. Agric. 25:850-855.

Barbetti, M. J. 1987. Root rot of s bterranean clover in Western Australia. p. 32-35. In B.S. Dear and W.J. Collins (ed.) Proc. of 3rd Natl. Subterranean Clover Workshop, Wagga Wagga, NSW. 31 Aug. - 2 Sept. Aust. Wool Corp., Melbourne.

Barbetti, M. J., and D. J. Gillespie. 1987. Fungal foliar diseases of subterranean clover. p. 26-28. In B.S. Dear and W.J. Collins (ed.) Proc. of 3rd Natl. Subterranean Clover Workshop, Wagga Wagga, NSW. 31 Aug. - 2 Sept. Aust. Wool Corp., Melbourne.

Barbetti, M. J., and K. Sivasithamparam. 1986. Fungal foliar diseases of subterranean clover. Rev. Plant Pathol. 65:513-521.

Barbetti, M. J., K. Sivasithamparam, and D. H. Wong. 1986. Root rot of subterranean clover. Rev. Plant Pathol. 65:287-295.

Barnes, D. K., S. A. Ostazeski, J. A. Schillinger, and C.H. Hanson. 1969. Effect of anthracnose (*Colletotrichum trifolii*) infection on stand and vigor of alfalfa. Crop Sci. 9:344-346.

Beale, P. E., and N. A. Goodchild. 1980. Genotype x environment interactions for reaction to *Kabatiella caulivora* in *Trifolium subterraneum* subspecies yanninicum. Aust. J. Agric. Res. 31:111-117.

Beale, P. E., and N. Thurling. 1980. Reaction of *Trifolium subterraneum* genotypes to different isolates of *Kabatiella caulivora*. Aust. J. Agric. Res. 31:89-94.

Bokor, A., and D. L. Chatel. 1973. `Scorch' disease of subterranean clover. Western Australian Department of Agriculture Bull. 3912.

Bray, R. A. 1988. Inheritance of rust resistance in *Macroptilium atropurpureum*. Plant Pathol. 37:88-95.

Bray, R. A., and J. A. G. Irwin. 1978. Selection for resistance to *Phytophthora megasperma* var. *sojae* in Hunter River lucerne. Aust. J. Exp. Agric. Anim. Husb. 18:708-713.

Bray, R. A., and J. A. G. Irwin. 1989. Recurrent selection for resistance to *Stemphylium vesicarium* within the lucerne cultivars Trifecta and Sequel. Aust. J. Exp. Agric. (in press).

Bretag, T. W. 1985. Fungi associated with root rots of annual *Medicago* spp. in Australia. Trans. Br. Mycol. Soc. 85:329-334.

Bretag, T. W., and J. F. Kollmorgen. 1986. Effect of trifluralin, benomyl and metalaxyl on the incidence and severity of root disease in annual *Medicago* spp. and evaluation of cultivars for resistance to root rot. Aust. J. Exp. Agric. 26:67-70.

Burgess, L. W., H. J. Ogle, J. P. Edgerton, L. L. Stubbs, and P. E. Nelson. 1973. The biology of fungi associated with root rot of subterranean clover in Victoria. Proc. R. Soc. Vic. 86:19-29.

Cameron, D. G. 1985. Tropical and subtropical pasture legumes. 5. Siratro (*Macroptilium atropurpureum*): the most widely planted sub-tropical legume. Queensl. Agric. J. 111:45-49.

Cameron, D. F., and J. A. G. Irwin. 1983. Inheritance of resistance to anthracnose disease in *Stylosanthes guianensis*. p.243-245. In C.J. Driscoll (ed.) Proc. of Eighth Australian Plant Breeding Conf., Adelaide. 14-18 Feb.

Chakraborty, S., D. F. Cameron, J. A. G. Irwin,and L. A. Edye. 1988. Evaluation of rate-reducing resistance to anthracnose (*Collectotrichum gloeosporioides*) in *Stylosanthes scabra*. Plant Pathol. (in press).

Clements, R. J. 1987. An Australian white clover breeding program - justification, objectives, timing and resources needed. p.5.1-5.5. In M.L. Curll (ed.) Proc. of National White Clover Improvement Workshop. University of New England, Armidale, 18-19 Aug. NSW Dep. Agric. and Aust. Wool Corp., Melbourne.

Clements, R. J., J. W. Turner, J. A. G. Irwin, P. W. Langdon, and R. A. Bray. 1984. Breeding disease resistant, aphid resistant lucerne for subtropical Queensland. Aust. J. Exp. Agric. Anim. Husb. 24:178-188.

Code, J. L., J. A. G. Irwin, and A. Barnes. 1985. Comparative etiological and epidemiological studies on rust diseases of *Phaseolus vulgaris* and *Macroptilium atropurpureum*. Aust. J. Bot. 33:147-157.

Collins, W. J. 1987. Screening for disease and insect pest resistance in pasture legume improvement - experience with subterranean clover. p. 4.17-4.20. In M.L. Curll (ed.) Proc. of National White Clover Improvement Workshop. University of New England, Armidale, 18-19 Aug. NSW Dep. Agric. and Aust. Wool Corp., Melbourne.

Dale, J.L., J. M. Manners, and J. A. G. Irwin. 1988. *Colletotrichum gloeosporioides* isolates causing two different anthracnose diseases on *Stylosanthes* spp. in Australia carry distinct double-stranded RNAs. Trans. Br. Mycol. Soc. 89 (in press).

Davis, R. D. 1987. Seedborne *Colletotrichum gloeosporioides* infection and fungicidal control in *Stylosanthes* spp. Seed Sci. Technol. 15:785-791.

Davis, R. D., and J. A. G. Irwin. 1982. Leaf lesions on *Stylosanthes guianensis* caused by *Curvularia eragrostidis* in North Queensland. Aust. Plant Pathol. 11:54.

Davis, R. D., J. A. G. Irwin, and D. F. Cameron. 1984. Variation in virulence and pathogenic specialization of *Colletotrichum gloeosporioides* isolates from *Stylosanthes scabra* cvv. Fitzroy and Seca. Aust. J. Agric. Res. 35:653-662.

Davis, R. D., J. A. G. Irwin, D. F. Cameron, and R. Shepherd. 1986. Yield losses in three species of *Stylosanthes* caused by *Colletotrichum gloeosporioides*. Aust. J. Exp. Agric. 27:67-72.
Davis, R. D., J. A. G. Irwin, D. F. Cameron, and R. K. Shepherd, 1987. Epidemiological studies on the anthracnose diseases of *Stylosanthes* spp. caused by *Colletotrichium gloeosporioides* in North Queensland and pathogenic specialization within the natural fungal populations. Aust. J. Agric. Res. 38:1019-1032.
Devine, T. E., C. H. Hanson, S. A. Ostazeski, and T. A. Campbell. 1971. Selection for resistance to anthracnose (*Colletotrichum trifolii*) in four alfalfa populations. Crop Sci. 11:854-855.
English, B. H., and J. M. Hopkinson. 1983. Effects of rust on Siratro seed production. Unpublished Queensland Dep. of Primary Industries Rep. WRS P67.2 MR.
Erwin, D. C. 1954. Root rot of alfalfa caused by *Phytophthora cryptogea*. Phytopathology 56:700-704.
Faris, M. A. 1985. Variability and interaction between alfalfa cultivars and isolates of *Phytophthora megasperma*. Phytopathology 75:390-394.
Frosheiser, F. I., and D. K. Barnes. 1973. Field and greenhouse selection for Phytophthora root rot resistance in alfalfa. Crop Sci. 13:735-738.
Gillespie, D. J. 1987. Root rot in subterranean clover. p. 36-39. In B.S. Dear and W.J. Collins (ed.) Proc. of 3rd Natl. Subterranean Clover Workshop, Wagga Wagga, NSW. 31 Aug. - 2 Sept. Aust. Wool Corp., Melbourne.
Greenhalgh, F. C., and S. Flett. 1987. Recent research on root diseases of subterranean clover in Victoria and its implications. p. 40-42. In B.S. Dear and W.J. Collins (ed.) Proc. of 3rd Natl. Subterranean Clover Workshop, Wagga Wagga, NSW. 31 Aug.-2 Sept. Aust. Wool Corp., Melbourne.
Greenhalgh, F. C., and P. R. Merriman. 1985. *Aphanomyces eutiches*, a cause of root rot of subterranean clover in Victoria. Australasian Plant Pathol. 14:34-37.
Greenhalgh, F. C., and P. A. Taylor. 1985. *Phytophthora clandestina*, cause of severe taproot rot of subterranean clover in Victoria, Australia. Plant Dis. 69:1002-1004.
Guyot, A. L. 1957. `Les Uredinees, Tome III. Genre Uromyces.' Encyclopedia Mycologique XXIX. (Paul Lechevalier, Paris.)
Hanson, C. H. 1972. Alfalfa science and technology. ASA, Madison, WI.
Hanson, E. W. 1950. Physiologic specialisation in *Kabatiella culivora*. Phytopathology 40:11.
Havey, M. J., D. P. Maxwell, and J. A. G. Irwin. 1987. Independent inheritance of genes conditioning resistance to *Phytophthora megasperma* from diploid and tetraploid alfalfa. Crop Sci. 27:873-879.
Helms, K. 1975a. Humidity, free water and light in relation to development of *Kabatiella caulivora* (Kirch) Karak in *Trifolium subterraneum* cv. Yarloop. Aust. J. Agric. Res. 26:511-520.
Helms, K. 1975b. Variation in susceptibility of cultivars of *Trifolium subterraneum* to *Kabatiella caulivora* and in pathogenicity of isolates of the fungus as shown in germination -inoculation tests. Aust. J. Agric. Res. 26:647-655.

Helms, K. 1987. Methods and some preliminary results from a national survey of virus diseases in subterranean clover 1984-1986. p. 20-21. In B.S. Dear and W.J. Collins (ed.) Proc. of 3rd Natl. Subterranean Clover Workshop, Wagga Wagga, NSW. 31 Aug. - 2 Sept. Aust. Wool Corp., Melbourne.

Irwin, J. A. G. 1972. Stagonospora root and crown rot of lucerne. Aust. Plant Pathol. Soc. Newsl. 1:29-30.

Irwin, J. A. G. 1974. Reaction of lucerne cultivars to *Phytophthora megasperma*, the cause of root rot in Queensland. Aust. J. Exp. Agric. Anim. Husb. 14:561-565.

Irwin, J. A. G. 1977. Factors contributing to poor lucerne persistence in southern Queensland. Aust. J. Exp. Agric. Anim. Husb. 17:998-1003.

Irwin, J. A. G. 1984. Etiology of a new *Stemphylium* incited leaf disease of alfalfa in Australia. Plant Dis. 68:531-532.

Irwin, J. A. G., and D. F. Cameron. 1978. Two diseases of *Stylosanthes* spp. caused by *Colletotrichum gloeosporioides* in Australia, and pathogenic specialization within one of the causal organisms. Aust. J. Agric. Res. 29:305-317.

Irwin, J. A. G., D. F. Cameron, R. D. Davis, and J. M. Lenne. 1986. Anthracnose problems with *Stylosanthes*. Trop. Grassl. Soc. Occas. Publ. 3:38-46.

Irwin, J. A. G., D. F. Cameron, and J. M. Lenne. 1984a. Responses of *Stylosanthes* to anthracnose. p. 295-311. In H.M. Stace and L.A. Edye (ed.) The biology and agronomy of *Stylosanthes*. Academic Press, Sydney, Australia.

Irwin, J. A. G., D. F. Cameron, and D. Ratcliff. 1984b. Influence of environmental factors on development of the anthracnose diseases of *Stylosanthes* spp. Aust. J. Agric. Res. 35:473-478.

Irwin, J. A. G., and J. L. Dale. 1982. Relationships between *Phytophthora megasperma* isolates from chickpea, lucerne and soybean. Aust. J. Bot. 30:199-210.

Irwin, J. A. G., and R. D. Davis. 1985. Taxonomy of some *Leptosphaerulina* spp. on legumes in Australia. Aust. J. Bot. 33:233-237.

Irwin, J. A. G., and R. M. Jones. 1977. The role of fungi and nematodes as factors associated with death of white clover (*Trifolium repens*) stolons over summer in south-eastern Queensland. Aust. J. Exp. Agric. Anim. Husb. 17:789-794.

Irwin, J. A. G., D. L. Lloyd, R. A. Bray, and P. W. Langdon. 1980. Selection for resistance to *Colletotrichum trifolii* in the lucerne cultivars Hunter River and Siro Peruvian. Aust. J. Exp. gric. Anim. Husb. 20:447-451.

Irwin, J. A. G., D. P. Maxwell, and E. T. Bingham. 1981a. Inheritance of resistance to *Phytophthora megasperma* in diploid alfalfa. Crop Sci. 21:271-276.

Irwin, J. A. G., D. P. Maxwell, and E. T. Bingham. 1981b. Inheritance of resistance to *Phytophthora megasperma* in tetraploid alfalfa. Crop Sci. 21:277-283.

Johnstone, G. R. 1987. Overview of virus disease problems affecting subterranean clover in Australia. p.16-19. In B.S. Dear and W.J. Collins (ed.) Proc. of 3rd Natl. Subterranean Clover Workshop, Wagga Wagga, NSW. 31 Aug. - 2 Sept. Aust. Wool Corp., Melbourne.

Johnstone, G. R., and M. J. Barbetti. 1987. Impact of fungal and virus diseases on pasture. p.235-248. In J.L. Wheeler *et al.* (ed.) Temperate pastures: their production, use and management. Australian Wool Corp./CSIRO, Melbourne.

Jones, R. J. 1982. The effect of rust (*Uromyces appendiculatus*) on the yield and digestibility of *Macroptilium atropurpureum* cv. Siratro. Trop. Grassl. 16:130-135.

Kellock, A. W., and E. Coster. 1968. Survey of bacterial wilt of lucerne. J. Agric., Vic. 66:242.

Lenne, J. M., and M. A. Calderon. 1984. Disease and pest problems of *Stylosanthes*. p. 279-294. In H.M. Stace and L.A. Edye (eds.). The biology and agronomy of *Stylosanthes*. Academic Press, Sydney, Australia.

Lu, N. S.-J., D. K. Barnes, and F. I. Frosheiser. 1973. Inheritance of Phytophthora root rot resistance in alfalfa. Crop Sci. 13:714-717.

Ludbrock, W. V., J. Brockwell, and D. S. Riceman. 1953. Bare-patch disease and associated problems in subterranean clover pastures in South Australia. Aust. J. Agric. Res. 4:403-413.

Manners, J. M. 1988. Transgenic plants of the tropical pasture legume, *Stylosanthes humilis*. Plant Sci. (in press).

Mears, P. T., and I. J. Partridge. 1986. Commercial usage of improved pastures in the Australian sub-tropics. Trop. Grassl. Soc. Occas. Publ. 3:119-127.

O'Brien, R. G., and W. Pont. 1977. Diseases of *Stylosanthes* in Queensland. Queensl. Agric. J. 103:126-128.

Ogle, H. J., J. A. G. Irwin, and R. A. Bray. 1988. Quantitative histological studies of compatible and incompatible interactions between accessions of *Macroptilium atropurpureum* and *Uromyces appendiculatus*. Plant Pathol. 37:96-103.

Ogle, H. J., J. A. G. Irwin, and D. F. Cameron. 1986. Biology of *Colletotrichum gloeosporioides* isolates from tropical pasture legumes. Aust. J. Bot. 34:537-550.

Ostazeski, S. A., and J. H. Elgin. 1982. Use of hypodermic inoculations of alfalfa for identifying host reactions and races of *Colletotrichum trifolii*. Crop Sci. 22:545-546.

Pont, W., and J. A. G. Irwin. 1976. Collectotrichum leaf spot and stem canker of *Stylosan hes* spp. in Queensland. Aust. Plant Pathol. Soc. News. No.5 (Supplement), Abstr. 35.

Purss, G. S. 1965. Diseases of lucerne. Queensl. Agric. J. 91:196-206.

Roberts, D. A. 1957. Observations on the influence of weather conditions upon severity of some disease of alfalfa and red clover. Phytopathology 47:626-628.

Rogers, V. E., J. A. G. Irwin, and G. Stovold. 1978. The development of lucerne with resistance to root rot in poorly aerated soils. Aust. J. Exp. Agric. Anim. Husb. 18:434-441.

Shipton, W. A. 1979. *Colletotrichum dematium* and *C. gloeosporioides* on *Stylosanthes*. Aust. Plant Pathol. Soc. Newsl. 8:45-46.

Simmonds, J. H. 1966. Host index of plant diseases in Queensland. Queensl. Dept. Primary Industries, Brisbane, Australia.

Staples, I. B. 1981. The technology of Townsville stylo (*Stylosanthes humilis*): A continuing need? J. Aust. Inst. Agric. Sci. 47:200-209.

Stovold, G. E. 1974. Root rot caused by *Pythium irregulare* Buisman, an important factor in the decline of established subterranean clover pastures. Aust. J. Agric. Res. 25:537-548.
Stovold, G. E., and A. Francis. 1988. Incidence of *Colletotrichum trifolii* on lucerne in New South Wales, its host range and reaction of lucerne cultivars to inoculation. Aust. J. Exp. Agric. 28:203-210.
Taylor, P. A. 1984. An unusual *Phytophthora* associated with root rot of subterranean clover. Plant Dis. 69:450.
Taylor, P. A., M. J. Barbetti, and D. H. Wong. 1985b. Occurrence of *Phytophthora clandestina* in Western Australia. Plant Protection Quarterly 1:57-58.
Taylor, P. A., I. G. Pascoe, and F. C. Greenhalgh. 1985a. *Phytophthora clandestina* sp. nov. in roots of subterranean clover. Mycotaxon 22:77-85.
Vinijsanun, T., D. F. Cameron, J. A. G. Irwin, and A. Barnes. 1987a. Phenotypic variation for disease resistance and virulence within naturalized populations of *Stylosanthes humilis* and *Colletotrichum gloeosporioides*. Aust. J. Agric. Res. 38:717-728.
Vinijsanun, T., J. A. G. Irwin, and D. F. Cameron. 1987b. Host range of three strains of *Colletotrichum gloeosporioides* from tropical pasture legumes, and comparative histological studies of interactions between Type B disease-producing strains and *Stylosanthes scabra* (non-host) and *S. guianensis* (host). Aust. J. Bot. 35:665-677.
Wong, D. H., M. J. Barbetti, and K. Sivasithamparam. 1984. Effects of soil temperature and moisture on the pathogenicity of fungi associated with root rot of subterranean clover. Aust. J. Agric. Res. 35:675-684.
Wong, D. H., K. Sivasithamparam, and M. J. Barbetti. 1986. Influence of soil temperature, moisture and other fungal root pathogens on pathogenicity of *Phytophthora clandestina* to subterranean clover. Trans. Br. Mycol. Soc. 86:479-482.

DISCUSSION

Brougham: Do we know of any examples of the use of gene pyramiding in legumes to overcome disease susceptibility etc.?

Irwin: No. But the technique has been used successfully in cereals.

Clements: Gene pyramiding has been used successfully in Australia for more than 30 yr to protect wheat against rust.

Clements: Is anything known yet about pathogen race formation in the fungus-causing subterranean clover scorch?

Irwin: Race formation does not seem to be a serious problem so far. The only variation reported has been in aggressiveness, not specificity towards different host genotypes.

Sheaffer: Are seed treatments used to reduce fungal attacks? For example, what about Apron (metalaxyl)?

Irwin: These might have use in some establishment situations. They are now widely used in subclover but not for other pasture legumes at present.

Hochman: Apron gives effective protection for 8 wk and is very cheap. However, because of pathogen dilution due to cultivation in the year of sowing the chance of establishment failure is low. It makes much better sense to use tolerant cultivars in most situations as this gives long-term protection.

R. Smith: What is the natural state of races in *Colletotrichum gloeosporioides* attacking *Stylosanthes* in the centre of origin (South America)?

Irwin: It would appear that the natural state is a mixture of host and pathogen biotypes and their interactions. There is complexity with respect to host-resistance genes and pathogen races.

Reed: How important is it for Australian national breeding programs to use a widespread mixture of races of a pathogen to which we wish to develop resistance?

Irwin: It is most important to have isolates from the whole range of the main regions where the cultivar will be grown for use in glasshouse screening programs, and field testing should be done over a range of locations.

Leath: In relation to the stability of resistance to disease in alfalfa, I wish to point out that in the USA, resistance to several diseases is still being selected at one geographic location and it holds up across the country, and resistance to bacterial wilt selected 40 yr ago is still effective. In the case of anthracnose, I do not believe that this is a case of resistance breakdown. Likely we merely presented Race 2 with a genotypic niche in which it could express itself.

Irwin: In general, I agree with your comments.

Marten: Do the Australian legume species that fix the most N_2 (that have the most effective host-rhizobial symbiosis) also have the greatest susceptibility to disease infection?

Irwin: White clover is generally disease resistant in Australia, but lucerne has shown disease susceptibility. Both could be classed as high N_2 fixers.

Kretschmer: How do you explain the lack of resistance to anthracnose in *Stylosanthes humilis* when genetic shifts in populations are rapid?

Irwin: Shifts probably don't include that for disease resistance.

ARTHROPOD PESTS AND THE PERSISTENCE OF PASTURE LEGUMES IN AUSTRALIA

P.G. Allen

SUMMARY

Various insects and mites are implicated in the decline in persistence of some legume pastures in Australia, especially temperate pastures and to a much lesser extent sub-tropical and tropical pastures. The types of damage and the losses which can be caused by these pests to different types of pasture are reviewed.

Insecticide is the most commonly used tactic for the control of pasture legume pests in Australia. Recently, the successful use of aphid-resistant lucerne and annual medics clearly demonstrated the value of resistant plants in reducing damage; resistance to various pests is now being sought in a number of pasture legume species. Numerous attempts have been made to implement biological control for pasture legume pests; a limited number of biological control agents have been successful in some regions but most have been ineffective.

Virtually all of the major pests damaging legume pastures have been accidently introduced to Australia. Contingency control strategies for potential new pests are discussed. Some future research directions are also discussed, in particular the need for damage assessment studies with pasture pests.

INTRODUCTION

Damage caused by insects and mites is implicated in the decline in persistence of some legume pastures in Australia, especially in temperate regions (e.g., Carter et al., 1982; Gillespie, 1983) and to a much lesser extent in tropical and sub-tropical regions (e.g., Lenne et al., 1980). Damage to Australian pastures by insects and mites has been reviewed for temperate, tropical, and sub-tropical regions by Wallace (1970); for the Northern Tablelands of New South Wales by Roberts et al. (1979); for temperate regions by Allen (1987). In addition, the report of Allsopp and Hitchcock (1987) on soil insect pests in Australia includes descriptions of the biology, damage, and control of a number of soil-dwelling pests of pasture which may influence legume persistence.

This paper summarizes the types of damage and the impact of the main insects and mites affecting pasture legume persistence, together with the tactics which are being either used or developed for their control. Contingency strategies for the control of new exotic pests in Australia and some future research directions are also discussed. The paper does not address arthropods affecting the establishment of new pastures because they can be easily controlled with insecticidal seed dressings or sprays at the time of establishment.

IMPACT OF ARTHROPODS ON PASTURE LEGUME PERSISTENCE

Pasture legumes in temperate regions are more likely to be damaged by arthropods than pasture legumes in sub-tropical or tropical regions of Australia, mainly because of (a) the larger number of species adapted to feeding on legumes and (b) the generally greater use of legumes for pasture improvement in temperate regions (Wallace, 1970). Plants can be damaged directly by pests consuming foliage, roots, sap or seed, and indirectly by pests transmitting plant pathogens or by predisposing plants to other pathogens. This paper is mainly confined to arthropods which cause direct damage to plants; vector-borne diseases of pasture legumes are discussed elsewhere in this publication.

Pasture pests are only one of many biotic and abiotic factors which can contribute to the poor persistence of legume pastures. At times they may be a major factor but their importance varies considerably and usually depends on other factors such as weather and pasture management (e.g., plant density, grazing and fertilizer practices, crop rotations, pest management). The interactions between the different factors reducing legume persistence are usually complicated and difficult to interpret.

Temperate Pastures

Annual Medic Pastures (*Medicago* spp.)

Annual medic pastures are susceptible to more major pest species than other legume pastures. The major pest species include redlegged earth mite (RLEM) (*Halotydeus destructor* (Tucker)); bluegreen aphid (BGA) (*Acyrthosiphon kondoi* Shinji); cowpea aphid (CPA) (*Aphis craccivora* Koch); sitona weevil (*Sitona discoideus* Gyllenhal); lucerne flea (*Sminthurus viridis* (Linnaeus)). Less important pest species include spotted alfalfa aphid (SAA) (*Therioaphis trifolii* (Monell) f. *maculata*); blue oat mite (*Penthaleus major* (Duges)). Many of these species are considered to contribute to the decline in quality of self-regenerating annual medic pastures across southern Australia (Carter, 1987; Carter et al., 1982).

Arthropod pests mainly influence the persistence of annual medic pastures by reducing the number of seedlings in autumn and by reducing seed yield in spring, but few studies have been conducted to quantify this damage.

Allen et al. (1986, unpublished data) recently demonstrated the relative importance of pests on the persistence of grazed *M. truncatula* cv. Hannaford pastures compared to other factors in South Australia. Establishment of high numbers of seedlings was influenced more by high initial plant densities (seed reserves), high rainfall in autumn and low amounts of cereal stubble than by arthropod damage. However, the densities of pests were generally low in their experimental pastures in one autumn, and the pests became most abundant after the seedlings had developed past the most susceptible cotyledon stage in another autumn. Synchrony between high densities of the pests and the cotyledon stage is necessary for high losses of plants to occur. This synchrony depends on weather and accordingly pests severely reduce plant numbers only in some years. Although arthropods may not always reduce plant density in autumn, their damage can reduce the vigor of the seedlings and consequently reduce the dry matter production of the plants in late autumn/early winter (Allen et al., 1987, unpublished data).

In spring, BGA and CPA regularly caused 30 to 50% reductions in seed yields in experimental pastures which yielded more than 150 kg/ha in the absence of these aphids (Allen et al., 1987, unpublished data). In a further series of experiments in grazed *M. truncatula* cv. Jemalong pastures, seed yield decreased linearly from about 700 kg/ha to 200 kg/ha as BGA densities increased to 50 per medic tip (Allen et al., 1987 unpublished data). The losses in yield were markedly increased when either RLEM or lucerne flea was present with BGA in spring; seed yields were then reduced from about 400 kg/ha to 50 kg/ha, even when BGA densities were as low as five per medic tip. Such low seed yields could lead to seed reserves in the soil less than the 200 kg/ha of seed suggested by Carter (1982) as being necessary for regeneration of good quality medic pastures.

Lodge and Greenup (1980) measured the influences of BGA and SAA on seedling survival and seed yields of *M. truncatula*, *M. polymorpha*, and *M. minima* in New South Wales using field plots which were either protected with insecticide or unprotected from aphids. Similar to Allen et al. (1987, unpublished data), they did not record high mortalities of seedlings in the untreated plots; they also postulated that high numbers of aphids had to coincide with the occurrence of cotyledons for plant losses to occur. Aphids were not present in sufficient numbers in their plots in early spring to reduce the seed yields of early flowering cultivars, but they were sufficiently abundant later to reduce the seed yield of the late flowering cv. Jemalong by 33%. In preliminary experiments in Western Australia, Gillespie (1988, personal communication) showed that RLEM can reduce seed yields of *M. murex* cv. Zodiac from 610 kg/ha to 140 kg/ha.

The deleterious effects of aphids, RLEM and lucerne flea on the persistence of annual medic pastures are shown in these studies, but further studies are required to elucidate the influence of factors such as medic cultivar, weather, plant nutrients and grazing management, on the losses caused by these pests in autumn and spring.

Annual Clover Pastures (*Trifolium* spp.)

Redlegged earth mite, BGA, and to a lesser extent lucerne flea are also implicated in the poor persistence of annual clovers, particularly subterranean clover (*T. subterraneum*) pastures, across southern Australia (Gillespie, 1983a; Stahle, 1987). Redlegged earth mite and lucerne flea can reduce seedling numbers in autumn and all three pests can reduce seed yields in spring. Again there are few quantitative data on the losses caused by these pests, although as early as 1944, Norris reported reductions in seed yields in subterranean clover pastures of up to 100 kg/ha (64%) which were caused by RLEM (Norris, 1944).

Grimm (1986) is measuring the influences of insect damage, fertilizer application, defoliation and weather on seedling numbers and seed yields in pastures with subterranean clover, mainly cv. Daliak, in Western Australia. The study is designed to examine the agronomy and phenology of subterranean pastures as well as the influence of insects on the pastures. To date, RLEM has only been a minor factor in plant losses in Grimm's experiments in autumn, but he warns that RLEM can seriously reduce seedling numbers in autumn when the mites hatch prior to the emergence of clover seedlings (also Gillespie et al., 1983a). Damage to seedlings in autumn may, however, markedly reduce herbage production in early winter, as with annual medic pastures (Nicholas & Hardy, 1976). In spring, BGA was more abundant in pasture which had not been defoliated than in defoliated pasture, and caused up to about 20% loss in seed yields in undefoliated pasture where seed yields were about 1000 kg/ha. Grimm (1986) suggests that BGA reduces seed yields in commercial pastures when grazing pressure is often light in spring because of abundant plant growth during that season.

Grimm (1983) showed that RLEM reduced seed yields in commercial Yarloop/Woogenellup subterranean clover pasture by about 44% to a yield of 440 kg/ha. The reduced yields were less than the 600 to 800 kg/ha proposed by Rossiter (1966) as being a realistic level for the success of a subterranean clover strain. Wolfe (1985) showed that BGA damage reduced seed yields of several subterranean clover strains in experimental micro-swards; the losses varied from about 16 to 66% amongst the strains where the seed yields of aphid-free micro-swards ranged from about 380 to 2600 kg/ha.

Black field cricket (*Teleogryllus commodus* (Walker)) can reduce the persistence of subterranean clover pastures grown on black, cracking soils. The cricket consumes clover seed reserves during summer and, at times, the losses can necessitate resowing (Mowatt & Birks, 1976). The crickets may also prune the roots of clover plants in late-autumn. The larvae of blackheaded pasture cockchafer (*Aphodius tasmaniae* Hope) are soil-dwelling but come to the soil surface to feed on green plant material during winter. Densities greater than about 150 to 200 larvae/m^2 can cause a sufficient loss of annual clover plants to change the composition of pastures towards weedy species e.g., *Erodium* spp. and *Arctotheca calendula* (Allen, 1986).

The larvae of redheaded pasture cockchafer (*Adoryphorus couloni* (Burmeister)) cut off the roots of pasture plants and can severely damage subterranean clover/grass pastures in south-eastern Australia (Hardy, 1981). This pest can reduce the persistence of pastures, and damage is accentuated when birds dig up the pasture to feed on larvae.

Similar to medic pastures, arthropod pests have been shown to cause losses in plant yields which could reduce the persistence of clover pastures. Grimm's (1986) combined entomological and agronomic study with pests in subterranean clover pastures is a positive step towards understanding the pest/host interaction together with some of the many factors influencing it.

Lucerne (*Medicago sativa*)

Prior to the accidental introduction of SAA and BGA to Australia in 1977 (Passlow, 1977a,b) insects had little influence onthe persistence of lucerne pastures. When SAA was first detected, virtually all of the lucerne grown in Australia was the highly susceptible cultivar, Hunter River. Within a year, SAA had spread to most lucerne growing areas in Australia (Lehane, 1982; Wilson et al., 1982) and was devastating irrigated and dryland rain-fed lucerne pastures to the same extent as in the USA when it first arrived there (Dickson et al., 1955). Regular spraying of insecticide was economic in irrigated stands for hay and seed production, but neither spraying nor other control tactics were economic in the large areas of low productivity, dryland pastures of Hunter River lucerne (Allen, 1984a). Within 3 yr of the first appearance of the aphids virtually all of the unprotected pastures were destroyed by the combined effects of SAA and dry seasons (e.g., Allen, 1982). SAA is now a minor pest of lucerne, mainly because of the wide-spread use of aphid-resistant cultivars of lucerne (I.D. Kaehne, 1988 personal communication).

BGA does not damage lucerne as severely as SAA. It can cause reduced dry matter yields with susceptible cultivars (Bishop, 1984), but is not implicated in the poor persistence of lucerne plants.

Larvae of white fringed weevil (*Graphognathus leucoloma* (Boheman)) feed on lucerne taproots and can kill plants, but severe damage is sporadic and is usually confined to patches in irrigated pastures (Berg, 1988). Wingless grasshopper (*Phaulacridium vittatum* (Sjostedt)) can occasionly reduce the persistence of dryland lucerne pastures grown on sandy soil; continuous grazing by this grasshopper during summer probably depletes the carbohydrate root reserves and induces plant mortality.

Perennial Clovers (*Trifolium* spp.)

Persistence of perennial clovers, *T. pratense*, *T. repens*, and *T. fragiferum*, is rarely influenced by arthropod pests on a large scale. Root-feeding scarabs (pasture whitegrubs) (*Sericesthis* spp., *Rhopaea* spp., and *Anoplognathus* spp.) can completely destroy perennial *Trifolium*/grass pastures in eastern Australia and allow invasion of weeds (Roberts et al., 1979). Larvae of these scarabs prefer to feed on grass roots, however, and pastures with a high legume content are less liable to severe attack. Similarly, high densities of the hepialids, (*Oncopera intricata* Walker, *O. rufobrunnea* Tindale, and *O. fasciculata* (Walker)) can reduce the persistence of perennial and annual clovers in clover/ grass pastures in south-eastern Australia; but lower numbers of hepialids mainly reduce the persistence of grasses and not clovers (McQuillan, 1985). Whitefringed weevil feed on the taproots of perennial clovers but do not usually cause economic damage to pastures.

Sub-Tropical and Tropical Pastures

Arthropods do not generally reduce the persistence of tropical and sub-tropical legume pastures in Australia, although their importance may increase with an increased use of legumes in improved pastures (Davies & Hutton, 1970). Most of the pasture legumes sown were first evaluated in field trials which were not protected from arthropods; so the cultivars used there have some resistance to potential pests (G. Strickland & R. Elder, 1988 personal communications).

Larvae of the amnemus weevils (*Amnemus superciliaris* (Pascoe) and *A. quadrituberculatus* (Boheman)) damage the roots of desmodium (*Desmodium uncinatum* cv. Silverleaf and *D. intortum* cv. Greenleaf) and glycine (*Neonotonia wightii*) plants. This damage can lead to the death of plants in late-winter, resulting in grass dominant pastures during the subsequent spring (Braithwaite & Rand, 1970; Mears, 1967). The effects of the damage to roots are particularly severe during dry periods (Jones, 1980, 1988); the larvae damage larger roots (>10 mm diam) which increases the susceptibility of plants to moisture stress. The combined effect of weevil damage and dry conditions caused severe losses in infested desmodium and glycine pastures during the 1964-65 drought in Queensland (Mears, 1967). In contrast, when conditions are favorable for plant growth, desmodium pastures can compensate for damage to at least 68% of roots; compensation occurs through the recovery of moderately damaged plants, through the development of new plants from stolons and (rarely) through volunteer seedlings (D. Smith, 1978). The persistence of desmodium pastures can also be reduced by the hepialids (*O. brachyphylla* Turner and *O. mitocera* (Turner)) (Quinlan et al., 1975).

The leucaena psyllid (*Heteropsylla cubana* Crawford) was first recorded in Queensland in 1986 and is now a major pest of leucaena (*Leucaena leucocephala*) in Queensland and the Northern Territory (Wilden et al., 1987). The psyllid is a sap-sucking insect which weakens plants by damaging successive young regrowth, causing a reduction in plant food reserves. Herbage production has been reduced by more than 50% (Wilden et al., 1987) and ungrazed plants in coastal regions of Queensland have been killed by the pest (D. Gramshaw, 1988, personal communication).

Damage to tropical legumes by other pests rarely reduces their persistence e.g., bean fly (*Ophiomyia phaseoli* (Tyron)) can severely damage seedlings of siratro (*Macroptilium atropurpureum*) but is rarely a problem with older plants (Cameron, 1985); whitefringed weevil and rough brown weevil (*Baryopadus corrugatus* Pascoe) damage the roots of Kenya white clover (*T. semipilosum*) but the clover plants usually compensate with new roots from root nodes (Cameron, 1986).

Other pasture legumes, such as *Centrosema virginianum*, *Vigna parkeri*, *D. heterophyllum*, *M. axillare*, *Lotononis baineseii* and *Stylosanthes* spp., are free of arthropod pests in Australia at present.

Economic Importance of Pests

The costs of reduced persistence of legume pastures caused by pests in Australia have not been evaluated, although estimates were made for losses caused by SAA and BGA to the lucerne industry. Lehane (1982) estimated that SAA and BGA destroyed a million ha of lucerne pasture, and cost the Australian grazing and hay-growing industries about $A200 m. M.V. Smith (1978) predicted that the complete loss of dryland Hunter River lucerne pastures in the Upper-South-East of South Australia would cost the region $A60 m over 5 yr; SAA and dry conditions had caused almost this loss by 1981 (Allen, 1982). The limited information on economic losses caused by pasture pests reflects the difficulties in measuring pasture losses and then relating the losses to livestock production.

CONTROL OF PASTURE LEGUME PESTS

The tactics for the control of pasture pests which are presently available or are being developed for application have not been integrated into pest management strategies for the most part, although an integrated approach, using insecticides, parasites, pathogens and aphid-resistant plants, was planned for the control of SAA (Lehane, 1982). Pasture pests are often not controlled perhaps because there is insufficient quantitative information on losses for farmers to make a rational economic decision to treat. Furthermore, where legume pastures are used in rotation with cereals, farmers tend to give priority to the management of cereals rather than to the management of pastures, to the detriment of proper pasture pest control (e.g., Cregan, 1985).

Chemical Insecticides

Insecticides are the most commonly used tactic for the control of pasture legume pests in Australia, and are likely to remain so in the near future (Goodyer et al., 1984). They are highly effective, can be applied quickly, are compatible with most farming systems, and are relatively cheap, although their cost can preclude their use in extensive, low-productive, dryland pastures. The insecticides currently registered for the control of legume pasture pests include a range of organophosphorus compounds and a carbamate (e.g., Hopkins, 1987; Swaine & Ironside, 1983). Organochlorine compounds are not permitted for the control of pasture pests in Australia because of residue contamination of livestock products.

The rates of application of insecticide recommended for the control of pasture legume pests are low; they range from 35 to 70 g active constituent (a.c.)/ha for RLEM and lucerne flea, to 40 to 150 g a.c./ha for legume aphids, to 300 to 500 g a.c./ha for larger pests, such as blackheaded pasture cockchafer and black field cricket. The costs for the insecticides, per se, range from about $A1 to $A9/ha, and the costs of application range from about $A4 to $A9/ha, depending on the method of treatment and the area treated.

Insecticide-induced resistance is generally not a problem with pests of legume pastures in Australia, probably because of low selection pressures for resistance. Insecticides are usually infrequently applied to pastures and only a small proportion of a pest population is treated at one time. An inexplicable exception to this generalization is the recent and extensive occurrence of insecticide-resistance with SAA. Populations of SAA developed resistance to a wide range of aphicides in New South Wales in 1979 (Walters & Forrester, 1979); in South Australia in 1981; in Victoria in 1983 and in Queensland in 1984 (V. Edge, 1986, personal communication). The type of resistance appears to be the same in all States, with the biotype still susceptible to chlorpyrifos and maldison.

Resistant Plants

The use of pest-resistant cultivars of pasture legumes appears to be the tactic most likely to provide low-cost, permanent control of many of the pasture legume pests. The potential value of resistant plants of pasture legumes became clearly evident with the successful use of aphid-resistant lucernes to prevent economic losses from SAA and BGA in Australia, New Zealand, and in the USA.

Lucerne (*M. sativa*)

Soon after the first detection of SAA in Australia, SAA-resistant cultivars of lucerne were imported from the USA (Walters, 1978); other SAA-resistant cultivars were then bred and released in Australia within a few years (Clements et al., 1984; Kaehne et al., 1983; Lance, 1980). Dry matter production, seasonal growth pattern and persistence of these cultivars were tested across Australia in the presence of aphids in glasshouses (Franzmann et al., 1979; Turner et al., 1981), and in the field in temperate (Hamilton et al., 1978; Ridland & Berg, 1981) and sub-tropical (Lloyd et al., 1985; Lowe et al., 1980) regions. Resistance to BGA and SAA is an essential character for all lucerne varieties released in Australia in the future (Kaehne et al., 1987) and all breeding material should be routinely screened for resistance to these pests (Kaehne et al., 1983). Aphid-resistant cultivars are virtually the only ones being used to resow the Hunter River lucerne pastures which were destroyed by SAA and drought during the late 1970s and early 1980s; many of the aphid-resistant cultivars have the added advantage that they are also resistant to a range of pathogens (Kaehne, 1985).

Aphid growth statistics have been used to assess aphid performance on host-plants; Wellings' (1985) data for BGA on lucerne suggest that these statistics should be obtained for different seasons and on different ages of lucerne to differentiate between resistance which can be attributed either to cultivar growth patterns or to resistant properties of the lucerne.

Kaehne et al. (1987) are also seeking resistance in lucerne to RLEM, lucerne flea, sitona weevil, and wingless grasshopper; sources of tolerance have been found for each of these pests but the tolerances are not yet present in commercial cultivars. Levels of tolerance can vary with the physiological age of the plants and some of the lines have insufficient tolerance to RLEM and lucerne flea to permit seedling establishment but are sufficiently tolerant as established plants. Future selection strategies may require the use of low rates of insecticide at plant emergence to protect the seedlings.

Annual Medics (*Medicago* spp.)

The successes with breeding for aphid-resistance in lucerne stimulated a search for pest-resistances in annual medic lines (Crawford et al., 1988). When SAA and BGA were first detected in Australia in 1977, cultivars of *M. rugosa* and *M. scutellata* were the only cultivars which had resistance to the aphids. These cultivars were not widely grown because their rates of seed softening were variable. Unfortunately, the widely used cultivars of *M. truncatula* were susceptible to the aphids. But, by 1982 Paraggio, a cultivar of *M. truncatula* which was moderately resistant to BGA and partly resistant to SAA, had been selected and released for commercial sowing (Crawford, 1987a). Paraggio was soon followed by Parabinga for sowing in low rainfall areas (Crawford, 1987b).

Resistance to insect pests is now an important criterion in annual medic breeding programs and is given high priority. The large genetic variability that exists in the Australian National Annual Medic Collection suggests a high potential for selecting for substantial levels of resistance to arthropod pests (Crawford, 1985). Screening for resistance to BGA, CPA, RLEM, lucerne flea and adult sitona weevil is in progress. Numerous aphid-resistant lines have now been distributed to agronomists for field evaluation in each State where annual medics are grown, as part of an informal National Annual Medic Improvement Program (e.g., Amor et al., 1986; Hochman, 1985). Backcrossing programs are expected to provide four new aphid-resistant cultivars for release by 1990. These cultivars will be alternatives for the four aphid-susceptible cultivars, Jemalong, Harbinger, Borung and Cyprus, which are most commonly used in Australia (Crawford et al., 1988). Encouraging results have been obtained from screening the annual *Medicago* gene pool for resistance to RLEM, lucerne flea and sitona weevil, but resistances to these pests are not yet included in commercial cultivars (A. Lake, 1988 personal communication).

Annual and Perennial Clovers (*Trifolium* spp.)

The arrival of BGA in Australia also stimulated screening for aphid resistance in subterranean clovers (Gillespie & Sandow, 1981). Seedlings of many lines in the National Subterranean Clover Collection have been screened for damage by BGA (Gillespie, 1983b). Most of the lines suffered reduced herbage yield, although many were better than the highly susceptible commercial cultivars. Gillespie found that the performance of lines is complicated by a large range of environmental and behavioral influences which may modify the plant's response to the aphid. The variable responses make field testing of lines lengthy and, to date, BGA-resistant cultivars of subterranean clover are not available.

Stahle (1987) demonstrated differences in the levels of resistance to RLEM amongst cultivars of white clover and suggested that the more resistant cultivars be grown in areas with a history of damage by RLEM.

Leucaena and Desmodium (*Leucaena* spp., *Desmodium* spp.)

Psyllid-resistant leucaena is considered to be necessary to minimize psyllid damage in leucaena pastures (Wilden et al., 1987). Bray and Woodroffe (1988) have already screened the seedlings of a number of lines from four species of leucaena for resistance to the biotype of the leucaena psyllid present in Australia. Seedlings in each line were damaged but the levels of damage varied amongst lines, indicating a range of resistance in seedlings. But, Bray and Woodroffe stressed the risks of extrapolating resistance of seedlings to that of mature plants.

Jones (1988) suggests that amnenus weevil resistance in desmodium could improve the persistence of infested desmodium pastures, particularly through periods of dry weather, however nothing is known about the variation in *Desmodium* spp. for this attribute.

Biological Control

Numerous attempts have been made to use parasites, predators, and pathogens for the control of pests of pasture legumes in Australia. Since 1979, 20 species of parasitic wasps and a predatory mite have been introduced for the control of seven different pests of pasture legumes (Field, 1984). These agents have been successful in some regions and ineffective in many others; nevertheless biological control with parasites and predators is still generally viewed as a potentially useful tactic against pasture pests.

The wasp (*Trioxys complanatus* (Quilis)) which parasitizes SAA is now successfully established across southern Australia as a result of intensive mass-rearing and release programs (Lehane, 1982; Wilson et al., 1982). Hughes et al. (1987) monitored SAA, the parasite and other factors in two irrigated Hunter River lucerne stands in New South Wales from 1978 to 1981. They concluded from their data and from a simulation model that SAA was biologically controlled by *T. complanatus*, and that this parasite was the main reason for SAA ceasing to be an economic problem within 6 yr of its first detection in Australia. In contrast, Allen (1984b) showed that *T. complanatus* did not effectively control SAA in dryland Hunter River lucerne pasture in South Australia. K. Walden and D. Wiltshire (1980 personal communication) also showed in South Australia that *T. complanatus* did not prevent damage by SAA to irrigated plants of lucerne in summer, but it did reduce numbers of SAA and damage in autumn. Factors limiting the general efficacy of *T. complanatus* in South Australia appear to be hot, dry weather during summer, predators which consume parasitized aphids, and secondary parasites (Samoedi, 1984).

The wasp parasite (*Aphidius ervi* (Haliday)) was introduced to Australia for the control of BGA and now occurs across southern Australia where it is considered to successfully control BGA in lucerne and subterranean clover pastures (e.g., Milne, 1986; Sandow, 1981). Milne and Bishop (1987) studied the impact of *E. ervi* and predators on BGA in irrigated Hunter River lucerne in New South Wales using selective insecticides to suppress the parasite and predators. They concluded that *E. ervi* and predators are potentially of considerable value in pest management programs in lucerne stands. Sandow (1981) considered that *E. ervi* was also successful in subterranean clover and annual medic pastures in Western Australia, but Allen et al., (1986, unpublished data) recently found that *A. ervi* occurred in annual medic pastures in low densities in South Australia but generally did not prevent BGA from reducing seed yields. BGA and *A. ervi* cannot survive in annual medic stubble during the summer months, and it appears that BGA colonizes subsequent pastures more effectively in autumn and winter than *A. ervi* in South Australia.

Microtonus aethiopoides (Loan), a parasitic wasp, was released in south-eastern Australia for the control of sitona weevil adults (Cullen & Hopkins, 1982). High levels of parasitism (>90%) by *M. aethiopoides* occur at times, but only after the adults have laid most of their eggs (D. Hopkins, 1986, personal communication). For this reason, *M. aethiopoides* is an ineffective parasite.

The predatory mites (*Anystis* sp. and *Neomolgus capillatus* (Kramer)) were introduced to Western Australia for the control of RLEM and lucerne flea, respectively (Wallace, 1981). They can markedly reduce the numbers of their respective hosts in restricted areas, but their rates of spread are extremely low and their real impact will take many years to evaluate. Another predatory mite (*Bdellodes lapidaria* (Kramer)) of European origin was noticed feeding on lucerne flea in Western Australia some years ago and it can restrict the increase in numbers of this pest (Wallace, 1981). This mite has been introduced to north-western Tasmania but was found to be climatically incompatible with the region (Ireson, 1984). J. Ireson (1988, personal communication) has successfully established a strain of *N. capillatus* from an area in France which is climatically similar to parts of Tasmania. The effectiveness of this strain is being assessed.

To date, pathogens are not effectively controlling pasture legume pests in Australia (Miller, 1984). The microsporidan (*Nosema locustae* Canning) in bran baits did not consistently control wingless grasshopper (Moulden & D'Antuono, 1984); the fungus (*Entomophaga grylii* (Fresenius) U.S. Pathotype I) was not effective against wingless grasshopper (Milner, 1985); the nematode (*Heterorhabditis heliothidis* (Khan, Brook & Hirschman)) and the fungus (*Beauvaria bassiana* (Balsamo)) could not be demonstrated to be effective against soil-dwelling sitona weevil larvae in the field (Bailey & Milner, 1985). *Zoophthera radicans* (Brefeld) Batko, a fungus introduced to control SAA (Milner et al., 1982) is not a major controlling factor of SAA, and pathogens also appear to have little influence on the density of BGA (Milne & Bishop, 1987).

Coles and Pinnock (1984) showed that inundative releases of the fungus (*Metarhizium anisopliae* (Metch.)) can cause high levels of control of blackheaded pasture cockchafer; they are currently obtaining efficacy data which are necessary for registration and commercialization of this fungus.

Berg et al. (1984) are studying the control of redheaded pasture cockchafer with the nematode (*Steinernema glasseri* (Steiner)) in Victoria. They found that low soil temperatures (<9-10°C) during May to August limits the level of control of larvae by the nematode to about 50%, which is not sufficient to prevent economic damage with most infestations. They are now searching in Tasmania for strains of the nematode which are more active than their current strain at low temperatures. Berg et al. (1987) have designed a drill which can successfully introduce the nematode into established pasture on a commercial scale, although the need to apply 1500 L/ha of water with the nematode may limit its use.

Some Tasmanian strains of *M. anisopliae* are also pathogenic to redheaded pasture cockchafer and have been found to grow at low soil temperatures (<10°C) (A. Rath, 1988 personal communication). Up to 75% control of larvae has been obtained with this fungus in a recent field experiment. Rath is now studying the duration of protection offered by this fungus when it is incorporated into the soil at the time of sowing new pastures. He considers that strains of *M. anisopliae* may offer at least 5 yr protection.

A major factor limiting the effectiveness of some of the above biological control agents appears to be incompatibility between the agent and the weather, especially as Australia has a wide range of climates. Agents are now being collected from regions both outside and within Australia which have climates similar to the region(s) where the pest occurs. With pathogens, the selection of more virulent strains than the strains tested may improve the level of control. Ineffectiveness of parasites or predators does not appear to be caused by insecticides; the overall application of insecticides to pastures is limited.

CONTINGENCY CONTROL STRATEGIES FOR POTENTIAL NEW PESTS

Most of the major pests damaging pasture legumes have been accidently introduced to Australia within the last 100 yr (Allen, 1987). More introduced insect pests are likely to be accidently introduced to Australia in the future, but the type and timing of them are difficult to predict. New pests will either be recognized legume pests, e.g., alfalfa weevil (*Hypera postica* Gyllenhal) and various *Sitona* spp., or species not currently recognized as pests, e.g., *Bruchidius trifolii* Motschulsky which feeds on *Trifolium* and *Medicago* seeds in Europe.

The above uncertainties make the planning of control strategies for potential new pests prior to their arrival in Australia difficult. In 1980, the Commonwealth plant quarantine authority in Australia compiled contingency plans for exotic pests, including pasture pests, based on the current knowledge of the pest and its control overseas at that time, but none of the pests has been detected in Australia yet. CSIRO is currently preparing a strategy for the management of the Russian wheat aphid (*Diuraphis noxia* (Mordwilko)) (R. Hughes, 1988, personal communication); the recent rapid spread of this pest to a number of countries including the USA suggests that it is likely to invade Australia. A comparable prediction can not be made for potential pasture pests.

Australian Governments and agricultural industries believe that the continuation of strict quarantine measures set for Australia to at least delay the accidental introduction of new pests is justified, although Smith (1983) argued that plant quarantine in Australia and New Zealand has been ineffective in preventing the entry of serious plant pests. He considered that resistant plants were the most successful means to control these pests and that joint international breeding projects should be encouraged with the rapid exchange of seed between countries. Screening of new Australian lines of pasture legumes against overseas pests has advantages but again the choice of pests for the screening is difficult. It may be more prudent to ensure that national collections of pasture legume species have the latest, complete range of genetic material with known resistance to pests. Screening new lines overseas and maintaining complete gene collections require international co-operation.

Accordingly, if an introduced pest is detected in Australia, a rapid assessment of the potential problem and rapid initiation of research and extension programs seems the most appropriate strategy to minimize production losses.

SOME FUTURE RESEARCH DIRECTIONS

"What density is worth treating?" is a question continually asked by farmers in reference to pasture pests. I believe that data from damage assessment studies are required to help answer this question. Economic thresholds may not be realistic for pasture pests (Kain & Atkinson, 1975), but the concept of thresholds ranging in complexity as postulated by Poston et al. (1983), or similar, would assist with decisions on the need to control pests. The application of such thresholds could be developed through expert systems (e.g. Bishop et al., 1987). Further selecting/breeding for plant resistance to pasture pests should be encouraged and projects should involve both plant breeders and entomologists, especially to integrate resistance with other control tactics. Pathogens have not been successful to date, but selection of more-virulent strains and development of improved strains with genetic engineering could enhance this tactic. Biological control with parasites and predators has had variable success with pasture pests in the past, but selection of better strains of existing species already in Australia may be of benefit. Parasites should be pursued for new pests, where possible. Insecticides will continue to be a major tactic for the control of pasture pests in the foreseeable future and research should be directed at improving their formulations, and at methods and timing of application. There is also a need to test low-toxicity synthetic pyrethroids for the control of pasture pests.

ACKNOWLEDGMENT

I wish to thank Dr. P.T. Bailey, Dr. S. C.McKillup, and Mr. E.J. Crawford for their useful comments on this paper and the following entomologists/agronomist for providing published and unpublished information - A. Bishop, G. Berg, R. Elder, M. Grimm, J. Ireson, R. Jones, P. McQuillan, A. Rath, P. Stahle, G. Berg, G. Strickland, and P. Wellings.

REFERENCES

Allen, P.G. 1982. The impact of spotted alfalfa aphid on dryland lucerne in South Australia. p. 17-23. In K.E. Lee (ed.) Proc. 3rd Australasian Conf. Grassl. Invert. Ecol., Adelaide, Australia.

Allen, P.G. 1984a. A control program for spotted alfalfa aphid in dryland lucerne pasture and its minimal implementation in South Australia. p. 114-120. In P. Bailey and D. Swincer (ed.) Proc. Fourth Aust. Appl. Entomol. Res. Conf., Adelaide, Australia.

Allen, P.G. 1984b. The management of spotted alfalfa aphid, *Therioaphis trifolii* (Monell) f. *maculata*, in dryland lucerne pasture in South Australia. Ph.D. thesis, University of Adelaide.

Allen, P. 1986. Blackheaded pasture cockchafer. Dep. Agric. South Australia Fact Sheet 3/86.

Allen, P.G. 1987. Insect pests of pasture in perspective. p. 211-225. In J.L. Wheeler et al. (ed.) Temperate Pastures; their production, use and management. Australian Wool Corp./ CSIRO, Melbourne.

Allsopp, P.G., and B.E. Hitchcock. 1987. Soil insect pests in Australia: Control alternatives to persistent organochlorine insecticides. SCA Tech. Rep. Ser. No. 21, CSIRO, Canberra, Australia.

Amor, R.L., P.E. Quigley, R.A. Latta, and J.W. Eales. 1986. Evaluation of new aphid resistant annual medics in north-west Victoria. J. Aust. Inst. Agric. Sci. 52:83-86.

Bailey, P., and R. Milner. 1985. *Sitona discoideus*: a suitable case for control with pathogens? p. 210-216. In R.B. Chapman (ed.) Proc. 4th Australasian Conf. Grassl. Invert. Ecol., Canterbury, New Zealand.

Berg, G. 1988. Weevils in pasture legumes. p. 4-7. In V. Chung (ed.) Crop Infor. Serv. Bull. 89. Dep. Agric. Rural Affairs, Victoria, Australia.

Berg, G.N., R.A. Bedding, P. Williams, and R.J. Akhurst. 1984. Developments in the application of nematodes for the control of subterranean pasture pests. p. 352-356. In P. Bailey and D. Swincer (ed.) Proc. Fourth Aust. Appl. Ent. Res. Conf., Adelaide, Australia.

Berg, G.N., P. Williams, R.A. Bedding, and R.J. Akhurst. 1987. A commercial method of application of entomopathogenic nematodes to pasture for controlling subterranean insect pests. Plant Prot. Quart. 2: 174-177.

Bishop, A.L. 1984. Damage to two varieties of lucerne by *Acyrthosiphon kondoi* Shinji in Australia. Gen. Appl. Entomol. 16:23-26.

Bishop, A.L., G.M. Lodge, and D.B. Waterhouse. 1987. A proposed expert system for the management of lucerne-LATIS. Rev. Marketing Agric. Econ. 55: 174-177.

Braithwaite, B.M., and J.R. Rand. 1970. The pest status of *Amnemus* spp. in tropical legume pastures in north coastal New South Wales. p. 676-681. In M.J.T. Norman (ed.) Proc. XI Int. Grassl. Congr., Surfers Paradise, Australia.

Bray, R.A., and T.D. Woodroffe. 1988. Resistance of some *Leucaena* species to the leucaena psyllid. Trop. Grassl. 22:11-16.

Cameron, D.G. 1985. Tropical and subtropical pasture legumes 5. Siratro (*Macroptilium atropurpureum*) the most widely planted subtropical legume. Queensl. Agric. J. 111:45-49.

Cameron, D.G. 1986. Tropical and subtropical pasture legumes 11. Kenya white clover (*Trifolium semipilosum*) - a tropical relative of white clover. Queensl. Agric. J. 112:109-113.

Carter, E.D. 1982. The need for change in making the best use of medics in the cereal-livestock farming systems of South Australia. p. 32. In Aust. Agron. Conf. Working Papers. Wagga Wagga, New South Wales, Australia.

Carter, E.D. 1987. Establishment and natural regeneration of annual pastures. p. 35-51. In J.L. Wheeler et al. (ed.) Temperate Pastures; their production, use and management. Australian Wool Corp./CSIRO, Melbourne.

Carter, E.D., E.C. Wolfe, and C.M. Francis. 1982. Problems of maintaining pastures in the cereal-livestock areas of southern Australia. p. 68-82. In Proc. Aust. Agron. Conf. Wagga Wagga, New South Wales, Australia.

Clements, R.J., J.W. Turner, J.A.G. Irwin, P.W. Langdon, and R.A. Bray. 1984. Breeding disease resistant, aphid resistant lucerne for subtropical Queensland. Aust. J. Exp. Agric. Anim. Husb. 24:178-188.

Coles, R.B., and D.E. Pinnock. 1984. Current status of the production and use of *Metarhizium anisopliae* for the control of *Aphodius tasmaniae* in South Australia. p. 357-361. In P. Bailey and D. Swincer (ed.) Proc. Fourth Aust. Appl. Ent. Res. Conf., Adelaide, Australia.

Crawford, E.J. 1985. Flowering response and centres of origin of annual *Medicago* species. New South Wales Dep. Agric. Tech. Bull. 32:7-11.

Crawford, E. 1987a. Paraggio. The first blue-green aphid resistant barrel medic. Dep. Agric. South Aust. Fact Sheet No. 5/87.

Crawford, E. 1987b. Parabinga. A moderately aphid-resistant barrel medic for low-rainfall cereal-growing areas. Dep. Agric. South Aust. Fact Sheet No. 6/87.

Crawford, E.J., A.W.H. Lake, and K.G. Boyce. 1988. Breeding annual *Medicago* species for semi-arid conditions in southern Australia. Adv. Agron. in press.

Cregan, P.D. 1985. Factors limiting farmer adoption of annual medic pasture improvement technology. New South Wales Dep. Agric. Tech. Bull. 32:55-56.

Cullen, J.M., and D.C. Hopkins. 1982. Rearing, releasing and recovery of *Microtonus aethiopoides* Loan (Hymenoptera : Braconidae) imported for the control of *Sitona discoideus* Gyllenhal (Coleoptera : Curculionidae) in south eastern Australia. J. Aust. Entomol. Soc. 21: 279-284.

Davies, J.G., and E.M. Hutton. 1970. Tropical and sub-tropical pasture species. p. 273-302. In R.M. Moore (ed.) Australian grasslands. Australian National University Press, Canberra.

Dickson, R.C., E.F. Laird, and G.R. Pesho. 1955. The spotted alfalfa aphid. Hilgardia 24:93-117.

Field, R.P. 1984. The use of parasites and predators in Australian biological control programs. p. 333-343. In P. Bailey and D. Swincer (ed.) Proc. Fourth Aust. Appl. Ent. Res. Conf., Adelaide, Australia.

Franzmann, B.A., W.J. Scattini, K.P. Rynne, and B. Johnson. 1979. Lucerne aphid effects on 18 pasture legumes in southern Queensland: a glasshouse study. Aust. J. Exp. Agric. Anim. Husb. 19:59-63.

Gillespie, D.J. 1983a. Pasture deterioration - causes and cures. J. Agric. W. Aust. Dep. Agric. 24:3-8.

Gillespie, D.J. 1983b. Developing clovers for disease and insect resistance. J. Agric. W. Aust. Dep. Agric. 24:14-15.

Gillespie, D.J., M.A. Ewing, and D.A. Nicholas. 1983. Subterranean clover establishment techniques. J. Agric. W. Aust. Dep. Agric. 24:16-20.

Gillespie, D.J., and J.D. Sandow. 1981. Selection for bluegreen aphid resistance in subterranean clover. p. 105-108. In J.A. Smith and V.W. Hays (ed.) Proc. XIV Int. Grassl. Conf., Kentucky. USA.

Goodyer, G.J., J.W. Watt, and V.E. Edge. 1984. Recent developments and future trends using insecticides, with particular reference to crop protection. p. 286-293. In P. Bailey and D. Swincer (ed.) Proc. Fourth Aust. Appl. Entomol. Res. Conf., Adelaide, Australia.

Grimm, M. 1983. Pasture insect research. p. 46-52. In Ann. Rep. 1982-83 Entomol. Branch, Western Australian Dep. Agric., Australia.

Grimm, M. 1986. Pasture entomology. p. 38-52. In Proc. Esperance Sandplain Research Seminar. Dep. Agric. Western Aust., Esperance, Australia.

Hamilton, B.A., L.R. Greenup, and G.M. Lodge. 1978. Seedling mortality of lucerne varieties in field plots subjected to spotted alfalfa aphid. J. Aust. Inst. Agric. Sci. 44:54-56.

Hardy, R.J. 1981. Some aspects of the biology and behaviour of *Adoryphorus couloni* (Burmeister) (Coleoptera : Scarabaeidae : Dynastidae). J. Aust. Entomol. Soc. 20: 67-74.

Hochman, Z. 1985. Early assessment of aphid tolerant accessions of annual medics for central western New South Wales. Dep. Agric. New South Wales Tech. Bull. 32:29-33.

Hopkins, D.C. 1987. A guide to the control of crop and pasture pests. Dep. Agric. South Aust. Bull. 1/81.

Hughes, R.D., L.T. Woolcock, J.A. Roberts, and M.A. Hughes. 1987. Biological control of the spotted alfalfa aphid, *Therioaphis trifolii* f. *maculata*, on lucerne crops in Australia by the introduced parasitic hymenopteran, *Trioxys complanatus*. J. Appl. Ecol. 24: 515-537.

Ireson, J.E. 1984. The effectiveness of *Bdellodes lapidaria* (Kramer) (Acari : Bdellidae) as a predator of *Sminthurus viridis* (L.) (Collembola : Sminthuridae) in North West Tasmania. J. Aust. Ent. Soc. 23: 185-191.

Jones, R.M. 1980. Persistence of greenleaf desmodium in established pastures. Trop. Grassl. 14:123-124.

Jones, R.M. 1988. Productivity and population dynamics of silverleaf desmodium(*Desmodium unicinatum*), greenleaf desmodium (*D. intortum*) and two *D.intortum* x *D. sanwicense* hybrids in coastal south east Queensland. Trop. Grassl. in press.

Kaehne, I. 1985. Lucerne variety recommendations for 1985. Dep. Agric. South Aust. Fact Sheet 19/85.

Kaehne, I.D., G.C. Auricht, E.T. Mayer, and J.A. Horsnell. 1987. Tolerance to pests and diseases in lucerne. p. 226-228. In J.L. Wheeler et al. (ed.) Temperate Pastures; their production, use and management. Australian Wool Corp./CSIRO, Melbourne.

Kaehne, I.D., A.W.H. Lake, and G.C. Auricht. 1983. Progress in lucerne breeding in South Australia. p. 280-281. In C.J. Driscoll (ed.) Proc. Eighth Australian Plant Breeding Conf., Adelaide, Australia.

Kain, W.M., and D.S. Atkinson. 1975. Problems of insect pest assessment in pastures. N.Z. Entomol. 6:9-13.

Lance, R. 1980. Breeding lucernes that resist aphids. Rural Research, CSIRO 106:22-27.

Lehane, L. 1982. Biological control of lucerne aphids. Rural Research, CSIRO 114:4-10.

Lenne, J.M., J.W. Turner, and F.D. Cameron. 1980. Resistance to diseases and pests of tropical pasture plants. Trop. Grassl. 14:146-151.

Lloyd, D.L., D. Gramshaw, T.B. Hilder, D.H. Ludke, and J.W. Turner. 1985. Performance of north American and Australian lucernes in the Queensland subtropics. 3. Yield, plant survival and aphid populations in rain grown stands. Aust. J. Exp. Agric. 25:91-99.

Lodge, G.M., and L.R. Greenup. 1980. The seedling survival, yield and seed production of three species of annual medics exposed to lucerne aphids. Aust. J. Exp. Agric. Anim. Husb. 20:457-462.

Lowe, K.F., J.W. Turner, and T.M. Bowdler. 1980. Preliminary assessment of the productivity and aphid resistance of ten American lucerne varieties and cv. Hunter River at Gatton, south-eastern Queensland. Trop. Grassl. 14:83-86.

McQuillan, P.B. 1985. An assessment of pasture damage caused by hepialid (*Oncopera intricata*) larvae in Tasmania. p. 7-17. In R.B. Chapman (ed.) Proc. 4th Australasian Conf. Grassl. Invert. Ecol., Canterbury, New Zealand.

Mears, P.T. 1967. The significance of the amnemus weevil in tropical legume-based pastures. J. Aust. Inst. Agric. Sci. 33:348-349.

Miller, L.A. 1984. Recent developments in the use of insect pathogens for pest control. p. 394-404. In P. Bailey and D. Swincer (ed.) Proc. Fourth Aust. Appl. Ent. Res. Conf., Adelaide, Australia.

Milne, W.A. 1986. The release and establishment of *Aphidius ervi* Haliday (Hymenoptera : Ichneumonoidea) in lucerne aphids in eastern Australia. J. Aust. Entomol. Soc. 25:123-130.

Milne, W.A., and A.L. Bishop. 1987. The role of predators and parasites in the natural regulation of lucerne aphids in eastern Australia. J. Appl. Ecol. 24: 893-905.

Milner, R.J. 1985. Field tests of a strain of *Entomophaga grylli* from the USA for biocontrol of the Australian wingless grasshopper, *Phaulacridium vittatum*. p. 255-261. In R.B. Chapman (ed.) Proc. 4th Australasian Conf. Grassl. Invert. Ecol., Canterbury, New Zealand.

Milner, R.J., R.S. Soper, and G.C. Lutton. 1982. Field release of an Israeli strain of the fungus *Zoophthora radicans* (Brefeld) Batko for biological control of *Therioaphis trifolii* (Monell) f. *maculata*. J. Aust. Entomol. Soc. 21:113-118.

Moulden, J.H., and M.F. D'Antuono. 1984. Evaluation of *Nosema locustae* for the control of wingless grasshoppers (*Phaulacridium* spp.) in Western Australia. p. 387-393. In P. Bailey and D. Swincer (ed.) Proc. Fourth Aust. Appl. Ent. Res. Conf., Adelaide, Australia.

Mowatt, P., and P. Birks. 1976. Field cricket control. Dep. Agric. Fish. South Australia, Fact Sheet No. 151/76.

Nicholas, D.A., and D.L. Hardy. 1976. Red legged earth mite cuts pasture production. J. Agric. W. Aust. 17: 33-34.

Norris, K.R. 1944. Experimental determination of the influence of the red-legged earth mite (*Halotydeus destructor*) on a subterranean clover pasture in Western Australia. CSIRO Australia, Bull. 183.

Passlow, T. 1977a. The spotted alfalfa aphid, a new pest of lucerne. Queensl. Agric. J. 103:329-330.

Passlow, T. 1977b. Blue-green aphid, a further new pest of lucerne. Queensl. Agric. J. 103:403-404.

Poston, F.L., L.P. Pedigo, and S.M. Welch. 1983. Economic injury levels: Reality and practicality. Bull. Entomol. Soc. Am. 29: 49-53.

Quinlan, T.J., R.J. Elder, and K.A. Shaw. 1975. Pasture dry matter reduction over the dry season due to *Oncopera* spp. (Lepidoptera : Hepialidae) on the Atherton Tableland, north Queensland. Aust. J. Exp. Agric. Anim. Husb. 15: 219-222.

Ridland, P.M., and G.N. Berg. 1981. Seedling resistance to pea aphid of lucerne, annual medic and clover species in Victoria. Aust. J. Exp. Agric. Anim. Husb. 21:507-511.

Roberts, R.J., T.J. Ridsdill-Smith, R.J. Milner, and P.J. Walters. 1979. Insect pests of pastures. p. 91-98. In R. Parkin (ed.) Pastoral Research on the Northern Tablelands of New South Wales. Dep. Agric. New South Wales, Australia.

Rossiter, R.C. 1966. The success or failure of strains of *Trifolium subterraneum* L. in a Mediterranean environment. Aust. J. Agric. Res. 17:425-446.

Samoedi, J. 1984. Factors affecting the efficiency of *Trioxys complanatus* (Quilis) a parasitoid of the spotted alfalfa aphid, *Therioaphis trifolii* (Monell) f. *maculata*, in South Australia. J. Aust. Inst. Agric. Sci. 50: 238.

Sandow, J.D. 1981. Can parasites and resistant plants control exotic lucerne aphids? J. Agric. W. Aust. 22: No. 2.

Smith, D. 1978. Evaluating long term control of *Amnemus superciliaris* (Pascoe) with pre-sowing insecticidal treatments in desmodium pastures. Queensl. J. Agric. Anim. Sci. 35:83-88.

Smith, H.C. 1983. International plant breeding: a more effective means of disease and pest control than quarantine. p. 218-219. In C.J. Driscoll (ed.) Proc. Aust. Plant Breeding Conf., Adelaide, Australia.

Smith, M.V. 1978. Existing Hunter River dryland lucerne stands - should they be protected? p. 253-255. Lucerne Aphid Workshop, Tamworth. New South Wales Dep. Agric., Australia.

Stahle, P.P. 1987. Resistance of four white clover varieties to attack by redlegged earth mite. p. 232-234. In J.L. Wheeler et al. (ed.) Temperate Pastures; their production, use and management. Australian Wool Corp./CSIRO, Melbourne.

Swaine, G., and D.A. Ironside. 1983. Insect pests of field crops. Queensl. Dep. Primary Industries Info. Series Q183006.

Turner, J.W., D.L. Lloyd, and T.B. Hilder. 1981. Effects of aphids on seedling growth of lucerne lines. 3. Blue-green aphid and spotted alfalfa aphid: a glasshouse study. Aust. J. Exp. Agric. Anim. Husb. 21:227-230.

Wallace, M.M.H. 1970. Insects of grasslands. p. 361-370. In R.M. Moore (Ed.) Australian grasslands. Aust. Natl. Univ. Press, Canberra.

Wallace, M.M.H. 1981. Tackling the lucerne flea and redlegged earthmite. J. Agric. W. Aust. 22: 72-74.

Walters, P.J. 1978. Lucerne Aphid Workshop Tamworth. New South Wales, Dep. Agric., Australia.

Walters, P.J., and N. Forrester. 1979. Resistant lucerne aphids at Tamworth. Agic. Gaz. Dep. Agric. New South Wales 90:7.

Wellings, P.W. 1985. Growth, development and survival of *Acyrthosiphon kondoi* (Homoptera: Aphididae) on five cultivars of lucerne. J. Aust. Entomol. Soc. 24:155-160.

Wilden, J.,G. Milne, and R. Elder. 1987. The leucaena psyllid in central Queensland. Central Queensl. Agnotes. Agdex No. 136/622. Queensl. Dep. Primary Industries, Australia.

Wilson C.G., D.E. Swincer, and K.J. Walden. 1982. The origins, distribution and host range of the spotted alfalfa aphid, *Therioaphis trifolii* (Monell) f. *maculata*, with a description of its spread in South Australia. J. Entomol. Soc. South Afr. 44:331-341.

Wolfe, E.C. 1985. Subterranean clover and annual medics - boundaries and common ground. New South Wales Dep. Agric. Tech. Bull. 32:23-28.

DISCUSSION

Buxton: Does plant species diversity in pastures reduce the impact of insects on pastures and keep insect pressure down?

Allen: Possibly, if the majority of plant species are not hosts to the insects. I suggest that the pressure of insects on the susceptible species in a diverse pasture would disadvantage those species to the extent that they may not persist.

Berberet: Mixed plantings of legumes and grasses may have lower populations of legume pests such as potato leafhopper than pure stands of legumes. This may serve to moderate some insect pest situations in mixed pastures.

Reed: How close is the alfalfa weevil to being in Australia and what impact might it have?

Allen: The alfalfa weevil is a threat to Australia, but it is difficult to predict how close it is and what impact it might have. Any weevil-resistant germplasm should be in our lucerne collections so that genes for resistance can be bred into new cultivars if the weevil arrives.

Reed: Does it attack other pasture legume species?

Berberet: There is evidence that it may attack annual *Medicago* species.

Brougham: Do you consider that a new insect pest would be detected more easily and quickly than a new disease?

Allen: Not necessarily - it would depend on the symptoms caused by the disease (some are obvious), and on the ease of detecting the insect, e.g., soil insects can be overlooked.

Irwin: I support this answer.

Hochman: What are the ecological factors explaining the phenomenon that when a pest like spotted alfalfa aphid (SAA) or bluegreen aphid first appears it has a devastating effect, yet after a few years growers can get away with returning to the susceptible cultivars?

Allen: The initial high densities of a new pest in a country have been shown with some pests to be due to the release of the pests from their predators and parasites. I suggest that growers in Australia would not get away with growing Hunter River lucerne again. This cultivar is highly susceptible to SAA and if this cultivar was widely grown again I do not believe that the introduced parasite and native predators would prevent further damage. I strongly advocate that the new aphid-resistant cultivars are grown.

Watson: Concerning the renewed use of Hunter River lucerne in Australia, the same is occurring with the local cv., Wairau, in New Zealand. This may be acceptable if buildup of effective biological control of aphids has been of greater significance than resistant cultivars.

Allen: Its future success may depend on the extent of growing the susceptible cultivar. I consider that, in reducing the damage caused by SAA, resistant cultivars have been more important overall than the introduced parasite in Australia. Furthermore, new aphid-resistant cultivars also have resistance to a number of important diseases.

Hoveland: In the southern USA, striped field crickets are generally a serious pest in destroying seedlings of arrowleaf, crimson, and white clover in grass sods. Do you have any serious pests on subterranean clover seedlings? Have you noted differences in damage among clover species in seedling damage?

Allen: Redlegged earth mite and lucerne flea can seriously reduce subterranean clover seedling numbers in autumn when the pests hatch prior to seedling emergence. It is a case of synchrony of occurrences of the pest and the most susceptible stages of the plant to damage. I do not know of differences in susceptibility between annual *Trifolium* species.

INITIATIVES IN PEST AND DISEASE CONTROL IN NEW ZEALAND TOWARDS IMPROVING LEGUME PRODUCTION AND PERSISTENCE

R. N. Watson, R. A. Skipp and B. I. P. Barratt

SUMMARY

Pest and disease research on forage legumes in New Zealand has been largely confined to white clover (*Trifolium repens* L.), red clover (*T. pratense* L.), and lucerne (*Medicago sativa* L.). Damage caused by insect pests, particularly those which feed on both grass and clover, is most easily recognized by farmers and is thus perceived as having greatest impact on pasture productivity. Costs have increasingly precluded the use of insecticide to control pasture pests. Integrated pest management, using a combination of chemical and biological control measures, is now showing promise. Nematode pests, virus diseases, and fungal root pathogens have more insidious effects on clover vigor and persistence. Successful control measures using host plant resistance or tolerance have been elusive in white clover. Developments in molecular biology will widen the scope for finding new sources of plant resistance.

Lucerne can outyield conventional pasture, especially in dry areas. Serious pest and disease outbreaks in the 1970s eroded farmer confidence in the crop, contributing to a dramatic decline in lucerne acreage. Resistant cultivars now available for most of the more serious pests and diseases make it appropriate to re-evaluate the place of lucerne in New Zealand agriculture.

INTRODUCTION

The extent of pest and disease research on forage legumes in New Zealand reflects their agronomic use and importance. Most work has been carried out on white clover (*Trifolium repens* L.) and red clover (*T. pratense* L.), the main legume components of permanent ryegrass (*Lolium perenne* L.)/clover pastures, and lucerne (*Medicago sativa* L.), the major forage legume grown as pure swards. Pest damage resulting directly in the removal of herbage, or death of plants through root grazing, is easily identified by farmers as a cause of lost production. Losses of up to 40% have been reported (East & Pottinger, 1984) and considerable effort has been made to develop suitable methods for the control of such pasture pests. Effects of nematode pests, root rotting fungi and systemic infection by viruses are less easy to detect and quantify. However, the insidious effects of these agents on productivity and persistence are now becoming better documented and various ways are being sought to reduce their influence.

This paper reviews some of the initiatives taken in New Zealand to assess the significance of pests and diseases and to find suitable methods to enhance forage legume vigor and persistence. It considers pest (insect and nematode) and disease (viral, fungal, and bacterial) problems affecting the establishment and persistence of pasture legumes and lucerne, respectively.

PESTS OF PASTURE LEGUMES

The major pasture pests range from those which affect clover only, through varying degrees of mixed feeding, to those which affect only grass (Table 1). Of these the endemic grass grub [*Costelytra zealandica* (White)] with national distribution, and porina (*Wiseana* spp.), with pest status in the southern North, and South Islands, are recognized as the most destructive pests affecting clover (East & Pottinger, 1984). As general feeders they have the greatest capacity for sporadic pasture damage. Porina and grass grub populations, and associated pasture damage, are directly enhanced by the presence of white clover in pasture (Farrell, 1976; Kain et al., 1979).

Table 1. Feeding preference of pasture pests in New Zealand.

Preference	Pest	Distribution	Pest status
CLOVER FEEDING ONLY	Root knot nematode (*Meloidogyne hapla*)	NI,nSI	Major,chronic
	Clover cyst nematode (*Heterodera trifolii*)	National	Major,chronic
	Blue butterfly (*Zizina otis labradus*)	National	Minor,localized
GENERAL FEEDERS (prefer populations enhanced by clover)	Lucerne flea (*Sminthurus viridis*)	National	Minor,localized
	Whitefringed weevil (*Graphognathus leucoloma*)	NI,nSI	Minor
	Stem nematode (*Ditylenchus dipsaci*)	National	Minor
	Grass grub (*Costelytra zealandica*)	National	Major,sporadic
	Striped chafer (*Odontria striata*)	sSI	Minor,v.localized
	Field slugs (*Deroceras* spp.)	National	Minor
	Porina caterpillar (*Wiseana* spp.)	sNI,SI	Major,sporadic
	Manuka beetle (*Pyronota* spp.)	National	Minor,v.localized
	Tasmanian grass grub (*Aphodius tasmaniae*)	NI,SI	Minor,v.localized
(prefer populations enhanced by grass)	Root lesion nematode (*Pratylenchus spp.*)	National	Minor
	Black field cricket (*Teleogryllus commodus*)	nNI	Major,sporadic
	Soldier fly (*Inopus rubriceps*)	nNI	Minor
	Black beetle (*Heteronychus arator*)	nNI	Minor,sporadic
GRASS FEEDING ONLY	Argentine stem weevil (*Listronotus bonariensis*)	National	Major

Perceptions of the importance of pests are influenced by the visual impact and timing of damage. Thus it is easily appreciated that total sward destruction during sporadic outbreaks by general feeders will directly effect stock feed reserves and immediate carrying capacity. The damaging effects on ryegrass of Argentine stem weevil [*Listronotus bonariensis* (Kuschel)] and black field crickets [*Teleogryllus commodus* (Walker)] have, however, only recently been widely appreciated, whereas the insidious effects on clover of whitefringed weevil [*Graphognathus leucoloma* (Bohemen)] (King & East, 1979), root feeding nematodes (Yeates et al., 1975; Risk & Ludecke, 1978) and slugs (Barker et al., 1985) are as yet recognized only by specialists. Subclinical pest damage was reported to cause considerable losses in pasture productivity, particularly of the clover component, and a marked reduction in N_2-fixation (Watson et al., 1985). Insect, mite and slug pests, and nematode pests of white clover have been reviewed by Gaynor and Skipp (1988) and Skipp and Gaynor (1988), respectively.

Pesticidal Controls

Short-term chemical controls are available for most major pasture pests (East & Pottinger, 1982, 1984). Exceptions, ironically, are for some pests which most directly affect clover. Root feeding clover nematodes cannot be economically controlled with nematicide. Whitefringed weevil larvae cannot be economically controlled with insecticide, although strategies to control surface active adults are feasible (East et al., 1975).

Grass grub is controlled by surface broadcasting of fensulfothion or isazophos granules in autumn to control larvae. Innovations such as banding granules, spraying, and solid stream liquid applications have generally not improved control (Trought, 1979; East et al., 1981). Drilling granules however, can improve insecticide persistence and penetration and enable the use of lower application rates, or cheaper but less residual alternatives to fensulfothion such as diazinon (Kain et al., 1982b; Stewart, 1984). Nevertheless reliable reduction of grass grub numbers from use of organophosphate insecticides has been difficult to achieve in many districts and soils (Trought, 1979; Carpenter et al., 1981; Lauren et al., 1982). Lindane, which is permitted for use in nondairy farming, acts more effectively against the adult than larval stages so that applications made in spring are more effective than traditional autumn applications (Lauren et al., 1982).

The cost of insecticide treatment for soil pests has become prohibitive. As a consequence registration of some promising insecticides for grass grub control, e.g., isofenphos and terbufos (East et al., 1981; Wrenn et al., 1985a) has not been pursued. Similarly pyrethroid insecticides, which require sophisticated timing of application using pheromone baited traps for effective adult control (Lauren et al., 1981; Henzell et al., 1985), have not been adopted into practice. Insecticides can disrupt the natural microbial control of grass grub (Miln & Carpenter, 1979), thus leading to increased insecticide use and reduced efficacy (Kain et al., 1982a; Thomson et al., 1985). Use of a variety of non-insecticidal ameliorants for grass grub such as stock treading, or N application to compensate for pasture damage, are more compatible with natural regulation and appear to be the best long-term alternative for grass grub management (Kain et al., 1982b, East, 1985).

Pests which feed on foliage such as porina, lucerne flea (*Sminthurus viridis* L.), black field cricket and Tasmanian grass grub (*Aphodius tasmaniae* Hope) are controlled more economically than soil pests. Cost efficiency has increased recently with improved insecticides, formulations, timing and rate of application, and the development of simple methods for determining action thresholds (Savage & French, 1981; Pottinger et al., 1985; Blank, 1984; East et al., 1980; Stewart et al., 1988). Insecticidal baits control porina (Haack et al., 1982; Henzell & Lauren, 1983) and cricket (Blank, 1984). Insect growth regulators, particularly difluron, could give economical control of whitefringed weevil (Henzell et al., 1979), porina or lucerne flea (Wrenn et al., 1985b; Wrenn & McGhie, 1986). The use of baits, more strategic use of insecticide, and more pest specific chemicals such as growth regulators, could reduce the deleterious impact of pesticides on natural pest control agents.

Biological Control

Vertebrate or insect predators do not regulate populations of the major pasture pests (East & Pottinger, 1975; Longworth, 1982), except for Australian soldier fly (East et al., 1986). The use of inundative applications of entomophagous nematodes, particularly *Heterorhabditis* and *Steinernema* spp. have shown initial promise in grass grub and porina control, but have not yet been used commercially (van der Mespel et al., 1986; Wright, 1988). Microbial pathogens exert a considerable influence on populations of major pasture pests (Wigley, 1985). Outbreak cycles of grass grub and porina are associated with epidemics of viral, bacterial, protozoan and fungal diseases. Wigley (1985) gave three principal control strategies involving microbial pathogens, (i) inundative release, (ii) on-farm management to conserve or enhance the activity of pathogens, (iii) release of pathogens into previously unexposed populations of the pest. Bacterial pathogens of grass grub, *Serratia* sp. and *Bacillus* sp. have the greatest prospects for commercial exploitation (Jackson, 1988). Baits inoculated with the fungus *Metarhizium anisopliae* (Metsch.) Sorokin can reduce populations of porina (Latch & Kain, 1983).

Microbial control can be disrupted by drought, cultivation, or pesticide use (Miln, 1982). These reduce effective levels of inoculum either directly or by reduction of the infected host population. Thus arable areas which include pasture crop rotations, and drought prone areas, are potentially the most susceptible to outbreaks of major pasture pests (Jackson et al., 1988). Establishment of pathogens into previously unexposed populations has not been fully exploited. There is considerable scope for exchange of pathogens of similar insect classes internationally.

Cultural Controls and Use of Alternative Legumes

Some management practices can reduce pest numbers. These include cultivation, grazing management, use of rollers, and pest tolerant plants. Cultivation prior to pasture establishment effectively reduces many pests but also is often followed by pest buildup to damaging densities (East & Kain, 1982; Fleming et al., 1982). Minimum tillage methods provide the means to upgrade pasture while conserving elements of natural pest control.

Although pest species vary in their response to increasing stock rates (East & Pottinger, 1983), intensive grazing can reduce levels of some pests. Stock rates up to 300 $cows^{-1}$ day^{-1} during winter block grazing reduce lucerne flea populations (Pottinger et al., 1985). Treading by mobs of cattle or sheep during autumn or early winter reduce grass grub populations when larvae are near the surface (East, 1979), and cause up to 90% reduction of slug densities (Ferguson & Barratt, 1986; Ferguson et al., 1988). Severe defoliation in spring causes high mortality of young porina larvae before they form burrows, and may increase the incidence of microsporidial infection and parasites (Stewart & Archibald, 1987). Pupae of soldier fly are similarly susceptible to desiccation (Robertson et al., 1979). Severe grazing during summer exposes grass grub to lethal soil temperatures (East & Willoughby, 1980), although such grazing intensity during summer can itself be detrimental to pasture. Mechanisms by which grazing reduces pest populations include direct treading, oxygen depletion or exposure of vulnerable stages to desiccation. The timing and application of grazing management has to be carefully optimized for each pest. The general intensity of year round grazing practised in New Zealand may contribute to the suppression of potential foliar pests such as aphids, leafhoppers, caterpillars, slugs, and thrips. Heavy rolling with 8-12 tonne rollers can also achieve effective control of grass grub (van Toor & Stewart, 1987). Use of a deeply grooved surface on the rollers increased control possibly by producing a shearing effect during soil compression. Part of the benefit of stock treading or rolling may be due to the compaction of soil assisting rerooting and recovery of pasture after grass grub damage (Kain et al., 1982a).

Pests can be sensitive to changes in pasture composition and structure, as influenced by different grazing systems (Brock, 1986) or other pests (King & East, 1979). Clover under rotational grazing gives a lower density of rooting nodes than under set stocking (Hay et al., 1988). This could make clover more vulnerable to attack by whitefringed weevil, grass grub, or nematodes. Nitrogen to compensate for loss of production by pests can be more cost effective than the use of insecticide to control them (Prestidge & East, 1984). It may be even more cost effective, however, to apply the N to undamaged pasture (Kain et al., 1982a). At high grass grub densities a combination of reduced rates of insecticide and N gave the most effective amelioration of pasture damage (van Toor et al., 1988).

Some forage legume species show resistance or tolerance to pasture pest attack (Biggs et al., 1986). For instance, grass grub numbers are effectively reduced in pasture containing *Lotus pedunculatus*, even in the presence of white clover (Kain et al., 1979). Establishment difficulties have restricted the uptake of *L. pedunculatus* by farmers. It is also susceptible to other pests such as whitefringed weevil (East, 1982) and the blue butterfly (*Zizina otis labradus*) (East et al., 1978).

Integrated Pest Management

Increasing pesticide costs and awareness of the need, and ability, to harmonize with natural pest regulation has led to a more soundly based approach to pest control. Integrated pest management (IPM) seeks to utilize all available pest control and ameliorating options in a rationalized manner (East & Pottinger, 1982; Kain et al.,

1982a; East et al., 1986). Integrated pest management is particularly suited to pasture, given its low monetary value and absence of requirement for cosmetic appearance. In New Zealand, IPM in pasture ranges from use of a single component (baiting) to control crickets if critical population densities are determined by a flushing method (Blank, 1984), to dual insecticide/grazing controls for lucerne flea (Pottinger et al., 1985), to a series of partially effective components for control of soil-inhabiting pests (Kain et al., 1982a, East & Pottinger, 1982).

Pest management requires the development of simple methods for rapid paddock by paddock assessment of pest populations by farmers, and a knowledge of damaging levels for each pest. Ideally pest management should become part of normal strategic planning to match seasonal feed supply with animal requirements, rather than a disruptive hiatus in the management schedule. Since the damage of pest populations is closely related to seasonal pasture growth conditions, both slow acting pests such as grass grub, and rapidly resurgent pests such as lucerne flea, require flexible options and an ongoing appraisal of pest effects on feed supply (Kain et al., 1982a,b). Stewart et al. (1988) have developed an insecticide treated strip method to provide early warning of pasture damage by grass grub. Farmers with higher stocking rates need to be the most sensitive to these requirements since they are less able to cushion the effects of pest damage through elasticity in pasture utilization or stock condition.

Plant Breeding for Pest Resistance or Tolerance

Plant resistance to pests has been a logical goal (Sutherland, 1979) but success in white clover has proven elusive. 'Grasslands Kopu', with resistance to stem nematode [*Ditylenchus dipsaci* (Kuhn)], is the first commercially available cultivar with pest resistance (van den Bosch et al., 1986; West & Steele, 1986). Current selection programmes are underway to improve resistance or tolerance to the more important clover cyst [*Heterodera trifolii* (Goffart)] and rootknot [*Meloidogyne hapla* (Chitwood)] nematodes (C. F. Mercer, 1988 personal communication). New Zealand white clover has been claimed to show levels of resistance to clover cyst nematode by overseas workers (Dijkstra, 1971) but there is no evidence to confirm that effective resistance occurs in New Zealand (Yeates et al., 1973). Similarly 'Huia' showed more resistance to slugs than ladino type clovers in Oregon (Anonymous, 1956) but its effective resistance to slugs has not been demonstrated in New Zealand (G.M. Barker, 1988 personal communication). Resistance to slugs and other pests is sometimes linked with cyanogenic properties of white clover.

Considerable attention has been given to selection in white clover for resistance to grass grub (van den Bosch & Gaynor, 1986). However, differences between clover lines in their ability to incite poor weight gains in feeding larvae have been inconsistent in laboratory screening trials (Wilson & Farrell, 1979). Field screening programmes (Kain et al., 1982b) have also been unproductive. Considerable research in determining the biochemical nature of resistance in other legumes to Scarabaeidae (Biggs et al., 1986) has not been of direct benefit to white clover breeding program. White clover plants with different root morphology gave no difference in grass grub feeding trials (van den Bosch & Gaynor, 1986). Morphology may however confer a level of tolerance in the field. Clover plants with a low density of rooting nodes on stolons could be more vulnerable to root feeding pests.

Hybridization of *T. repens* with other species of *Trifolium*, e.g., *T. uniflorum* to increase tap-rootedness, and *T. ambiguum* for virus resistance has produced regenerants using embryo culture methods (Williams & Williams, 1981). Other hybrids are being evaluated in connection with resistance to root nematodes (Mercer, 1988). The agronomic value of such hybrids has yet to be demonstrated.

Advances in molecular biology widen the scope for realization of pest and disease resistance in plants. Much of the molecular biology research effort in New Zealand is devoted to improving white clover. Plant cell regeneration techniques and a system for introducing cloned genes into white clover has been developed (White & Greenwood, 1986). This opens the way for introduction of desirable genes, including those conferring pest resistance (White, 1988). Tissue-specific promoter genes would allow expression of resistance only in target plant tissue, e.g., grass grub resistance in rhizobial root nodules or roots.

Legume Establishment

There are greater potential threats to legume establishment, including pest attack, when conventional tillage has not been used before sowing (Clements, 1986; Barker, 1986). Poor clover establishment has commonly resulted from both hill country oversowing, and lowland undersowing methods in New Zealand (Charlton & Thom, 1984; Ledgard et al., 1988).

In oversowing trials in upland Otago 60-70% of legume seeds failed to reach the cotyledon stage, largely through poor seed-soil contact and desiccation (Barratt, 1986 unpublished data). Following emergence, protection of white clover seedlings by insecticide applications for 3 months after sowing increased establishment from 28 to 60%. Plant density was increased from 28 to 71/m^2, and significant increases in white clover dry matter production persisted for up to 4 yr after the initial 3 month protection.

Plant competition from existing vegetation can have an overriding effect on seedling growth in non-tillage establishment (Watson et al., 1986). Severe growth retarding effects of pests as demonstrated for grass grub (Wilson, 1978), nematodes (Yeates et al., 1975), slugs (Ferguson, 1984) and endemic adult weevils (Barratt, 1985a) in white clover add to these competitive disadvantages. In addition to seedling attack black field crickets were also responsible for considerable removal and consumption of oversown seed directly (Blank et al., 1980).

Nematicide treatment of white clover seed gave large top and root growth responses at 4 and 8 wk after sowing in laboratory studies using field infested soil, even although nematode infestations were not greatly different from untreated seedlings at that stage (Watson, 1987 unpublished data). The implication was that nematode infestation as early as the cotyledon stage determined the future growth course of the seedling, probably by its effect on taproot growth and nodulation (Skipp & Watson, 1987). Barratt (1985a) showed that cotyledon damage by adult weevils reduced nodulation and subsequent seedling growth of white clover oversown in rangeland. Seed protection may thus be important in clover establishment in no tillage systems. Appropriate methods of seed protection need to be evaluated and may require various components of molluscicidal, insecticidal, nematicidal, and fungicidal activity.

Barker et al. (1988) found that the method of vegetation control used during pasture renovation with ryegrass altered the way pests reacted to seedlings. Total sward desiccation with herbicide gave greater seedling growth rates than no herbicide. Herbicide banded over drill rows gave an intermediate effect. Greater attraction of pests onto drill rows occurred on the blanket treatment, and least with no herbicide. Pesticide used at planting, however, gave greater reduction of the total pest population in the blanket treatment because of the more effective attraction into the toxic zone. Increasing seed rates to compensate for pest attack has not been shown to increase herbage yields and may result in greater attraction of pests onto the drill rows (Clements, 1986; Robertson et al., 1981; Barratt & Johnstone, 1984; Barker,1986 unpublished data).

Over large areas of New Zealand clover persists by natural regeneration from seed or stolons. Pest effects on this process have been demonstrated (Watson et al., 1985). Effects were most marked after severe drought had eliminated clover from the pasture (Watson et al.,1985, unpublished data). Regeneration after autumn rain restored clover to predrought levels by the following spring on treated plots, but this took a further year in untreated plots where whitefringed weevil, grass grub, and nematode pests were present.

Ferguson and Barratt (1986) showed that mob-stocking prior to direct drilling gave over 90% mortality of grey field slugs [*Deroceras reticulatum* (Müller)] which can severely damage establishing clover (Charlton, 1979). Development of a trap for absolute determination of slug densities (Ferguson, 1988, unpublished data) will enable strategic use of stock treading in slug control.

LUCERNE PESTS

Lucerne was promoted in the 1960s and 1970s as a high-yielding forage crop particularly suited to semiarid environments, with resistance to major pasture pests. Only grass grub (in seedling crops), whitefringed weevil and stem nematode were considered potentially important amongst 49 pests recorded (Pottinger & McFarlane, 1967). Since that time blue-green lucerne aphid (*Acyrthosiphon kondoi* Shinji), pea aphid [*A. pisum* (Harris)], spotted alfalfa aphid (*Therioaphis trifolii f maculata* Buckton) and sitona weevil (*Sitona discoideus* Gyllenhal) have become widely established and whitefringed weevil has extended its range into major lucerne growing areas (Kain & Trought, 1982; Douglas, 1986).

Kain and Trought (1982) reviewed research on the effects of blue-green and pea aphids in lucerne production and discussed chemical and biological control possibilities. Dunbier et al. (1977, 1979) reported on the performance of lucerne cultivars in relation to blue-green aphid and stem nematode. Lucerne cultivars AS14R, WL311, WL318 from USA, and 'Rere' were more resistant to pea aphid and spotted alfalfa aphid than the standard New Zealand cv. 'Wairau' (Rohitha et al., 1985). All cultivars tested were susceptible to blue-green lucerne aphid. 'Grasslands Oranga' has resistance to blue-green lucerne aphid (Easton & Cornege, 1984). Pea aphid and blue-green aphid caused considerable damage to lucerne in the seasons after their first appearance. Since then severe infestations have not occurred. There has been renewed grower interest in returning to Wairau, considered to persist better than 'Rere' in Canterbury (S.L. Goldson, 1988 personal communication).

Sitona weevil has been a more intransigent problem, particularly in Canterbury on free draining soils which have low N retention (S.L. Goldson, 1988 personal communication). Intensive studies of the biology, pest status, and control of sitona weevil have been centred in this region (Goldson et al., 1984, 1985, 1988). Defoliation by adult weevils in summer can result in yield reductions or loss of a complete harvest. Larval damage is, however, considered to have more serious implications for lucerne productivity and persistence through destruction of root nodules and pruning of root hairs (Goldson & Proffitt, 1985). Larval density thresholds above which significant yield losses occurred were characteristically sharp, at about 2000/m^2 in 1 yr of their study. The re-establishment of nodules on lucerne roots after larval attack appears to be inhibited, possibly by a fungal interaction, thus accentuating the run-down in lucerne vigour (S.L. Goldson, 1988, personal communication). Insecticide controls as foliar sprays can be directed against the adult prior to egg laying in May (Goldson, 1984). Soil (T. E. T. Trought, 1982, personal communication) or downward translocating foliar application of systemic insecticides have not effectively controlled larvae (Barratt, 1985b). Stock treading effects in winter have not been shown to be beneficial and may directly damage the lucerne (Goldson & Wynn-Williams, 1984).

A wasp parasite of sitona adults, (*Microctonus aethiopoides* Loan), was released in Canterbury in 1982 (Stufkens et al., 1987). Although well established and showing initial promise, its ability to control sitona weevil has not been fully determined (Goldson et al., 1988).

Stem nematode can reduce production and stand life of susceptible lucerne cultivars. Resistant cultivars from the USA have been advocated, although these have not always given resistance to New Zealand races of the nematode (Dunbier et al., 1979; Greenwood et al., 1984). Root nematodes of lucerne, particularly *Meloidogyne hapla* and *Pratylenchus* spp. occur widely but their pest status in lucerne has not been established. Grandison (1976) found that vigor and rooting of seedlings improved when they were protected against root knot nematode.

Methods of establishing lucerne without cultivation have not been developed successfully enough to become widely practised (Douglas, 1986). As lucerne is normally sown into a finely prepared seedbed, pest problems during early seedling growth are minimized. Roots remain susceptible to grass grub and whitefringed weevil during the establishment year before becoming resistant and tolerant, respectively (Pottinger & McFarlane, 1967; East, 1982). In many parts of Otago where lucerne is grown, hemivoltine grass grub populations can severely deplete seedling numbers after sowing in spring if cultivation has been ineffective. Granular insecticide at planting can provide effective protection of seedling pests including whitefringed weevil (East & Parr, 1977). Lucerne probably remains a superior yielding forage plant in areas prone to drought and pasture pests (Douglas et al., 1987). The stabilization of recently arrived pest populations, and development of cultivars resistant to pests and diseases, means that it is probably timely to re-evaluate the place of lucerne in appropriate farming districts.

DISEASES OF PASTURE LEGUMES

Diseases of white and red clover have been well assessed qualitatively, but definitive information on their effect on pasture production or clover persistence is lacking. Fungicide treatments are not applied to pasture for the control of clover diseases in commercial agriculture, although they have been used extensively in northern regions for the control of epidemic outbreaks of the animal mycotoxic disease, facial eczema. The diseases of other pasture legumes have received little attention.

Virus Diseases

White clover mosaic virus (WClMV) (Fry, 1959), soybean dwarf virus (SDV) (Ashby et al., 1979a = subterranean clover red leaf virus, Johnstone & McClean, 1987), alfalfa mosaic virus (AMV) (Fry, 1952), clover yellow vein virus (CYVV) (Forster & Musgrave, 1985), red clover necrotic mosiac virus (RCNMV) (Morris-Krsinich et al., 1983), and a white clover strain of lucerne Australian latent virus (LALV), (Forster & Morris-Krsinich, 1985) have been found in white clover in New Zealand. The WClMV, SDV, and particularly in lucerne-growing regions, AMV, are widespread in New Zealand. Pastures often contain a high proportion of plants infected by one or more of these viruses (Fry, 1959; Ashby et al., 1979b, Forster et al., 1985). The CYVV and LALV are restricted to the North Island (Forster & Musgrave, 1985; Forster & Morris-Krsinich, 1985) while the incidence of RCNMV is low (R. L. S. Forster, 1988, personal communication).

The WClMV has been found to reduce dry matter production of white clover in the glasshouse by up to 25% (Fry, 1959). Effects may be more severe in mixed clover/grass swards (Catherall, 1987). Forster (1984 personal communication) recorded reductions in N_2 fixation (acetylene reduction) by white clover plants infected with WClMV, alone or in combination with other viruses, of 34% (WClMV), 51% (WClMV + AMV), and 49% (WClMV + CYVV) in field plots. Resistance to WClMV has not been found in white clover (Clark & Barclay, 1972; Catherall, 1987) but the resistance of *T. ambiguum* and *T. fragiferum* could be utilized through interspecific hybridization (Scott, 1982a; D. R. Musgrave, 1982, personal communication). Plant Diseases Division, DSIR, is investigating incorporation of subgenomic coat protein RNA from WClMV into the white clover genome to protect white clover from WClMV (Forster et al., 1987).

The AMV and CYVV affect growth, nodulation, and N_2-fixation of white clover (Akita, 1982; Barnett & Diachun, 1985; Forster & Musgrave, 1985). Control of AMV might also be obtained using a virus coat protein gene, or interspecific hybrids. Resistance to CYVV is available within white clover clones selected in the USA (D. R. Musgrave, 1983, personal communication).

The WClMV, bean yellow mosaic virus (BYMV) (= pea mosaic virus, Barnett & Diachun, 1985), SDV and AMV, but not RCNMV, have been found in red clover in New Zealand (Fry, 1952; Latch & Proctor, 1966; Ashby et al., 1979a; R. L. S. Forster, 1988, personal communication). The WClMV is reported to reduce growth (Fry, 1959; Scott, 1982b) and N_2 fixation (Khadhair et al., 1984) of red clover.

Foliage Diseases

Leafspots caused by *Leptosphaerulina trifolii* (Rost.) Petr. (pepper spot), *Mycosphaerella killianii* Petr. [= *Cymadothea trifolii* (Killian) Wolf] (sooty blotch), *Pseudopeziza trifolii* (Biv.) Fckl. (pseudopeziza leafspot), *Uromyces trifolii* (Hedwig. f. *ex* DC.) Fckl. (rust), may seasonally reach a high incidence on white clover in pastures (Skipp & Lambert, 1984). Other fungal pathogens, including *Cercospora zebrina* Pass., *Stagonospora meliloti* (Lasch.) Petr., *Ascochyta* spp., and *Phoma* spp. cause more general leaf necrosis. Increased yields of pasture and seed crops following application of fungicides (Skipp & Lambert, 1984; Lewis & Asteraki, 1987; Watson et al., 1985; P.T.P. Clifford,1985 personal communication) suggest that foliage diseases can affect clover growth in the field.

Leaf diseases caused by *L. trifolii*, *M. killianii*, *P. trifolii*, *C. zebrina*, *Phoma medicaginis* Malb. & Roum. (spring black stem and leafspot), and *Uromyces fallens* (Arthur) Barth. (rust) are present on red clover in New Zealand (Dingley, 1969; Skipp & Nan, 1986 unpublished data).

Soilborne Diseases

Roots of white clover seedlings can be destroyed by soilborne fungi which cause post-emergence damping-off. Although *Pythium* spp. (particularly *P. irregulare* Buis.) are abundant in pasture soils in New Zealand, *Thanatephorus cucumeris* (Frank) Donk (= *Rhizoctonia solani* Kuhn) probably causes the greatest loss of white clover seedlings (Skipp & Watson, 1987).

Roots of pasture plants develop a characteristic internal microflora of soilborne, root-invading fungi (Skipp et al., 1985). The fungi most commonly found in white clover roots in a survey of New Zealand pastures were: mycorrhizal fungi, *Bimuria novae-zelandiae* D. Hawksw., Chea & Sheridan, *Codinaea fertilis* Hughes & Kendrick, *Fusarium oxysporum* Schlecht., *Phoma chrysanthemicola* Hollos, species of *Chrysosporium*, *Cylindrocarpon*, and *Colletotrichum*, and fungi with sterile mycelium (Skipp & Christensen, 1983). Most of these fungi invaded undamaged roots of inoculated plants (Skipp & Christensen, 1982). These "weak pathogens" would be expected to be most damaging when plants were subject to stress e.g., drought, shading, defoliation, nutritional imbalance, and damage caused by pests and diseases (Leath et al., 1971; Menzies, 1973; Latch & Skipp, 1987). Fungi which can cause extensive root lesions [*Chalara elegans* Nag Raj & Kendrick = *Thielaviopsis basicola* (Berk. & Br.) Ferrais] and/ or stolon rotting (*Thanatephorus cucumeris*) have been detected in pastures where persistence of white clover has been poor (Skipp & Watson, 1987).

Outbreaks of clover rot caused by *Sclerotinia trifoliorum* Eriks. can occur during winter but damage is rarely severe even in pure white clover stands. In an evaluation of nine cultivars, "Grasslands Kopu" white clover showed the least clover rot. "Grasslands Pitau" and "Huia" were susceptible (Watson, 1988). The effect of *S. trifoliorum* on red clover in pasture has not been investigated, but significant reductions in plant density in the year after establishment were noted by Ledgard et al. (1988). In pure stands on experimental stations clover rot is generally more damaging to red clover than to white clover.

Evaluation of the relative persistence of overseas and New Zealand red clovers has been carried out by DSIR Grasslands Division in an experimental block maintained as a red clover monoculture (to increase inoculum of soilborne pathogens). Pot and field experiments (Skipp, 1986; Skipp et al., 1986; 1983, unpublished data) showed that establishment of red clover in this block was affected by *P. ultimum* and other *Pythium* spp. Treatment of seed and soil with metalaxyl, but not benomyl, increased numbers of surviving seedlings. Few plants which became established in the block lived for more than 2 yr. Plants developed severe cortical rot and stele necrosis of tap roots associated with infection by *Chalara elegans*, *Fusarium solani* (Mart) Sacc., *Cylindrocladium scoparium* Morgan, and *Verticillium dahliae* Kleb. Some Swiss lines of red clover which had persisted well in the block had significantly less root rot and stele necrosis than plants of the New Zealand cultivars Hamua, Turoa, and Pawera. *Chalara elegans* and *F. solani* have been found in pasture red clover in warm North Island areas, and *V. dahliae*, in a South Island seed producing region (1986 unpublished data).

Species of *Phoma* (often *P. medicaginis*), *Acremonium*, *Gliocladium*, *Cylindrocarpon*, *Fusarium*, and other fungi, have frequently been isolated from apparently healthy stele tissue of red clover plants from pastures (1985 unpublished data). They may have entered the tap root through splits in the crown caused by internal breakdown (Leath, 1985), hoof damage, and expansion of the crown. These fungi could contribute to tap root decay and ultimately cause plant death.

LUCERNE DISEASES

Virus Diseases

The AMV, a lucerne strain of LALV, and lucerne transient streak virus (LTSV) are widespread on lucerne in New Zealand (Ashby et al., 1979a) and SDV has also been reported (Wilson & Close, 1973). The incidence of LALV and LTSV has remained fairly constant. However, AMV in lucerne in North Island stands has increased dramatically since the mid-1970s due to the entry and spread of two vectors, the blue-green aphid and pea aphid, and the importation of infected seed from the USA (Forster et al., 1985).

Reported effects of AMV on growth and nodulation of lucerne (Forster et al., 1985; Ohki et al., 1986) indicate that control measures are desirable. Breeding lines have been identified in New Zealand in which a large proportion of plants has remained free of AMV for up to 5 yr in the field (Forster, 1988 personal communication).

Foliage Diseases

Leafspots caused by *Pseudopeziza medicaginis* (Lib.) Sacc. (common leafspot), *Leptosphaerulina briosiana* (Poll.) Graham and Luttrell (pepper spot), *Phoma medicaginis*, and *Stemphylium botryosum* Wallr. (stemphylium leafspot), and to a lesser extent *Pseudopeziza jonesii* Nannf. (yellow leaf blotch), *Uromyces striatus* Schroet. (rust) and *Stagonospora meliloti* (stagonospora leafspot and root rot) damage lucerne crops in New Zealand (Hart & Close, 1976; Harvey & Martin, 1980; Close et al., 1982).

Soilborne Diseases

Infection of roots of lucerne seedlings by *Pythium* spp. causes post-emergence damping-off. *Pythium irregulare* is common in pasture soils (Skipp & Watson, 1987), *P. ultimum,* and other *Pythium* spp., in soils from lucerne stands (1985 unpublished data). Lucerne establishment can be improved by treatment of seed with systemic fungicides (Falloon & Skipp, 1982).

Bacterial wilt caused by *Corynebacterium michiganense* pv. *insidiosum* (McCulloch) Dye & Kemp is widespread in New Zealand and can affect persistence of susceptible cultivars (Dunbier et al., 1981). The use of resistant cultivars from the USA and New Zealand ('Rere' and 'Oranga') has reduced the significance of the disease (Dunbier & Sanderson, 1976; Close et al., 1982). However, these cultivars are susceptible to verticillium wilt (caused by *Verticillium albo-atrum* Reinke & Berth.). Verticillium wilt is widespread in New Zealand and can reduce yield and persistence of stands (Dunbier et al., 1981; Hawthorne, 1983). Variation in wilt susceptibility can be found within and between cultivars (Hawthorne, 1983, 1987). Current DSIR lucerne breeding programmes include selection for resistance to verticillium wilt.

Crown rot and root decay can be an important cause of declining productivity in lucerne stands (Stephen et al., 1982). Soilborne pathogenic fungi found associated with the crown rot syndrome include *Thanatephorus cucumeris, Stagonospora meliloti, Phoma medicaginis, Chalara elegans, Sclerotinia trifoliorum,* and species of *Fusarium, Diaporthe* and *Cylindrocarpon* (Close et al., 1982; Hawthorne, 1983 personal communication).

CONCLUSION

In comparison with USA and Australia, New Zealand pastoral agriculture is characterized by a narrow diversity of climatic influences, pasture species in commercial use, and major pests and diseases. Pests and diseases do, however, cause acute or subclinical damage in pasture. These types of damage have distinctive features in relation to clover persistence.

Acute damage is largely the result of insect attack. It is generally sporadic, involves one pest, and affects both grass and clover in pasture. The short duration of pest attack, and neutral effect on grass-clover competitive balance, means that even after severe damage, clover usually recovers along with the rest of the pasture. Acute damage by insects is easily recognized, and quantifiable in terms of lost pasture availability or animal production. Intensive research to ameliorate insect losses in pasture has resulted in multiple options for control, in the form of pest management packages. Current research is directed towards development of control strategies which are compatible with natural regulation of pests. Areas where understanding of individual pests is limited are outlined by Pottinger (1985).

Chronic, subclinical damage may result in even greater loss of productive potential in pasture than that caused by sporadic insect attack. Such damage, however, is difficult to quantify and commonly is the result of a complex interaction among pests (e.g., root-feeding nematodes), pathogens (e.g., viruses or root-rotting fungi) and environmental factors. The damaging agents may act disproportionately against clover, reducing its competitive vigor. The development of more effective means to quantify effects of pest and disease complexes, and identification of sources of resistance or tolerance provide major challenges and opportunities for progress in the search for greater persistence of pasture legumes.

Multidisciplinary studies are needed to quantify and integrate the interactive effects between plants, animals, pests, diseases and the environment in determining grass-clover balance.

Cooperative potential between countries exists for all levels of pest and disease research towards improving legume persistence. In particular the development of standardized quantitative assessment methods, biological control, plant breeding and methods for improving legume establishment could benefit. Comparative studies of the epidemiology of specific pests or diseases may reveal natural controls or managerial practices which could be mutually exploited. Exchange of plant ecotypes for screening against nematodes or diseases may find genetic material which has resistance or tolerance to local races of pathogen.

REFERENCES

Akita, S. 1982. Effects of alfalfa mosaic virus on the growth of swards of white clover. Bull. Natl. Grassl. Res. Inst. 21:54-66.

Anonymous. 1956. A legume that slugs leave alone. Oregon's Agric. Progr. 3:7-8.

Ashby, J. W., R. L. S. Forster, J. D. Fletcher, and P. B. Teh. 1979a. A survey of sap-transmissible viruses of lucerne in New Zealand. N.Z. J. Agric. Res. 22:637-640.

Ashby, J. W., P. B. Teh, and R. C Close. 1979b. Symptomatology of subterranean clover red leaf virus and its incidence in some legume crops, weed hosts, and certain alate aphids in Canterbury, New Zealand. N.Z. J. Agric. Res. 22:361-365.

Barker, G. M. 1986. Biology of pest slugs and their significance in conservation-tillage systems. p. 83-106. *In* R. R. Hill et al. (ed.) Proc. Int. Symp. on Establishment of Forage Crops by Conservation-Tillage: Pest Management. U.S. Regional Pasture Res. Lab., University Park, PA.

Barker, G. M., P. J. Addison, and R. P. Pottinger. 1985. Biology and pest status of slugs (Mollusca) in two Waikato pastures. p. 18-25 *In* R.B. Chapman (ed.) Proc. 4th Australasian Conf. on Grassl. Invert. Ecol. Caxton Press, Christchurch.

Barker, G. M., L. N. Robertson, R. N. Watson, and B. E. Willoughby. 1988. Pasture renovation: interactions of vegetation control on pest infestations. Proc. 5th Australasian Conf. on Grassl. Invert. Ecol. August 1988, Melbourne. (In press.)

Barnett, O. W., and S. Diachun. 1985. Virus diseases of clovers. p. 235-268. *In* N. L. Taylor (ed.) Clover science and technology. ASA, CSSA, and SSSA, Madison, WI.

Barratt, B. I. P. 1985a. Effect of cotyledon damage on nodulation and growth of white clover oversown into native grassland in Central Otago, New Zealand. p. 127-132. *In* R. B. Chapman (ed.) Proc. 4th Australasian Conf. on Grassl. Invert. Ecol. Caxton Press, Christchurch.

Barratt, B. I. P. 1985b. An attempt to control *Sitona discoideus* larvae with systemic insecticides. Proc. N.Z. Weed Pest Control Conf. 38:35-37.

Barratt, B. I. P., and P. D. Johnstone. 1984. Effects of insects, seeding rate and insecticide seed dressings on white clover and Maku lotus in tussock grassland. N.Z. J. Agric. Res. 27:13-18.

Biggs, D. R., G. A. Lane, G. B. Bussell, D. L. Gaynor, and O. R. W. Sutherland. 1986. Biochemical mechanisms of pest resistance in pasture legumes. Proc. Plant Breeding Symp. DSIR 1986. Agron. Soc. N.Z. Spec. Publ. 5:316-321.

Blank, R. H. 1984. Adoption by Northland farmers of a pest management package to control black field cricket, *Teleogryllus commodus* (Orthoptera:Gryllidae). N.Z. Entomol. 8:37-41.

Blank, R. H., D. S. Bell, and H. M. Morgan. 1980. Black field crickets consume oversown pasture seed. Proc. N.Z. Weed Pest Control Conf. 33:37-40.

Brock, J. L. 1986. Some observations of pasture management effects on grass grub, porina and earthworm populations. Proc. N.Z. Grassl. Assoc. 47:273-278.

Carpenter, A., T. K. Wyeth, and P. J. T. Allan. 1981. Chemical control of grass grub in the southern North Island: Results of 1977-79 trials. Proc. N.Z. Weed Pest Control Conf. 34:225-228.

Catherall, P. L. 1987. Reaction of some white clover (*Trifolium repens*) varieties to inoculation with white clover mosaic virus. Tests of Agrochemicals and Cultivars 8 Ann. Appl. Biol. 110(Suppl.):142-143.

Charlton, J. F. L. 1979. Effects of slugs during establishment of oversown legumes in box experiments. p. 253-255. *In* T. K. Crosby and R.P. Pottinger (ed.) Proc. 2nd Australasian Conf. on Grassl. Invert. Ecol. Government Printer, Wellington.

Charlton J. F. L., and E. R. Thom. 1984. Establishment and persistence of new herbage species and cultivars. N.Z. Agric. Sci. 18:130-135.

Clark. M. F., and P. C. Barclay. 1972. The use of immuno-osmophoresis in screening a large population of *Trifolium repens* L. for resistance to white clover mosaic virus. N.Z. J. Agric. Res. 15:371-375.

Clements, R. O. 1986. The impact of insect pests during the establishment of forage crops and some possible solutions to the problems. p. 7-22. *In* R. R. Hill et al. (ed.) Proc. Int. Symp. on Establishment of Forage Crops by Conservation Tillage:Pest Management. U.S. Regional Pasture Res. Lab., University Park, PA.

Close, R. C., I. C. Harvey, and F. R. Sanderson. 1982. Lucerne diseases in New Zealand and their control. p. 61-69. *In* R. B. Wynn-Williams (ed.) Lucerne for the 80's. Spec. Publ. 1. Agron. Soc. N.Z.

Dijkstra, J. 1971. Breeding for resistance to *Heterodera* in white clover. Euphytica 20:36-46.

Dingley, J. M. 1969. Records of plant diseases in New Zealand. Bull. N.Z. Dept. Sci. Indust. Res. 192.

Douglas, J. A. 1986. The production and utilization of lucerne in New Zealand. Grass Forage Sci. 41:81-128.

Douglas, M. H., D. W. Brash, B. I. P. Barratt, and J. M. Keoghan. 1987. Successful lucerne growing in inland Otago. Proc. N.Z. Grassl. Assoc. 48:193-197.

Dunbier, M. W., R. C. Close, and T. J. Ellis. 1976. Disease- and pest-resistant lucerne cultivars for New Zealand. Proc. N.Z. Weed Pest Control Conf. 29:46-49.

Dunbier, M. W., W. M. Kain, and K. B. McSweeney. 1977. Performance of lucerne cultivars resistant to blue-green lucerne aphid in New Zealand. Proc. N.Z. Weed Pest Control Conf. 30:155-159.

Dunbier, W. M., T. P. Palmer, T. J. Ellis, and P. A. Burnett. 1979. Field evaluation of lucerne cultivars for *Ditylenchus dipsaci* (Nematoda:Tylenchidae) and *Acyrthosiphon kondoi* (Hemiptera:Aphididae) p. 99-102. *In* T. K. Crosby and R. P. Pottinger (ed.) Proc. 2nd Australasian Conf. on Grassl. Invert. Ecol. Government Printer, Wellington.

Dunbier, M. W., and F. R. Sanderson. 1976. Lucerne cultivars resistant to disease. N.Z. J. Agric. 132(2):10-12.

Dunbier, M. W., F. R. Sanderson, and T. J. Ellis. 1981. Effect of bacterial wilt and verticullium wilt on six lucerne cultivars. Proc. N.Z. Weed Pest Control Conf. 34:21-24.

East, R. 1979. Effects of grazing management on *Costelytra zealandica* populations (Coleoptera:Scarabaeidae). p. 180-184. *In* T. K. Crosby and R. P. Pottinger (ed.) Proc. 2nd Australasian Conf. on Grassl. Invert. Ecol. Government Printer, Wellington.

East, R. 1982. Interactions between whitefringed weevil *Graphognathus leucoloma* and legume species in the northern North Island. N.Z. J. Agric. Res. 25:131-140.

East, R. 1985. A review of progress in the development of pasture pest managements in the northern North island. p. 353-358. *In* R. B. Chapman (ed.) Proc. 4th Australasian Conf. in Grassl. Invert. Ecol. Caxton Press, Christchurch.

East, R., and W. M. Kain. 1982. Prediction of grass grub, *Costelytra zealandica* (Coleoptera:Scarabaeidae) populations. N.Z. Entmol. 7:222-227.

East, R., and J. Parr. 1977. Chemical control of whitefringed weevil in lucerne. Proc. N.Z. Weed Pest Control Conf. 30: 50-55.

East, R., and R. P. Pottinger. 1975. Starling (*Sturnus vulgaris* L.) predation on grass grub (*Costelytra zealandica* (White), Melolonthinae) populations in Canterbury. N.Z. J. Agric. Res. 18:417-452.

East, R., and R. P. Pottinger. 1982. Integrated pasture pest management in New Zealand. p. 102-113. *In* P. J. Cameron et al. (ed.) Proc. Australasian Workshop on Development and Implementation of IPM. Government Printer, Wellington.

East, R., and R. P. Pottinger. 1983. Use of grazing animals to control insect pests of pasture. N.Z. Entomol. 7:352-359.

East, R., and R. P. Pottinger. 1984. The cost of pasture pests. N.Z. Agric. Sci. 18:136-140.

East, R., R. A. Prestidge, and L. N. Robertson. 1986. Recent advances in pasture pest management in the northern North Island. N.Z. J. Ecol. 9:101-109.

East, R., R. N. Watson, and R. D. Welsh. 1978. Damage to pasture legumes by larvae of the common blue, *Zizina otis labradus* (Lepidoptera:Lycaenidae). N.Z. Etomol. 6:390-391.

East, R., R. D. Welsh, and C. M. Miller. 1975. Control of whitefringed weevil adults with insecticides. Proc. N.Z. Weed Pest Control Conf. 28:213-216.

East, R., and B. E. Willoughby. 1980. Effects of summer defoliation on grass grub (*Costelytra zealandica*) populations. N.Z. J. Agric. Res. 23:547-562.

East, R., B. E. Willoughby, N. A. Haigh, and N. R. Wrenn. 1980. Chemical control of Tasmanian grass grub in the northern North Island. Proc. N.Z. Weed Pest Control Conf. 33:155-157.

East, R., B. E. Willoughby, and N. R. Wrenn. 1981. Evaluation of the solid stream application technique for grass grub control in the North Island. Proc. N.Z. Weed Pest Control Conf. 34:229-233.
Easton, H. S., and E. Cornege. 1984. 'Grasslands Oranga' lucerne (*Medicago sativa*). N.Z. J. Exp. Agric. 12:283-286.
Falloon, R. E., and R. A. Skipp. 1982. Fungicide seed treatments improve lucerne establishment. Proc. N.Z. Weed Pest Control Conf. 35:127-129.
Farrell, J. A. K. 1976. Field microplot trials on plant resistance to porina. Proc. N.Z. Weed Pest Control Conf. 29:168-171.
Ferguson, C. 1984. Slug feeding on seeds and seedlings of ryegrass and white clover. Proc. N.Z. Weed Pest Control Conf. 36:212-215.
Ferguson, C. M., and B. I. P. Barratt. 1986. Damage assessment and control of slugs in direct-drilled pastures in southern New Zealand. p. 113. *In* R. R. Hill et al. (ed.) Proc. Int. Symp. on Establishment of Forage Crops by Conservation Tillage: Pest Management. U.S. Regional Pasture Res. Lab., University Park, PA. (abstract only).
Ferguson, C. M., B. I. P. Barratt, and P. A. Jones. 1988. Control of the grey field slug [*Deroceras reticulatum* (Müller)] by stock management prior to direct drilled pasture establishment. J. Agric. Sci. (In press.)
Fleming, S., R. Archibald, K. Stewart, and J. Kalmakoff. 1982. Natural regulation of *Wiseana* spp. by an enzootic disease. p. 283-298. *In* K.E. Lee (ed.) Proc. 3rd Australasian Conf. on Grassl. Invert. Ecol. S.A. Government Printer, Adelaide.
Forster, R. L. S., P. J. Guilford, and D. V. Faulds. 1987. Characterization of the coat protein subgenomic RNA of white clover mosaic virus. J. Gen. Virol. 68:181-190.
Forster, R. L. S., and B. A. M. Morris-Krsinich. 1985. A distinct strain of lucerne Australian latent virus in white clover in New Zealand. Ann. Appl. Biol. 107:449-454.
Forster, R. L. S., B. A. M. Morris-Krsinich, and D.R. Musgrave. 1985. Incidence of alfalfa mosaic virus, lucerne Australian latent virus, and lucerne transient streak virus in lucerne crops in the North Island of New Zealand. N.Z. J. Agric. Res. 28:279-282.
Forster, R. L. S., and D. R. Musgrave. 1985. Clover yellow vein virus in white clover (*Trifolium repens*) and sweet pea (*Lathyrus odoratus*) in the North Island of New Zealand. N.Z. J. Agric. Res. 28:575-578.
Fry, P. R. 1952. Occurrence of lucerne-mosaic virus in New Zealand. N.Z. J. Sci. Tech., Sec.A 34:320-326.
Fry, P. R. 1959. A clover mosaic virus in New Zealand pastures. N.Z. J. Agric. Res. 2:971-981.
Gaynor, D. L., and R. A. Skipp. 1988. Pests. p. 461-492 In M. J. Baker and W. M. Williams (eds.) White clover. CAB International, Wallingford, England.
Goldson, S. L. 1984. Sitona weevil in lucerne: biology and control. MAF, AgLink FPP548.
Goldson, S. L., C. B. Dyson, J. R. Proffitt, J. R. Frampton, and J. A. Logan. 1985. The effect of *Sitona discoideus* Gyllenhall (Coleoptera:Circulionidae) on lucerne yields in New Zealand. Bull. Entmol. Res. 75:429-442.
Goldson, S. L., E. R. Frampton, B. I. P. Barratt, and C. M. Ferguson. 1984. The seasonal biology of *Sitona discoideus* Gyllenhall Coleoptera:Circulionidae), an introduced pest of New Zealand lucerne. Bull. Entomol. Res. 74:249-259.

Goldson, S. L., E. R. Frampton, and J. R. Proffitt. 1988. Population dynamics and larval establishment of *Sitona discoideus* Coleoptera:Circulionidae) in New Zealand lucerne. Ann. Appl. Biol. 112:177-195.

Goldson, S. L., and J. R. Proffitt. 1985. Measurement of the impact of different larval densities of *Sitona discoideus* Gyllenhall on Canterbury lucerne. p. 26-34. In R. B. Chapman (ed.) Proc. 4th Australasian Conf. on Grassl. Invert. Ecol. Caxton Press, Christchurch.

Goldson, S. L., and R. B. Wynn-Williams. 1984. The effect of winter mob-stocking on sitona larval populations in lucerne. Proc. N.Z. Weed Pest Control Conf. 37:110-112.

Grandison, G. S. 1976. Root knot and stem nematodes of lucerne. Proc. N.Z. Weed Pest Control Conf. 29:31-34.

Greenwood, P. B., G. W. Yeates, and G. W. Sheath. 1984. Effect of nematicide on stem nematode (*Ditylenchus dipsaci*) populations and productivity, and survival of 3 lucerne cultivars in North Otago. N.Z. J. Agric. Res. 27:557-562.

Haack, N. A., W. M. Kain, and B. P. Springett. 1982. The use of insecticidal baits for the control of porina (*Wiseana* spp). p. 349-354. *In* K. E. Lee (ed.) Proc. 3rd Australasian Conf. on Grassl. Invert. Ecol. S.A. Government Printer, Adelaide.

Hart, R. I. K., and R. C. Close. 1976. Control of leaf diseases of lucerne with benomyl. Proc. N.Z. Weed Pest Control Conf. 29:42 45.

Harvey, I. C., and R. J. Martin. 1980. Leaf spot diseases on lucerne cultivars. N.Z. J. Exp. Agric. 8:295-296.

Hawthorne, B. T. 1983. Variation in pathogenicity among isolates of *Verticillium albo-atrum* from lucerne. N.Z. J. Agric. Res. 26:405-408.

Hawthorne, B. T. 1987. Qualitative and quantitative assessments of the reaction of lucerne plants to verticillium wilt. N.Z. J. Agric. Res. 30:349-359.

Hay, M. J. M., J. L. Brock, V. J. Thomas, and M. V. Knighton. 1988. Seasonal and sheep grazing management effects on branching structure and dry weight of white clover plants in mixed swards. Proc. N.Z. Grassl. Assoc. 49:197-201.

Henzell, R. F., and D. R. Lauren. 1983. Conventional and microencapsulated spray and bait formulation for porina control. Proc. N.Z. Weed Pest Control Conf. 36:190-194.

Henzell, R. F., D. R. Lauren, and M. R. Briscoe. 1985. The relationship between grass grub beetle mating and pheromone trap catches. p. 35-38. *In* R.B. Chapman (ed.) Proc. 4th Australasian Conf. on Grassl. Invert. Ecol. Caxton Press, Christchurch.

Henzell, R. F., D. R. Lauren, and R. East. 1979. Effect on the egg hatch of whitefringed weevil of feeding lucerne treated with the insect growth regulator diflubenzuron. N.Z. J. Agric. Res. 22:197-200.

Jackson, T. A. 1988. Development of *Serratia entomophila* as an inundative biological control agent for grass grub (*Costelytra zealandica*). *In* Proc. 5th Australasian Conference on Grassl. Invert. Ecol. August 1988, Melbourne. (In press.)

Jackson, T. A., N. D. Barlow, J. F. Pearson, and R. A. French. 1988. The effects of paddock rotation policies on grass grub (*Costelytra zealandica*) damage in Canterbury. *In* Proc. 5th Australasian Conference on Grassl. Invert. Ecol. August 1988, Melbourne. (In press.)

Johnstone, G. R., and G. D. McClean. 1987. Virus diseases of subterranean clover. Ann. Appl. Biol. 110:421-440.

Kain, W. M., D. S. Atkinson, and R. K. Darwin. 1982a. A development and implementation of integrated pest management of pasture pests in intensive dairy systems in the Dannevirke district of southern Hawkes Bay. p. 91-101. *In* P. J. Cameron et al. (ed.) Proc. Australasian Workshop on Development and Implementation of IPM. Government Printer, Wellington.

Kain, W. M., R. East, and J. A. Douglas. 1979. *Costelytra zealandica* - pasture species relationships on the pumice soils of the central North Island of New Zealand (Coleoptera:Scarabaeidae). p. 86-90. *In* T. K. Crosby and R. P. Pottinger (ed.) Proc. 2nd Australasian Conf. on Grassl. Invert. Ecol. Government Printer, Wellington.

Kain, W. M., N. A. Thomson, I. C. Plunkett, D. S. Atkinson, and J. R. Malcolm. 1982b. Development and implementation of integrated management of pasture pests in intensive dairy farms of the southern North island region of New Zealand. p. 339-348. *In* K. E. Lee (ed.) Proc. 3rd Australasian Conf. on Pasture Insect Ecol. S.A. Government Printer, Adelaide.

Kain, W. M., and T. E. T. Trought. 1982. Insect pests of lucerne in New Zealand. p. 49-60. *In* R.B. Wynn-Williams (ed.). Lucerne for the 80's. Spec. Pub. No. 1. Agron. Soc. N.Z.

Khadhair, A. H., R. C. Sinha, and J. F. Peterson. 1984. Effect of white clover mosaic virus on various processes to symbiotic N_2 fixation in red clover. Can. J. Bot. 62:38-43.

King, P. D., and R. East. 1979. Effects of pasture composition on the dynamics of *Heteronychus arator* and *Graphognathus leucoloma* populations (Coleoptera:Scarabaeidae and Circulionidae). p. 79-83. *In* T. K. Crosby and R. P. Pottinger (ed.) Proc. 2nd Australasian Conf.l on Grassl. Invert. Ecol. Government Printer, Wellington.

Latch, G. C. M., and W. M. Kain. 1983. Control of porina caterpillar (*Wiseana* spp.) in pasture by the fungus *Metarhizium anisophliae.* N.Z. J. Exp. Agric. 11:351-354.

Latch, G. C. M., and C. H. Proctor. 1966. Incidence of some viruses on *Trifolium pratense* L. in New Zealand pastures. N.Z. J. Agric Res. 9:726-728.

Latch, G. C. M., and R. A. Skipp. 1987. Diseases. p. 421-460. *In* M. J. Baker and W. M. Williams (ed.) White clover. CAB International, Wallingford, England.

Lauren, D. R., R. F. Henzell, and N. R. Wrenn. 1981. Comparison of two spray data prediction techniques for control of grass grub beetles. Proc. N.Z. Weed Pest Control Conf. 34:243-246.

Lauren, D. R., R. F. Henzell, and N. R. Wrenn. 1982. A comparison of lindane broadcast and drilled in spring for grass grub beetle control. Proc. N.Z. Weed Pest Control Conf. 35:94-97.

Leath, K. T. 1985. General diseases. p. 204-233. *In* N. L. Taylor (ed.) Clover science and technology. ASA, CSSA, and SSSA, Madison, WI.

Leath, K. T., F. L. Lukezic, H. W. Crittenden, E. S. Elliott, P. M. Halisky, F. L. Howard, and S. A. Ostazeski. 1971. The *Fusarium* root rot complex of selected forage legumes in the northeast. Bull. Penn. Agric. Exp. Stn. 777.

Ledgard, S. F., G. J. Briar, and R. N. Watson. 1988. New cultivars for Waikato dairy pasture: establishment, production and nitrogen fixation during the first year. Proc. N.Z. Grassl. Assoc. 49:207-211.

Lewis, G. C., and E. J. Asteraki. 1987. Incidence and severity of damage by pests and diseases to leaves of twelve cultivars of white clover. Tests of Agrochemicals and Cultivars 8 Ann. Appl. Biol. 110(Suppl.):140-141.

Longworth, J. F. 1982. Pathogen, parasites and predators in pest control. p. 137-146. *In* K. E. Lee (ed.) Proc. 3rd Australasian Conf. on Grassland Invert. Ecol. S.A. Government Printer, Adelaide.

Menzies, S. A. 1973. Root rot of clover caused by *Codinaea fertilis*. N.Z. J. Agric. Res. 16:239-245.

Mercer, C. F. 1988. Reaction of some *Trifolium* species to *Meloidogyne hapla* and *Heterodera trifolii*. *In* Proc. 5th Australasian Conf. on Grassld. Invert. Ecol. August 1988, Melbourne. (In press.)

Mespel, G. J. van der, W. M. Kain, and W. M. Wouts. 1986. Evaluation of an entomophagous nematode *Steinernema bibionis*) for control of porina larvae (*Wiseana* spp.) in the Otaki district. Proc. N.Z. Weed Pest Control Conf. 39:114-116.

Miln, A. J. 1982. The effects of cultivation on disease incidence in grass grub populations. Proc. N.Z. Weed Pest Control Conf. 35:86-89.

Miln, A. J., and A. Carpenter. 1979. Relationship between pathogen and insecticide in grass grub. Proc. N.Z. Weed Pest Control Conf. 32:92-95.

Morris-Krsinich, B. A. M., R. L. S. Forster, and D. W. Mossop. 1983. Translation of red clover necrotic mosaic virus RNA in rabbit reticulocyte lysate: Identification of the virus coat protein cistron on the larger RNA strand of the bipartite genome. Virology 124:349-356.

Ohki, S. T., W. T. Leps, and C. Hiruki. 1986. Effects of alfalfa mosaic virus infection on factors associated with symbiotic N_2 fixation in alfalfa. Can. J. Plant. Pathol. 8:277-281.

Pottinger, R. P. 1985. The future of pasture pest research in relation to New Zealand grassland agriculture. p. 359-366. *In* R. B. Chapman (ed.) Proc. 4th Australasian Conf. on Grassld. Invert. Ecol. Caxton Press, Christchurch.

Pottinger, R. P., and R. P. McFarlane. 1967. Insect pests and nematodes in lucerne. p. 229-247. *In* R. H. M. Langer (ed.) The Lucerne Crop. Reed, Wellington.

Pottinger, R. P., N. R. Wrenn, and R. A. McGhie. 1985. Timing treatments for control of lucerne flea (*Sminthurus viridus*) in the northern North Island. p. 133-140. *In* R. B. Chapman (ed.) Proc. 4th Australasian Conf. Grassl. Invert. Ecol. Caxton Press, Christchurch.

Prestidge, R. A., and R. East. 1984. Use of fertilizer nitrogen to manipulate pasture plant quality and compensate for damage by grass grub, *Costelytra zealandica* (Coleoptera:Scarabaeidae). N.Z.Entomol. 8:24-29.

Risk, W., and T. E. Ludecke. 1978. Some factors affecting pasture in Southland. Proc. N.Z. Grassl. Assoc. 40:51-59.

Robertson, L. N., R. H. Davison, and A. C. Firth. 1981. Insecticide rates, seeding rates and Australian soldier fly control in pasture. Proc. N.Z. Weed Pest Control Conf. 34:161-163.

Robertson, L. N., R. P. Pottinger, P. J. Gerard, and G. M. Dixon. 1979. *Inopus rubriceps* control in pasture by grazing management (Diptera:Stratiomyidae). p. 185-188. *In* T. K. Crosby and R. P. Pottinger (ed.) Proc. 2nd Australasian Conf. on Grassl. Invert. Ecol. Government Printer, Wellington.

Rohitha, B. H., R. P. Pottinger, and A. C. Firth. 1985. Lucerne aphid resistance in five lucerne cultivars grown in the Waikato. p. 80-84. *In* R. B. Chapman (ed.) Proc. 4th Australasian Conf. on Grassl. Invert. Ecol. Caxton Press, Christchurch.

Savage, M. J., and R. A. French. 1981. Porina chemical control: determination of application time. MAF, AgLink FPP588.

Scott, S. W. 1982a. Tests for resistance to white clover mosaic virus in red and white clover. Ann. Appl. Biol. 100:393-398.

Scott, S. W. 1982b. The effects of white clover mosaic virus infection on the yield of red clover (*Trifolium pratense* L.) in mixtures and in pure stands. J. Agric. Sci. 98:455-460.

Skipp, R. A. 1986. Fungal root pathogens in pasture soils and effects of fungicides. Proc. N.Z. Weed Pest Control Conf. 39:60-64.

Skipp, R. A., and M. J. Christensen. 1982. Invasion of white clover roots by fungi and other soil micro-organisms. III. The capacity of fungi isolated from white clover roots to invade seedling root tissue. N.Z. J. Agric. Res. 25:97-101.

Skipp, R. A., and M. J. Christensen. 1983. Invasion of white clover roots by fungi and other soil micro-organisms. IV. Survey of root-invading fungi and nematodes in some New Zealand pastures. N.Z. J.Agric. Res. 26:151-155.

Skipp, R. A., M. J. Christensen, and Z.-B. Nan. 1986. Invasion of red clover (*Trifolium pratense*) roots by soilborne fungi. N.Z. J. Agric. Res. 29:305-315.

Skipp, R. A., and D. L. Gaynor, 1988. Pests - nematodes. p 493-512. *In* M. J. Baker and W. M. Williams (ed.) White clover. CAB International, Wallingford, England.

Skipp, R. A., and M. G. Lambert. 1984. Damage to white clover foliage in grazed pastures caused by fungi and other organisms. N.Z. J. Agric. Res. 27:313-320.

Skipp, R. A., G. C. M. Latch, and M. J. Christensen. 1985. Fungal invasion of clover and grass roots in New Zealand pasture soils. p. 66-68. *In* C. A. Parker et al. (ed.) Ecology and management of soilborne plant pathogens. APS, St. Paul, MN.

Skipp, R. A., and R. N. Watson. 1987. Pot experiments with pasture soils to detect soilborne pathogens of white clover and lucerne, and effects of field application of fungicides. N.Z. J. Agric. Res. 30:85-93.

Stephen, R. C., D. J. Saville, I. C. Harvey, and J. Hedley. 1982. Herbage yields and persistence of lucerne (*Medicago sativa* L.) cultivars and the incidence of crown and root diseases. N.Z. J. Exp. Agric. 10:323-332.

Stewart, K. M. 1984. Control of grass grub by drilling insecticide granules into pasture. Proc. N.Z. Weed Pest Control Conf. 37:117-120.

Stewart, K. M., and R. D. Archibald. 1987. The effects of pasture management on population density and diseases of porina (Lepidoptera:Hepialidae). N.Z. J. Exp. Agric. 15:375-379.

Stewart, K. M., R F. van Toor, and B. I. P. Barratt. 1988. An 'insecticide test plot' technique for early detection of pasture damage by grass grub, *Costelytra zealandica* (White) (Coleoptera:Scarabaeidae). N.Z. J. Agric. Res. (In press.)

Stufkens, M. W., J. A. Farrell, and S. L. Goldson. 1987. Establishment of *Microctonus aethiopoides*, a parasitoid of the sitona weevil, in New Zealand. Proc. N.Z. Weed Pest Control Conf. 40:31-32.

Sutherland, O. R. W. 1979. Invertebrate-plant relationships and breeding pest-resistant plants. p. 84-88. *In* T. K. Crosby and R. P. Pottinger (ed.) Proc. 2nd Australasian Conf. on Grassl. Invert. Ecol. Government Printer, Wellington.

Thomson, N. A., M. R. Lawrence, A. J. Popay, J. F. Lagan, and W. M. Kain. 1985. The effect of long term use of insecticide for grass grub control on grass grub numbers and milkfat production. p. 305-312. *In* R. B. Chapman (ed.) Proc. 4th Australasian Conf. on Grassl. Invert. Ecol. Caxton Press, Christchurch.

Toor, R. F. van, R. S. Littlejohn, and K. M. Stewart. 1988. Fensulfothion control of grass grub [*Costelytra zealandica* (White)] in combination with nitrogen fertilizer to ameliorate pasture damage. N.Z. J. Agric. Res. (In press.)

Toor, R. F. van, and K. M. Stewart. 1987. Comparison of a grooved and smooth roller for control of grass grub. Proc. N.Z. Weed Pest Control Conf. 40:191-193.

Trought, T. E. T. 1979. Review of chemical control of *Costelytra zealandica* in New Zealand, 1968-78 (Coleoptera:Scarabaeidae). p. 160-164. *In* T. K. Crosby and R. P. Pottinger (eds.) Proc. 2nd Australasian Conf. on Grassl. Invert. Ecol. Government Printer, Wellington.

van den Bosch, J., and D. L. Gaynor. 1986. Screening white clover for grass grub resistance. p. 311-315. *In* T. A. Williams and G. S. Wratt (ed.) Plant breeding Symp. DSIR 1986. Spec. Publ. 5. Agron. Soc. N.Z.

van den Bosch, J., J. A. Lancashire, B. M. Cooper, T. B. Lyons, and W. M. Williams. 1986. G18 white clover - a new cultivar for intensive lowlands. Proc. N.Z. Grassl. Assoc. 47:173-178.

Watson, R. N. 1988. Infection of nine white clover cultivars by *Sclerotinia trifoliorum* and *Ditylenchus dipsaci*. Proc. N.Z. Weed Pest Control Conf. 41:121-125.

Watson, R. N., G. M. Barker, L. N. Robertson, R. S. Marsden, and E. A. Wilson. 1986. Nematicidal protection of ryegrass seedlings in pasture renovation trials. Proc. N.Z. Weed and Pest Control Conf. 39:41-44.

Watson, R. N., G. W. Yeates, R. A Littler, and K. W. Steele. 1985. Responses in nitrogen fixation and herbage production following pesticide applications on temperate pastures. p. 103-113. *In* R. B. Chapman (ed.) Proc. 4th Australasian Conf. on Grassl. Invert. Ecol. Caxton Press, Christchurch.

West, C. P., and K. W. Steele. 1986. Tolerance of white clover cultivars to stem nematode (*Ditylenchus dipsaci*). N.Z. J. Exp. Agric. 14:227-229.

White, D. W. R. 1988. Use of cell and molecular genetic manipulation to improve pasture plants. Proc. N.Z. Grassl. Assoc. 49:67-72.

White, D. W. R., and D. Greenwood. 1986. Development of a transformation system in white clover using modified Ti-plasmid vectors. p. 349-354. *In* T. A. Williams and G. S. Wratt (ed.) Proc. Plant Breeding Symp. DSIR 1986. Spec. Publ. No. 5. Agron. Soc. N.Z.

Wigley, P. J. 1985. An evaluation of strategies for using insect pathogens to control pasture pests in New Zealand. p. 276-285. *In* R.B. Chapman (ed.) Proc. 4th Australasian Conf. on Grassl. Invert. Ecol. Caxton Press, Christchurch.

Williams, W. M., and E. G. Williams. 1981. Use of embryo culture with nurse endosperm for interspecific hybridization in pasture legumes. p. 163-165. *In* J. Allan Smith and V. W. Hays (ed.) Proc. 14th Int. Grassl. Congr., Lexington, KY. Westview Press, Boulder, CO.

Wilson, E. R. L. 1978. Effect of grass grub on establishment of legume seedlings in pots. N.Z. J. Agric. Res. 21:727-731.

Wilson, E. R. L., and J. A. K. Farrell. 1979. Some problems encountered in screening white clover for resistance to *Costelytra zealandica* (Coleoptera:Scarabaeidae). p. 99-94. *In* T. K. Crosby and R. P. Pottinger (ed.) Proc. 2nd Australasian Conf. on Grassl. Invert. Ecol. Government Printer, Wellington.

Wilson, J., and R. C. Close. 1973. Subterranean clover red leaf virus and other legume viruses in Canterbury. N.Z. J. Agric. Res. 16:305-310.

Wrenn, N. R., and R. A. McGhie. 1986. An evaluation of the insect growth regulators diflubenzuron and teflubenzuron for control of lucerne flea (*Sminthurus viridus*). Proc. N.Z. Weed Pest Control Conf. 39:27-32.

Wrenn, N. R., R. A. McGhie, and R. P. Pottinger. 1985a. Evaluation of terbufos for grass grub control. Proc. N.Z. Weed Pest Control Conf. 38:23-26.

Wrenn, N. R., R. A. McGhie, and R. P. Pottinger. 1985b. Bioassay and field experiments for evaluation of difluron for porina and caterpillar control in pasture. p. 286-292. *In* R. B. Chapman (ed.) Proc. 4th Australasian Conf. in Grassl. Invert. Ecol. Caxton Press, Wellington.

Wright, P. J. et al. 1988. Nematode control of pests in New Zealand. *In* Proc. 5th Australasian Conf. on Grassld. Invert. Ecol. August 1988, Melbourne. (In press.)

Yeates, G. W., W. B. Healy, and J. P. Widdowson. 1973. Screening of legume varieties for resistance to the root nematodes *Heterodera trifolii* and *Meloidogyne hapla*. N.Z. J. Agric. Res. 16:81-86.

Yeates, G. W., W. B. Healy, J. P. Widdowson, N. A. Thomson, and B. N. MacDiarmid. 1975. Influence of nematicides in growth of plots of white clover on a yellow-brown loam. N.Z. J. Agric. Res. 128:411-416.

DISCUSSION

Leath: I was pleased that you mentioned "cumulative stress load." Rarely does a single pest take out a stand. Usually it is the "cumulative" stress of many factors.

Irwin: Do you feel that current methodologies allow us to accurately quantify losses due to soil pathogens?

Watson: It is R. A. Skipp's impression that there is a definite need for new initiatives in this area. Qualitatively , it is known which organisms infect white clover in New Zealand, but relative importance within complexes has not been quantified.

J. Hay: Although it is suggested that the way to make progress combatting insect damage in white clover is by breeding for resistance, efforts within New Zealand have not been at all profitable. Could you comment?

Watson: I feel it is important that the reasons for this should be elucidated. Perhaps other strategies for survival mean the plant has not had to develop chemically based mechanisms for example. Open pollination of white clover, and high buried seed loads into which improved clover may be sown, may make the whole idea of plant resistance somewhat suspect.

Sheaffer: I noticed that you observed a response to seed treatment with metalaxyl. This contrasts with much of our work from the northern USA.

Watson: R. A. Skipp has shown good responses to the fungicide in controlling damping off from *Pythium spp.* from seed and soil applied treatments for red clover and lucerne, but not for white clover.

J. Hay: Stem nematode is not mentioned as a pest of white clover. 'Pitau' is susceptible.

Watson: Stem nematode is mentioned in Table 1 as a pasture pest as it has a wide host range, including grasses. It becomes a problem in spaced white clover plants and pure species evaluations, but has not been demonstrated as a pest in mixed pasture in New Zealand. It is regarded as important in Europe so that resistance is an important criterion where seed exports are concerned.

Gramshaw: Soil mineral N should be able to carry grass plants over short-term denodulation of clover by pests.

Watson: Our responses to pesticide application showed that N_2 fixation was chronically depressed and thus would be expected to eventually limit the productive potential of pasture.

DISEASES AND FORAGE STAND PERSISTENCE IN THE UNITED STATES

K.T. Leath

SUMMARY

Diseases are a major cause of premature stand decline. Often, several diseases are active in a field at the same time and interact with other stresses to produce a cumulative stress load. Fungi, bacteria, nematodes, and viruses are the causal agents of disease, and disease problems worsen as the age of a stand increases. With perennial forage species, diseases, too, are perennial. The root and crown rot complex, an insidious, chronic disease caused by a complex of organisms, debilitates plants and renders stands unprofitable.

Fungicides are not applied to forages except as a seed treatment, so, in essence, diseases are not really controlled. They are managed. Disease losses are minimized through sound crop management. Sound management is also the key to stand persistence, because this keeps all stresses on the plants to a minimum. Minimally stressed plants maintain healthy roots longer, and healthy roots are the key to sustained forage production. Higher levels of resistance to root-rot pathogens, foliar diseases and viruses, coupled with improved management techniques are needed to provide enhanced long-term performance of forage legumes in the future.

Diseases are a major cause of premature stand decline. More than 400 pathogens have been described on forage grasses and legumes (O'Rourke, 1976), and all plant organs are subject to attack. Rarely do diseases act independently to shorten stand life. They frequently act as merely one component of a group of biotic and abiotic agents, which interact to produce a cumulative stress load. Often several diseases will be active in stands and on individual plants at the same time, and only occasionally can a stand failure be attributed to a single disease. The effects of a disease outbreak can reduce performance of a stand long after symptoms disappear. With perennial forge crops, it must be remembered that diseases, too, are often perennial. Diseases occur simultaneously and sequentially and a succession of chronic diseases, which vary geographically, from year to year, and within a single season, contribute greatly to those ill-defined phenomena referred to as "winterkill" and "lack of persistence."

Diseases are frequently categorized by the portion of the plant on which symptoms are manifested, e.g., leaf diseases, root rot, crown and stem blight, however their effects are never limited to single organs and should be considered as entire plant diseases. Leaf diseases affect root function and vice versa. It often is the subtle effects of minor diseases acting as a complex that results in premature stand decline.

Regardless of the cause, the failure of a forage stand to persist can be attributed to the failure of individual plants to maintain healthy root systems. It is this concept of "a healthy root system" that should be the focus and goal of forage producers and agricultural advisors, when assessing the problem of long-term maintenance of forage stands.

The contribution of diseases to stand decline is not easily quantified. Estimates of annual yield loss from diseases are available, but a monetary value is difficult to assign to the gradual reduction in productivity that occurs in forages and the shortened proration of seeding costs.

DISEASES AND INTERACTIONS

Diseases of forage species having particular local importance are listed and described in state crop production guides and manuals. More extensive coverage is available for clovers (Leath, 1985), alfalfa (Graham et al., 1979; Leath et al., 1988), and grasses (Braverman et al., 1986). A good general reference on forage crop diseases is that by O'Rourke (1976), and the Alfalfa Analyst available from the Certified Alfalfa Seed Council., Inc., Woodland, CA, is most useful for diagnosing alfalfa problems. Specific diseases and their relative importance at various locations in the USA and Canada are given for red clover (Table 1), white clover (Table 2), crimson clover (Table 3), arrowleaf clover (Table 4), subterranean clover (Table 5), and birdsfoot trefoil (Table 6). Data in these tables are from a 1983 survey by the author as obtained from research and extension personnel at the locations listed.

Most diseases can become severe under extenuating circumstances, but they can be classified as acute or chronic with a fair level of surety. Losses to Sclerotinia rot have increased over the past few years in many areas of the USA, probably due to increased practice of fall seeding by conservation tillage methods. Typical of acute diseases would be wilts, anthracnoses, damping-off, Phytophthora root rot, and Sclerotinia crown and stem rot. Typical of chronic diseases would be leaf spots, mildews, rusts, viroses, and most root rots. High levels of resistance have been developed in response to the threat posed by the acute diseases, and such diseases should not be major considerations in the premature stand decline syndrome. It is the chronic diseases that contribute most to the gradual reduction in stand productivity. The chronic diseases exert a low-level, continuous drain on plant performance, and teamed with each other or with climatic or edaphic pressures or mismanagement, cause individual plant death and ultimately collective death of plants or stand failure.

Table 1. Diseases of red clover (Trifolium pratense L.) in the USA and Canada.[1]

	United States																Canada				
Disease	Arkansas[2]	Colorado	Illinois	Indiana	Iowa	Kentucky	Louisiana	Minnesota	Mississippi	Missouri	North Carolina	Oregon	Pennsylvania	Virginia	Washington	Wisconsin	Alberta	New Brunswick	Ontario	Prince Edward Island	Quebec
Anthracnose																					
Northern			1-2	3	2-3	1		3		3			3			4	2	1	3	1	2
Southern	3		3	2	3	2	2	2		3	2		2	2							
Black root rot							2				1										
Black stem																					
Spring	2		2	2	2-4	1	2	2					2	2		2	2	1	2	2	
Summer	2		2	2	2	1		2			2		2	2		2					
Common leaf spot	2		2	2		1		1					2	1				3	1	3	2
Downy mildew																					1
Lepto leaf spot								1					1				1				1
Nematodes																					
Clover cyst													2								
Lance											2										
Root knot											2									2	
Root lesion																				3	
Spiral											2										
Stubby root											2										
Phyllody																				2	2
Powdery mildew	3	2		1	2-3	2	3	1-2		1	3		2-3	2	1	3	2	3	3	2	2
Root rot complex	4		4	4	3-4	4		3-4		4			4	4	1	4	4	4	4	4	4
Rust	1		2	1	1-2	1		1		1	1		1			1			3	1	1
Sclerotinia rot									3		2		2			1		2		2	
Sooty blotch											1									1	2
Southern blight									1												
Stagonospora leaf spot																	3				1
Stemphylium leaf spot	2		2	2	2	1		2			1		2	1		3	1		2	2	2-3
Viruses																					
Alfalfa mosaic												2	1								
Bean yellow mosaic	2		4	4		2		3				2	3-4			?	1			1	
Clover yellow vein													1								
Peanut stunt													1								
Red clover vein mosaic								3				2	2								
Unidentified					1			1							2				2		2
Wisconsin pea streak																	1				
Winter crown rot																	4				

[1] Disease severity ratings: 1=minor importance; 2=occasionally severe; 3=frequently severe; 4=limits yield and life of stand; ?=occurs but severity not known.

[2] Information received from more than one individual at a location was combined.

Table 2. Diseases of white clover (<u>Trifolium</u> <u>repens</u> L.) in the USA and Canada.[1]

Disease	United States									Canada	
	Arkansas	Florida	Georgia	Louisiana	Mississippi	North Carolina	Oregon	Pennsylvania	South Carolina	Alberta	Prince Edward Island
Anthracnose											
Southern						3			1		
Black stem											
Summer		3				1					
Common leaf spot		3							2		2
Curvularia leaf spot	1	3				3					
Damping-off			2								
Leaf spot complex				1					1		
Lepto leaf spot	2		2		1	2-3		1	2	1	1
Nematodes											
Clover cyst						1					
Complex		3	2			2			2		2
Lance						2					
Root knot		3			2	2			3		
Root lesion											2
Spiral		3				2					
Sting		3									
Stubby root						2					
Phyllody											3
Powdery mildew				2					1		
Root rot complex		3				3		3	2-3	2	3
Rust	1							2	1	2	2
Sclerotinia rot			2	2	2	3			1-2		
Sooty blotch	2					1		2	1	2	2
Southern blight					2	2			1-2		
Stagonospora leaf spot										1	
Viruses											
Alfalfa mosaic									2		
Bean yellow mosaic	4						1				
Clover yellow mosaic							2				
Clover yellow vein									3		
Complex		3			3	3-4		2	4		
Peanut stunt									3		
Red clover vein mosaic									1		
White clover mosaic							2		2		
Winter crown rot										4	

[1]Disease severity ratings: 1=minor importance; 2=occasionally severe; 3=frequently severe; 4=limits yield and life of stand; ?=occurs but severity not known.

Table 3. Diseases of crimson clover (<u>Trifolium incarnatum</u> L.) in the USA.[1]

Disease	United States					
	Florida	Georgia	Louisiana	Mississippi	North Carolina	Oregon
Anthracnose						
Northern					1	
Southern					2	
Damping-off		2				
Fusarium wilt				2		
Nematodes	2	2				
Powdery mildew			1			
Pythium rot		2				
Root and crown rot complex				2		
Sclerotinia crown and stem rot		1	2	3	2	
Sooty blotch		1	2		2	
Viruses						
Unidentified	2			2		
Pea enation mosaic						2

[1]Disease severity ratings: 1=minor importance; 2=occasionally severe; 3=frequently severe; 4=limits yield and life of stand; ?=occurs but severity not known.

The root-rot complex, or Fusarium root rot as it is often called, is of particular interest regarding lack of persistence. It begins very early in the life of a stand (O'Rourke & Millar, 1966), progresses slowly or perhaps even goes latent temporarily, and increases in severity as the plant ages. Roots and crowns of forage legumes are attacked by many different fungi. <u>Rhizoctonia</u>, <u>Stagonospora</u>, and other fungi cause crown and root rots and reduce stand persistence when environmental conditions are favorable. <u>Fusarium</u> species have been repeatedly isolated from roots of forage legumes by different people in diverse geographic areas, which suggests a major role for this group of fungi in the root rot complex.

<u>Fusarium</u> species penetrate roots of forage legumes directly (Chi et al., 1964) and colonize roots of plants shortly after seeding (O'Rourke & Millar, 1966). The species of <u>Fusarium</u> most often associated with root rots of forage legumes are: <u>F</u>. <u>oxysporum</u>, <u>F</u>. <u>avenaceum</u>, <u>F</u>. <u>solani</u>, <u>F</u>. <u>acuminatum</u>, <u>F</u>. <u>moniliforme</u>, and <u>F</u>. <u>tricinctum</u>. The pathogenicity of <u>Fusarium</u> species varies considerably among and within species. Most isolates attack seedlings but are far less virulent against older, more mature

Table 4. Diseases of arrowleaf clover (*Trifolium vesiculosum* L.) in the USA.[1]

Disease	United States: Florida	Georgia	Mississippi	Texas
Bean yellow mosaic virus		2		
Internal crown breakdown			3[2]	
Leaf spot unidentified	1			
Lepto leaf spot		1		
Nematodes				
Complex	2			
Root knot			2	
Pythium		2		
Phytophthora root rot			2	
Red leaf complex		2		
Rhizoctonia root rot and damping-off		2		
Root rot complex	3			2
Sclerotinia crown and stem rot			2	
Southern blight		2	1	
Stemphylium leaf spot		1	?	
Viruses	3	1	3[2]	

[1]Disease severity ratings: 1=minor importance; 2=occasionally severe; 3=frequently severe; 4=limits yield and life of stand; ?=occurs but severity not known.

[2]Occurred in research plots.

plant roots (Leath & Kendall, 1978). Some *Fusaria* in conjunction with soil-borne *Pseudomonas* bacteria cause more severe root rot than either organism causes individually (Leath et al., 1986).

Roots are colonized by *Fusarium* species soon after emergence from the seed, but disease symptoms may not appear for some time. This delay is generally attributed to the "weak" pathogenicity of root rotting *Fusaria*, which cause more severe rot when plants are under stress. Evidence supports that such a relationship does exist (Leath et al., 1971), and that these stresses come in many forms. Inadequate K fertility, poor soil drainage, incorrect soil pH, too frequent harvests, adverse winter conditions, insect injury, and chronic diseases are major stress factors contributing to the increased proneness of alfalfa to root and crown rots. Fertility, soil pH, and harvest frequency are controllable to a large degree. The other stress factors are less manageable but

Table 5. Diseases of subterranean clover (Trifolium subterraneum L.) in the USA.[1]

Disease	United States: Florida	Georgia	Mississippi	North Carolina
Cercospora leaf spot			2	
Damping-off		2		2
Leaf and stem spots	3			
Nematodes	3			
Powdery mildew		2		
Pythium root rot		2		
Root and crown rot complex	3	2		
Sclerotinia crown and stem rot			2	
Stemphylium leaf spot	3			
Virus complex	3		2	
White mold		2		

[1]Disease severity ratings: 1=minor importance; 2=occasionally severe; 3=frequently severe; 4=limits yield and life of stand; ?=occurs but severity not known.

can be ameliorated. For example, poorly drained fields can be avoided, planted to some other crop, or else tiled and drained. Winter injury can be minimized by maintaining adequate fertility, cutting early enough in the fall, and leaving stubble of sufficient height.

Root feeding insects, especially the clover root curculio (Sitona hispidulus L.), injure roots and enhance infection by root pathogens. Fusarium oxysporum, F. solani, and F. avenaceum are common inhabitants of curculio feeding sites (Dickason et al., 1968; Leath et al., 1963; Leath and Hower, 1987). It is quite likely that gains in reducing the severity of the root rot complex will not precede similar gains in reducing curculio injury.

Insect injury can also interact as a stress factor. Aphids and leafhoppers, for example, feed on the tops of the plants and reduce the nutrients available for root growth and function. Fusarium root rot in alfalfa subjected to aphid and leafhopper feeding was nearly twice as severe as that of plants grown without stress (Leath & Byers, 1977). Insecticides reduce insect stress, but resistant cultivars would provide a better, long-term solution to the insect stress problem.

Foliar diseases pose a problem similar to that of insects. Fungicides are not generally used on forages, and although the economics of using them has improved (Broscious & Kirby, 1985),

Table 6. Diseases of birdsfoot trefoil (<u>Lotus</u> <u>corniculatus</u> L.) in the USA and Canada.[1]

Disease	United States					Canada	
	Maryland	Missouri	Pennsylvania	Vermont	Wisconsin	Alberta	Prince Edward Island
Anthracnose <u>Colletotrichum</u> sp.			2				
Cercospora leaf spot	1						
Charcoal rot	3						
Fusarium wilt		1		4			
Leaf spot unidentified				1-2	?		
Mycoleptodiscus rot	3						
Phomopsis stem blight	2						
Pythium seedling rot			1	2			
Rhizoctonia							
Root rot	1		3				
Aerial blight	3	3					
Root lesion nematode							3
Root rot complex	4	4	4		?	4	3
Southern blight	2						
Stemphylium leaf spot and stem canker	1		2			2	
Viruses							
Tobacco ringspot	1						
Tomato ringspot	1						
Winter crown rot						4	

[1]Disease severity ratings: 1=minor importance; 2=occasionally severe; 3=frequently severe; 4=limits yield and life of stand; ?=occurs but severity not known.

the concern over residues still limits their potential use for foliar disease control in the near future. Some resistance to foliar diseases is found in various cultivars, but no cultivar has high levels of resistance to the several diseases that prevail over a major part of our growing area. Here again, however, resistant cultivars seem to be the best long-term solution.

There is no question that virus (Leath & Barnett, 1981; McLaughlin & Boykin, 1988; Tu & Holmes, 1980) and nematode diseases contribute to stand decline, but research in this area has been limited. The life of clover plants is significantly shorter when virus and <u>Fusarium</u> spp. infect the plant simultaneously than when either pathogen acts alone (Leath et al., 1971). An extensive survey of nematode injury to alfalfa has not been made, but the importance of stem (Griffin, 1968), lesion (Nelson et al., 1985), and clover cyst (Leath, 1985) nematodes in reducing stand life of forage legumes needs further investigation. Evidence indicates that nematode feeding injury favors the development of root rot in forage legumes (McGuire et al., 1958). Pathogens, such as viruses and nematodes, greatly reduce the vigor of a plant. Likely, they also reduce the ability of a plant to withstand and recover from crown and root rots, and the converse is also probably true, that crown and root rots make plants more susceptible to stress injury (Hwang & Flores, 1987).

Much of the discussion thus far relates to forage legumes more than to forage grasses, and it is quite likely that diseases play a stronger role in the early demise of legume stands than they do for grasses. In any case, it should not be assumed that grasses do not suffer from diseases in much the same way as do legumes. Forage grasses in the USA do not suffer from the acute diseases, e.g., wilts, that afflict legumes, nor do their roots seem to be as severely debilitated by disease. Although roots of grasses can be attacked rather severely by some of the same Fusaria that attack legume roots (Leath & Kendall, 1980, unpublished data), these fungi are not known to cause problems in production. Perhaps grasses replace rotted roots more quickly than do the legumes, or perhaps the loss of a major root of a grass plant does not have the impact of a tap root loss to a legume.

Grasses certainly have an abundance of leaf diseases, and these do become sufficiently severe to slow growth and significantly stress the plants. Grass viruses are known (Braverman et al., 1986), but their impact has not really been assessed. Nematodes, however, likely cause grass stands to become unprofitable. This has been observed in Pennsylvania with a Cricconemoides sp. in orchardgrass. In this instance, the age of the stand was estimated to be in excess of 8 yr. The use of recommended crop rotations would minimize such problems.

MINIMIZING DISEASE IMPACT

At present in forage crop production, fungicide use is primarily as seed treatments. Therefore, diseases are not really controlled; they are managed. It is not possible to separate the management of a crop from the management of its diseases. Disease management is a part of crop management and should not be considered otherwise. Crop management is the master system under which pest management is included.

Disease management is not new to forages. Farmers have been empirically adopting practices that reduced diseases ever since the inception of forage production. It is the degree of refinement, based on accumulated knowledge about host and pathogen, that makes the disease management of today a new approach, and the success of the disease management program is only as good as the completeness and accuracy of the information on which it is based. Data are sufficient to permit the development of forage growth models (Bula & Massengale, 1972; Fick, 1977), but epidemiological data are insufficient for any of the numerous diseases afflicting forages. Until these data are available, disease management in forages will be imprecise.

A forage stand consists of thousands of individual plants/ha growing en masse in spite of biological and physical stresses and because of a competent management system. Such a system manipulates stress factors to favor the crop species over the weed species. These forages are harvested several times during each growing season and then are expected to survive over the winter and produce again the following year. Superimposed on the stress of harvest is that caused by diseases. When such stresses become too great, individual plants die, stands are thinned, and production decreases. Forages are intensively managed over a period of several years, therefore, there are many opportunities to implement management strategies that reduce disease losses. Root rot begins shortly after planting and progresses gradually, increasing in severity with the age of the stand. It is the rate of disease increase that management can change. There is a close relationship between disease and management. Minimizing disease loss depends upon the correct and timely application of management strategies.

Many fungi, bacteria, viruses, nematodes, and air pollutants contribute to the disease picture. Pathogens interact with each other and with other biotic and physical stresses to produce a cumulative overall effect. Stresses need not occur at the same time to interact. Stresses separated in time can and often do interact. For example, an autumn epidemic of foliar disease might limit the accumulation of carbohydrates stored in roots making plants more prone to winterkill and causing slower growth the following spring, or devastation by spring pathogens. Another example might be that of insect feeding in a young stand which could limit root growth and make plants less able to tolerate root rot or drought.

Many forages are perennials and some of their diseases are also. Root rots, viruses, and wilts are not transient. Once infection has taken place, the plant remains diseased. The impact of that disease and the rate at which it develops are functions of climate and management. Because climate must be tolerated, management becomes the prime strategy. Some crop-management strategies acting directly or indirectly to reduce disease losses are listed in Table 7. The application of management strategies to reduce disease losses begins before the crop is planted. These early strategies can only be implemented once. If the opportunity to implement them is missed, the consequences must be suffered. Awareness and careful planning are needed to maximize the effects of these strategies. Often some of the accepted management practices are not appreciated from a disease-management viewpoint. Weed control, fertilization, and insect control are all disease-management strategies, because their application results in less loss from disease. Weeds not only compete with forage plants for growth requirements but also serve as alternative hosts for pathogens. The maintenance of K at levels that permit vigorous plant growth also results in reduced populations of potential root pathogens (O'Rourke & Millar, 1966), and the control of leafhoppers reduces the impact of root disease. Many other practices that are recognized mainly as production practices also minimize disease losses.

Table 7. Management practices that directly or indirectly affect disease development in forage legumes.

	Stage of crop development	
Preplant	Planting	Production
Crop rotation	Planting date	Insect control
Weed control	Weed control	Disease monitoring
Fertility	Soil preparation	Crop sanitation
Liming	Type of seeding	Weed control
Mixed vs. pure stand	Fungicide	Timely harvests
Species	Seeding rate	pH maintenance
Variety	Seeding depth	P and K fertility

Much of our disease management is preventive rather than therapeutic, therefore, of necessity it must be continuous. Even a brief break in its application may result in increased disease severity and larger losses. Against many pathogens there exists some inherent host-plant resistance, but this resistance does not constitute immunity, and the expression of this resistance can be modified by management practices (Leath, 1981; Leath & Ratcliffe, 1974).

A forage grower usually has two main objectives--production and stand persistence. These objectives are not necessarily attained by the same management system. Usually, the management system used represents a compromise that permits a satisfactory yield and a satisfactory stand life. Managing perennial forages is complex compared to managing an annual crop where only yield is of concern.

GENERAL APPRAISAL

Most forage crops are generally expected to produce satisfactorily for several years. If perennial, they should produce indefinitely, however, in reality, their productive lives are typically more like 3 to 5 yr for legumes and probably slightly longer for grasses.

Why don't stands last indefinitely? More than likely it is because the plants' root systems are not functioning adequately. Plant roots quite often are asked to perform in what is most certainly an adverse and at times even hostile environment. They routinely encounter biological and physical factors that put continuous and often severe stress on their ability to carry on normal functions. Diseases are part of this stress complex.

Maintaining healthy roots is highly desirable, but achieving this is not simple. Because it is quite difficult to alter many aspects of the soil--the root's environment--one approach has been to alter the plant through genetics, and in this way much progress has been achieved. For example, plants resistant to pathogens causing wilts and Phytophthora root rot have been bred, and varieties with multiple pest resistance are commonplace. Plant breeding, however, has not provided answers to all the problems affecting alfalfa roots.

At present there is no resistance to the root rot complex. Therefore crop management remains the major strategy against the problem. A stand can be expected to reach an unprofitable state sooner under poor management than under good management, because root health is directly related to management quality. Root rot will become limiting as early as 2 yr after seeding in a mismanaged alfalfa stand, or as long as 5 to 7 yr in a well-managed stand. Under poor management the root rot complex is a fairly rapid, acute disease; under good management it is a slow, chronic disease. Roots anchor plants in the soil, fix nitrogen, absorb water and nutrients, and store carbohydrates for winter survival. Their state of health is critical to plant survival and productivity.

PROSPECTIVE

The root rot complex is likely to plague producers of forage legumes for a long time. Certainly, it is not easy to resolve. Because of the extreme variability of the fungi associated with root and crown rots of alfalfa, the prospects of breeding a cultivar with a high level of resistance that is effective over a broad geographic area are not bright. This is not to say that progress in selection for more persistent alfalfa will be in vain, however. Selection to reduce stand loss from root disease may well be accomplished indirectly through selection of alfalfas that produce and store carbohydrates more efficiently, grow roots faster, and tolerate stresses better. Evidence in the clovers and also in alfalfa indicates that when insect and foliar disease stresses are reduced, so is the rate of root rot development. Good management practices will continue to be essential to reducing stand loss. Sound management is critical to the realization by any alfalfa cultivar of its inherent disease resistance.

Research is needed to determine the impact that low-temperature fungi are having on forage persistence over much of the northern half of the United States. The activity of such fungi on turf over much of this area is obvious and most certainly contributes to "winterkill" of forages.

The interactions of rhizosphere organisms have received only cursory attention in forage species, and increased examination should reveal both beneficial and detrimental relationships that are presently unknown.

The immediate future is not likely to bring major changes in approaches to reduce losses in forage crops. Our present system is effective and economical, and it causes minimal disturbance to the environment, but it can be improved. In the future, present strategies will be intensified; new varieties will have resistance to more pathogens; and levels of resistance will be higher. Greater use of grass-legume mixtures, improved adaptation of crop to site, new formulations and methods of application of fertilizer, and increased efficiency of insect control all will contribute to improved persistence of forage stands. Rapid evaluation of forage quality and the attendant emergence of forages as a market commodity will place protein content in its proper perspective and thus increase interest in the control of foliar disease, perhaps employing the use of fungicides. Biocontrol of pathogens in other crops looks promising and should be evaluated for potential application in forage systems.

With forage crops, contrary to many other crops, there is not the knowledge needed to apply established principles of disease management in an optimum manner. Reliable crop-loss data are needed to determine the economic feasibility of applying control strategies. A great deal of fundamental research in pathogen biology and disease development must be completed before models for forage-crop diseases can be developed and incorporated into master models of crop production. Although disease losses in forage crops are kept to a practical level through host-plant resistance and the judicious application of sensible crop-management strategies, there is still a need to develop a more effective disease-management system. A coordinated and sustained research program at commercial, state, and federal levels is the only solution to current forage disease problems and to those of the future.

REFERENCES

Braverman, S.W., F.L. Lukezic, K.E. Zeiders, and J.B. Wilson. 1986. Diseases of forage grasses in humid temperate zones. Pa. Agric. Exp. Stn. Bull. 859.

Broscious, S.C., and H.W. Kirby. 1985. Fungicides and application timing in relation to leaf spot disease and yield of alfalfa. Phytopathology 75:1346 (Abstr.).

Bula, R.J., and M.A. Massengale. 1972. Environmental physiology. p. 167-183. In C.H. Hanson (ed.) Agronomy 15.

Chi, C.C., W.R. Childers, and E.W. Hanson. 1964. Penetration and subsequent development of three Fusarium species in alfalfa and red clover. Phytopathology 54:434-437.

Dickason, E.A., C.M. Leach, and A.E. Gross. 1968. Clover root curculio injury and vascular decay of alfalfa roots. J. Econ. Entomol. 61:1163-1168.

Fick, G.W. 1977. The mechanism of alfalfa regrowth: A computer simulation approach. Search Agric. 7(3):1-28. Cornell Univ., Ithaca, NY.

Graham, J.H., F.I. Frosheiser, D.L. Stuteville, and D.C. Erwin. 1979. A compendium of alfalfa diseases. Am. Phytopathol. Soc., St. Paul.

Griffin, G.D. 1968. The pathogenicity of Ditylenchus dipsaci to alfalfa and the relationship of temperature to plant infection and susceptibility. Phytopathology 58:929-932.

Hwang, S.F., and G. Flores. 1987. Effects of Cylindrocladium gracile, Fusarium roseum and Plenodomus meliloti on crown and root rot, forage yield and winterkill of alfalfa in north-eastern Alberta. Can. Plant Dis. Surv. 67:31-33.

Leach, C.M., E.A. Dickason, and A.E. Gross. 1963. The relationship of insects, fungi and nematodes to the deterioration of roots of Trifolium hybridum L. Ann. Appl. Biol. 52:371-385.

Leath, K.T. 1981. Pest management--alfalfa diseases. In D. Pimentel (ed.) Handbook of pest management in agriculture. CRC Ser. in Agric. CRS Press, Boca Raton, FL.

Leath, K.T. 1985. General diseases. p. 205-233. In N.L. Taylor (ed.) Clover science and technology. Agronomy 25.

Leath, K.T., and O.W. Barnett. 1981. Viruses infecting red clover in Pennsylvania. Plant Dis. 65:1016-1017.

Leath, K.T., and R.A. Byers. 1977. Interaction of fusarium root rot with pea aphid and potato leafhopper feeding on forage legumes. Phytopathology 67:226-229.

Leath, K.T., D.C. Erwin, and G.D. Griffin. 1988. Diseases and nematodes. p. 621-670. In A.A. Hanson et al. (ed.) Alfalfa and alfalfa improvement. Agronomy 29.

Leath, K.T., and A.A. Hower. 1987. Activity of Fusarium species in feeding sites of clover root curculio larvae in roots of alfalfa. Phytopathology 77:1616 (Abstr.).

Leath, K.T., and W.A. Kendall. 1978. Fusarium root rot of forage species: pathogenicity and host range. Phytopathology 68:826-831.

Leath, K.T., F.L. Lukezic, H.W. Crittenden, E.S. Elliott, P.M. Halisky, F.L. Howard, and S.A. Ostazeski. 1971. The Fusarium root rot complex of selected forage legumes in the Northeast. Pa. State Univ. Bull. 777.

Leath, K.T., F.L. Lukezic, B.W. Pennypacker, and W.A. Kendall. 1986. Fusarium avenaceum and Pseudomonas viridiflava interact to cause necrosis in roots of red clover. Int. Congr. Microbiol. Proc. 14:285 (Abstr.).

Leath, K.T., and R.H. Ratcliffe. 1974. The effect of fertilization on disease and insect resistance. p. 481-502. In D. Mays (ed.) Forage fertilization. ASA, Madison, WI.

McGuire, J.M., H.J. Walters, and D.A. Slack. 1958. The relationships of root-knot nematodes to the development of Fusarium wilt in alfalfa. Phytopathology 48:344.

McLaughlin, M.R., and D.L. Boykin. 1988. Virus diseases of seven species of forage legumes in the Southeastern United States. Plant Dis. 72:539-542.

Nelson, D.L., D.K. Barnes, and D.H. MacDonald. 1985. Field and growth chamber evaluations for root-lesion nematode resistance in alfalfa. Crop Sci. 25:35-39.

O'Rourke, C.J. 1976. Diseases of grasses and forage legumes in Ireland. The Agricultural Institute, Dublin, Ireland.

O'Rourke, C.J., and R.L. Millar. 1966. Root rot and root microflora of alfalfa as affected by potassium nutrition, frequency of cutting, and leaf infection. Phytopathology 56:1040-1046.

Tu, J.C., and T.M. Holmes. 1980. Effect of alfalfa mosaic virus infection on nodulation, forge yield, forage protein, and overwintering of alfalfa. Phytopathol. Z. 97:1-9.

DISCUSSION

Clements. You described a conceptual relationship between the percentage of plants resistant to a given disease and the change in plant population with time, as the disease kills individual plants. A figure of 50% resistance was said to confer adequate field resistance to most diseases. If several disease are operating at once, is 50% resistant plants for each disease sufficient to ensure an adequate stand?

Leath. Yes. It seems to be an adequate number. With some diseases a lower percentage of resistant plants would suffice. Disease pressure in a field situation often is not as severe as those employed during the selection program.

Irwin. I concur that the "cumulative stress load" that you mentioned is important in considering persistence. I raise the question as to whether the phenomenon of acquired resistance has a role in control of diseases of pasture legumes? For example, infection by one pathogen, or an avirulent strain, protecting the plant from subsequent infection.

Leath. Yes. It is an important concept, in a sense "plant vaccination," and is worthy of research. At present, however, it is a long way from field application.

Matches. Do you see resistance breaking down as a result of changes or variation in pathogenic races?

Leath. No. Resistance selected at one location holds up at other locations, and resistance selected many years ago is still as good as ever. In general, races have not posed serious problems in forage legumes.

Barnes. What is the potential for the use of biotechnology in providing disease resistance? Are there any examples?

Leath. Application of biotechnological techniques have indeed enhanced disease resistance in some non-forage crops, however, I believe that successful applications in forage crop disease are still to come.

Hoveland. Sometimes legume persistence can be improved quite well by recurrent selection in hostile environments. Florida 77 alfalfa does not have a high level of resistance to anything specific, however, it does have fairly good tolerance to a wide range of pests, such as nematodes, root rots, and leaf spots.

INSECTS THAT REDUCE PERSISTENCE AND PRODUCTIVITY OF FORAGE LEGUMES IN THE USA

R. C. Berberet and A. K. Dowdy

SUMMARY

In general, there is limited information on effects of insect infestations over the stand-life of alfalfa and other forage legumes. Within the last 10 yr, particularly, researchers have been emphasizing studies of longer duration and providing some important results to aid in management of perennial forage crops to extend productive stand life. The necessity for research on interactions among insect pest species, as well as, interrelationships of insects, pathogens, and weeds has been demonstrated. The response of scientists in terms of multidisciplinary studies has produced some excellent results and has served to emphasize that much more information is needed. As we learn more about the numerous stresses imposed on perennial forage crops due to pest infestations and environmental factors, we also realize that effective management strategies must be developed to help maintain healthy and productive stands for harvesting or grazing purposes.

Forage legumes provide habitats and nutrient sources for large numbers of insect species. Extensive entomological research has been conducted on bionomics of key pests and beneficial species. Traditionally, studies have had relatively specific objectives of determining seasonal life histories or control strategies for individual pest species. Few of these studies have been designed as long-term efforts to describe effects of pests as contributors to stand decline in perennials such as alfalfa, *Medicago sativa* L. Interactions among insect pests or between insects and other groups, such as weeds and pathogenic fungi, have not been studied extensively. Of the papers published in the USA on entomological research in legumes, the greatest number have dealt with alfalfa. Far fewer have involved research on clovers and limited studys has been devoted to maintenance of legumes in pastures.

Emphases on integrated pest management (IPM) have resulted in productive research efforts, particularly with alfalfa. Studies have involved multiple-pest interactions and combined effects of several pest species in reducing productivity. Entomologists, weed scientists, and plant pathologists are working cooperatively to learn how pest stresses contribute to stand decline over a period of several years. Synergistic effects among pests, such as the importance of root-feeding insects in providing points of entry for pathogenic fungi, are being described. The relationship of insect damage to increasing weed interference in stands is being investigated. The result of these efforts is an improved understanding of the relationships of pests with other aspects of legume production systems. Current research results relating to the importance key insect pests in reducing productivity and persistence of major forage legumes in the USA are reviewed in this chapter.

SOIL INSECTS

Clover Root Curculio

The clover root curculio, *Sitona hispidulus* (F.), has been known to occur in the USA since 1876 (Webster, 1915). The species has spread widely throughout this country and Canada. All *Trifolium* spp. as well as various *Medicago* spp. are known to be hosts for this pest, (Jackson, 1922). The clover root curculio has been cited as a particularly serious pest of red clover, *T. pratense* L., (Newton & Graham, 1960), white clover, *T. repens* L., (Powell et al., 1983), and alfalfa (Underhill et al., 1955). There are a number of *Sitona* spp. which infest legumes, and confusion exists as to the identity of species causing damage in some instances. Manglitz et al. (1963) have discussed possible misidentifications that exist in the literature concerning descriptions of damage caused by *S. hispidulus* and by the sweet-clover weevil, *S. cylindricollis* Fahraeus, on *Melilotus* spp.

The clover root curculio completes one generation/year, with eggs deposited on the soil surface in close proximity to host plants during part or all of the time period from October to May, depending upon temperature in winter months. Small larvae must enter the soil to find root nodules on which to feed in order to survive on alfalfa (Quinn & Hower, 1986). However, Byers and Kendall (1982) reported that nodules were not essential for survival of larvae in laboratory tests on red clover. Larger larvae feed on side roots and rootlets. The fourth and fifth instars, in particular, feed on the taproots (Dintenfass & Brown, 1986). Most of the injury to roots of forage legumes occurs from March through June in the USA, a time called the "critical injury period" by Underhill et al. (1955).

As a result of feeding by larger larvae on taproots, the clover root curculio has been implicated as an agent contributing to serious stand declines in alfalfa and clovers (Newton & Graham, 1960; James et al., 1980). Dickason et al. (1968) found damage to roots as deep as 70 cm in the soil and reported that damage was most severe within 25 cm of the soil surface. They also found a relatively consistent pattern of cumulative damage by the clover root curculio over the life of alfalfa stands in Oregon. Cranshaw (1985) observed that the incidence of extensive root scarring increased greatly from the second to fifth year after establishment in Minnesota.

Despite evidence of injury to alfalfa, researchers have obtained conflicting results regarding the potential for yield reductions due to clover root curculio. Dickason et al. (1968) found no significant reductions in forage yields in spite of relatively severe injury to root systems of alfalfa plants. However, Godfrey (1984) reported consistent yield reductions due to infestations of this pest in field plots. Godfrey and Yeargan (1985) found that under a dry soil regime in a greenhouse test, alfalfa yield decreased with increasing larval population densities. Lamp et al. (1985) reported that clover root curculio can cause significant yield reductions under dry to normal soil moisture conditions in established stands.

Although the clover root curculio may reduce yields when causing serious injury to root systems of plants, its role in promoting infections of pathogenic fungi and bacteria may be of greater importance relative to stand loss. In studies conducted with alsike clover, _T. hybridum_ L., an association was found between injury by the insect and vascular decay in plant roots (Leach et al., 1961; Dickason et al., 1968). The most common fungal species isolated from roots was _Fusarium oxysporum_ Schl. sensu Snyd. & Han. Newton and Graham (1960) observed a high degree of association between injury by the clover root curculio and/or additional root boring species such as the clover root borer, _Hylastinus obscurus_ (Marsham), and root rot of red clover. In a study where insect injury and occurrence of root rot was found on less than 50% of the plants in the first year, injury by root boring insects was detected on 80-90% of the plants and root rot occurred on 100% in the second season. The interaction of the insect and disease agents had arrested plant growth by June of the second year and had killed two-thirds of the plants by October. In addition to reports from the USA, interaction of these agents has been cited as an important cause for stand loss in red clover in eastern Canada (Thompson & Willis, 1967). Greenhouse studies have shown that injury by the clover root curculio also leads to increased incidence of _Fusarium_ root rot in white clover (Graham & Newton, 1960).

Dickason et al. (1968) observed frequent association of clover root curculio injury and infection by _F. oxysporum_ in alfalfa. The fact that damage by this insect species could lead to higher incidence of _Fusarium_ root rot was confirmed in greenhouse studies by Hill et al. (1969). Studies on alfalfa have also indicated an association between occurrence of bacterial wilt, _Corynebacterium insidiosum_ (McCull.) H. L. Jens. and injury by clover root curculio (Hill et al., 1971).

Clover Root Borer

The clover root borer, _H. obscurus_, has been described as a serious pest, particularly in red clover, throughout much of the USA and in Canada. This species completes one generation/year and injury is caused to host plants by tunneling of both adults and larvae within roots (Rockwood, 1926). However, the economic importance of this species as a pest in red clover has not been consistently demonstrated. In some studies involving use of insecticides, there have been either increases in yield or increases both in stand retention and yield after reduction of clover root borer populations (Woodside & Turner, 1956; Dickason & Terriere, 1961). Other research has shown no association between satisfactory control of the pest and increased yields of forage or stand longevity in red clover (Gyrisco & Marshall, 1950; Koehler et al., 1961). It has been concluded that this insect is but one of many species that may cause damage in red clover and that its effective control may give measureable returns only if other insect and disease factors are not limiting.

As has been the case with the clover root curculio, the clover root borer has been implicated as a contributor to greater incidence of root rot infections in host plants (Koehler et al., 1961). However, it is difficult to determine the individual role of the borer either in possible mechanical transmission of pathogens as suggested by Deane and Morrison (1957), or in opening plants to infection by feeding on roots. It is common to find a complex of insects and pathogenic species in damaged roots (Newton & Graham, 1960; Waters, 1964). Whereas, it appears to be usual for damage by the clover root curculio to occur prior to infection by pathogens such as Fusarium spp., some studies have indicated that the clover root borer may seek out weakened or diseased plants in which to oviposit, and thus not be important in initiating root rot (Pruess, 1959; Leath & Byers, 1973). Newton and Graham (1960) stated that the borers did not feed on the decaying tissues, but rather they fed in adjacent healthy tissue and contributed to additional stress on the plants. Despite questions about individual species effects, stresses caused by the entire complex of insects and disease agents that attack the roots undoubtedly cause mortality of plants and reduce stand density and productivity of red clover.

Alfalfa Weevil Complex

After introduction during 1904 (Titus, 1910) of what has become known as the western strain of the alfalfa weevil, Hypera postica (Gyllenhal), and a subsequent introduction in 1951 (Poos and Bissell, 1953) designated the eastern strain, this species has become a perennial pest of considerable importance to alfalfa producers throughout the USA. A third introduction which was detected about 1939 (Wehrle, 1939), has been identified as the Egyptian alfalfa weevil, H. brunneipennis (Boheman). This population has been found in Arizona and California. However, recent findings summarized by Manglitz and Ratcliffe (1988) indicate that this population does not differ from the eastern strain of H. postica.

The alfalfa weevil does feed on Trifolium spp. and in field tests of various clovers has shown preference for white, alsike, and mike clovers (Hoveland & Bass, 1963; Byrne & Blickenstaff, 1968). However, its preference for alfalfa has far exceeded that for clovers. Research emphases related to crop losses due to H. postica have centered on alfalfa and little information has been developed on damage by this species to clovers, particularly from long-term studies such as are necessary to describe effects on stand retention.

In most alfalfa-growing areas of the USA, the larval stage of H. postica has caused serious damage, and at times, total defoliation of alfalfa. The severity of the damage has led to large reductions in plantings of alfalfa in the southern states. Numerous studies have described effects of feeding by larvae and adults of H. postica on alfalfa plants. Liu and Fick (1975) reported reductions in seasonal yields exceeding 1 Mg/ha with larval populations of 1.1-2.7/stem. Lower yields have resulted, in part, from the fact that plant height and root reserves as measured by total nonstructural carbohydrates (TNC) are reduced in association with larval attack (Fick & Liu, 1976). In addition, damage in the first crop of alfalfa has resulted in decreased rates of regrowth after harvest (Fick, 1976). Berberet and McNew (1986) described effects of variable levels of infestation up to 17 larvae/stem in reducing plant growth, maturation, and yield in the first crop. The highest infestation levels caused complete defoliation of alfalfa and reduced growth and yield by 80-90%.

As part of the emphasis on IPM in alfalfa, determination of economic threshold and economic injury levels for H. postica has received considerable emphasis. An essential requirement for determination of these levels is knowledge of the relationship of the weevil infestation and alfalfa yield reduction. Hintz et al. (1976) found that the relationship of these variables changed depending upon the growth stage of alfalfa when damage occurred, with smaller plants being much less able to tolerate weevil feeding. Berberet et al. (1981) reported that, when plants were heavily damaged before attaining a height of 15-20 cm, reduction in forage yield at first harvest averaged 188 kg/ha for each increase of one larva/stem in the population density. In addition, residual effects on the second crop, which was not infested, averaged 155 kg/ha for each increase of one larva/stem. Residual effects resulted, to some extent, from reduced stem density as well as from reduced regrowth rate. In this study and one conducted by Wilson et al. (1979), it appeared that effects of a single instance of defoliation due to the alfalfa weevil, despite causing temporary reduction in stem density, did not result in death of plants and permanent loss of stand.

Extensive experimentation has been conducted in Oklahoma from 1981 to 1987 for the purpose of determining the combined effects of alfalfa weevil and weed species such as cheat, Bromus secalinus L., crabgrass, Digitaria sp., and cupgrass, Eriochloa sp., on stand retention and productivity of alfalfa. In two studies conducted from 1981 to 1986 (Berberet et al., 1987), insecticide and herbicides were used individually and in combination to regulate natural infestations of alfalfa weevils and weeds. For one treatment, alfalfa weevils were controlled by application of carbofuran in the first crop, while in a second treatment, weeds were controlled with a dormant season application of terbacil (3-tert-butyl-5-chloro-6-methyluracil) plus oryzalin (3,5-dinitro-N_4,N_4-dipropysulfanilamide). A third treatment involved use of both types of pesticides (pest-free) and the fourth was unsprayed. The first study was initiated on a stand in its third year of production and continued for 3 yr. The second was begun on a stand in its second year and continued for 4 yr. Data were collected on pest infestation levels, stand densities, and yields (three-five harvests) during each season.

In the first study, peak weevil numbers averaged 6.2 larvae/stem without insecticide. Weed infestations which were greatest in the unsprayed treatment, consisted primarily of warm-season species (crabgrass and cupgrass). Weeds comprised up to 42% of the harvested forage in this treatment during August of the first year (1981), and up to 73% of forage in the same month of 1982. During the period when weed infestation levels were highest, they comprised just 3-4% of forage harvested in the two treatments which received herbicides. In spite of stresses imposed by pest infestations, the beginning stand density of 23-25 stems/0.1 m^2, was not appreciably reduced in any treatment until the second year (Table 1). In the third year of the study (1983), the combined effects of alfalfa weevil and weed infestations (unsprayed) resulted in greatly reduced stem density relative to other treatments (Table 1). Also, yield losses relative to the pest-free treatment were much greater in the unsprayed alfalfa than in either treatment infested by only one pest type (Table 2).

In the second study which began in 1983, peak weevil populations averaged 6.4 larvae/stem in unsprayed plots. Cool-season grasses such as cheat were the predominate weed species. In the fourth year (1986), the weed content of first harvest forage in this treatment reached 72% compared to 7% in the treatment which received insecticide only. Where the herbicides were used, the weed content was held to 4% or less. Observations on stand retention were similar to those in the first study except that consistently reduced stem densities were not apparent until the third year (1985) (Table 3). Consistent reductions in stem counts resulted after repeated infestations by both weevils and weeds. Little effect on yields was seen in this study due to weed infestation alone (Table 4). Losses exceeded 1.5 Mg/ha in each season due to damage by weevil larvae and were much higher where both types of pests occurred. It seemed apparent that yield and stand reductions resulting from combined pest infestations were much greater than the sum of the effects of the pests occurring individually.

A third, more comprehensive study was begun in fall of 1982 with the objective of describing the individual and combined effects of the alfalfa weevil and weed species in three cultivars ('WL318', 'Arc', and a registered Oklahoma common designated 'OK08') with varied harvest management practices (Dowdy 1988). In

Table 1. Alfalfa stem densities after treatment with insecticides, herbicides, or both, 1981-83.

	1981		1982		1983
Treatment	Crop 1	Crop 3	Crop 1	Crop 3	Crop 1
	No./0.1m^2				
Insect. + herb.	22a*	18a	24a	20a	16a
Insecticide	22a	20a	23a	16b	13ab
Herbicides	21a	18a	20b	19ab	17a
Unsprayed	20a	18a	19b	9c	9b

*Numbers within columns followed by the same letter are not significantly different (P>0.05; Duncan's MRT).

Table 2. Seasonal forage yield reductions with insecticide, herbicide, or unsprayed treatments in Oklahoma, 1981-82.[1]

Treatment	1981 Total	1981 Alfalfa	1982 Total	1982 Alfalfa	2-yr total (alfalfa)
	Mg/ha				
Insecticide	(0.3)[2]	0.8	(0.8)[2]	1.1	1.9
Herbicides	1.6	1.7	1.3	1.4	3.1
Unsprayed	0.5	2.5	1.7	4.6	7.1

[1]Reductions relative to the pest-free treatment (total = weeds + alfalfa).
[2]Yield increase.

this experiment, cultivars comprised main plots, harvest treatments consisting of late fall harvest (mid-November), winter-grazing with cattle (December and January), and no fall or winter harvest (last cut in mid-September) were randomized on subplots, and the four pesticide options described for previous studies were applied on sub-subplots within each harvest treatment. Treatments were applied annually from 1982 to 1987. Data were collected each year on weevil and weed infestation levels, stem densities, and yields. At the termination of the experiment, plants were undercut and counted from a 2.5 m^2 area in each sub-subplot.

Peak populations of alfalfa weevil averaged 4.4 larvae/stem in the first crop on plots where fall growth was left unharvested and no insecticide was applied. Numbers typically were somewhat lower in treatments where fall growth (and habitat for overwintering weevil adults) was removed by cutting or grazing. The weed component of the forage was relatively light in all treatment combinations through 1985. Due to declining stands, particularly in OK08, and effects of repeated weevil infestation, average weed content of forage reached 80-90% at first harvest of 1987 in some unsprayed treatment combinations. The accumulated effects of pest

Table 3. Alfalfa stem densities after treatment with insecticide, herbicides, or both, 1984-86.

Treatment	1984 Crop 1	1984 Crop 3	1985 Crop 1	1985 Crop 3	1986 Crop 1
	No./0.1m^2				
Insect. + herb	22a*	25a	28a	23a	23a
Insecticide	21a	25a	28a	23a	22a
Herbicides	19a	22a	22b	22a	16b
Unsprayed	13b	22a	18c	19b	12c

*Numbers within columns followed by the same letters are not significantly different (P>0.05; Duncan's MRT).

Table 4. Seasonal forage yield reductions with insecticide, herbicide, or unsprayed treatments in Oklahoma, 1984-86.[1]

Treatment	1984 Alfalfa	1985 Total	1985 Alfalfa	1986 Total	1986 Alfalfa	3-yr total (alfalfa)
	Mg/ha					
Insecticide	(0.4)[2]	(0.3)[2]	0	0	0.4	0
Herbicides	2.5	1.6	1.7	2.7	2.9	7.1
Unsprayed	3.1	1.5	3.1	2.4	5.1	11.3

[1]Reductions relative to pest-free treatment (total = weeds + alfalfa).
[2]Yield increase.

stress over the 5-yr period on productivity of the alfalfa were evident in seasonal yield data for 1987 (Table 5). Means are given for WL318 and OK08 only as they represent the highest and lowest levels. Over the course of the study, there were no consistent differences in forage yields due to harvest treatments within cultivars. Overall, it appeared that winter-grazing, in particular, may have value in the southern USA for reduction of alfalfa weevil and weed infestations without causing any acceleration of stand decline.

Table 5. Seasonal forage yields in the fifth year that harvest and pesticide treatments were imposed in Oklahoma.

Cultivar	Harvest treatment: Fall cut, No herb.	Fall cut, Herb.	Winter grazed, No herb.	Winter grazed, Herb.	Unharvested, No herb.	Unharvested, Herb.
	Mg/ha					
	alfalfa + weeds					
WL318						
no insect.	11	14	13	15	11	13
insect.	15	14	15	15	17	16
OK08						
no insect.	6	9	8	11	8	9
insect.	9	12	9	14	9	11
	alfalfa only					
WL318						
no insect.	5	11	6	12	4	10
insect.	10	13	9	14	11	14
OK08						
no insect.	<1	3	1	4	<1	4
insect.	3	7	3	9	3	5

LSD (total) for cultivar = 3; for harvest = 2; for pesticide = 0.5
LSD (alfalfa) for cultivar = 4; for harvest = 3; for pesticide = 0.6 (P = 0.05)

Table 6. Alfalfa stem densities after imposition of harvest and pesticide treatments over a 5-yr period in Oklahoma.

	Harvest treatment					
	Fall cut		Winter grazed		Unharvested	
Cultivar	No herb.	Herb.	No herb.	Herb.	No herb.	Herb.
	No./0.1m^2					
WL318						
no insect.	7	13	11	13	9	15
insect.	12	15	14	16	16	16
OK08						
no insect.	5	5	2	5	2	5
insect.	3	8	4	8	5	4

LSD for cultivar = 5; for harvest = 4; for pesticide = 1(P = 0.05)

Final stem densities counted in July of 1987 illustrated the value of the improved cultivar (WL318) for stand retention in comparison to OK08 (Table 6). Consistently higher plant populations were recorded for WL318 at the termination of the study, as well (Table 7). There was convincing evidence of the role that pest stresses have in promoting stand decline as stem and plant numbers were consistently lower without pesticides. It was also evident that stress imposed on plants as a result of defoliation by weevil larvae reduced the competitiveness of alfalfa against weeds.

Studies conducted in California have shown minimal long-term effects on productivity of alfalfa resulting from poor alfalfa weevil control, but indicated that lack of controls for winter annual weeds resulted in significant losses that continued to escalate over the life of the stand. Yield reductions were greatest with combined alfalfa weevil and weed stresses (Lamp et al., 1985). An interesting aspect of these studies related to effects of weeds in reducing egg and larval populations of the

Table 7. Final alfalfa plant counts after imposition of harvest and pesticide treatments over a 5-yr period in Oklahoma.

	Harvest treatment					
	Fall cut		Winter grazed		Unharvested	
Cultivar	No herb.	Herb.	No herb.	Herb.	No herb.	Herb.
	No./m^2					
WL318						
no insect.	22	47	30	45	23	37
insect.	45	42	41	57	28	50
OK08						
no insect.	6	18	5	17	6	20
insect.	11	25	9	25	21	20

LSD for cultivar = 21; for harvest = 17; for pesticide = 4(P=0.05)

alfalfa weevil. Numbers of eggs were considerably lower in weedy compared to weed-free stands and this difference became more pronounced with each successive year. The effects of weed infestations in reducing larval populations of the weevil had previously been reported by Norris et al. (1984). These authors stated that the differences in populations were not altered by the type of herbicide used and therefore must be attributed to changes in vegetation. Wolfson and Yeargan (1983) also found greater alfalfa weevil population densities in areas where weeds had been controlled with a herbicide.

Summers and McClellan (1975) reported that a significant amount of the defoliation of alfalfa grown in California, previously attributed to the Egyptian alfalfa weevil was actually due to foliar diseases. Lamp et al. (1985) discussed the need for research data on effects of both insect pests and pathogens over the life of a stand. In studies conducted from 1980 to 1983, it was concluded that pathogenic fungi were much more important than insect pests in contributing to stand decline. Insects such as alfalfa weevil were more instrumental in reducing the amount of foliage produced by plants. There appeared to be no synergistic interaction between pathogens and foliar insect pests. However, stress resulting from either of these pest types tended to increase weed encroachment and lead to rapid loss of stand. In conclusion, all long-term studies that have included members of the alfalfa weevil complex show that consideration of pest interactions is much more important in describing stand decline than concentration on insect species alone.

APHIDS

Of the numerous aphid species that infest forage legumes, we have included only those which are perennial pests throughout much of the USA. A common problem in working with aphids is the frequency of infestations by more than one species and the difficulty in determining the effects of any single species on productivity and stand retention. The emphasis of research has primarily been on developing resistant varieties of alfalfa and clovers (Manglitz, 1985; Manglitz & Ratcliffe, 1988), and relatively little information is available on yield and stand reductions resulting from aphid infestations.

Pea Aphid

The pea aphid, Acyrthosiphon pisum (Harris), occurs throughout the USA and much of Canada. Its host range includes virtually all forage legumes that are grown in North America (Ellsbury & Nielson, 1981), although it is best known as a pest of alfalfa and several Trifolium spp. The pea aphid is a cool-season pest whose populations typically reach economic injury levels in the spring and fall. A number of researchers have reported that this aphid has caused serious reductions in growth and yield of alfalfa (Franklin, 1953; Harvey et al., 1971). A severe infestation of this pest is reported to have caused the loss of half of the first crop yield in Kansas during 1959 (Hackerott et al., 1963).

Actual losses suffered as a result of pea aphid infestations in alfalfa appear to be somewhat dependent upon soil moisture levels. Hobbs et al. (1961) found that pea aphid population densities of 1400 to 1800/sweep did not cause a yield reduction in an irrigated stand. By contrast, Cuperus and Radcliffe (1982) estimated the economic threshold in alfalfa to be ca. 70/sweep or 1.2/stem within 2 wk of harvest in nonirrigated fields where moisture for plant growth was limited. A serious side-effect of pea aphid injury in alfalfa has been observed in Alberta, Canada by Harper and Lilly (1966) and Harper and Freymen (1979, 1983). In both greenhouse and field experiments, heavy infestations of this aphid decreased cold-hardiness. The possibility of greater winter killing where pea aphids have occurred may result in stand reductions over much of the USA and Canada.

In an experiment involving individual and combined infestations of the pea aphid and alfalfa weevil in the first crop, Wilson and Quisenberry (1986) found that each pest type reduced yield of alfalfa at first harvest, but the effects did not carry over to the second crop. The greatest yield reductions resulted with combined infestations and residual effects of these infestations were evident at second harvest. Although this description points out the potential for increased damage when pea aphids occur along with another insect species, an even greater threat to stand longevity may exist with combined occurrence of aphids and plant pathogens. Leath and Byers (1977) reported that there was a significant interaction between this aphid and *Fusarium* spp. that accelerated development of root rot disease symptoms and death of plants in both alfalfa and red clover. These studies provide an additional example where insect infestation and disease incidence should not be treated as separate, discrete problems.

Blue Alfalfa Aphid

Since its discovery in California during 1974, the blue alfalfa aphid, *A. kondoi* Shinji, has spread over much of the USA. Like the pea aphid, this species occurs primarily in cool, spring weather on the first crop of alfalfa (Stern et al., 1980; Berberet et al., 1983). Its host range does include several legume genera that are used for forage production among which *Medicago* and *Melilotus* appear to be most suitable for its reproduction and survival (Ellsbury & Nielson, 1981). Damage to alfalfa appears to start when blue alfalfa aphid populations reach 20/stem if top growth is short. Taller plants (>25 cm) can withstand population levels of 40-50/stem (Stern et al., 1980). When populations of 80-90/stem have been allowed to persist for more than a short time, serious yield losses resulted (Summers & Coviello, 1984). The blue alfalfa aphid has caused up to 40-cm reduction in plant height and greater than 2 Mg/ha reduction in forage yield. Although extensive studies have not been conducted to determine the potential for loss of stands due to this pest, our observations in Oklahoma indicate that it causes stunting and death of plants more rapidly than the pea aphid.

Spotted Alfalfa Aphid and Yellow Clover Aphid

In 1954, damage to alfalfa by what was thought to be the yellow clover aphid, Therioaphis trifolii (Monell), was reported over New Mexico, Arizona, and California in the southwestern USA (Tuttle & Butler, 1954). During 1955, widespread outbreaks of this aphid were reported in several states including Oklahoma (Bieberdorf & Bryan, 1956). Within a short time of the first reports of damage in the Southwest, questions arose regarding the proper identity of the aphid. By comparison of anatomical characteristics and preferred hosts, it was identified as the spotted alfalfa aphid, T. maculata (Buckton), a species that is closely related to the yellow clover aphid (Dickson et al., 1955; Dickson, 1959). The yellow clover aphid had previously been known as a minor pest of red clover in the eastern USA. Peters and Painter (1957) found that the preferred hosts of the spotted alfalfa apahid included most Medicago spp. that were tested, as well as several species of Melilotus. The best-suited hosts for the yellow clover aphid were found in the genus Trifolium, and few Medicago spp. were acceptable. In a recent review of the taxonomic status of these aphids, Manglitz and Ratcliffe (1988) indicated that T. trifolii and T. maculata may be forms of the same species T. trifoli.

Nielson and Barnes (1961) reported that the greatest population densities of the spotted alfalfa aphid occurred between April and October in Arizona. Over the last 15 years in Oklahoma, outbreaks of the aphid have occurred in late fall (November) and in spring, with the last widespread, heavy infestations observed during February and March of 1981 (Berberet et al., 1983). In both locations, conditions of limited rainfall and low relative humidities appeared to be optimal for reproduction and survival of the aphid. Mittler and Sylvester (1961) theorized that the spotted alfalfa aphid may be a more serious pest than the pea aphid because it is better adapted to drought conditions and it disperses much more slowly from damaged host plants.

The spotted alfalfa aphid is quite damaging to susceptible plants and causes chlorosis and necrosis of leaves in the presence of high population densities. The aphid also causes a toxic reaction that may result in death of plants and thinning of stands. Kindler et al. (1971) artifically infested alfalfa in field cages and found that, over 2 yr, spotted aphids reduced yields of susceptible alfalfa cultivars by an average of 38% and stands by 6%. Dickson et al. (1955) reported that a single, heavy infestation may ruin a stand of alfalfa. In Oklahoma, we have observed 70-80% reduction in stands over a period of several weeks during early spring when population densities ranged up to 800/stem on alfalfa 5-10 cm tall. Somewhat in contrast to these results, Burkhardt (1959) reported that control of heavy infestations of spotted alfalfa aphid in fall did not increase yields in the following season.

A great deal of success has been achieved in breeding alfalfa with resistance the spotted alfalfa aphid (Nielson & Lehman, 1980). Although use of resistant cultivars greatly reduces the chances of yield and stand loss due to this pest, emergence of adapted strains or biotypes may negate the effectiveness of resistance. Nielson et al. (1970) reported the existence of a new biotype of spotted alfalfa aphid that severely damaged previously resistant cultivars in Arizona and California. There is a great need for additional research on effects of this aphid on stand persistence in alfalfa, especially for studies relating to interactions of improved cultivars and various biotypes within the pest species.

POTATO LEAFHOPPER

The potato leafhopper, Empoasca fabae (Harris), is regarded currently as the most serious insect pest of alfalfa throughout much of the central USA. Populations of this species become established each year after an initial period of migration from overwintering areas near the Gulf of Mexico. Typically, damage becomes most evident on the second and third crops of alfalfa. The leafhopper feeds by removing plant fluids using its piercing-sucking mouthparts and causes a characteristic chlorosis (hopper-burn) beginning at the tips of leaves and spreading over entire leaves as feeding continues. The same type of symptom is produced in red clover and has been called "yellow top" (Hollowell et al., 1927). Studies have been conducted for many years to relate this symptom with other evidence of plant injury. Among the first attempts was that by Poos and Johnson (1936), who showed that potato leafhopper infestation resulted in reduction of both plant height and yield in alfalfa and red clover.

More recent studies have been concentrated on the effects of potato leafhopper on productivity of alfalfa. Kouskolekas and Decker (1968) reported that the most dramatic effect of infestations on alfalfa was stunting of growth. These authors also observed that reductions in plant growth did not always cause proportional reductions in yield. Hower and Flinn (1986) concurred in stating that correlations between plant height and yield were not consistent. They found that alfalfa height was lowered by 6-54% with leafhopper populations of one to eight/stem and yield was reduced from 8-70% with the same infestation levels. Several researchers have found that losses due to the leafhopper were greatest where fertility and/or soil moisture levels were low (Shaw et al., 1986; Oloumi-Sadeghi et al., 1988).

Another critical concern related to damage by the potato leafhopper is reduced forage quality. Shaw and Wilson (1986) conducted a study in which crude protein of forage was reduced from 23 to 18% due to infestation by this leafhopper. Hower and Flinn (1986) calculated that the percentage reduction in protein yield (% crude protein X dry weight) was usually greater than the percentage reduction in total dry weight of forage produced. Thus, the reduction in total nutrient production/ha is a critical consideration in assessing the importance of the leafhopper as a pest in legumes.

The timing of infestations after harvest was important in determining the extent of losses in studies of Kouskolekas and Decker (1968), who reported that early infestations caused much greater yield losses than those which occurred 2-3 wk after harvest. Cuperus et al. (1983) calculated economic threshold levels from 0.3 to 0.5 potato leafhoppers/sweep in alfalfa regrowth ranging from 5-17 cm in height.

High leafhopper population densities occurring soon after harvest resulted in greater than 50% reduction stem counts according to Kouskolekas and Decker (1968). Although this did not necessarily mean that plants had been killed, it showed that plant vigor had been reduced by the insect infestation. Shaw and Wilson (1986) found that the ability of alfalfa plants to accumulate carbohydrate root reserves was impaired by the potato leafhopper. This slowed the rate of regrowth in the heavily damaged plots and could have reduced the capability of alfalfa plants to compete with weeds. Oloumi-Sadeghi et al. (1988) found that stem densities decreased by 40% from 1 yr to the next in plots where pesticides were not applied for insect (primarily potato leafhopper) or weed control. Apparently, the pest stresses had resulted in increased mortality of plants during winter. An additional factor that may be important in assessing the damage potential of potato leafhopper on alfalfa is the effect of weed type and infestation densities on the insect populations. Presence of high densities of grass weeds has tended to reduce leafhopper numbers relative to those in weed-free stands, while infestations of broadleaf weeds has led to increased numbers (Lamp et al., 1984; Oloumi-Sadeghi et al., 1988).

REFERENCES

Berberet, R. C., D. C. Arnold, and K. M. Soteres. 1983. Geographical occurrence of Acyrthosiphon kondoi Shinji in Oklahoma and its seasonal incidence in relation to Acyrthosiphon pisum (Harris), and Therioaphis maculata (Buckton) (Homoptera: Aphididae). J. Econ. Entomol. 76:1064-1068.

Berberet, R. C., and R. W. McNew. 1986. Reduction in yield and quality of leaf and stem components of alfalfa forage due to damage by larvae of Hypera postica (Coleoptera: Curculionidae). J. Econ. Entomol. 79:212-218.

Berberet, R. C., R. D. Morrison, and K. M. Senst. 1981. Impact of the alfalfa weevil, Hypera postica (Gyllenhal) (Coleoptera: Curculionidae), on forage production in nonirrigated alfalfa in the southern plains. J. Kans. Entomol. Soc. 54:312-318.

Berberet, R. C., J. F. Stritzke, and A. K. Dowdy. 1987. Interactions of alfalfa weevil (Coleoptera: Curculionidae) and weeds in reducing yield and stand of alfalfa. J. Econ. Entomol. 80:1306-1313.

Bieberdorf, G. A., and D. E. Bryan. 1956. Research on the spotted alfalfa aphid. Oklahoma A. & M. Coll. Bull. B-469.

Burkhardt, C. C. 1959. Effects of heavy fall infestations of spotted alfalfa aphids on subsequent spring growth of alfalfa in Kansas. J. Econ. Entomol. 52:642-643.

Byers, R. A., and W. A. Kendall. 1982. Effects of plant genotypes and root nodulation on growth and survival of Sitona spp. larvae. Environ. Entomol. 11:440-443.

Byrne, H. D., and C. C. Blickenstaff. 1968. Host-plant preference of the alfalfa weevil in the field. J. Econ. Entomol. 61:334-335.

Cranshaw, W. S. 1985. Clover root curculio injury and abundance in Minnesota alfalfa of different stand age. Great Lakes Entomol. 18:93-95.

Cuperus, G. W., and E. B. Radcliffe. 1982. Economic injury levels and economic thresholds for pea aphid, Acyrthosiphon pisum (Harris), on alfalfa. Crop Prot. 1:453-463.

Cuperus, G. W., E. B. Radcliffe, D. K. Barnes, and G. C. Marten. 1983. Economic injury levels and economic threshold for potato leafhopper (Homoptera: Cicadellidae) on alfalfa in Minnesota. J. Econ. Entomol. 76:3141-3149.

Deane, B. C., and F. O. Morrison. 1957. The distribution and importance of the clover root borer Hylastinus obscurus (Marsh) (Coleoptera: Scolytidae) in Quebec. Can. J. Plant Sci. 37:26-33

Dickason, E. A., C. M. Leach, and A. E. Gross. 1968. Clover root curculio injury and vascular decay of alfalfa roots. J. Econ. Entomol. 61:1163-1168.

Dickason, E. A., and L. C. Terriere. 1961. Insecticide residues on red clover after clover root borer control with aldrin and heptachlor granules. J. Econ. Entomol. 54:1058-1059.

Dickson, R. C. 1959. On the identity of the spotted alfalfa aphid in North America. Ann. Entomol. Soc. Am. 52:63-68.

Dickson, R. C., E. F. Laird, Jr., and G. R. Pesho. 1955. The spotted alfalfa aphid. Hilgardia 24:93-118.

Dintenfass, L. P., and G. C. Brown. 1986. Feeding rate of larval clover root curculio, Sitona hispidulus (Coleoptera: Curculionidae), on alfalfa taproots. J. Econ. Entomol. 79:506-510.

Dowdy, A. K. 1988. Population densities of the alfalfa weevil, Hypera postica (Gyllenhal), in alfalfa, Medicago sativa L., as influenced by late fall harvest, winter grazing, and weed control. Ph.D. diss. Oklahoma State Univ., Stillwater.

Ellsbury, M. M., and M. W. Nielson. 1981. Comparative host plant range studies of the blue alfalfa aphid, Acyrthosiphon kondoi Shinji, and the pea aphid, Acyrthosiphon pisum (Harris) (Homoptera: Aphididae). USDA Tech. Bull. 1639. U.S. Gov. Print. Office, Washington, DC.

Fick, G. W. 1976. Alfalfa weevil effects on regrowth of alfalfa. Agron. J. 68:809-812.

Fick, G. W., and B. W. Y. Liu. 1976. Alfalfa weevil effect on root reserves, developmental rate, and canopy structure of alfalfa. Agron. J. 68:595-599.

Franklin, W. W. 1953. Insecticidal control plot tests for pea aphids in relation to alfalfa hay yields. J. Econ. Entomol. 46:462-467.

Godfrey, L. D. 1984. Effects and interactions of the clover root curculio, Sitona hispidulus (F.), alfalfa weevil, Hypera postica (Gyllenhal) and root rot fungi in the alfalfa agroecosystem. Ph.D. diss. Univ. of Kentucky, Lexington (Diss. Abstr. AAC8416135).

Godfrey, L. D., and K. V. Yeargan. 1985. Influence of soil moisture and weed density on clover root curculio, Sitona hispidulus, larval stress to alfalfa. J. Agric. Entomol. 2:370-377.

Graham, J. H., and R. C. Newton. 1960. Relationship between injury by the clover root curculio and incidence of fusarium root rot in ladino white clover. Plant Dis. Rep. 44:534-535.

Gyrisco, G. G., and D. S. Marshall. 1950. Further investigations on the control of the clover root borer in New York. J. Econ. Entomol. 43:82-86.

Hackerott, H. L., E. L. Sorensen, T. L. Harvey, E. E. Ortman, and R. H. Painter. 1963. Reactions of alfalfa varieties to pea aphids in the field and greenhouse. Crop Sci. 3:298-301.

Harper, A. M., and S. Freyman. 1979. Effect of the pea aphid, Acyrthosiphon pisum (Homoptera: Aphididae), on cold-hardiness of alfalfa. Can. Entomol. 111:635-636.

Harper, A. M., and S. Freyman. 1983. Cold-hardiness of 1-, 2-, and 3-year-old alfalfa infested with the pea aphid (Homoptera: Aphididae). Can. Entomol. 115:1243-1244.

Harper, A. M., and C. E. Lilly. 1966. Effects of the pea aphid in southern Alberta. J. Econ. Entomol. 59:1426-1427.

Harvey, T. L., H. L. Hackerott, and E. L. Sorensen. 1971. Pea aphid injury to resistant and susceptible alfalfa in the field. J. Econ. Entomol. 64:513-517.

Hill, Jr., R. R., J. J. Murray, and K. E. Zeiders. 1971. Relationships between clover root curculio injury and severity of bacterial wilt in alfalfa. Crop Sci. 11:306-307.

Hill, Jr., R. R., R. C. Newton, K. E. Zeiders, and J. H. Elgin, Jr. 1969. Relationships of the clover root curculio, Fusarium wilt, and bacterial wilt in alfalfa. Crop Sci. 9:327-329.

Hintz, T. R., M. C. Wilson, and E. J. Armbrust. 1976. Impact of alfalfa weevil larval feeding on the quality and yield of first cutting alfalfa. J. Econ. Entomol. 69:749-754.

Hobbs, G. A., N. D. Holmes, G. E. Swailes, and N. S. Church. 1961. Effect of the pea aphid, Acyrthosiphon pisum (Harris) (Homoptera: Aphididae), on yields of alfalfa hay on irrigated land. Can. Entomol. 93:801-804.

Hollowell, E. A., J. Monteith, Jr., and W. P. Flint. 1927. Leafhopper injury to clover. Phytopathology 17:399-404.

Hoveland, C. S., and M. H. Bass. 1963. Susceptibility of mike clover (Trifolium michelianum Savi) to alfalfa weevil. Crop Sci. 3:452-453.

Hower, A. A., and P. W. Flinn. 1986. Effects of feeding by potato leafhopper nymphs (Homoptera: Cicadellidae) on growth and quality of established stand alfalfa. J. Econ. Entomol. 79:779-784.

Jackson, D. J. 1922. Bionomics of weevils of genus Sitona injurious to leguminous crops in Britian, Part II. Sitona hispidula F., S. sulcifrons Thun. and S. crinita Herbst. Ann. Appl. Biol. 9:93-115.

James, J. R., L. T. Lucas, D. S. Chamblee, and W. V. Campbell. 1980. Influence of fungicide and insecticide applications on persistence of ladino clover. Agron. J, 72:781-784.

Kindler, S. D., W. R. Kehr, and R. L. Ogden. 1971. Influence of pea aphids and spotted alfalfa aphids on the stand, yield of dry matter, and chemical composition of resistant and susceptible varieties of alfalfa. J. Econ. Entomol. 64:653-657.

Koehler, C. S., K. D. Fezer, H. H. Neunzig, and G. G. Gyrisco. 1961. The economic importance of the clover root borer. J. Econ. Entomol. 54:631-635.

Kouskolekas, C., and G. C. Decker. 1968. A quantitative evaluation of factors affecting alfalfa yield reduction caused by the potato leafhopper attack. J. Econ. Entomol. 61:921-927.

Lamp, W. O., R. J. Barney, E. J. Armbrust, and G. Kapusta. 1984. Selective weed control in spring-planted alfalfa: effects on leafhoppers and planthoppers (Homoptera: Auchenorrhyncha) with emphasis on potato leafhopper. Environ. Entomol. 13:207-213.

Lamp, W. O., K. V. Yeargan, R. F. Norris, C. G. Summers, and D. G. Gilchrist. 1985. Multiple pest interactions in alfalfa. *In* R. E. Frisbie and P. L. Adkisson (ed.) Integrated pest management on major agricultural systems. Texas A&M Univ., College Station.

Leach, C. M., E. A. Dickason, and A. E. Gross. 1961. Effects of insecticides on insects and pathogenic fungi associated with alsike clover roots. J. Econ. Entomol. 54:543-546.

Leath, K. T., and R. A. Byers. 1973. Attractiveness of diseased red clover roots to the clover root borer. Phytopathology 63:428-431.

Leath, K. T., and R. A. Byers. 1977. Interaction of *Fusarium* root rot with pea aphid and potato leafhopper feeding on forage legumes. Phytopathology 67:226-229.

Liu, B. W. Y., and G. W. Fick. 1975. Yield and quality losses due to alfalfa weevil. Agron. J. 67:828-832.

Manglitz, G. R. 1985. Insects and related pests. *In* N. L. Taylor (ed.) Clover science and technology. ASA, CSSA, and SSSA, Madison, WI.

Manglitz, G. R., D. M. Anderson, and H. J. Gorz. 1963. Observations on the larval feeding habits of two species of *Sitona* (Coleoptera: Curculionidae) in sweetclover fields. Ann. Entomol. Soc. Am. 56:831-835.

Manglitz, G. R., and R. H. Ratcliffe. 1988. Insects and mites. p. 671-704. *In* A. A. Hanson et al. (ed.) Alfalfa and alfalfa improvement. ASA, CSSA, and SSSA, Madison, WI.

Mittler, T. E., and E. S. Sylvester. 1961. A comparison of the injury to alfalfa by the aphids, *Therioaphis maculata* and *Macrosiphum pisi*. J. Econ. Entomol. 54:615-622.

Newton, R. C., and J. H. Graham. 1960. Incidence of root-feeding weevils, root rot, internal breakdown, and virus and their effect on longevity of red clover. J. Econ. Entomol. 53:865-867.

Nielson, M. W., and O. L. Barnes. 1961. Population studies of the spotted alfalfa aphid in Arizona in relation to temperature and rainfall. Ann. Entomol. Soc. Am. 54:441-448.

Nielson, M. W., and W. F. Lehman. 1980. Breeding approaches in alfalfa. *In* F. G. Maxwell and P. R. Jennings (ed.) Breeding plants resistant to insects. John Wiley and Sons, New York, NY.

Nielson, M. W., W. F. Lehman, and V. L. Marble. 1970. A new severe strain of the spotted alfalfa aphid in California. J. Econ. Entomol. 63:1489-1491.

Norris, R. F., W. R. Cothran, and V. E. Burton. 1984. Interactions between winter annual weeds and Egyptian alfalfa weevil (Coleoptera: Curculionidae) in alfalfa. J. Econ. Entomol. 77:43-52.

Oloumi-Sadeghi, H., L. R. Zavaleta, S. J. Roberts, E. J. Armbrust, and G. Kapusta. 1988. Changes in morphological stage of development, canopy structure, and root nonstructural carbohydrate reserves of alfalfa following control of potato leafhopper (Homoptera: Cicadellidae) and weed populations. J. Econ. Entomol. 81:368-375.

Peters, D. C., and R. H. Painter. 1957. A general classification of available small seeded legumes as hosts for three aphids of the "yellow clover aphid complex." J. Econ. Entomol. 50:231-235.

Poos, F. W., and T. L. Bissell. 1953. The alfalfa weevil in Maryland. J. Econ. Entomol. 46:178-179.

Poos, F. W., and H. W. Johnson. 1936. Injury to alfalfa and red clover by the potato leafhopper. J. Econ. Entomol. 29:325-331.

Powell, G. S., W. V. Campbell, W. A. Cope, and D. S. Chamblee. 1983. Ladino clover resistance to the clover root curculio (Coleoptera: Curculionidae). J. Econ. Entomol. 76:264-268.

Pruess, K. P. 1959. Effect of host condition on the clover root borer. J. Econ. Entomol. 52:1143-1145.

Quinn, M. A., and A. A. Hower. 1986. Multivariate analysis of the population structure of Sitona hispidulus (Coleoptera: Curculionidae) in alfalfa field soil. Can. Entomol. 118:517-524.

Rockwood, L. P. 1926. The clover root borer. USDA Bull. 1426. U.S. Gov. Print. Office, Washington, DC.

Shaw, M. C., and M. C. Wilson. 1986. The potato leafhopper: scourge of leaf protein! - and root carbohydrates too? p. 152-160. M.C. Wilson (ed.) In Breeding quality into alfalfa. Proc. Eleventh Natl. Alfalfa Symp., Purdue Univ., IN. 5-6 Mar. 1986.

Shaw, M. C., M. C. Wilson, and C. L. Rhykerd. 1986. Influence of phosphorus and potassium fertilization on damage to alfalfa, Medicago sativa L., by the alfalfa weevil, Hypera postica (Gyllenhal) and potato leafhopper, Empoasca fabae (Harris). Crop Prot. 5:245-249.

Stern, V. M., R. Sharma, and C. Summers. 1980. Alfalfa damage from Acyrthosiphon kondoi and economic threshold studies in southern California. J. Econ. Entomol. 73:145-148.

Summers, C. G., and R. L. Coviello. 1984. Impact of Acyrthosiphon kondoi (Homoptera: Aphididae) on alfalfa: field and greenhouse studies. J. Econ. Entomol. 77:1052-1056.

Summers, C. G., and W. D. McClellan. 1975. Interaction between Egyptian alfalfa weevil feeding and foliar disease: impact on yield and quality in alfalfa. J. Econ. Entomol. 68:487-489.

Thompson, L. S., and C. B. Willis. 1967. Distribution and abundance of Sitona hispidula (F.) and the effect of insect injury on root decay of red clover in the maritime provinces. Can. J. Plant Sci. 47:435-440.

Titus, E. G. 1910. The alfalfa weevil. Utah State Coll. Agric. Exp. Stn. Bull. 110.

Tuttle, D. M, and G. D. Butler, Jr. 1954. The yellow clover aphid - a new alfalfa pest in the southwest. J. Econ. Entomol. 47:1157.

Underhill, G. W., E. C. Turner, Jr., and R. G. Henderson. 1955. Control of the clover root curculio on alfalfa with notes on life history and habits. J. Econ. Entomol. 48:184-187.

Waters, N. D. 1964. Effects of Hypera nigrirostris, Hylastinus obscurus, and Sitona hispidula populations on red clover in southwestern Idaho. J. Econ. Entomol. 57:907-910.
Webster, F. M. 1915. Alfalfa attacked by the clover root curculio. USDA Farmers' Bull. 649. U.S. Gov. Print. Office, Washington, DC.
Wehrle, L. P. 1939. A new insect introduction. Bull. Brooklyn Entomol. Soc. 34:170.
Wilson, H. K., and S. S. Quisenberry. 1986. Impact of feeding by alfalfa weevil larvae (Coleoptera: Curculionidae) and pea aphid (Homoptera: Aphididae) on yield and quality of first and second cuttings of alfalfa. J. Econ. Entomol. 79:785-789.
Wilson, M. C., J. K. Stewart, and H. D. Vail. 1979. Full season impact of the alfalfa weevil, meadow spittle bug, and potato leafhopper in an alfalfa field. J. Econ. Entomol. 72:830-834.
Wolfson, J. L., and K. V. Yeargan. 1983. The effects of metribuzin on larval populations of alfalfa weevil, Hypera postica (Coleoptera: Curculionidae). J. Kans. Entomol. Soc. 56:40-46.
Woodside, A. M., and E. C. Turner, Jr. 1956. Control of the clover root borer in Virginia. J. Econ. Entomol. 49:640- 642.

DISCUSSION

Buxton. Is there likely to be success in breeding for resistance to potato leafhopper in the future?

Berberet. I am confident that success in breeding for resistance to the potato leafhopper will be improved in the future. However, a concern in the USA is the apparent reduction of emphasis on breeding of pest resistant legumes for leafhopper and other insects.

Sheaffer. Will it be possible to find resistance to alfalfa weevil in an agronomically acceptable cultivar?

Berberet. While we have not found resistance in the form of antibiosis or antixenosis for alfalfa weevil, we have achieved a high degree of tolerance in cultivars such as 'Cimarron', which is an excellent commercial cultivar. Eventually, other forms of resistance will be available.

Sheaffer. What factors impart resistance (tolerance)?

Berberet. The plants tend to have large, vigorous terminals that withstand some weevil feeding and tend to produce extensive lateral branching in response to feeding and can compensate for some defoliation.

R. Smith. What is the status of biological control of the alfalfa weevil in the USA?

Berberet. In the northeastern USA, several parasitic species have been established and a highly successful control program has resulted. In other portions of the country more limited biological control has been achieved, especially where few parasitic species have been established. In Oklahoma for example, just one parasite is widely established despite 15 yr of releases.

Watson. Is renodulation by lucerne limited, after feeding by *Sitona* sp. larvae has resulted in destruction of these structures? This appears to be a problem in New Zealand following damage by larvae of *Sitona discoideus*.

Berberet. I am not aware of studies in the USA that have addressed this problem.

Matches. Do you know of any entomologists who are working extensively with pasture legumes in the USA?

Berberet. No, not at this time. To my knowledge, the amount of research related to entomology in grazing systems would be quite limited at present. It should be given consideration for greater emphasis in the future.

Reed. Is berseem clover (T. *alexandrinum* L.) known in the USA as a host which harbors aphids that vector viral diseases in lupines and beans?

Berberet. I am aware of no information on this topic.

Buxton. Do mixed stands of grasses and legumes have an effect on insect populations relative to what would be seen in pure stands of legumes?

Berberet. It has been reported that alfalfa stands which have a fairly high grass component tend to be less heavily infested by the potato leafhopper than pure stands. Conversely, grass infestation tends to cause no change in population of alfalfa weevil larvae on a per stem basis. On a per area (m^2) basis, larval numbers are lower because alfalfa plant numbers would tend to be lower.

GENERAL DISCUSSION OF
MAJOR PESTS AND DISEASES

Gramshaw: The presentations on diseases and insect pests from the three countries suggest that there are a number of specific organisms that are of economic importance and a common problem in two or more countries. This provides an opportunity for cooperative research.

Berberet: Pathogens and insects should be considered as a complex rather than individual organisms. Combined disease and pest stresses can greatly reduce legume persistence even though the organisms individually incur only minor damage.

Gramshaw: One disease complex often referred to is the root rot complex. Is there commonality among causal pathogens in the three countries, and is there adequate research methodology to effectively deal with this complex?

Leath: The root rot complex and viruses in clovers are universal in their occurrence and it is quite possible that virus effects go unnoticed. Without an appropriate assay it is impossible to elucidate the virus situation.

Irwin: There is a need for etiological studies to identify the components within root rot complexes, and the components will vary with location. Methodologies are required that allow quantification of losses due to root pathogen complexes.

Hoveland: Superior performance of some genotypes is often difficult to explain in terms of specific known resistances to diseases and insect pests. For example, Florida 77 lucerne.

Gramshaw: Conversely, the recognition of the importance of Phytophthora and Colletotrichum on lucerne in the Queensland subtropics led to specific breeding objectives and the successful development of resistant cultivars.

Irwin: This is an example of advances that can be made when the etiology of the disease or diseases is quantified.

Leath: The components of any root rot complex vary geographically and even within the same season. Methodology would be a common area for cooperation between researchers in different countries. Methods used to study complexes certainly need refining.

Irwin: A suggestion on methods could be the use of multiple regression or methyl bromide fumigation techniques.

Reed: Furadan was being used in several U.S. states in the early 1980s and gave improved pasture establishment where no pests were detected. Has this led to the identification of new problems and has it become a commercial practice?

Leath: The effects cannot be explained and Furadan is used commercially.

Watson: Unexplained responses could be due to very early protection of seedlings against nematodes or other pests that are important in subsequent seedling growth. Assessments of organisms are often made well after establishment when the early differences in pest burdens have dissipated.

Rotar: What is happening in the rhizosphere of young seedling roots?

Caradus: In newly formed nodal roots of white clover the first invaders are nematodes which then allow potentially pathogenic fungi to enter.

Woomer: Are there any biocontrol micro-organisms that can be included with rhizobial inoculant to give early disease and pest protection?

Leath: We have a project evaluating Bacillus spp. for protecting red clover from 'damping off' organisms. A commercial product is already available for protecting peanuts and I believe that the potential for further success is quite good.

Irwin: Do legume breeders endorse the view that there is a need for diversity in legume cultivars to combat a variable pathogen?

R. Smith: For perennial species it would be best to have diversity to withstand varying stresses over numerous environmental sequences. With annuals, uniform lines may be adequate since these are subjected to fewer stresses.

Sheath: The need for diversity in a legume species or community would assume greater importance where stress is irregular or unpredictable.

Allen: Most plant breeders select or breed for high levels of resistance to insects and diseases. Such high levels of resistance may not always be necessary because lower levels can be integrated with other control tactics. High levels of resistance may lead to lines with good agronomic characters being discarded, and may lead to selection of pest biotypes which can circumvent the resistance in the field. I would be interested to hear of the attitude towards developing cultivars with high levels of resistance.

Berberet: The use of high levels of resistance has been successful with lucerne in the past. Has it caused problems in Australia?

Allen: No, not yet, but biotypes of spotted alfalfa aphid were selected in the field in California which damaged resistant cultivars.

R. Smith: Selecting for high levels of resistance is the easiest method. There are two types of resistance, physical and chemical resistance. What type of resistance do you think occurs in aphid-resistant medics?

Allen: It seems to be antibiosis related to chemical resistance.

Watson: Why has pest and disease resistance in white clover proven so elusive? With the open-pollination of the plant and high buried seed loads, is it a realistic goal to attempt selection for resistance? Alternatively, what is the likely field persistence of introduced genotypes in old white clover pastures, especially those with a high level of perpetuation through seed?

Allen: The solution is not easy, but similar problems appear to be occurring with the selection of aphid-resistant subterranean clovers in Western Australia.

Gramshaw: Can we identify disease and pest research areas of common interest?

Clements: Australian scientists are beginning to study the incidence and effects of viruses of white clover. New Zealand scientists may need to do the same. We could draw on experience obtained in the USA. This is a potential area for immediate cooperation.

Allen: In preparation for the accidental introduction of new pests and diseases to the three countries, I suggest this workshop recommends that:

1. Cooperation among the countries occurs to ensure that collections of the various germplasms in each country contain all of the sources of known resistance to pests and diseases in pasture legumes.

2. Arrangements be developed whereby a breeder in one country can have lines tested against exotic pests and diseases in another country which has the pest or disease.

Leath: This concept should be endorsed as a highly beneficial area for cooperative effort.

DEVELOPING PERSISTENT PASTURE LEGUME CULTIVARS FOR AUSTRALIA

R.J. Clements

SUMMARY

More than 160 pasture legume cultivars have been developed in Australia by the direct use of natural ecotypes and by plant breeding, but relatively few are widely sown. Persistence has been sought commonly, using mainly empirical methods. Progress towards more persistent cultivars of lucerne, subterranean clover, annual medics, white clover and *Stylosanthes* species is reviewed briefly. Cultivars have also been developed from other *Trifolium* species, from other temperate legume genera, and from many tropical genera, notably *Macroptilium*, *Centrosema*, *Leucaena*, and *Cassia*. Breeding has usually been commenced prematurely, but has been successful in the long run. Disease and pest resistance are now common improvement objectives. There is still ample scope for developing new species and extending the area of use of those already domesticated.

THE VARIETAL SCENE

Australia's sown pasture legumes are all exotics. Their deliberate cultivation on farms commenced almost 200 yr ago, when lucerne (*Medicago sativa* L.) and white clover (*Trifolium repens* L.) were sown in eastern Australia. During the 50 yr from 1889-1939, two key groups of Mediterranean annuals, subterranean clover (*T. subterraneum* L.) and the annual medics (*Medicago* spp.) were domesticated. Representatives of each group were first sown on farms during the 1890s. Deliberate on-farm cultivation of tropical pasture legumes in the genus *Stylosanthes* commenced during the 1930s (Humphreys, 1967). These five groups of legumes remain those most widely sown today. Of the 165 cultivars of pasture legumes described in the Register of Australian Herbage Plant Cultivars to February 1988 (Barnard, 1972; Mackay, 1982; updates), 91 belong to these groups. Apart from a group of fodder crops, the remainder mainly occupy niches in the subtropics, high rainfall and alpine areas or on heavy or poorly drained soils. Indeed it has been the failure of the main suite of legumes to persist in these areas that has led frequently to a search for alternative species.

In Australia, most cultivars have been developed by collecting, testing, and commercializing natural ecotypes. Not only has this process usually preceded plant breeding, it has continued to provide new cultivars up to the present time. Despite a significant investment in plant breeding in recent years, only one-third of the cultivars released since 1960 have resulted from breeding programs, and there is no sign that this trend is changing (Williams & Clements, 1986). There is no white clover breeding program (Clements, 1987), and all of the registered *Stylosanthes* cultivars are naturalized or introduced ecotypes. Only for lucerne has plant breeding been the predominant source of new cultivars. The phrases "plant improvement" and "variety development" in this paper include both plant breeding and the commercialization of naturally occurring genotypes.

Public organizations and farmers have dominated the development of pasture legume cultivars in Australia. Recently, the private sector has marketed more than 20 proprietary lines of lucerne, mainly from the USA, but their share of the present market is small. All Australian State Departments of Agriculture, several CSIRO Divisions and several Universities have been involved in cultivar development, and vigorous programs are continuing in all states. At present, the main plant improvement centres for pasture legumes are in Brisbane (tropical legumes and lucerne), Perth (subterranean clover, *Lupinus* spp. and *Ornithopus* spp.), Adelaide (lucerne, annual medics, and *Onobrychis viciifolia* Scop.) and Yanco, NSW (lucerne). Smaller centres exist elsewhere, and a national white clover improvement program is under consideration. Testing of the products of plant introduction and breeding is decentralized; for example, the Queensland Department of Primary Industries maintains a network of 12 major screening centres over a latitudinal range of 11 degrees (1400 km). This regional testing plays a key role in plant improvement programs.

Despite the plethora of commercial cultivars, and despite an emerging trend towards sowing mixtures of cultivars, new plantings are dominated by only a few popular lines. For example, only two of the seven registered white clover cultivars (cv. Haifa and Grasslands Huia) and 3 of the 11 *Stylosanthes* cultivars (*S. guianensis* (Aubl.) Sw. var. *intermedia* (Vog.) Hassler cv. Oxley, *S. hamata* (L.) Taub. cv. Verano, and *S. scabra* Vog. cv. Seca) are sown extensively at present. About 40 lucerne cultivars are available commercially, but six cultivars provide more than three-quarters of the newly sown area; indeed, cv. Hunter River, which until 1977 was effectively the only sown cultivar, is regaining a substantial popularity. Annual medic plantings are dominated by two cultivars, and only 6 of the 26 registered cultivars are sown extensively. Only for subterranean clover is there a large number of extensively sown cultivars,and even here 11 cultivars provided 96% of the certified seed produced in 1986/87 (S.G. Clark and K.F.M. Reed, 1988, personal communication).

Good persistence has been a characteristic feature of many of Australia's pasture legumes. Most of the key legumes are in fact intrinsically persistent, and this has been a significant factor in their widespread use. The earliest cultivars of subterranean clover and annual medics were locally adapted races which had persisted on farms for very many years. The popularity of Hunter River lucerne was mainly due to its greater persistence than other cultivars, several of which were more productive in the short term. Townsville stylo (*S. humilis* Kunth.) had a history of persistence and spread for about 30 yr before the first commercial seed was harvested from locally adapted races. White clover spreads naturally in humid parts of eastern Australia onto newly cleared ground, and an ecotype which developed in the Clarence valley region of New South Wales was shown to be more persistent than most introduced lines (O'Brien, 1970). Extensive areas in south-eastern Australia have been colonized by annual species of *Trifolium* and *Medicago* that have never been commercialized (Donald, 1970); at times, these species dominate the pastures. Clearly, there has been a long history of persistence of pasture legumes in this country. If there is now a problem with persistence of the key legumes, it can only be because either the environment (in the widest sense) or the genotypes have changed unfavorably. Have our expectations increased? Were the early successes in the "easiest" environments?

CULTIVAR DEVELOPMENT 1950-1988

The modern era of pasture plant improvement in Australia began after World War II. It has been characterized by increasingly sophisticated plant introduction and breeding programs, commencing with the pioneering plant collecting missions to tropical southern America (1948), southern Africa (1952) and the Mediterranean region (1951-54), and culminating in the use of genetic engineering technologies during the 1980s.

The following review of selection criteria for pasture legumes since 1950, and of progress achieved, is based substantially on information provided in the Register of Australia Herbage Plant Cultivars (*loc. cit*) and the references included therein.

Lucerne

Because lucerne rarely regenerates naturally from fallen seed in Australian pastures, persistence depends upon the survival of plants established at sowing. The slow rate of population turnover, unrestricted transport of seed within Australia, repeated importation of commercial seed from overseas until about 1930, and concentration of seed production in favorable rather than marginal areas have minimized the evolution of locally adapted races, although some regional types were well recognized by 1920 (Cameron, 1973). Lucerne therefore provides the best Australian example of a sustained effort to develop more persistent varieties by breeding, and it is worthwhile reviewing progress in some detail.

Hunter River Lucerne, the standard cultivar until 1977, is a semi winter-dormant variety. Introduction and testing of a range of overseas varieties over a period of many years led to the release of the winter-active African (1962), Siro Peruvian (1967), and Demnat (1972). (Small areas of Peruvian lucerne have been grown commercially in Australia for more than 50 yr). Although these cultivars are higher yielding than Hunter River in the short term, especially during the mild Australian winters, they lack persistence and have not been planted on large areas. The cv. Paravivo, registered in 1971, was derived by intercrossing surviving plants from a 6-yr-old stand of cv. African. In subsequent testing (Barnard, 1972), cv. Paravivo was shown to be more persistent than cv. African, but it has never been widely sown.

Resistance to continuous grazing has been sought mainly by exploiting escape mechanisms associated with low spreading crowns and more or less procumbent stems. During the 1960s, breeding programs utilizing "creeping rooted" genotypes were conducted at Canberra (ACT), Deniliquin (NSW), and Brisbane (Q). Despite success in transferring the creeping-rooted character to a locally adapted genetic background (Daday, 1968; Bray, 1969), none of the bred lines was commercially successful. Persistence was variable (often inferior to cv. Hunter River, even under continuous grazing), and productivity was unacceptably low. The creeping - rooted character itself has a moderate to high heritability, and in spaced plants is positively correlated genetically with persistence, but varies in expression between environments (Heinrichs & Morley, 1962; Daday, 1968; Bray, 1969).

Wild, "spreading" ecotypes from Spain, with low, broad crowns and prostrate branches were found to be more persistent than erect cultivars when continuously grazed by sheep, under both Mediterranean and subtropical conditions (Leach, 1970; Leach at al., 1982). The wild Spanish lines were used extensively in the breeding of cv. Sheffield, registered in 1980, which shows superior persistence under grazing (Kaehne, 1978). Priority is now being given to upgrading the vigor and productivity of cv. Sheffield, and to increasing its resistance to pests and diseases (I.D. Kaehne, 1988, personal communication).

Most lucerne cultivars will not tolerate prolonged waterlogging. During 1961-1974, repeated cycles of selection for survival on heavy, waterlogged soils near Deniliquin, NSW resulted in the development of cv. Falkiner and other elite breeding material (Rogers, 1977; Rogers et al., 1978). The improved persistence of these lines was shown to be associated with resistance to root rot (*Phytophthora megasperma*).

Lucerne stands in the Australian subtropics persist for only a few years (Cameron, 1973). Poor persistence is due to a complex of factors, many of which are poorly understood, including diseases, waterlogging under high temperatures, poor cutting and grazing management, and invasion by summer-growing grasses (Leach & Clements, 1984). Selection for resistance to *P. megasperma* and *Colletotrichum trifolii* from 1973-1977 led to the development of elite lines which were much more persistent than their parents in Queensland (Bray & Irwin, 1978; Irwin et al, 1980).

Following these results in NSW and Queensland, disease resistance is now a standard selection criterion in Australian lucerne breeding programs.

The arrival in 1977 of the spotted alfalfa aphid *Therioaphis trifolii* f. *maculata* and the blue-green aphid *Acyrthosiphon kondoi* led to rapid devastation of Hunter River lucerne stands and to a new phase of lucerne improvement. An immediate reaction was the importation of aphid-resistant cultivars from the USA, but by 1980, six Australian-bred, aphid-resistant cultivars had been registered. The importance of aphid resistance in lucerne persistence in Australia is well-documented (Allen, these Proceedings).

The present emphasis in breeding programs is to produce cultivars with resistance to multiple diseases and pests. The first such cultivar, Trifecta, was registered in 1983, the second, Sequel, in 1985, and the third, Aurora, in 1986. Trifecta and Sequel are among the most persistent and highest-yielding cultivars developed so far for the Australian subtropics (Clements et al., 1984; Lowe et al., 1987), but the practical achievement has been to extend the haymaking lifetime of lucerne stands in Queensland by only 1 or 2 yr (Lowe et al., 1988). This will not change the trend in Australia for lucerne persistence to increase with latitude (Leach, 1978), but it does show that worthwhile gains are achievable.

This brief survey shows that Australian plant breeders have increased the persistence of lucerne in unfavorable situations (water-logging soils, subtropics, continuous grazing) and have maintained the persistence of lucerne in more favorable situations in the face of emerging threats (aphids and diseases). Much remains to be done. Resistance to the newly described pathogen Acrocalymma medicaginis (Alcorn & Irwin, 1987) and to pests such as redlegged earth mite (Halotydeus destructor) are likely to be key objectives (Kaehne et al., 1985).

Subterranean Clover

Francis and Gladstones (1983), Collins and Gladstones (1985), and Collins and Stern (1987) have recently reviewed Australian subterranean clover improvement programs. The first known artificial cross was made by Trumble in 1929, and subsequently genetic and breeding studies were conducted in South Australia and Victoria during the 1930s and 1940s, but no cultivars resulted.

Systematic plant improvement commenced in 1950 in Perth, WA and 1954 in Canberra, ACT. In retrospect, breeding objectives prior to 1963 are now seen to have been poorly conceived; indeed, the registration description of the first bred cultivar, Howard (first certified in 1964), states that it lacked persistence in some areas. No other cultivar resulted from this early work, but cv. Howard was a useful parent in subsequent breeding.

During the 1960s, the key improvement objective was to reduce the content of oestrogenic isoflavones, especially formononetin, which cause infertility in sheep. Three low-formononetin cultivars were released in 1967 and a fourth in 1972. Low formononetin content is now a standard selection criterion and a prerequisite of all new cultivars.

In 1967, the Perth program was reorganized, and has since evolved into the National Subterranean Clover Improvement Program. Breeding objectives reflect the complex array of environments and management systems in which subterranean clover is grown, but long-term persistence is regarded as the ultimate criterion of field success (Collins & Gladstones, 1985). There is a lack of emphasis on herbage yield and on animal production trials. Instead, there is a concentration on environment-dependent persistence characters such as flowering and maturity dates, high seed yield and hardseededness (especially for early and midseason cultivars), good burr burial, and resistance to diseases such as clover scorch (Kabatiella caulivora) and root rots (Phytophthora, Pythium and Rhizoctonia spp). Field testing extends over several states. Eleven cultivars have been registered since 1975, mostly flowing from the Perth-based program.

Subterranean clover cultivars differ in persistence, but because of the huge area and multitude of niches within the general region of adaptation, all cultivars are able to persist somewhere. For example, cv. Woogenellup has many weaknesses which reduce its persistence in some regions: rapid breakdown of hard seeds, non-burial of seeds, susceptibility to scorch and root diseases, more specific Rhizobium requirement. Yet Woogenellup is among the most popular cultivars in Australia, and on the north-western slopes of NSW is more persistent than most cultivars (Archer et al., 1987). In strongly Mediterranean climates in Western Australia, the following cultivars are reputed to be relatively persistent: Daliak, Dinninup, Seaton Park, Dalkeith and Junee (Rossiter, 1977; W.J. Collins, 1988, personal communication). Seaton Park and Daliak are also productive and persistent in southern NSW (Dear & Loveland, 1985; Dear et al., 1987), while Mt. Barker is still persistent in more humid areas of NSW. In higher rainfall areas of Victoria, the cultivars Larisa and Trikkala of subspecies yanninicum persist well (Reed et al., 1985), as do mid-season and late-flowering cultivars of ssp. subterraneum such as Mt. Barker and Karridale (Hotton & Curnow, 1987; Clark & Hirth, 1987).

Annual Medics

Several annual medics, especially M. polymorpha L. (burr medic) and M. minima (L.) Bart. (woolly burr medic), have become naturalized over vast areas of southern Australia, partly because of their spiny burrs which contaminate the wool of sheep. However, only since 1975 have three cultivars of burr medic been registered, selected primarily for having spineless burrs, while there is no commercial cultivar of M. minima. The commercialization of burr medic was long preceded by that of M. truncatula Gaertn., Fruct. & Semin. (barrel medic) and M. scutellata (L.) Miller (snail medic) during the 1930s. Barrel medic is now the most widely sown annual medic, and its utilization led to the development of ley farming systems in the drier pairs of the eastern Australian Wheat Belt.

The development of new and improved medics for Australia has been reviewed by Crawford (1983). Seven species, including spineless burr medic, are now cultivated deliberately, and there are 26 registered cultivars. The reason for the large number of species is that, although they are nearly all adapted to alkaline soils, there is a wide range in adaptation to soil texture and structure. Medicago rugosa Desr. (gama medic), for example, is adapted to heavy black soils and calcareous clay loams, while M. littoralis Rohde ex Loisel. (strand medic) and M. tornata (L.) Miller (disc medic) are adapted to light sandy soils. There is also much variation in the length of the growing season in southern Australia. Thus the development of new cultivars has emphasized adaptation to soils and climate. During the 1950s, emphasis was placed on developing an early flowering medic for light, calcareous, sandy soils; the resulting cultivar, Harbinger strand medic (released in 1959), is still widely planted. Another widely sown cultivar is Jemalong barrel medic, also developed during the 1940s and early 1950s, which is adapted to a wide range of soil types in eastern Australia but which has a restricted adaptation in Western Australia. Thus, the barrel medic cultivars Cyprus (released in 1959), Cyfield and Ghor were developed in Western Australia to extend the use of this very valuable species in that state, while cv. Borung, Akbar, and Ascot were developed in the eastern states to provide cultivars better suited to the Wimmera region (Victoria) and the alkaline soils of north-western NSW.

Recognition during the 1950s and 1960s of the value of gama medic on heavy clays led to the development of cv. Paragosa (1966).

Subsequently, a second cv. Paraponto (1978), with greater hardseededness was released. Similarly, three cultivars of disc medic were developed for alkaline sandy soils. The process of commercializing new species for particular niches is continuing. The most recent example is the release of *M. murex* Willd. cv. Zodiac, which is adapted to acid soils (Gillespie, 1987). Current extension of the burr medic cv. Serena onto acid soils has relied in part on the development of new, acid-tolerant strains of *Rhizobium meliloti* (Howieson & Ewing, 1986).

During the late 1970s and early 1980s, the advent of the spotted alfalfa aphid and the blue-green aphid placed the continued cultivation of annual medics at risk. Eight aphid-resistant cultivars of four species were registered between 1979-1988, but in many environments they are inferior agronomically to the cultivars they were intended to replace (A.W.H. Lake, 1988, personal communication). Multiple aphid resistance is rare among introduced lines of the commercially sown medics, so there is little opportunity to develop new cultivars directly from naturally occurring ecotypes. Instead, aphid resistance is being transferred by backcrossing to Harbinger strand medic and Jemalong, Cyprus and Borung barrel medic cultivars (Lake et al., 1987).

Crawford (1983) lists the main selection criteria for new medic cultivars as follows: seedling vigor, herbage yield during winter, time of flowering and seed maturation, seed production, acceptable pod spininess, appropriate hardseededness, and resistance to pests and diseases (especially aphids and the sitona weevil, *Sitona discoideus*). The emphasis on regeneration and persistence is striking, both in Crawford's review and in the registration statements of many individual cultivars. Crawford singles out one particular cultivar which ..... "only underwent a 6-yr selection and evaluation program", and notes that it was found subsequently to be non-persistent. It will be interesting to see whether the first wave of aphid-resistant cultivars released since 1977 with less time for field-testing prove to be as persistent as the cultivars of the 1950s and 1960s.

White Clover

In view of the economic importance of white clover in Australia, there has been a surprisingly small local effort to develop improved cultivars (Clements, 1987). Only three of the seven white clover cultivars registered in Australia were developed in this country. The first of these, cv. Irrigation, is an ecotype which originated in the irrigation districts of northern Victoria and was commercialized during the 1930s. The other two, cv. Haifa and Siral, are ecotypes from Israel and Algeria, respectively. Haifa is by far the most popular of these three cultivars. It was selected for persistence, herbage yield, heat tolerance and seed production. Its persistence relies heavily on regeneration from seed. In contrast, cv. Siral, which was also selected for persistence and herbage yield, persists characteristically by perennation of individual plants.

Future Australian white clover improvement programs are likely to concentrate on persistence and herbage yield (e.g., Curll, 1987). The importance of diseases and pests in reducing persistence and productivity in Australia needs to be determined.

Stylosanthes

Several recent reviews have described the history of development and improvement of Stylosanthes in Australia (e.g., Humphreys, 1967; Edye & Grof, 1983; Cameron et al., 1984). During the 1930s and 1940s, first S. humilis (Townsville stylo) and then S. guianensis var. guianensis (stylo) were commercialized through the pioneering efforts of a few scientists and farmers. The first systematic study of variation among naturalized populations of Townsville stylo was carried out by D.F. Cameron during the 1960s, leading to the registration of the three named cultivars of this species in 1968-1969.

Deliberate introduction of Stylosanthes to Australia commenced during the 1930s and was greatly accelerated from 1965 onwards. Extensive and widespread testing led to the registration of three key cultivars: Oxley fine-stem stylo (S. guianensis var. intermedia), in 1969; Verano Caribbean stylo (S. hamata), in 1975; and Seca shrubby stylo (S. scabra) in 1977. Each has limitations but they have proved to be strongly persistent in suitable environments in northern Australia. Fine-stem stylo has the narrowest adaptation, being suited to light-textured, free-draining soils in the subtropics. It is also low-yielding. All introductions of this botanical variety have similar limitations. Attempts to improve the performance of fine-stem stylo by hybridization with the more tropical S. guianensis var. guianensis commenced in 1972, but it has proved extraordinarily difficult to combine the persistence of fine-stem stylo with the desirable attributes of the tropical parents. Some early experimental pastures of Oxley fine-stem stylo have now persisted for more than 20 yr.

Shrubby stylo and Caribbean stylo are both more tropical in adaptation, but there are good prospects for the development of cultivars better suited to the subtropics. Indeed, the shrubby stylo cultivar Fitzroy, registered in 1980, showed great promise in the subtropics before anthracnose disease (Colletotrichum gloeosporioides) reduced its usefulness. Recent collections of S. hamata from elevated parts of Venezuela may be more cold-tolerant than cv. Verano (L.A. Edye, 1988 personal communication).

Anthracnose disease is the major threat to the continued use of Stylosanthes in tropical pastures. Originally recorded more than 50 yr ago in Brazil, it was first seen in Australia in 1973. By 1984, two separate types of anthracnose (A and B) had been identified, each consisting of several races, and Townsville stylo had been virtually eliminated from thousands of hectares in northern Australia. All registered cultivars of S. humilis and S. guianensis var. guianensis are susceptible under field conditions. Therefore, disease resistance is the prime objective in Australian plant improvement programs. Because the pathogen is so variable, a wide range of breeding technologies is being used, including strategies to reduce infection rate and to introduce resistance genes from alien species. A more detailed review of anthracnose disease and the development of resistant cultivars is provided by Irwin in these Proceedings.

Other Legumes

In addition to those in the main groups described above, 46

cultivars of other temperate/Mediterranean legumes and 28 cultivars of other tropical legumes have been registered. They represent a significant development effort which cannot be reviewed in detail here. Some broad generalizations can be made. First, 17 cultivars belong to species of Lupinus, Vicia, Vigna, and Lablab sown principally as fodder crops or grain legumes. Persistence for more than one growing season is not a major selection criterion for such species. Second, there is a large group of Trifolium cultivars (9 species, 24 cultivars) adapted mainly to specialized niches. An example is Caucasian clover (T. ambiguum M.B., 6 cultivars), adapted to alpine and sub-alpine conditions. Selection objectives for these Trifolium species have varied widely. Although restricted in distribution, the areas established in favored regions are noteworthy, e.g., 400 000 ha sown to strawberry clover (T. fragiferum L.) in South Australia (Donald, 1970). The remaining temperate legumes are from genera such as Ornithopus (serradella), Astragalus (milk vetch) and Onobrychis (sainfoin).

The tropical legume group includes a disproportionately large number of cultivars selected originally in subtropical or elevated tropical regions. Despite early indications of success, persistence in these regions has generally been disappointing, especially under heavy grazing, and efforts are continuing to find legumes adapted to the subhumid subtropics. Cassia rotundifolia Pers. cv. Wynn, registered in 1984, is in this category. It must be emphasized that persistence is not the sole criterion of success; Bargoo jointvetch (Aeschynomene falcata (Poir) DC.), registered in 1973, has persisted at almost every subtropical site at which it has been sown, for periods now exceeding 30 yr (Wilson et al., 1982), but difficulties of seed production have greatly limited its acceptance. There are many herbaceous legumes which might persist in the subtropics but which are unsuitable because they are toxic, unpalatable or spiny (Williams, 1983).

Three tropical legume genera, Macroptilium, Leucaena, and Centrosema, deserve special mention because attempts to develop cultivars from them have been determined and sustained. Siratro (M. atropurpureum (DC.) Urb.), released in 1960, has persisted under moderate grazing for up to 20 yr in the subtropics (Jones et al., 1983), and was sown on ca. 220 000 ha in Queensland by 1982 (Walker, 1983). However, it will not persist under close grazing and is susceptible to a rust, Uromyces appendiculatus. The area of persistent siratro pastures is now small. A second cultivar has not been released despite continuous breeding from 1956-1977 (Hutton & Beall, 1977). The selection objectives included herbage yield and acid soil tolerance, but did not include persistence, which was not then recognized as a major problem, or resistance to rust, which was not observed until 1978.

Leucaena leucocephala (Lam.) deWit, a small tree or shrub, was commercialized during the 1960s. Despite its strong persistence, it was not well accepted because it is difficult to establish and contains the toxic amino acid mimosine. Breeding commenced in 1956 and has continued almost uninterrupted. A high-yielding cultivar (cv. Cunningham) was registered in 1976, but breedng for reduced mimosine content was unsuccessful (Bray, 1985). During the 1980s, rumen bacteria able to degrade the breakdown products of mimosine were introduced to Australia and isolated (Jones, 1985), so that mimosine derivatives are no longer a problem for ruminants.

However, the leucaena psyllid, *Heteropsylla cubana*, now threatens the persistence and future use of this legume, and breeding for psyllid resistance may be necessary.

Centrosema pubescens Benth. (centro) was one of the first tropical legumes sown in Australian pastures, and a stable area of ca. 20 000 ha exists on the humid tropical coast of north Queensland. Plant improvement programs with *Centrosema* were conducted from 1966-1983. In 1971, the cv. Belalto of *C. schiedeanum* (Schlecht.) Williams & Clements was registered (Grof & Harding, 1970). It is known to be more persistent than centro under close grazing, but has not been successful commercially because no seed industry has been developed. Another species, the annual *C. pascuorum* Mart. ex Benth., has been developed for pastures in the Top End (i.e., the northern third) of the Northern Territory. The cv. Cavalcade, registered in 1984, was bred for increased herbage and seed yield (Clements et al., 1986). Its persistence so far has been good. A later-flowering ecotype, cv. Bundey, was selected for its superior growth on seasonally flooded tropical lowlands (McCosker, 1987). Efforts to breed a cultivar of *C. virginianum* (L.) Benth. for the subhumid subtropics have been unsuccessful, despite substantial improvements in persistence and herbage yield (Clements & Thomson, 1983).

LESSONS LEARNED

Phases in the development of pasture legumes in Australia exist, but they overlap and their margins are blurred. A 'crude matching of plants to places' (Donald, 1970) has led to the systematic collection and evaluation of ecotypes and to the development of cultivars for particular regions or purposes. The recognition of 'second generation' limitations and of opportunities for novel genetic improvement of the main legumes within their present areas of adaptation has led to a third phase of greater scientific sophistication, but there is a continuing effort to find 'new' legumes and to extend the use of the main legumes beyond their present limits.

A common thread in the history of improvement of pasture legumes in Australia has been premature plant breeding. It has been premature on two counts. First, almost invariably, breeding was initiated before representative genetic resources had been assembled or evaluated, so that later introductions frequently performed better than bred lines. Second, breeding objectives were poorly defined and in retrospect are seen to have been poorly conceived. This is particularly true of plant persistence, where early ideotypes were often woefully inadequate. Thus, the earliest breeding programs usually either failed to produce cultivars or produced cultivars which were not accepted commercially.

It can be remarkably difficult to decide at what stage breeding should be initiated (Cameron, 1983). It is almost as if there is a 'learning phase' which has to be experienced before progress is possible. If so, it may reflect in part the need for a good deal of background genetic information which is usually generated by plant breeding and is rarely available beforehand. It may also reflect the need for breeders to educate themselves and to stimulate research by others. For example, although the early years of subterranean clover breeding were disappointing in terms of released cultivars, during this time breeders and agronomists developed a conceptual and descriptive framework and an information base of lasting value. Significant advances in knowledge were reported, notably on the genetic control of flowering and hardseededness, the inheritance of oestrogenic isoflavones, the sub-specific taxonomy of T. subterraneum and its implications for cross compatibility, and the ecology of subterranean clover-based pastures. Major reviews by Morley (1961), Donald (1963), and Rossiter (1966) were published. The work led to a better definition of breeding objectives and strategies, and eventually to the release of successful bred cultivars.

Cameron (1983) has argued that in developing highly persistent cultivars, an empirical approach has been most successful because the plant attributes underlying persistence are so varied and complex. Jones and Clements (1981) in discussing ecophysiology and adaptation of pasture plants came to a similar conclusion. The Australian experience generally bears out this view. Many authors have attempted to describe the key elements of ecological success or failure, but their descriptions tend to be assertive rather than experimentally derived. In practice, the factors contributing to success are species-, site- and management-dependent. For example, seed yield (itself very much a derived character), burr burial and hardseededness are prime selection criteria in subterranean clover improvement programs because of their likely contribution to persistence (Francis & Gladstones, 1983; Collins & Gladstones, 1985; Collins & Stern, 1987). However, a cursory survey of the relevant literature (12 references) reveals that the importance of these characters varies tremendously with climate, land use (grazing vs. ley farming), grazing pressure and probably soil type and companion grass species. This is well recognized by the breeders, who have developed a range of selection objectives to suit particular environments (Collins & Gladstones, 1985). The successful cultivars themselves vary considerably with respect to these characters. Despite these difficulties, modelling approaches such as that used by Rossiter et al. (1985) will be increasingly useful aids to plant improvement.

One factor that is clearly implicated in the recent failure of previously persistent cultivars is the advent of 'new' diseases and pests. It is in the development of pest- and disease-resistant cultivars that plant breeders (as distinct from plant developers) have made the greatest contribution to improving persistence. It is also in this area that the new 'genetic engineering' technologies are most likely to contribute to persistence, although they have not yet done so in Australia. These technologies are already being used to improve simply inherited traits unrelated to persistence, and their application to disease resistance breeding in Stylosanthes is the subject of intensive research (J.M. Manners, 1988, personal communication).

In conclusion, it should be emphasized again that persistence is not the sole criterion of merit. Legumes are grown in pastures to fix N and provide high-quality feed for animals. In order to do this well, they must produce large quantities of herbage. Farmers and graziers will grow high-quality fodder crops and legumes such as lucerne even though they do not persist; they will not sow persistent, low-yielding legumes.

ACKNOWLEDGEMENT

I am grateful to the following colleagues from southern states for their helpful comments: W.J. Collins. I.D. Kaehne, A.W.H. Lake, J.W. Read, and K.F.M. Reed.

REFERENCES

Alcorn, J.L., and J.A.G. Irwin. 1987. Acrocalymma medicaginis gen. et sp. nov. causing root and crown rot of Medicago sativa in Australia. Trans. Br. Mycol. Soc. 88: 163-167.

Archer, K.A., G.M. Lodge, R.S. Wetherall, and D.B. Waterhouse. 1987. Legume evaluation and selection in northern New South Wales. p. 283-286. In J.L. Wheeler et al. (ed.) Temperate pastures : their production, use and management. Australian Wool Corp./CSIRO, Melbourne.

Barnard, C. 1972. Register of Australian herbage plant cultivars. CSIRO, Melbourne.

Bray, R.A. 1969. Variation in and correlations between yield and creeping-rootedness in lucerne. Aust. J. Agric. Res. 20:47-55.

Bray, R.A. 1985. Breeding leucaena. p. 317-322. In R.F. Barnes et al. (ed.) Forage legumes for energy-efficient animal production. USDA-ARS, CSIRO & NZDSIR, Washington, DC.

Bray, R.A., and J.A.G. Irwin. 1978. Selection for resistance to Phytophthora megasperma var. sojae in Hunter River lucerne. Aust. J. Exp. Agric. Anim. Husb. 18: 708-713.

Cameron, D.F. 1983. To breed or not to breed. p. 237-250. In J.G. McIvor and R.A. Bray (ed.) Genetic resources of forage plants. CSIRO, Melbourne.

Cameron, D.F., E.M. Hutton, J.W. Miles, and J.B. Brolmann. 1984. p. 589-606. In H.M. Stace and L.A. Edye (ed.) The biology and agronomy of Stylosanthes. Academic Press, Sydney.

Cameron, D.G. 1973. Lucerne (Medicago sativa) as a pasture legume in the Queensland subtropics. J. Aust. Inst. Agric. Sci. 39: 98-108.

Clark, S.G., and J.R. Hirth. 1987. Growth and persistence of Mediterranean genotypes of midseason-late maturing subterranean clover (Trifolium subterraneum) in Victoria. Aust. J. Exp. Agric. 27: 551-557.

Clements, R.J. 1987. An Australian white clover breeding program - justification, objectives, timing and resources needed. p. 5.1-5.5. In M.L. Curll (ed.) National white clover improvement. Australian Wool Corp./NSW Department of Agriculture, Glen Innes, Australia.

Clements, R.J., and C.J. Thomson. 1983. Breeding Centrosema virginianum in subtropical Queensland. CSIRO Australia, Div. Tropical Crops & Pastures, Tech. paper no. 26.

Clements, R.J., J.W. Turner, J.A.G. Irwin, P.W. Langdon, and R.A. Bray. 1984. Breeding disease resistant, aphid resistant lucerne for subtropical Queensland. Aust. J. Exp. Agric. Anim. Husb. 24: 178-188.

Clements, R.J., W.H. Winter, and C.J. Thomson. 1986. Breeding Centrosema pascuorum for northern Australia. Trop. Grassl. 20: 59-65.

Collins, W.J., and J.S. Gladstones. 1985. Breeding to improve subterranean clover in Australia. p. 308-315. In R.F Barnes et al. (ed.) Forage legumes for energy-efficient animal production. USDA-ARS, CSIRO & NZDSIR, Washington, D.C.

Collins, W.J., and W.R. Stern. 1987. The national subterranean clover improvement program - progress and directions. p. 276-278. In J.L. Wheeler et al. (ed.) Temperate pastures : their production, use and management. Australian Wool Corp./CSIRO, Melbourne.

Crawford, E.J. 1983. Selecting cultivars from naturally occurring genotypes : evaluating annual Medicago species. p. 203-215. In J.G. McIvor and R.A. Bray (ed.) Genetic resources of forage plants. CSIRO, Melbourne.

Curll, M.L. 1987. A proposal for describing and testing material. p. 4.1-4.5. In M.L. Curll (ed.) National white clover improvement. Australian Wool Corp./NSW Department of Agriculture, Glen Innes, Australia.

Daday, H. 1968. Heritability and genotypic and environmental correlations of creeping root and persistency in Medicago sativa L. Aust. J. Agric. Res. 19:27-34.

Dear, B.S., P.D. Cregan, and Z. Hochman. 1987. Factors restricting the growth of subterranean clover in New South Wales and their implications for further research. p. 55-57. In J.L. Wheeler et al. (ed.) Temperate pastures : their production, use and management. Australian Wool Corp./CSIRO, Melbourne.

Dear, B.S., and B. Loveland. 1985. A survey of seed reserves of subterranean clover pastures on the southern tablelands of New South Wales. p. 214. In J.J. Yates (ed.) Proc. 3rd Aust. Agron. Conf., Hobart. Australian Soc. of Agron., Parkville, Vic.

Donald, C.M. 1963. Competition among crop and pasture plants. Adv. Agron. 15: 1-118.

Donald, C.M. 1970. Temperate pasture species. p. 303-320. In R.M. Moore (ed.) Australian grasslands. Australian National University Press, Canberra.

Edye, L.A., and B. Grof. 1983. Selecting cultivars from naturally occurring genotypes : evaluating Stylosanthes species. p. 217-232. In J.G. McIvor and R.A. Bray (ed.) Genetic resources of forage plants. CSIRO, Melbourne.

Francis, C.M., and J.S. Gladstones. 1983. Exploitation of the genetic resource through breeding : Trifolium subterraneum. p. 251-260. In J.G. McIvor and R.A. Bray (ed.) Genetic resources of forage plants. CSIRO, Melbourne.

Gillespie, D.J. 1987. Murex, a new medic for acid soils. p. 172-174. In J.L. Wheeler et al. (ed.) Temperate pastures : their production, use and management. Australian Wool Corp./CSIRO, Melbourne.

Grof, B., and W.A.T. Harding. 1970. Yield attributes of some species and ecotypes of Centrosema in north Queensland. Queensl. J. Agric. Anim. Sci. 27: 237-243.

Heinrichs, D.H., and F.H.W. Morley. 1962. Quantitative inheritance of creeping-root in alfalfa. Can. J. Genet. Cytol. 4: 79-89.

Hotton, G.B., and B.C. Curnow. 1987. Persistence of Mt. Barker subterranean clover. p. 164. In T.G. Reeves (ed.) Proc. 4th Aust. Agron. Conf., Melbourne. Australian Soc. of Agron., Parkville, Vic.

Howieson, J.G., and M.A. Ewing. 1986. Acid tolerance in the Rhizobium meliloti - Medicago symbiosis. Aust. J. Agric. Res. 37: 55-64.

Humphreys, L.R. 1967. Townsville lucerne : history and prospect. J. Aust. Inst. Agric. Sci. 33: 3-13.

Hutton, E.M., and L.B. Beall. 1977. Breeding of Macroptilium atropurpureum. Trop. Grassl. 11: 15-31.

Irwin, J.A.G., D.L. Lloyd, R.A. Bray, and P.W. Langdon. 1980. Selection for resistance to Colletotrichum trifolii in the lucerne cultivars Hunter River and Siro Peruvian. Aust. J. Exp. Agric. Anim. Husb. 20: 447-451.

Jones, R.J. 1985. Leucaena toxicity and the ruminal degradation of mimosine. p. 111-119. In A.A. Seawright et al. (ed.) Plant toxicology. Queensland Poisonous Plants Committee/Queensland Department of Primary Industries, Brisbane.

Jones, R.M., and R.J. Clements. 1981. Ecophysiology and adaptation of pasture plants. p. 232-249. In D.E. Byth and V.E. Mungomery (ed.) Interpretation of plant response and adaptation to agricultural environments. Queensland Branch, Australian Institute of Agricultural Science, Brisbane.

Jones, R.M., J.C. Tothill, and L.'t Mannetje. 1983. CSIRO agronomic research on Siratro-based pastures in south-east Queensland. p. 45-57. In R.F. Brown (ed.) Siratro in south-east Queensland. Queensland Department of Primary Industries, Brisbane.

Kaehne, I.D. 1978. The performance under intensive continous grazing of second generation bulk populations derived from crosses between wild and exotic alfalfas and cultivated non-hardy varieties. Rep. 26th Alfalfa Improvement Conf., S. Dakota State Univ., p. 47-48.

Kaehne, I.D., G.C. Auricht, A.W.H. Lake, and E.T. Meyer. 1985. A progress report on lucerne breeding in South Australia. p. 207. In J.J. Yates (ed.) Proc. 3rd Aust. Agron. Conf., Hobart. Australian Soc. of Agron., Parkville, Vic.

Lake, A.W.H., K.E. Sfreddo, and B.G. Baron. 1987. The breeding and selection of an aphid-resistant Harbinger-type strand medic by backcrossing. p. 345. In J.J. Yates (ed.) Proc. 4th Aust. Agron. Conf., Melbourne. Australian Soc. of Agron., Parkville, Vic.

Leach, G.J. 1970. An evaluation of lucerne lines at the Waite Agricultural Research Institute, South Australia. Aust. J. Exp. Agric. Anim. Husb. 10: 53-61.

Leach, G.J. 1978. The ecology of lucerne pastures. p. 290-308. In J.R. Wilson (ed.) Plant relations in pastures. CSIRO, Melbourne.

Leach, G.J., and R.J. Clements. 1984. Ecology and grazing management of alfalfa pastures in the subtropics. Adv. Agron. 37: 127-154.

Leach, G.J., D. Gramshaw, and F.H. Kleinschmidt. 1982. The survival of erect and spreading lucerne under grazing at Lawes and Biloela, southern Queensland. Trop. Grassl. 16: 206-213.

Lowe, K.F., D. Gramshaw, T.M. Bowdler, R.L. Clem, and B.G. Collyer. 1987. Yield, persistence and field disease assessment of lucerne cultivars and lines under irrigation in the Queensland subtropics. Trop. Grassl. 21: 168-181.

Lowe, K.F., B.L. Bartholomew, and T.M. Bowdler. 1988. Hay production of lucerne cultivars in the Lockyer Valley, south-east Queensland. Trop. Grassl. 22: (in press).

Mackay, J.H.E. 1982. Register of Australian herbage plant cultivars. Supplement to the 1972 edition. CSIRO, Melbourne.

McCosker, T.H. 1987. Agronomic and grazing evaluation of 3 lines of Centrosema pascuorum under seasonally flooded conditions in the

Northern Territory. Trop. Grassl. 21: 81-91.
Morley, F.H.W. 1961. Subterranean clover. Adv. Agron. 13: 58-123.
O'Brien, A.D. 1970. White clover (Trifolium repens L.) in a subtropical environment on the east coast of Australia. p. 165-168. In M.J.T. Norman (ed.) Proc. XI Int. Grassl. Congr., Surfers Paradise. University of Queensland Press, Brisbane.
Reed, K.F.M., P.M. Schroder, J.W. Eales, R.M. McDonald, and J.F. Chin. 1985. Comparative productivity of Trifolium subterraneum and T. yanninicum in south-western Victoria. Aust. J. Exp. Agric. 25: 351-361.
Rogers, V.E. 1977. Breeding for field resistance to root rot of lucerne. p. 14(a)1-14(a)4. In R.W. Downes (ed.) Proc. 3rd Congr. Soc. Adv. Breeding Res. in Asia & Oceania (SABRAO). SABRAO, Canberra.
Rogers, V.E., J.A.G. Irwin, and G. Stovold. 1978. The development oflucerne with resistance to root rot in poorly aerated soils. Aust. J. Exp. Agric. Anim. Husb. 18: 434-441.
Rossiter, R.C. 1966. Ecology of the Mediterranean annual-type pasture. Adv. Agron. 18: 1-56.
Rossiter, R.C. 1977. What determines the success of subterranean clover strains in south-western Australia? Proc. Ecol. Soc. Aust. 10: 76-88.
Rossiter, R.C., R.A. Maller, and A.G. Pakes. 1985. A model of changes in the composition of binary mixtures of subterranean clover strains. Aust. J. Agric. Res. 36: 119-143.
Walker, B. 1983. Introduction. p. 1-3. In R.F. Brown (ed.) Siratro in south-east Queensland. Queensland Department of Primary Industries, Brisbane.
Williams, R.J. 1983. Tropical legumes. p. 17-31. In J.G. McIvor and R.A. Bray (ed.) Genetic resources of forage plants. CSIRO, Melbourne.
Williams, R.J., and R.J. Clements. 1986. The future role of plant introduction in the development of tropical pastures in Australia. p. 20-28. In G.J. Murtagh and R.M. Jones (ed.) Proc. 3rd Aust. Conf. Tropical Pastures, Rockhampton. Tropical Grassl. Soc. of Australia, Brisbane.
Wilson, G.P.M., R.M. Jones, and B.G. Cook. 1982. Persistence of jointvetch (Aeschynomene falcata) in experimental sowings in the Australian subtropics. Trop. Grassl. 16: 155-156.

DISCUSSION

Forde: What is the morphological basis of the grazing tolerance of the new lucerne cv. Sheffield?

Clements: It has a broad, protected crown, but is not creeping-rooted.

Reed: It was based on collections from Spain and Afghanistan which had undergone grazing for a long period.

Sheaffer: What is the dormancy reaction of Sheffield?

Clements: It is not winter-dormant, and is not derived from M. falcata germplasm. Its winter activity is similar to that of cv. Hunter River, which in Australia is classed as semi-winter-active.

Hoveland: Has Sheffield been well accepted by farmers?

Clements: The reception has been mixed. There is a good supply of commercial seed, but Sheffield is not one of the main cultivars being planted at present. The main obstacle to acceptance seems to be the rather low yield. Ian Kaehne in South Australia is working to upgrade Sheffield.

Reed: I endorse Bob's comments on farmer acceptance.

Buxton: How long would you like to have a grazing alfalfa plant stand?

Clements: It varies upon the needs of the farmer, from 2 to 3 yr in ley pastures grown in rotation with crops to many years for permanent pastures. As a generalization, we'd like a stand to be capable of lasting 5 yr or more under near-continuous grazing.

(Unknown): In your written paper you referred to premature plant breeding in Australia. Would you expand on this point, please?

Clements: It is not just an Australian phenomenon, but we certainly have a long history of premature plant breeding. In the written paper I used subterranean clover and lucerne as examples, but I could just as easily have used the annual medics or most of the breeding programs with tropical legumes. The two main difficulties have been inadequate breeding objectives and/or inadequate genetic resources at the time that breeding commenced. The result has been either a failure to produce a variety at all, or the production of a variety which has not been successful in commerce. There are a few exceptions, and with time breeding has become much more successful, especially with lucerne and subterranean clover. In my opinion, pasture plant breeders are limited not by a lack of smart technology, but mainly by a lack of well-defined breeding objectives. This is particularly true for legume persistence.

Allen: You mentioned that you had some concerns with the objective of multi-disease and multi-pest resistance in the lucerne breeding projects. What are those concerns?

Clements: My concern is not with multiple pest and disease resistance as an objective, but with the preoccupation with that as a sole objective. For example, in Australia a great deal of lucerne is grazed, and there is a need to build on the success achieved with the cv. Sheffield. We need to develop a range of grazing tolerant lucernes that are more productive. We also need lucernes that are more tolerant of soils high in Al. Some people believe we should improve haymaking characteristics.

I am concerned that no commercial company would be prepared to find the resources for the "long haul" research needed to obtain the grazing tolerant cultivars I mentioned. The trend towards short-term projects and commercialization of public research will make it harder for public breeders to do the job.

Hochman: I am concerned that the new aphid-tolerant medics are considered not to be as well adapted as cv. Jemalong. This observation is based on only a few years' data. It is important that an analysis of those seasons be made and a longer term climatic analysis be undertaken before breeding aphid tolerance into Jemalong, which was widely but not particularly well adapted before the arrival of aphids.

Clements: My information came from Andrew Lake, the breeder of annual medics, based in South Australia.

Kretschmer: In your verbal presentation, you failed to mention the successful tropical legumes such as centro, calopo, buffalo (Alysicarpus), and lotononis. Why not spend more effort on calopo and Alysicarpus introductions instead of so much effort on stylos?

Clements: I have made some brief comments on centro in the written paper, but the others are not treated in detail. In tropical Australia, Stylosanthes has been found to be very well adapted in the extensive semi-arid parts, and is receiving most attention. However, in the subtropics we have no proven widely adapted legumes at present, and an extensive range of legumes is being evaluated. The test locations extend into the tropics, and the range of plants includes representatives of most of the genera you mention.

R. Smith: How do you see the commercial interests entering into the cultivar development scene in Australia?

Clements: I believe there will be increased commercial emphasis if the potential exists for economic gain. Direction of public resources will depend upon policy. Plant Variety Rights have only just been introduced in Australia, and we are presently in a phase of negotiation with commercial companies, looking at opportunities for collaboration.

BREEDING FOR LEGUME PERSISTENCE IN NEW ZEALAND

J.R. Caradus and W.M. Williams

SUMMARY

Persistence has a high priority in legume breeding programmes in New Zealand. Most work in this area has been done with white clover (*Trifolium repens* L.) though other species receiving consideration include subterranean clover (*T. subterraneum* L.), *Lotus* species, alfalfa (*Medicago sativa* L.) and red clover (*T. pratense* L.). Four studies are reported examining characters associated with persistence. The most persistent white clover material in New Zealand tends to be of New Zealand origin, have a moderate to high cyanogenesis level and high stolon densities.

An outline is given of breeding strategies for improving persistence in dryland environments, with interspecific competition, in high country environments, under nutrient deficiencies and toxicities, and in nematode infested soil.

INTRODUCTION

Persistence is the bottom line of adaptation and production and as such has a high priority in breeding programmes. Legume persistence can be adversely affected by grazing and management effects, diseases and pests, competition from other herbage species, deficiencies and toxicities of nutrients and climatic factors.

The present paper aims to outline methods being used to breed for improved persistence in a number of legume species. Plant characters that are related to persistence will be identified and the cost of persistence under grazing and in difficult environments will be discussed. The main species of interest are white clover, red clover, *Lotus* species, alfalfa and subterranean clover. However, white clover is by far the most important legume species in New Zealand and most work has been done with this species.

SCREENING FOR PERSISTENCE

Many factors are associated with persistence and, because of this, the problem is genetically complex. Plant characters important for persistence in one environment, e.g., drought, may not be involved in another environment, e.g., pest attack. Because persistence is so basic to adaptation and productivity, it is important for the plant breeder to be able to prioritize the genetic needs and to select according to these priorities. The most efficient way of integrating the chief priorities is to select in the target environment, although special circumstances may justify selection under artificial conditions, e.g., artificial disease epidemics, frost chambers, etc.

SPECIES COMPARISONS

It is important to distinguish the morphological requirements for persistence of different legume species. Plants of several major legumes such as red clover, alfalfa, and birdsfoot trefoil (*Lotus corniculatus* L.), depend for their survival on the persistence of the plant crown and its supporting tap-root. Stoloniferous and rhizomatous species, e.g., white clover, depend on tap-root and crown survival only in the seedling stage. Because most organs last little more than 1 yr, these species later depend for survival on efficient replacement of roots and shoots. The annual species, and to some extent the perennials, also depend on seedling re-establishment from in situ seed dissemination. In this paper we will deal with vegetative persistence only.

The major pasture legumes differ in their adaptations and this, to some extent, is related to morphology. For example, the tap-rooted species tend to be better adapted to drought than the stoloniferous and rhizomatous species.

SELECTION CRITERIA FOR PERSISTENCE

There are inherent difficulties in the assessment of persistence, the greatest problem being the length of time required to determine it. Identification of specific plant characters that might ensure increased persistence would be advantageous.

Plant characters usually associated with persistence in white clover will clearly vary with the nature of the environment. However, for several stress environments, including low soil fertility, winter cold and heavy grazing pressure a set of common characters is apparent - namely small leaf size, and dense spreading habit. For summer drought, the ability to produce thickened roots may be more important. Against this background a number of recent studies of persistence in white clover are evaluated below. Scrutiny of these enables evaluation of the models and modification if necessary.

Because of the creeping nature of white clover in swards, dead plants are overgrown by adjacent survivors. In spaced plant nurseries, however, this is not permitted and so white clover spaced plants provide useful information on individual plant persistence that is not easily obtainable from swards. The recent spaced plant studies are considered below in relation to plant characters and plant breeding models.

Study 1

An evaluation of 254 lines from 32 countries, including unselected introductions through to finished cultivar material was carried out by Williams and Cooper (1980). Measurements were made of leaf size and density after the 1st yr and growth and number of plants surviving after 18 months. An attack of stem nematode (*Ditylenchus dipsaci* Kuhn.) occurred early in the 2nd yr and shortened the duration of the trial.

Survival data were analyzed by leaf size class (1=very small to

6=large), density class (1=very open to 6=very dense) and breeding category (either unselected, some selection or cultivar). Persistence increased with leaf size (Table 1a) and was highest for moderately dense lines (Table 1b). Unselected material had inferior persistence to selected material, whether or not selections had culminated in a finished cultivar (Table 1c).

Table 1. The effect of (a) leaf size class, (b) density class and (c) breeding category on yield and persistence of 251 white clover lines grown as spaced plants at Palmerston North for 18 months.

(a)	Leaf size	No. of lines	Density score (1-6)	Final growth score (0-5)	Persistence (%)
1	Very small	6	3.7	0.9	56
2	Small	31	3.6	0.9	57
3	Medium-small	61	3.3	1.1	68
4	Medium	98	2.8	1.2	70
5	Medium-large	31	2.5	1.7	84
6	Large	12	2.3	2.2	90
	P		***	***	***
	LSD 0.05		0.8	0.3	13

(b)	Density class	No. of lines	Final growth score (0-5)	Persistence (5)
1	Very open	11	1.0	53
2	Open	72	1.2	65
3	Moderately open	65	1.2	71
4	Moderately dense	68	1.3	75
5	Dense	23	1.2	78
6	Very dense	3	1.1	68
	P		n.s.	***
	LSD 0.05		-	20

(c) Breeding category	No. of lines	Leaf size score(1-6)	Density score(1-6)	Final growth score	Persistence (%)
Unselected	165	3.2	3.0	1.1	66
Some Selection	41	4.3	2.7	1.5	77
Cultivar	45	3.9	3.1	1.4	76
P		***	n.s.	**	***
LSD 0.05		0.3	-	0.1	6

The superior persistence of large leaved moderately dense plants in this study clearly is inconsistent with the usual model for stressed environments. However, results are readily interpreted in relation to the relative resistance of the plant materials to stem nematode. Resistance to this pest is present mainly in large leaved ladino material and is simply inherited (Williams, 1972). Further, lines selected in the local environment would have already undergone selection for resistance and would thus be showing superior persistence to unselected introductions. This study provides an example of an environment in which pest resistance over-rides other considerations in white clover persistence. In this case grazing pressure was lax, soil fertility high and soil moisture adequate, and so larger leaved materials were able to express their potential.

Study 2

This study was an evaluation of 109 cultivars and 16 elite breeding lines from 24 countries at Palmerston North, in which measurements were made of leaf size, height, cyanogenesis (%) in the first year, and growth and persistence after 2.75 yr.

In this study, persistence in a spaced plant nursery has shown only a weak relationship with small leaf size and a tendency toward cyanogenesis among the persistent plants (Table 2). Agronomically successful white clover cultivars in New Zealand have all been found to be moderate to high in cyanogenesis but no clear physiological basis for this has been established. The weak relationship with leaf size probably reflects the relatively benign environment in relation to soil fertility and moisture and grazing pressure.

The insertion of clover plants into a grass sward is expected to give a more realistic indication of agronomic performance than spaced plants. The following two trials compared white clover lines in small plots (50 x 50 cm or 50 x 75 cm) in a grass sward under rotational grazing with 9 or 10 grazings a year.

Study 3

An evaluation was made at Palmerston North of 109 cultivars and 16 breeding lines of white clover from 24 countries in small plots

Table 2. Correlation between leaf size, height, cyanogenesis (%), growth and persistence after 2.75 yr of 125 white clover cultivars grown as spaced plants at Palmerston North.

	Leaf size	Height	Cyanogenesis	Growth score
Height	0.92***	-		
Cyanogenesis	0.05	0.03	-	
Growth score	-0.26**	-0.21*	0.39**	-
Persistence	-0.19*	-0.12	0.42**	0.69***

in a grass sward (Caradus et al., 1988). Measurements were made of stolon and flower head numbers prior to each grazing, and clover and grass dry matter yields in late spring and autumn for 2 yr. In the first year measurements were also made of leaf size, clover canopy height and cyanogenesis. Yearly means were calculated for stolon number, flower number and proportion of clover in the sward.

As is common in such studies, tall, large leaved materials grew well in the first year but by Year 2 had begun to give way to the smaller leaved, more densely branching types (Table 3). Again the trend toward superior persistence of cyanogenic lines was apparent and, in this case there was also a significant but very weak tendency for profuse flowering in the first year to be detrimental to second year performance.

Cultivars with the highest stolon numbers in Year 2 were generally very small leaved and prostrate with a low harvest index and therefore contributed little to total harvested dry matter. These cultivars included Barbian, Kent Wild White, N.Z. hill country selection (Isolation V), selection from Southland (N.Z.) sheep farms and Whatawhata early flowering (a selection from dry hill country, N.Z.) (Fig. 1).

Table 3. Correlations of leaf size, cyanogenesis %, clover canopy height measure in the first year, with mean stolon number, mean flower number and mean proportion of clover in sward in Years 1 and 2, of 125 white clover cultivars grown in small plots in a grazed grass sward.

	Leaf size	Cyano-genesis	Height	Stolon no. Yr.1	Stolon no. Yr.2	Flower no. Yr.1	Flower no. Yr.2	Propn clover Yr.1
Cyanogenesis	0.17	-						
Height	0.84***	0.21*	-					
Stolon no. Yr.1	-0.66***	0.04	-0.64***	-				
Stolon no. Yr.2	-0.46***	0.31**	-0.47***	0.77***	-			
Flower no. Yr.1	-0.16	-0.20*	-0.18*	0.16	-0.01	-		
Flower no. Yr.2	-0.15	0.15	-0.14	0.23**	0.36**	0.51***	-	
Prop^n clover Yr.1	0.57***	0.31**	0.55***	-0.36**	-0.08	-0.19	0.08	-
Prop^n clover Yr.2	0.27**	0.58***	0.28**	-0.13	0.23**	-0.30**	0.21*	0.56***

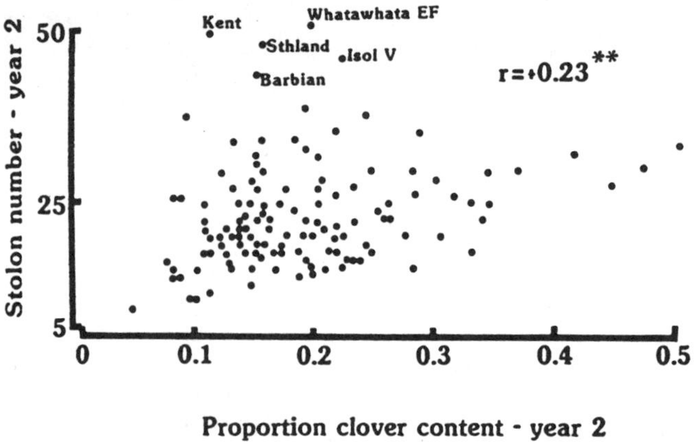

Fig. 1. Correlation between mean stolon number (per 190 cm^2) and mean proportion of clover in sward in Year 2, for 125 white clover cultivars grown in a grazed grass sward (study 3).

The success of white clover in pastoral agricultural systems is undoubtedly due to its stoloniferous nature and its ability to form roots at nodes. Maintenance of a high stolon population in the sward is obviously an important requirement for long term persistence. Stolon number is often strongly correlated with plant type, such that the apparently more productive large leaved, upright cultivars have fewer stolon tips than the apparently less productive small leaved, prostrate cultivars. Figure 2 indicates the emphasis

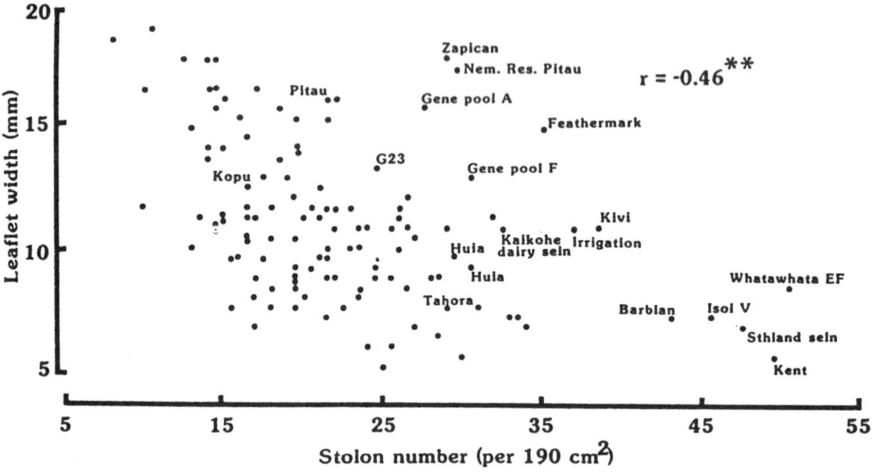

Fig. 2. Correlation of leaflet size (mm) in Year 1 with mean stolon number (per 190 cm^2) in the second year of study 3, comparing 125 white clover cultivars grown in a grazed grass sward. With the exception of Kent (U.K.), Irrigation (Australia), Barbian (Netherlands), and Zapican (Uruguay) all named points represent cultivars or breeding lines of New Zealand origin.

which has been placed on selection in New Zealand for high stolon densities associated in many cases with larger than average leaves. Along with Irrigation from Australia, most of the New Zealand lines clearly fall outside the general relationship between leaf size and stolon density.

Study 4

An evaluation was made of 27 lines which had performed better than Huia and/or Pitau during previous New Zealand trials. Cultivars were grown in small plots in a grass sward and measurements made of number of stolon tips, flower number, and yield of grass and clover before each grazing (nine grazings per year). Leaf size was calculated by dividing leaf weight by leaf number at each harvest. Stolon and flowering material harvested were weighed separately and will not be considered here. Changes in correlations between stolon number, leaf number, leaf size and proportion of clover in sward (as an indicator of persistence) with time give some indication of the importance of these plant characters for long-term persistence in this type of experiment (Table 4). With time the correlation between stolon numbers and proportion of clover becomes more positive (suggesting that the most persistent lines are those with high stolon numbers); the correlation between leaf size and proportion of clover becomes more negative (suggesting that the most persistent lines are those with smaller leaves); the correlation between leaf number and proportion of clover remains highly positive; the correlation between leaf number and stolon number is consistently positive; and the correlation between leaf size and stolon number is consistently negative.

Table 4. The effect of time on correlation coefficients between stolon number, leaf number, leaf size (weight), and proportion of clover in sward for 27 lines of white clover grown in small plots in a grass sward and rotationally grazed by sheep.

Grazing	Harvest	Date	Stolon no./ Propn clover	Leaf wt./ Propn clover	Leaf no./ Propn clover	Leaf no./ Stolon no.	Leaf wt./ Stolon no.
5	1	Feb. 1987	-0.27	n.m.[1]	n.m.	n.m.	n.m.
6	2	March 1987	-0.21	0.61	0.61	0.06	-0.63
7	3	May 1987	+0.10	0.56	0.51	0.72	-0.50
8	4	July 1987	-0.22	0.53	0.61	0.25	-0.72
9	5	Aug. 1987	-0.19	0.30	0.70	0.16	-0.31
10	6	Sept. 1987	0.42[2]	0.11	0.67	0.91	-0.69
11	7	Oct. 1987	0.35	-0.05	0.65	0.82	-0.68
12	8	Dec. 1987	0.09	-0.03	0.62	0.62	-0.68
13	9	Jan. 1988	0.43	-0.19	0.71	0.75	-0.70

[1] n.m. = not measured
[2] Coefficients underlined are significant $P < 0.05$, df = 25.

PERSISTENCE AND GERMPLASM ORIGIN

There is good evidence to show that in New Zealand local germplasm is generally more persistent than introduced germplasm. In two spaced plant trials (studies 1 and 2 mentioned earlier), New Zealand cultivars had the highest persistence and final growth scores (Table 5). The next most persistent cultivars originated from Argentina and Poland (Study 1), and from Britain, Australia, Ireland and Japan (Study 2). Consistently poor material in both trials came from Brazil and Israel.

In a study comparing 10 cultivars and elite selections (five from New Zealand and five from overseas) at six sites for at least 2 yr (Chapman et al., 1986) New Zealand material showed the greatest adaptability for yield when data were re-calculated using a

Table 5. Persistence and final growth score of bred cultivars grown in spaced plant trials at Palmerston North, analysed by (a) country of origin (Study 1) and (b) area of origin (Study 2). Standard errors are given.

Country or area of origin	No. of cultivars	Persistence, %	Growth score a (0-5) b (0-10)
(a) Argentina	1	90	1.2
Belgium	1	80	1.3
Brazil	1	65	0.9
Britain	5	78 ± 7	1.2 ± 0.1
Czechoslovakia	3	50 ± 6	0.7 ± 0.1
Denmark	6	73 ± 4	1.0 ± 0.1
Israel	1	70	1.0
New Zealand	2	95 ± 5	2.1 ± 0.6
Poland	3	90 ± 3	1.9 ± 0.2
Sweden	2	75 ± 0	1.1 ± 0.2
U.S.A.	6	74 ± 9	1.7 ± 0.3
USSR	1	50	0.8
(b) Australia	3	89 ± 3	3.5 ± 0.1
British Isles	18	91 ± 2	3.6 ± 0.1
Eastern Europe	12	81 ± 2	3.0 ± 0.2
Japan	2	89 ± 9	3.3 ± 0.5
Mediterranean	3	78 ± 7	2.8 ± 0.5
New Zealand	13	97 ± 1 (94 ± 3)†	4.3 ± 0.3
North America	19	77 ± 2	2.9 ± 0.1
Scandinavia	23	81 ± 1	3.2 ± 0.1
South Africa	1	87	2.6
South America	4	82 ± 3	3.2 ± 0.2
Western Europe	21	81 ± 2	3.3 ± 0.1

† Four bred cultivars.

genotype-environment regression analysis (Finlay & Wilkinson, 1963) (Fig. 3a). Slopes of all New Zealand material were greater than 1.3 and for overseas material less than 0.9. All regressions were significant except that for 'Clarence', which showed the poorest adaptability. Genotype-environment analysis of stolon density (Fig. 3b) showed increasing improvement (increasing slope) as leaf size decreased and, with the exception of Kent, as New Zealand germplasm content increases. Overall the least adaptable line for stolon density was the largest leaved overseas line studied, Haifa, and the most adaptable line the smallest leaved New Zealand cultivar, Tahora. All regressions shown in Fig. 3b were significant.

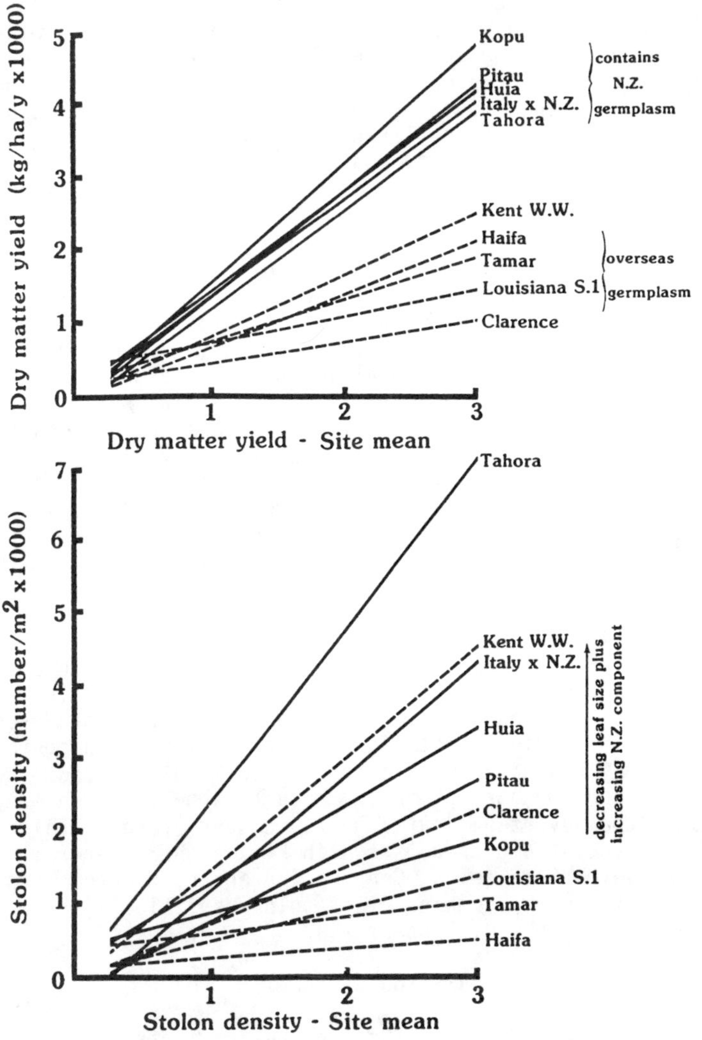

Fig. 3. Regression of line means on site means for (a) dry matter yield and (b) stolon density of 10 white clover lines (adapted from Chapman et al., 1986).

A similar study has been undertaken for nine cultivars of subterranean clover, all of Australian origin, at eight dry hill country sites (Chapman et al., 1986). The major adaptive characters were found to be days to flowering and tolerance of water-logged soils. For most (6) of the hill country sites late flowering was the predominant factor in subclover persistence. However, at the driest site mid-season material was most suitable and at the wettest site cultivars of subspecies yanninicum with water-logging tolerance proved to be superior to all cultivars of subspecies subterraneum.

THE COST OF PERSISTENCE

Genotypes of white clover that persist under intensive set-stocking managements tend to be small leaved with a low harvest index; they have a high stolon density and probably low stolon elongation rates. In contrast genotypes that perform best under lax defoliation treatments, such as infrequent cutting or rotational grazing by cattle tend to be large leaved with high harvest index; they have a low stolon density and high stolon elongation rates.

Williams et al. (1982) showed that under continuous sheep grazing a small leaved hill country selection produced twice as many stolons as medium leaved Huia, giving a dry matter advantage of 50%. The small leaf size and prostrate habit of the hill country line resulted in a low harvest index (Caradus & Williams, 1981).

The leaf sizes of genotypes of six populations of white clover collected from areas that had received differing frequencies and intensities of cutting for several years were measured at time of collection and after 2 yr growth in pots in a glasshouse (Caradus, 1984). Correlation of leaf size of populations between measurements was significant, $r = +0.90^{*}$. Leaf size after 2 yr in pots decreased with increasing intensity of defoliation received in the field.

In a pot study, Caradus (1986) showed that large leaved white clover genotypes had high harvest index and that genotypes from, and presumably adapted to, intensively grazed hill country had low harvest indices and were low yielding. Also, the genotypes with a low harvest index fixed the lowest amount of N per unit of P absorbed ($r = +0.56^{**}$, for correlation of proportion of dry matter in shoot vs. N/P ratio of the whole plant). Therefore selection for a low harvest index, to ensure persistence under intensive grazing, may result in genotypes with low N_2-fixation per unit of P absorbed.

Despite indications that high stolon density which confers persistence may have a price attached in terms of reduced productivity and low harvest index, results show that, in many situations, the advantages outweigh the disadvantages. This is reflected in the strong trend shown in Fig. 2 where several New Zealand selections (e.g., Gene pool A and F, Nematode Resistant Pitau and Feathermark) combine high stolon densities with leaf sizes consistent with high yield potentials.

BREEDING STRATEGIES

Persistence in Dryland Environments

Breeding programmes related to white clover persistence in summer dry areas of New Zealand have focused on (i) drought escape, (ii) drought tolerance and (iii) selection for taprootedness.

Drought escape is related to the ability of a plant to flower early and set seed before the onset of summer dry conditions. Following periods of summer drought, there is an opportunity for plant populations to re-establish themselves from soil seed reserves in the autumn (Macfarlane & Sheath, 1986). A selection has been made for this purpose and is known as 'Whatawhata Early Flowering'. This selection has been evaluated for production and persistence against eight other white clover cultivars at two summer dry hill sites. At Whatawhata Hill Country Research Station 'Whatawhata Early Flowering' had similar production to other New Zealand material of the same morphology (i.e., Tahora that was bred from a collection of white clovers from moist hill country). At Wairakei, Whatawhata Early Flowering out produced the other cultivars. While a large percentage of plants were lost each year due to the very drought prone nature of the trial site, the selection produced two to three times the number of flowers and seeds compared with Huia and Tahora (Macfarlane & Sheath, 1986).

Drought tolerance is achieved by plants persisting vegetatively through drought periods. More than 100 lines of white clover have been evaluated for vegetative survival at two dryland hill sites, over a 2 yr period, which included one severe summer drought. There was a definite trend for populations collected from dryland sites to have greater vegetative survival than other New Zealand material (Woodfield & Caradus, 1987). However, even within the dryland population collection there was considerable variation, with two populations collected from North Canterbury having 20 and 28% survival, and 10 populations from Canterbury, Marlborough, Wairarapa and Southern Hawke's Bay over 10% survival.

Selection of genotypes with increased taprootedness is based on the rationale that improved moisture uptake may be achieved by plants with more extensive root systems and increased root to shoot ratio (Smith & Morrison, 1983). Studies have shown that dry hill country white clover populations appear to be more taprooted than populations from moist hill country (Caradus & Woodfield, 1986; Woodfield & Caradus, 1987). This is partly due to the larger leaf size of dry hill country populations but also to a larger taproot diameter and higher number of taproots developing at nodes. Dryland populations and survivors after drought have similar root morphologies to that of Huia but are smaller leaved and possibly lower yielding (Woodfield & Caradus, 1987). Root characters that show reasonable heritabilities are taproot diameter and proportion of taproot to total root weight (Caradus & Woodfield, 1989; Woodfield & Caradus, 1989).

Lotus corniculatus lines have also been evaluated for persistence in dryland areas (Charlton et al., 1978; Forde & de Lautour, 1978). At the driest site the most persistent lines after three summers came from Europe, the Mediterranean region, and South America. At a moister site the South American material did not persist.

Competition

Plants growing in mixtures, as is the case for most herbage legumes other than lucerne, may compete for any environmental factors for which they have a common requirement but which is in short supply. The most important of these environmental factors are light, nutrients and water.

Interspecific competition is incorporated at some stage into all New Zealand breeding programs with legumes. In the past this used to be towards the end of the program whereas more recently experimental lines are being evaluated for agronomic performance in a grass sward at a much earlier stage of selection. There have been no breeding programs specifically selecting for ability of a legume species to compete with a particular grass species.

Persistence in High Country

Persistence of legumes in New Zealand high country is affected by both low temperature and low moisture (Scott et al., 1985), Scott (1985) evaluated the persistence of 24 herbage legume species and lines within most of these species in the Mackenzie Basin. Although there was considerable variation for persistence within some species, the most persistent species include white clover, alsike clover (*T. hybridum* L.), red clover, *Lotus corniculatus*, crown vetch (*Coronilla varia* L.), and subterranean clover. Other studies have shown alsike clover to be more frost tolerant than white clover (Clifford, 1973). Scott and Sutherland (1984) evaluated 38 alsike lines at three sites in the Mackenzie basin for yield, adaptability, and persistence. Yield and adaptability of lines appeared to be completely unrelated to persistence. In another study Scott et al. (1974) compared five to eight lines of red clover, alfalfa, white clover, alsike clover and *Lotus* species in the Mackenzie Country and found that the most persistent species were alfalfa and red clover. No lines of white clover, alsike clover, and lotus were as persistent as the worst line of alfalfa.

Nutrient Deficiencies and Toxicities

Breeding programs under way in this area include selection for:

1. Tolerance of low P soils.
2. Efficiency of P use on high-P soils.
3. Tolerance of high Al levels in soil.
4. Tolerance of high soil-N levels.

Despite the fact that response to added P in glasshouse studies do not correlate well with those in the field (Caradus & Snaydon, 1986) the strategy at present is to identify white clover germplasm that differs in some aspect of its P nutrition, whether it be P response (Mackay et al., 1988) or the way in which it adjusts the ratio of inorganic P to total P when P stressed (Caradus & Snaydon, 1987). Once differences are clearly identified, and it is shown that the differences are heritable, then the selected germplasm will be field tested.

Selection for Al-tolerance is being carried out in Al-amended soil in the glasshouse. At least two screenings have been shown to be required before final selections can be made. Apparent genotypic differences within white clover for Al-tolerance have been detected (Caradus et al., 1987) and heritability for putative Al-tolerance/intolerance is at present being determined.

Fifteen cultivars of white clover, including all those that have been consciously bred for high-N tolerance overseas, are being evaluated for N-tolerance in a field trial under grazing to determine whether germplasm differences for this character are manifested in New Zealand. Clovers have been sown with perennial ryegrass (Lolium perenne L.) and there will be four rates of N applied. Persistence under high N with grass competition and grazing of the cultivars will be determined.

Grazing and Management Effects

Cultivars that have resulted from attempts to breed for differing grazing managements include Tahora white clover for set-stocked and rotationally grazed sheep pastures (Williams et al., 1982) and Kopu white clover for dairy and cattle pastures (van den Bosch et al., 1986).

Pest and Disease Resistance

Diseases and pests that can reduce white clover persistence in New Zealand include nematodes, namely root knot (Meloidogyne spp.), clover cyst (Heterodera trifolii Goffart), and stem nematode (Ditylenchus dipsaci), grass grub (Costelytra zealandica (White)), Sclerotinia trifoliorum Eriksson and a range of viruses. Studies are in hand to determine variability within white clover for resistance/tolerance to each of these pests and diseases.

Very little variation has been observed within white clover for resistance to grass grub; most differences can be explained by the nutritional status of the root material eaten.

Efforts to increase the nematode resistance of white clover cultivars used in New Zealand began with screening for stem nematode resistance (Williams, 1972). Ladino cultivars were most resistant and crosses between susceptible Grasslands Pitau and Regal ladino resulted in the more resistant cultivar Grasslands Kopu.

Other screening studies are now under way to identify resistance to root knot nematode (Meloidogyne hapla (Chitwood)) and clover cyst nematode (Heterodera trifolii). Two approaches have been taken, (i) blanket screening and recurrent selection in nematode infested soil for resistance/tolerance to both nematode species and (ii) screening in sterilised soil to which has been added inoculum of either root knot or clover cyst nematodes. There is evidence that tolerance to both nematode species has been achieved using method (i) (Mercer & Cooper, 1987, unpublished data).

CONCLUDING COMMENTS AND FUTURE DIRECTIONS

Future direction should maintain emphasis on improving persistence in dryland, nutrient deficient, and pest-infested areas but increase awareness of importance in selecting for (i) resistances to viruses, (ii) competitive ability with specific grass species, such as prairie grass, tall fescue and phalaris and include programs comparing the effect of high and low endophyte ryegrass on persistence, (iii) Sclerotinia resistance and (iv) persistence under a wider range of environments and animal classes, e.g., deer (Cervus elaphus L.) and goats (Caprus hircus L.).

REFERENCES

Caradus, J.R. 1984. The phosphorus nutrition of populations of white clover (Trifolium repens L.). Ph.D Reading University.

Caradus, J.R. 1986. Variation in partitioning and percent nitrogen and phosphorus content of the leaf, stolon and root of white clover genotypes. N.Z. J. Agric. Res. 29: 367-379.

Caradus, J.R., A.C. Mackay, J. van den Bosch, and D.R. Woodfield. 1988. The relationship of yield and plant morphology of a world collection of white clover cultivars. p.69-70. In Proc. 9th Australian Plant Breeding Conf. Wagga Wagga. 27 June-1 July, 1988. Agricultural Research Institute, Wagga Wagga, NSW, Australia.

Caradus, J.R., A.D. Mackay, and M.W. Pritchard. 1987. Towards improving the aluminium tolerance of white clover. Proc. N.Z. Grassl. Assoc. 48: 163-169.

Caradus, J.R., and R.W. Snaydon. 1987. Aspects of the phosphorus nutrition of white clover populations. I. Inorganic phosphorus content of leaf tissue. J. Plant Nutri. 10: 273-286.

Caradus, J.R., and W.M. Williams. 1981. Breeding for improved white clover production in New Zealand hill country. p.163-168. In C.E. Wright (ed.) Plant physiology and herbage production. Occas. Symp. 13. Br. Grassl. Soc., Berkshire, U.K.

Caradus, J.R., and D.R. Woodfield. 1986. Evaluation of root type in white clover genotypes and populations. p.322-325. In T.A. Williams and G.S. Wratt (ed.) Plant Breed. Symp. DSIR 1986. Spec. Publ. 5. Agron. Soc. N.Z., Crop Res. Div., DSIR, Christchurch.

Caradus, J.R., and D.R. Woodfield. 1989. Estimates of heritability for, and relationships between, root and shoot characters of white clover. I. Replicated clonal material. (in manuscript).

Chapman, D.F., G.W. Sheath, M.J. Macfarlane, P.J. Rumball, B.M. Cooper, G. Crouchley, J.H. Hoglund, and K.H. Widdup. 1986. Performance of subterranean and white clover varieties in dry hill country. Proc. N.Z. Grassl. Assoc. 47: 53-62.

Charlton, J.F.L., E.R.L. Wilson, and M.D. Ross. 1978. Plant introduction trials. Performance of Lotus corniculatus introductions as spaced plants in Manawatu. N.Z. J. Exp. Agric. 6: 201-206.

Clifford, P.T.P. 1973. Alsike clover in South Island high country. Tussock Grassl. and Mountain Lands Inst. Rev. 27: 18-21.

Finlay, K.W., and G.N. Wilkinson. 1963. The analysis of adaptation in a plant-breeding programme. Aust. J. Agric. Res. 14: 742-754.

Forde, M.B., and de Lautour, G. 1978. Plant introduction trials. Classification of Lotus introductions. N.Z. J. Exp. Agric. 6: 293-297.

Macfarlane, M.J., and G.W. Sheath. 1986. The development of Whatawhata early flowering white clover. Ann. Rep. 1985/86 Whatawhata Hill Country Research Station, ARD, MAF. pp.74-75.

Mackay, A.D., J.R. Caradus, J. Dunlop, G.S. Wewala, M.C.H. Mouat, M.G. Lambert, A.L. Hart, and J. van den Bosch. 1988. Response to phosphorus of a world collection of white clover cultivars. Proc. 3rd International Symposium on Genetic aspects of plant mineral nutrition. Braunschweig, F.R.G. (in press).

Scott, D. 1985. Plant introduction trials : genotype-environment analysis of plant introductions for the high country. N.Z. J. Exp. Agric. 13: 117-127.

Scott, D., J.M. Keoghan, G.G. Cossens, L.A. Maunsell, M.J.S. Floate, B.J. Wills, and G. Douglas. 1985. Limitations to pasture production and choice of species. In R.E. Burgess and J.L. Brock (ed). Using herbage cultivars. Grassl. Res. Practice Series 3: 9-15.

Scott, D., G.C. Stringer, K.F. O'Connor, and P.T.P. Clifford. 1974. Growth of legume varieties on a yellow-brown high country soil. N.Z. J. Exp. Agric. 2: 251-259.

Scott, D., and B.L. Sutherland. 1984. Plant introduction trials : comparison of 38 alsike clover lines in the Mackenzie basin. N.Z. J. Exp. Agric. 12: 203-207.

Smith, A., and A.R.J. Morrison. 1983. A deep rooted white clover for South African conditions. Proc. Grassl. Soc. South Africa 18: 50-52.

van den Bosch, J., J.A. Lancashire, B.M. Cooper, T.B. Lyons, and W.M. Williams. 1986. G.18 white clover - a new cultivar for lowland pastures. Proc. N.Z. Grassl. Assoc. 47: 173-177.

Williams, W.M. 1972. Laboratory screening of white clover for resistance to stem nematode. N.Z. J. Agric. Res. 15: 363-370.

Williams, W.M., and B.M. Cooper. 1980. Plant introduction trials. Evaluation of white clover (*Trifolium repens* L.) introductions at Palmerston North. N.Z. J. Exp. Agric. 8: 259-265.

Williams, W.M., M.G. Lambert, and J.R. Caradus. 1982. Performance of a hill country white clover selection. Proc. N.Z. Grassl. Assoc. 43: 188-195.

Woodfield, D.R., and J.R. Caradus. 1987. Adaptation of white clover to moisture stress. Proc. N.Z. Grassl. Assoc. 48: 143-149.

Woodfield, D.R., and J.R. Caradus. 1989. Estimates of heritability for, and relationships between, root and shoot characters of white clover. II. Regression of progeny on mid-parent (in manuscript).

DISCUSSION

Hoveland: What mechanisms are associated with aluminium tolerance in white clover?

Caradus: This has as yet not been determined. Our major aim at present is to identify aluminium-tolerant germplasm and establish that aluminium tolerance is heritable in white clover.

Matches: What is the correlation between stolon density and herbage yield?

Caradus: In the grazed sward trial, that I mentioned, highest yields were obtained in the first year from lines having low stolon densities. However, in the second year, the medium to high stolon density lines had the highest yields, presumably due to better persistence of these lines.

Leath: How are you screening for resistance to *Sclerotinia*?

Caradus: We have used both mycelium and ascospore inoculation. Use of mycelium has not been very productive and use of ascospores is more laborious, but in the end may be more important. We have found it necessary to make sure that plants are not damaged during handling prior to inoculation as this may predispose them to infection.

Watson: Can you identify reasons for the yield advantages that New Zealand cultivars and first generation crosses with overseas cultivars have in spread plantings?

Caradus: I am not entirely confident that the reasons have been clearly identified. It may be, however, that it is the more stoloniferous nature of the New Zealand material as shown in Fig. 2 that gives them an advantage.

Buxton: Were the more drought resistant lines of white clover more taprooted?

Caradus: There was a tendency for this to be the case.

Sheaffer: Were the results from the small plot study realistic?

Caradus: There is always a need to compromise between what is desirable and what is possible. Despite the small size of the plots used, we are confident that the results are realistic.

Smith: Please comment on your good correlation between proportion of clover in the first and second years.

Caradus: While this relationship was reasonable for the first 2 yr, current information, as we now proceed into the 4th yr of this trial, would suggest that this relationship declines.

Brougham: What plant characters ensure persistence to extreme cold?

Caradus: The most obvious is the winter dormancy of white clover cultivars bred in Scandinavia and colder parts of northern Europe. These cultivars also tend to be small to medium leaved and moderately prostrate.

Curll: Your ranking of the performance of New Zealand and Australian cultivars in terms of persistence and stolon density illustrates the importance of environment when evaluating cultivars. Therefore, ranking of cultivars in New Zealand is quite different to that in Australia. Hence the need to have an improvement program specifically for each country, targeting the particular environments, and objectives for that country.

Sheath: What have been the difficulties encountered in breeding for grass grub resistance in white clover?

Caradus: Methodology has been a major problem, but there does not seem to have been a great deal of variation within white clover anyway. Genotypes with large taproots have shown greater tolerance and nutritive value of the root tissue may also be involved.

BREEDING AND GENETICS OF LEGUME PERSISTENCE

R.R. Smith and A.E. Kretschmer, Jr.

SUMMARY

The major limitation to the use of temperate forage legumes for hay or pasture is their relative short life-span in swards. The depletion of stands (lack of persistence) is a direct function of stresses imposed by climatic, pest, edaphic, management, and other factors and their interactions. While the primary objective of most forage legume breeding programs is to improve persistence, little information is available on the genetics and breeding for persistence per se in temperate forage legumes. Numerous cultivars and germplasm lines have been released with an indication of improved persistence. Phenotypic recurrent selection is the most frequently used breeding procedure to develop this germplasm. Examples of improved persistence using this procedure are presented for several legumes. The effect of selection for resistance or tolerance to specific stress factors on forage legume persistence is dependent upon the availability of genetic variability, the complexity of inheritance, and the effectiveness of procedures used to identify resistant genotypes. Interspecific hybridization between cultivated species and their wild relatives, especially using new biochemical procedures, offers promise of improving persistence by the transfer of desirable characteristics from the wild species.

INTRODUCTION

The major limitation in the use of temperate forage legumes for hay or pasture is their relative short life-span in swards. The stand depletion becomes evident as soon as harvesting, grazing or cutting commences. The loss of stand continues progressively, depending on the interaction of climatic, pest, edaphic, management, and physiological factors as well as competition among species. The plants response to these factors and their ability to persist under hay or grazing conditions is dependent upon the degree of stress imposed. In the field, the environment is continually changing, imposing an ever changing set of stress conditions.

Numerous cultivars and germplasm lines have been released with an indication of improved persistence or longevity. However, a review of the literature did not reveal any reported incidences where these releases are the result of studies designed to evaluate the genetics of persistence. Therefore, while it is the intent of this paper to address the breeding and genetics of legume persistence per se as a complex tract, we will attempt to report progress from selection for persistence. Examples for improved persistence as the result of selection for specific stresses, on occasion, will be cited.

PERSISTENCE FROM A PLANT BREEDERS PERSPECTIVE

What does persistence mean? How do you define it? Perhaps it could be best described as the survival of plant material against specific stresses unique to the existing environment. This sounds simple, but as mentioned earlier, numerous factors may impose stresses in any one environment, or one factor may impose opposite stresses in different environments. For example, red clover (*Trifolium pratense* L.) survival in northern USA is a function of tolerance to low temperatures, but in the mid-south and southern USA tolerance to high temperatures is necessary.

Generally, the longevity or age of a stand is a function of the severity of the stress. The actual impact of a stress may not be expressed until a later date. For example, stress imposed in the first harvest of the first year may not be apparent until the second harvest of the second year. In red clover, stands are more difficult to maintain as the incidence of root rot organisms increase, and the problem is compounded further when unadapted cultivars are used or other stresses such as physiological factors are imposed (Taylor & Smith, 1973). As the numbers of stress factors and their interaction increase, often imposing opposite pressures, the effect on plant growth is impossible to predict. Therefore, the plant breeder, with the aid of specialists in other disciplines, must define those factors that effect plant growth in their specific environment. Once these factors are defined, the objectives are to improve plant performance under specific stress conditions in a sequential manner and to address interactions between stresses as they occur. Improvement may be rapid if the response to the stress is controlled by a few genes, such as pest resistance, but much slower for more complexly inherited traits such as persistence. Perhaps the entire genome of a species is involved with persistence.

SELECTION FOR PERSISTENCE

In perennial legumes, the most frequently used breeding procedure is phenotypic recurrent selection (PRS). Desirable phenotypes are chosen and intercrossed, generally in a polycross isolation, to produce the subsequent generation. Such polycross seed can be used for progeny testing of the original selections, for testing progress made with initial selections, and/or as a source for further selection. Other breeding procedures will be discussed as warranted.

In an attempt to develop persistent germplasm legume breeders have generally used PRS. Surviving plants in old swards or yield trails are chosen and intercrossed. This has been one of the primary objectives of the red clover improvement program at the University of Wisconsin for the past 40 years. Emphasis has primarily been to select surviving, reasonably healthy plants from 3- or 4-yr-old yield trials rather than from pasture or hay fields. Management and germplasm history are available for yield trial plots, but often lacking for pasture or hay fields. The development of the red clover cultivars released jointly by the USDA-ARS and the University of Wisconsin between 1950 and 1988 is presented in Table 1. 'Lakeland' was released in 1961 and the base germplasm was selected for disease resistance from a mixture of hybrids of lines and cultivars (Hollowell, 1961). Subsequent to this, 'Arlington' was released in 1973 (Smith et al., 1973). Germplasm in this cultivar had been subjected to three cycles of persistence in 3-yr-old yield trials. 'Marathon' (officially released in 1987 but not currently registered) was developed using PRS for survival in 4-yr-old yield trials followed by one cycle of selection for 3-yr persistence to wet, acid (pH 5.7) soil conditions. Both Arlington and Marathon were continually screened for resistance to the foliar disease, northern anthracnose [causal agent -- Kabatiella caulivora (Kuchn.) Karak].

The performance of Wisconsin-released cultivars in Wisconsin is presented in Table 2. Substantial improvement in forage dry matter yield and percent stand at the end of the 3rd year has been achieved. In the past, red clover stands declined rapidly between the 2nd and 3rd years, often severe enough to discontinue the sward. This problem was somewhat reduced by the release of Arlington and the current performance of Marathon suggests even further progress. Currently, we are continuing to select within these cultivars to improve pest resistance using a combination of PRS, progeny testing, and family selection.

Other examples of selection for persistence should also be cited. Schonhorst et al. (1980) developed AZ-Ron alfalfa germplasm (Medicago sativa L.) which was selected for tolerance to frequent herbage removal. The cultivar Moapa was harvested 36 times over a 4-yr period at the early bud stage. Sixty of the most vigorous survivors were selected and intercrossed to produce AZ-Ron. AZ-Ron consistently out yields its parents, Moapa, and all other test entries in the 3rd and 4th year because of its greater persistence and vigor. Cooke & Sonmor (1977) developed ML-48-65 white clover (Trifolium repens L.) germplasm by intercrossing 16 surviving plants from a 4-yr-old spaced-plant

Table 1. Red clover cultivars developed in Wisconsin, 1950-88.

Cultivar	Release date	Method of development
Lakeland	1961	Line hybridization followed by six cycles PRS for disease resistance
Arlington	1973	Three cycles of PRS for persistence and disease resistance
Marathon	1987	Two cycles of PRS for persistence in old stand followed by one cycle of selection in wet, acid soils plus maintenance of disease resistance.

Table 2. Performance of red clover cultivars developed in Wisconsin.

Cultivar	Performance in third year			
	Yield			
	Rel. Lake[1]	Loss from Yr 2 to Yr 3	Fall stand	NA Rest.[2]
	%	%	%	%
Lakeland	100	35	32	18
Arlington	127	20	48	37
Marathon	137	4	67	40

[1]Forage dry matter yield relative to Lakeland (5.29 mg ha^{-1}), average of 12 tests in Wisconsin.
[2]Percent plants resistant to northern anthracnose (NA).

nursery. ML-48-65 is very winter-hardy and withstands close mowing and grazing. Miller et al. (1983) developed NC-83 birdsfoot trefoil (Lotus corniculatus L.) germplasm by selecting 30 surviving clones from breeding programs at Illinois, Missouri, Iowa, and Minnesota which were superior in forage yield, seed production, pest resistance and persistence. NC-83 is persistent and has excellent yield. Pedersen et al. (1986) developed the birdsfoot trefoil cultivar AU Dewey which was selected for its rhizomatous nature and good vigor. AU Dewey's persistence in southern USA pastures is attributed to its high reseeding capacity. Numerous other examples of germplasm release could be cited but like those discussed, none provided information on the genetics of persistence.

Taylor et al. (1968) examined the feasibility of breeding red clover for persistence utilizing polycross progeny testing. Twenty clones were selected and intercrossed from 1500 clones subjected to viruses (primarily bean yellow mosaic virus). These 20 selected clones were subjected to polycross progeny testing at two locations. Persistence of the progeny were correlated with persistence of the clones. These authors concluded that the use of polycross tests in breeding red clover for persistence was effective. In the evaluation of the alfalfa cultivar Trifecta, Clements et al. (1984) observed that polycross progeny testing indicated that both yield and persistence could be further improved by selection and reported a heritability estimate of 0.86 for persistence.

Most alfalfa cultivars in the USA are developed as "hay" types, and numerous high yielding, persistent, pest resistant hay cultivars are available. However, these hay type alfalfas are generally not well suited to grazing and thus do not persist well under these conditions. To overcome this problem, cultivars with rhizomatous or creeping-rooted plants were developed (Heinrichs, 1963). Perhaps through appropriate management, alfalfa stands can be maintained under continuous grazing conditions (Iversen, 1967). Counce et al. (1984) screened and characterized 22 alfalfa cultivars for persistence under mowing and continuous grazing conditions. Both hay and creeping-rooted type cultivars were included. The cultivars did not differ in performance under mowing but did under grazing conditions. Persistence of the cultivars under the different conditions was not correlated. Top growth and total nonstructural carbohydrates in the tap root were less for persistent cultivars than nonpersistent cultivars. These authors concluded that the prospect for selecting alfalfa for persistence under grazing conditions are good, but they suggest productivity would be sacrificed unless selection for productivity and persistence was practiced. Subsequently, Smith & Bouton (personal communication) have undertaken a project to use recurrent phenotypic selection to develop an alfalfa population which is productive and persists under grazing conditions. Progress has been achieved after two cycles of selection in upright and creeping-rooted types from both existing cultivars and accessions from the Plant Introduction collection. The selected population was superior in yield and persistence after 6 months of continuous grazing in each of 2 yr.

It is generally agreed that the performance of legume germplasm as measured in spaced-plant nurseries is of little value to predict the performance in a dense stand. The exception is for some qualitatively inherited characteristics. Therefore, due to the complexity of persistence, it would appear that the best opportunity for improving persistence would be to select in dense stands where individual plants are subjected to intra-and inter-plant competition. Rowe (1988) suggested that competition among alfalfa genotypes eliminates some of the less vigorous genotypes. He examined the effects of density, and thus competition, at three levels with respect to rate of plant mortality and changes in plant and plot yields. The plants were space planted in the field to stand 10, 17.3, and 30 cm on centers with a relative ratio of 9:3:1 per unit area. The rates of thinning were linear over the life of the experiment (2 years) but different for each density. Further, he suggests that selection through competition could be effective in a space planted nursery with plants at 17.3 cm centers or less if the trait called competitiveness is heritable.

SELECTION FOR SPECIFIC STRESS TRAITS

Improvement of general plant health through selection for pest resistance and favorable edaphic factors has led to improved persistence in forage legumes. Excellent reviews on specific responses to selection at the whole plant level on such stress factors effecting clovers and alfalfa are presented by Leath (1985), Barnett & Diachun (1985), Manglitz (1985), Leath et al. (1988), Manglitz & Ratcliffe (1988), Elgin et al. (1988), and Sorensen et al. (1988). This topic is also addressed by others at this conference so will not be discussed further by the current authors. However, the classic example of the development of alfalfa cultivars resistant to bacterial wilt [causal agent -- <u>Corynebacterium insidiosum</u> (McCull.) H.L. Jens.] warrants comment. The release of these cultivars in the late 1930's marked the beginning of breeding for alfalfa disease resistance and substantially improved the persistence of alfalfa at the time (Elgin et al. 1988).

Most research on selection for resistance to stress traits has been accomplished using screening techniques imposed at the whole plant level. However, the regeneration of plants from tissue culture in the early 1970s opened a whole new field of potential selection procedures which could be imposed at the cellular level. Much of the research in this area has been limited to alfalfa and the clovers. While specific screening procedures have been effective in identifying alfalfa cells resistant to toxic substances, Bingham et al. (1988), point out that the reaction must also be transmitted to the whole plant to be effective. The process has been effective for some qualitative inherited characteristics such as disease resistance. However, for more quantitatively inherited traits such as yield or persistence it may be difficult, if not impossible, to select at the cellular level. Current excellent reviews of tissue culture activities in alfalfa and the clovers are presented by Bingham et al. (1988) and Rupert & Collins (1985), respectively.

Selection for resistance to a specific stress trait is dependent upon the effectiveness of identifying resistant genotypes in the population. Thus, it is very difficult to achieve progress if the breeder must rely on natural conditions occurring in spaced plant nurseries or broadcast plots. Therefore, the most effective procedures are to impose the stress conditions under controlled environmental conditions. Currently, most selection procedures employed to identify genotypes resistant to pests or edaphic stress are conducted under controlled conditions with the assumption that the levels of resistance achieved will also be observed under field conditions. Also, these stress conditions are often imposed on germplasm for several cycles before field testing occurs. We were concerned with this problem in our red clover breeding program in Wisconsin as we attempted to improve disease resistance in our relatively persistent germplasm (Smith, 1980). Our procedure was to select for disease resistance in the greenhouse during the winter months, intercross in isolation during the summer months and initiate the next cycle of selection the following winter. Selection pressure was approximately 5% each cycle with the retention of a minimum of 100 plants each cycle. After four cycles of such a selection scheme, the effects on other agronomic characteristics were measured by testing the four cycles and the base population under field conditions for 3 yr. The annual selection procedure did not alter cycle means for green matter yield or persistence (Table 3). Plants tended to become more upright but no significant change in maturity was evident. Genetic variability existed in the fourth cycle for yield, maturity, and growth type but not for persistence (data not presented). Even with no evidence of significant changes between cycles of selections, we believe it appropriate to field test the selections after two or three cycles of annual selection. This provides an opportunity to evaluate the germplasm for agronomic worth and to perhaps obtain an evaluation for field tolerance to the stress trait.

Table 3. Mean disease reaction, green matter yield, growth type, maturity, and persistence of red clover germplasm representing four cycles of selection for disease resistance.

Cycle	Disease[1] NA	TS	Yield (3rd yr)[2] Cut I	Cut II	Growth Type[3]	Maturity[4]	Persistence 3rd yr[5]
0	3.0	4.6	335	123	3.0	6.6	51
1	3.2	4.5	339	128	3.4	7.1	54
2	1.8	4.1	352	152	3.4	7.0	57
3	1.7	3.9	313	147	3.6	6.8	61
4	1.3	3.5	296	131	3.7	6.7	58
LSD (5%)	0.3	0.3	NS	NS	0.5	0.3	NS

[1]NA=northern anthracnose; TS=target spot: 1=resistant, 5=susceptible.
[2]grams green weight per plant.
[3]1=decumbent, 5=upright, at first harvest.
[4]1=vegetative, 8=green seed stage, at first harvest.
[5]percent surviving plants.

EFFECTS OF PLOIDY LEVEL ON PERSISTENCE

In the 1940s, it was believed that one method to improve crop productivity was to polyploidize certain crop plants. Red clover, normally a diploid (2x=14), appeared ideally suited for polyploidization. As a result, the first tetraploid (4x=28) red clover cultivar, Ulva, was released by the Swedish Seed Association in 1959 (Bingefors & Ellerstrom, 1964). The performance of Ulva and other tetraploid cultivars developed throughout Europe have demonstrated the superiority of tetraploid red clover over diploids. The tetraploids are higher yielding and more persistent (Frame, 1976). No direct genetic explanation is given; however, improved disease resistance, greater frost tolerance, and improved drought tolerance have been suggested as factors contributing to improved yield and persistence (Anderson, 1973; Crowley, 1975; Frame, 1976). The fact that tetraploids have four alleles at each locus in contrast to only two in the diploids provides for increased opportunity for intra-and inter-allelic interactions to exist and provide additional additive, dominance, and epistatic genetic effects not present in the diploid forms.

The effects of ploidy level on physiological and agronomic traits in alfalfa are summarized by McCoy & Bingham (1988). The tetraploid forms exceed the diploid for several agronomic and physiological traits but no studies report a change in persistence due to ploidy level.

INTERSPECIFIC HYBRIDIZATION

Genetic variability for stress factors is generally observed within the plant species concerned and resistance to these factors can be achieved using appropriate screening procedures and breeding schemes. On occasion, appropriate genetic variability is lacking in species but is available in other species of the genus. However, to transfer the desirable trait from one species to the other depends upon the ability to hybridize the species of concern. When the two species are sexually compatible, it is generally easy to make the transfer and has been used successfully in many genera.

The problem arises when desirable genes are in wild relatives of the domestically used species and normal sexual hybridization between the species is extremely difficult if not impossible to make. Recent developments in tissue culture techniques have provided new procedures which hopefully will overcome some of these interspecific hybridization barriers (Bingham et al., 1988; Rupert & Collins, 1985).

Kura clover, _T. ambiguum_ Bieb., is resistant to the most serious virus diseases of white clover, _T. repens_ L. Also, kura clover has a strong taproot and rhizome system which could improve the persistence of white clover. Natural crosses between these species have been unsuccessful, but recently the hybrid between these two species has been attained (Williams, 1978; Williams & Verry, 1981). The interspecific hybrid was accomplished using a specialized tissue culture technique, embryo culture with transplanted endosperm (Williams & De Lautour, (1980). The initial hybrid was sterile, however, subsequent hybrids showed sufficient fertility to obtain backcross and F_2 progeny. Research on this hybrid is currently being conducted in the USA and New Zealand.

Embryo rescue techniques were employed by Phillips et al. (1982) to obtain the interspecific hybrid between _T. sarosiense_ Hazsl. and _T. pratense_, a sexually incompatible cross. These investigators were successful in transferring the rhizomatous root characteristic from _T. saroclense_ to the hybrid, but the hybrid was sterile. Subsequent attempts to produce the hybrid have been successful but at present the hybrids are sterile in spite of numerous efforts to restore fertility.

The development of improved techniques for manipulatory excised embryos and recent accomplishments in somatic cell fusion offer greater possibilities for making wide crosses and for obtaining desirable characters from other species (Bingham et al., 1988). It must be pointed out, however, that the use of cell fusion or cell suspension techniques is dependent upon regeneration of whole plant tissue. The capacity for regeneration from callus appears to be quite genotypic dependent, and germplasm which readily regenerates is lacking in most legume species other than alfalfa.

CONCLUDING REMARKS

Selection and breeding for persistence in forage legumes is a complex task since numerous stress factors are interacting simultaneously on plant growth. Progress from selection for persistence will be slow unless a single stress factor is identified which impacts significantly on persistence. Then progress from selection can be rapid but only if genetic variability exists for the stress trait and the inheritance of such a trait is relatively simple.

Identification of plant genotypes resistant to stress factors is a direct function of the effectiveness of the screening procedure employed. Cooperation of scientists from many disciplines (agronomy, genetics, physiology, entomology, pathology, etc.) in both public institutions and private industry is of essence in order to appropriately characterize the stress factors and to establish procedures to identify resistant germplasm. Conventional procedures have been successful but new methodology in the area of genetic engineering and molecular biology will provide greater opportunities for improvement.

Current research on stress resistance must continue; however, scientists must place increased emphasis on physiological and edaphic traits, intra-and inter-specific competition, and climatic factors as they effect persistence of legumes. Financial support is always limiting, but as scientists, we must use that support most effectively and encourage increased support through appropriate channels.

REFERENCES

Anderson, L.B. 1973. Relative performance of the late-flowering tetraploid red clover 'Grassland 4706', five diploid red clovers, and white clover. N.Z.J. Exp. Agric. 1:233-237.

Barnett, O.W., and S. Diachun. 1985. Virus diseases of clover. _In_ N.L. Taylor (ed.) Clover science and technology. Agronomy 25:235-268.

Bingefors, S., and S. Ellerstrom. 1964. Polyploidy breeding in red clover. The tetraploid variety Svalof's Ulva compared with some diploid and tetraploid varieties. Z. Pflanzenzuechtg. 51:315-334.

Bingham, E.T., T.J. McCoy, and K.A. Walker. 1988. Alfalfa tissue culture. _In_ A.A. Hanson et al. (ed.) Alfalfa and alfalfa improvement. Agronomy 29:903-929.

Clements, R.J., J.W. Turner, J.A.G. Irwin, P.W. Langdon, and R.A. Bray. 1984. Breeding disease resistant, aphid resistant lucerne for subtropical Queensland. Aust. J. Exp. Agric. Anim. Husb. 24:178-188.

Cooke, D.A., and L.G. Sonmor. 1977. Registration of ML-48-65 white clover germplasm. Crop Sci. 17:189.

Counce, P.A., J.H. Bouton, and R.H. Brown. 1984. Screening and characterizing alfalfa for persistence under mowing and continuous grazing. Crop Sci. 24:282-285.

Crowley, J.G. 1975. Red clover -- a new look at an old crop. Farm Food Res. 6:38-40.

Elgin, J.H., Jr., R.E. Welty, and D.B. Gilchrist. 1988. Breeding for disease and nematode resistance. _In_ A.A. Hanson et al. (ed.) Alfalfa and alfalfa improvement. Agronomy 29:827-858.

Frame, J. 1976. The potential of tetraploid red clover and its role in the United Kingdom. J. Brit. Grassl. Soc. 31:139-152.

Heinrichs, D.H. 1963. Creeping alfalfas. Adv. Agron. 15:317-337.

Hollowell, E.A. 1961. Registration of varieties of red clover. Agron. J. 53:403.

Iversen, C.E. 1967. Grazing management of lucerne. p. 129-133. In R.H.M. Langer (ed.) The lucerne crop. A.H. and A.W. Reed Wellington, New Zealand.

Leath, K.T. 1985. General diseases. In N.L. Taylor (ed.) Clover science and technology. Agronomy 25:205-233.

Leath, K.T., D.C. Erwin, and G.D. Griffin. 1988. Diseases and nematodes. In A.A. Hanson et al. (ed.) Alfalfa and alfalfa improvement. Agronomy 29:621-670.

Manglitz, G.R. 1985. Insects and related pests. In N.L. Taylor (ed.) Clover science and technology. Agronomy 25:269-294.

Manglitz, G.R., and R.H. Ratcliffe. 1988. Insects and mites. In A.A. Hanson et al. (ed.) Alfalfa and alfalfa improvement. Agronomy 29:671-704.

McCoy, T.J., and E.T. Bingham. 1988. Cytology and cytogenetics of alfalfa. In A.A. Hanson et al. (ed.) Alfalfa and alfalfa improvement. Agronomy 29:737-776.

Miller, D.A., P.R. Beuselich, I.T. Carlson, and L.J. Elling. 1983. NC-83 birdsfoot trefoil germplasm. Crop Sci. 23:1017.

Pedersen, J.F., R.L. Haaland, and C.S. Hoveland. 1986. Registration of 'AU Dewey' birdsfoot trefoil. Crop Sci. 26:1081.

Phillips, G.C., G.B. Collins, and N.L. Taylor. 1982. Interspecific hybridization of red clover (*Trifolium pratense* L.) with *T. sarosiense* Hazsl. using in vitro embryo rescue. Theor. Appl. Genet. 62:17-24.

Rowe, D.E. 1988. Alfalfa persistence and yield in high density stands. Crop Sci. 28:491-494.

Rupert, E.A., and G.B. Collins. 1985. Tissue culture. In N.L. Taylor (ed.) Clover science and technology. Agronomy 25:405-443.

Schonhorst, M.H., A.K. Dobrenz, R.K. Thompson, and M.A. Massengale. 1980. Registration of AZ-Ron alfalfa germplasm. Crop Sci. 20:831.

Smith, R.R. 1980. Effect of selection on annual basis for disease resistance on agronomic characteristics in red clover. p.70. In Agronomy abstract.

Smith, R.R., D.P. Maxwell, E.W. Hanson, and W.K. Smith. 1973. Registration of Arlington red clover. Crop Sci. 13:771.

Sorensen, E.L., R.A. Byers, and E.K. Horber. 1988. Breeding for insect resistance. In A.A. Hanson et al. (ed.) Alfalfa and alfalfa improvement. Agronomy 29:859-902.

Taylor, N.L., W.A. Kendall, and W.H. Stroube. 1968. Polycross progeny testing of red clover (*Trifolium pratense* L). Crop Sci. 8:451-454.

Taylor, N.L., and R.R. Smith. 1973. Breeding for pest resistance in red clover. p. 125-127. In Proc. Southern Pasture and Forage Crops Improvement Conf. Sarasota, FL 13-14, June. ARS-USDA, New Orleans, LA.

Williams, E. 1978. A hybrid between *Trifolium repens* and *T. ambiguum* obtained with the aid of embryo culture. N.Z.J. Bot. 16:499-506.

Williams, E.G., and G. De Lautour. 1980. The use of embryo culture with transplanted nurse endosperm for the production of interspecific hybrids in pasture legumes. Bot. Gaz. 14:252-257.
Williams, E.G., and I.M. Verry. 1981. A partial fertile hybrid between *Trifolium repens* and T. *ambiguum*. N.Z.J. Bot. 19:1-7.

DISCUSSION

Buxton: As opposed to grazing situations, is there generally a positive relation between persistence and yield among genotypes with hay production?

R. Smith: Yes, in our program in Wisconsin our first year yields are similar and improved yield is expressed in subsequent years.

Irwin: Earlier you were selecting for *Fusarium* resistance, have you made progress for this?

R. Smith: Progress was made in the greenhouse but the improvement was not realized as greater performance and persistence in the field. I will ask this same question of Ken Leath.

Leath: We did not take the germplasm to the field for evaluation. Our objective was to ascertain if resistance to a specific *Fusarium* was a hereditary trait and it was. Alfalfa researchers in Quebec, Canada, using some of our isolates and methods improved resistance to *Fusarium* root rot in alfalfa that was experienced in the field.

Caradus: Do you know if increasing ploidy levels in legume species other than red clover increases persistence?

R. Smith: No. Reports in alfalfa suggest improved performance for some characters but not persistence.

Forde: Perhaps one gets the biggest advantage just going from diploid to the tetraploid level.

Matches: I have not found that the improved red clovers have been any better than common types under grazing. To my knowledge U.S. breeders have managed red clover breeding programs under a hay management system. Is it true that most red clover germplasm has not been evaluated under grazing?

R. Smith: Yes, however, from the success of the Georgia Program with alfalfa germplasm I believe their procedure should be used on red clover. The procedure is very simple and easy to apply.

Sheafer: With the slow progress made in red clover breeding due to diseases, do you think that it is worthwhile to continue breeding red clover and possibly white clover.

R. Smith: First, I believe we have been successful in the improvement of red clover. The improvement in alfalfa has not been of large magnitude since the incorporation of bacterial wilt resistance in the 1950s. Second, red clover is adapted to considerable acreage in the USA that is not suited for the growth of alfalfa. Finally, I believe the USA would be in a very vulnerable state if we restricted our germplasm base to one species.

Sheath: Could you provide us the programs in the USA conducting research and/or selection on aluminum tolerance in legumes?

R. Smith: To my knowledge only two programs in alfalfa are selecting for aluminum tolerance, Beltsville, MD and Athens, GA. However, red clover cultivars have been characterized for aluminum tolerance at the Hill Country Research Center, Beckly, WV.

Marten: In our environment in northern USA we have evidence that one can afford to sacrifice up to 15-20% in yield to gain legume persistence in a legume-grass alternative situation to maximize economic yield of animal product per unit area. Do scientists from Australia and New Zealand find a similar situation in their environments?

Hochman: Yes, data from grazing experiments in Australia would support this statement.

Jones: How successful has the white clover cultivar, Osceola, been in the southeastern USA? This cultivar was selected for better resistance to pests and diseases and increase stolon survival.

Kretschmer: Osceola is a poor seeding Ladino type as compared to the naturalized Louisiana type. In the south it has not persisted well because of poor seed production.

GENERAL DISCUSSION OF GENETICS AND BREEDING FOR PERSISTENCE

Scott: Would you conclude that Australian scientists have sometimes engaged in premature plant breeding?

Clements: Yes, Australia has a history of premature plant breeding for both temperate and tropical legumes. In almost every case, the earliest plant breeding programs either failed to produce cultivars, produced cultivars that were unsuccessful commercially, or produced cultivars that were inferior to germplasm that was subsequently introduced. The problem often has been poor definition of breeding objectives, not lack of breeding technology. Plant breeding with sound objectives is worthwhile.

Sheath: Could someone from Australia indicate the past performance and likely future of plant introduction programs?

Clements: The Australian plant introduction program has been active for many years and has contributed much to new cultivar development. We expect this program to continue and to be used extensively in future breeding programs.

Hoveland: The discussion thus far suggests that in the USA the federal government is responsible for plant breeding. Actually, more plant breeding is supported at state agricultural experiment stations with state funds than by the USDA. This varies greatly among states with some providing excellent support for breeding of forage legumes and others providing little or nothing.

COLLABORATIVE RESEARCH AMONG SCIENTISTS IN AUSTRALIA, NEW ZEALAND, AND THE UNITED STATES OF AMERICA

Robert F Barnes

To collaborate means to work jointly with others, especially in an intellectual endeavor, such as a project involving scientific writing or research to be jointly accredited. In essence, it means to labor together. Efforts may involve communications, simply an exchange of information; or they may involve cooperation, which is to act or work with another or others for mutual benefit. The terms cooperate and collaborate are synonymous.

Effective collaboration/cooperation cannot be legislated or easily achieved by administrative edict. It is dependent upon the creation of an appropriate atmosphere and is driven by factors such as:

1. Identification of researchable problems. If the research area involved is too difficult for the state of scientific techniques and no definitive approach is available, it will preclude the research from being successful. Problems need to be defined.

2. Personal scientific interests. One of the best ways to get people to cooperate is to have them agree to work together on a specific problem through the exchange of ideas and resources.

3. Compatible personalities and backgrounds of individual scientists. Informal cooperative efforts depend upon the willingness of individual scientists to pool resources and expertise in striving to attain common goals. Personnel who have established themselves in their own research area are often best suited to become involved in interdisciplinary and collaborative research activities because many research organizations emphasize early independent research in their reward system.

4. Good communications. Special care needs to be taken to assure good communications. It is highly desirable for cooperating researchers to hold frequent discussions and share information. Modern telecommunication systems will enhance the sharing of information and reduce the need for structured meetings.

5. Leadership. The leaders of a collaborative team effort must have a vision of the goal to be completed and understand the process of interdisciplinary cooperative efforts to achieve that goal. If not, the team will get off track and become nonproductive quickly.

6. Administrative support. The role of the administrator should be to create an optimistic and positive research atmosphere and to chart the future course of action. Administrative decrees generally are not effective, but a little encouragement can go a long way.

7. Funding and budget control. Adequate resources allocated in a definable time period, appropriate for the problem being addressed, are important. Where no funds are involved the effort becomes, in essence, a labor of love.

8. Promise of reward. Potential for publication and recognition for accomplishment should be defined as soon as possible, authorship should be decided early and should be based upon assumption of responsibility.

The key ingredient to any research and education program, particularly collaborative/cooperative programs, is people. Excellence in organizational performance can be achieved only through excellence of individual performance.

HISTORICAL PERSPECTIVE

An active cooperative effort has been underway among scientists in Australia, New Zealand, and the USA for some time in the area of forage-livestock research, particularly emphasizing legumes. Formal program activities have been under the auspices and support of the Australian/U.S. Agreement for Scientific and Technical Cooperation and the New Zealand/U.S. Science and Technology Agreement.

A cooperative binational (Australian-U.S.) workshop on forage evaluation was held in Armidale, New South Wales, Australia, 27 to 31 Oct. 1980 (Wheeler & Mochrie, 1981). At the conclusion of that workshop, an Australian-U.S. Committee on Forage Evaluation and Utilization was formed. Priority research areas were identified for cooperative research (Barnes, 1981) and included:

1. The nutrition-behavior interface or the plant-animal interface.
2. Legume breeding, adaptation, and management.
3. Instrumental analysis of forage quality.

A binational New Zealand-U.S. agricultural cooperative workshop was held in June 1981, in conjunction with the XIV International Grassland Congress held at Lexington, KY, USA. Potential collaborative opportunities were identified for pasture legumes, involving three primary areas: (i) management,

(ii) species improvement through breeding, and (iii) species-management interactions.

At that time, the committee established in 1980 at Armidale, was expanded to include New Zealand personnel. It was renamed the Committee for Cooperation on Forage Evaluation and Utilization. Areas identified for possible future workshops included:

1. Alfalfa-pasture interface.
2. Legume adaptation and maintenance in pastures.
3. Grazing animal behavior.
4. Techniques for forage quality evaluation.
5. Fiber in forage plants.
6. Modeling-economics-forage evaluation.
7. Forages for dairy cattle.
8. Plant-animal interactions at an early stage in forage evaluation.

Following the XIV International Grassland Congress, a decision was made to organize a trilateral workshop on the topic, "Forage Legumes for Energy-Efficient Animal Production." The workshop was held in Palmerston North, New Zealand, 30 Apr. to 4 May 1984. The Proceedings were published and distributed by the U.S. Department of Agriculture, Agricultural Research Service (USDA-ARS) in 1985 (Barnes et al., 1985). At the conclusion of the trilateral workshop, participants identified eight areas of recommended activities for future collaborative efforts. National coordinators, or more appropriately "communicators," were identified from each country to serve in maintaining contact and sharing future information.

Two other recent conferences involved forage-livestock researchers, educators, and practitioners from Australia, New Zealand, and the USA.

The first of these led to publication of the compendium of papers originally presented in an abbreviated form at a special session of the XV International Grassland Congress, 30 Aug. 1985, Kyoto, Japan, entitled, "Grazing-Lands Research at the Plant-Animal Interface" (Horn et al., 1987). This special session provided a forum for scientists conducting comprehensive plant-animal interface research to present research results and exchange ideas. The published Proceedings serve as a frame of reference to chart future progress toward understanding the complex plant-animal interface of forage and animal productivity. Eight delegates from the USA and two each from Australia, New Zealand, and the United Kingdom participated in the session.

The second recent conference was an international symposium entitled, "Establishment of Forage Crops by Conservation-Tillage: Pest Management," held 15 to 19 June 1986, in State College, PA, USA. The symposium commemorated the 50th anniversary of the U.S. Regional Pasture Research Laboratory, USDA-ARS. An emphasis was placed on identifying knowledge gaps and research needs associated with four classes of pests -- insects, weeds, plant pathogens, and slugs -- that adversely affect forage crop establishment (Hill et al., 1986).

SUMMARY OF RECENT COLLABORATIVE/COOPERATIVE WORK

In June 1988, an inquiry was sent to each individual designated as a national communicator at the 1984 Trilateral Workshop. A summary of those responding is outlined below.

1. Legume persistence under grazing

Title: Evaluation of invertebrate pest complex in direct drilled forage crops in the seedling stage.

Countries: New Zealand and USA

Personnel and location:

B.I.P. Barratt
Invermay Agric. Res. Centre
P.B. Mosgiel
New Zealand

R.A. Byers
USDA-ARS
U.S. Reg. Pasture Res. Lab.
University Park, PA 16802 USA

Activity:

B.I.P. Barratt is serving as a visiting scientist for 6 months (April through October 1988) at the U.S. Regional Pasture Research Laboratory. Field and greenhouse studies are underway with insect pests and slugs and their impact upon the establishment of no-till forage crops. A technique developed in New Zealand for quantitative slug density estimation is being evaluated under U.S. conditions.

2. Nitrogen relationships

Title: Plant nutrition -- Regulation of nutrient acquisition and transport.

Countries: Australia and USA

Personnel and location:

Frank Smith
CSIRO
Plant Nutrition Lab.
306 Carmody Rd.
St. Lucia, Queensland 4067
Australia

W.A. Jackson
Dep. of Soil Science
North Carolina State Univ.
Raleigh, NC 27695 USA

Activity:

Exchange visits of scientists in 1981 and 1983 involving studies on plant nutrition. Emphasis was placed on the influence of N nutrition on P absorption in studies at North Carolina State University during Frank Smith's 1981 visit. Two manuscripts resulted from these studies on N enhancement of phosphate transport (Smith & Jackson 1987a, b). Regulation of P uptake and transport within the plant under various levels of ambient P concentration were studied with stylo during W.A. Jackson's visit to CSIRO in 1983. Three manuscripts are in preparation from these studies:

(1) F.W. Smith, W.A. Jackson, and P.J. Vanden Berg. Phosphorus nutrition of *Stylosanthes hamata*. I. Internal phosphorus flows during development of phosphorus stress.

(2) F.W. Smith and W.A. Jackson. Phosphorus nutrition of *Stylosanthes hamata*. II. Regulation of phosphorus uptake.

(3) F.W. Smith and W.A. Jackson. Phosphorus nutrition of *Stylosanthes hamata*. III. Regulation of phosphorus translocation.

3. Mechanisms of microbial nodulation

a. Title: Distribution and symbiotic effectiveness of *Rhizobium meliloti* in rangeland soils of the intermountain west.

Countries: New Zealand and USA

Personnel and location:

W.L. Lowther
Invermay Agric. Res. Centre
P.B. Mosgiel
New Zealand

M.D. Rumbaugh
D.A. Johnson
USDA-ARS, Crops Res. Lab.
UMC 63, Utah State Univ.
Logan, UT 84322 USA

D.D. Dwyer
Dep. of Range Science
Utah State Univ.
Logan, UT 84322

Activity:
W.L. Lowther spent 11 months as a visiting scientist at the Logan, UT, USA location starting in January 1985. A field and laboratory program was undertaken to survey the distribution and to assess the symbiotic effectiveness of naturalized populations of *Rhizobium meliloti* in rangeland soils of the U.S. intermountain west. The objective was to determine their influence on establishment and symbiotic N_2 fixation of alfalfa.

Results from these studies were used to formulate inoculation recommendations and to illustrate the potential for increased symbiotic N_2 fixation and, hence, alfalfa (*Medicago sativa* L.) yield by introducing highly effective competitive strains of rhizobia (Lowther et al., 1987a, b).

b. Title: Host microbial interaction in N_2 fixation by *Lotus pedunculatus*.

Countries: New Zealand and USA

Personnel and location:

C.P. Vance
USDA-ARS
Univ. of Minnesota
411 Borlaug Hall
1991 Upper Buford Circle
St. Paul, MN 55108 USA

C.E. Pankhurst
DSIR
Palmerston North
New Zealand

Activity:

C.P. Vance and C.E. Pankhurst conducted collaborative research from January 1984 to December 1987 on N_2 fixation relative to host microbial interactions in trefoil (*Lotus pedunculatus*) (Vance et al., 1987).

c. Title: Biological nitrogen fixation in *Trifolium* and *Medicago* germplasm.

Countries: Australia and USA

Personnel and location:

K.A. Lawson
7 Penrhyn Ave.
Beecroft
New South Wales, 2119
Australia

M.D. Rumbaugh
D.A. Johnson
USDA-ARS, Crops Res. Lab.
UMC, Utah State Univ.
Logan, UT 84322 USA

Activity:

K.A. Lawson, microbiologist, did post-doctoral work from March 1986 to March 1988 on projects associated with the selection and evaluation of *Trifolium* germplasm and *Rhizobium* strains; and biological N_2 fixation in co-evolved *Rhizobium* and *Medicago* ecotypes. Manuscripts are being prepared for journal publication, but the following are in print: Lawson et al., 1986, 1987; and Rumbaugh et al., 1987.

4. Efficiency of protein and energy utilization by ruminants

Title: Efficiency of protein utilization by ruminants.

Countries: Australia and New Zealand

Personnel and location:

T.N. Barry
Massey Univ.
Palmerston North
New Zealand

P. Larkin
D. Spencer
CSIRO, Plant Industry Div.
Canberra
Australia

Activity:

No formal projects have been initiated, but useful informal contact has been maintained between Australia and New Zealand scientists. As a result of the 1984 trilateral meeting, CSIRO Plant Industry Division has set up a group, headed by P. Larkin, to introduce condensed tannin into alfalfa using molecular biology techniques. T.N. Barry has visited Canberra twice to study the progress of the program. The same laboratory, under the direction of D. Spencer, has established a program to introduce genes coded for the production of proteins of low rumen degradability into forage plants for grazing.

5. Categorizing legume germplasm to minimize edaphic limitations

Title: Informal germplasm exchange.

Countries: Australia, New Zealand, and USA

Personnel and location:

M.B. Forde
Plant Introduction Officer
Grassland Division
DSIR
Private Bag
Palmerston North
New Zealand

S. Christiansen
U.S. AID-Casablanca
State Department
Washington, DC 20523 USA

H.G. Bishop
B. Walker
Dep. of Primary Industries
Mackay, Queensland
Australia 4740

W.J. Collins
Dep. of Agriculture
South Perth
W. Australia

A.E. Kretschmer, Jr.
Agric. Res. & Educ. Ctr.
Univ. of Florida
Box 248
Ft. Pierce, FL 34954-0248 USA

R.J. Clements
CSIRO Cunningham Laboratory
Div. of Tropical Crops & Pastures
306 Carmody Road
St. Lucia, Queensland
Australia 4067

Activity:

a. White clover (*Trifolium repens* L.) germplasm collected in southwest Europe (1986) and in Australia (1985) is currently under evaluation at Palmerston North. The material will be made available to the Australian white clover breeding program at Glen Innes.

b. A collection of late-flowering subterranean clover (*T. subterraneum* L.) accessions from France and Sardinia was received by DSIR, Palmerston North, in 1987 from W.J. Collins. The accessions are undergoing preliminary evaluation for use in a breeding program in New Zealand.

c. New Zealand has been invited to join a U.S./Yugoslavia/International Board for Plant Genetic Resources project for forage germplasm collection in Yugoslavia. The collection activities will involve the continental climate areas in 1988 and potentially the coastal areas in 1989. (New Zealand is particularly interested in the latter.)

d. Cooperative collection trips have been made by A.E. Kretschmer with R. J. Clements, CSIRO, in southeastern Texas, Louisiana, and Florida, USA and with R. Reid, formerly CSIRO, Davies Laboratory, Townsville, in Mexico. B. Walker and H.B. Bishop, Dep. of Primary Industries, were supplied with 125 *Aeschynomene* germplasm accessions which were tested at the Mackay Research Center, Queensland. After thorough evaluation of the total germplasm on hand, a joint-vetch (*A. americana* L.) ecotype was released. Kretschmer collaborated with CSIRO, CIAT, and EMBRAPA (Brazil) in the preparation of a World Centrosema Catalogue. Publications pertinent to tropical forage legume collection and exchange include Kretschmer et al. (1985, 1986).

e. S. Christiansen (formerly USDA-ARS) visited Australia in October to November 1984. This trip was prompted by Clements' visit to El Reno, OK, USA in late 1983. Christiansen visited pasture legume research centers in Australia and obtained valuable accessions of subterranean clover, annual medics, and other species. Christiansen also participated in a native pastures workshop at the Narayen Research Station from 23 to 25 Oct. 1984, where recommendations for future research on native grasslands were drawn up.

6. Shrub and browse legumes

Title: Genetic improvement in the genus *Leucaena*, and on-site adaptability evaluations of other woody legumes of tropical/subtropical origin.

Countries: Australia and USA

Personnel and location:

R.A. Bray
CSIRO
306 Carmody Road
St. Lucia, Queensland 4067
Australia

J.L. Brewbaker
Horticulture Dep.
Univ. of Hawaii
3190 Maile Way
Honolulu, HI 96822 USA

Activity:

a. *Leucaena* species. International Leucaena Psyllid Trials (LPT) were initiated in 1987 with collaboration between R.A. Bray and J.L. Brewbaker, supported by the Nitrogen Fixing Tree Assoc. (NFTA), with funds from USDA and USAID. Twelve varietal trials were held throughout Asia in 1987 and a second series sent in 1988 to collaborators, largely in southeast Asia. The leucaena psyllid (*Heteropsylla cubana*) entered the Pacific/Asia/Australia region in 1984 and has caused significant yield loss. Genetic resistance was immediately evident in Hawaii and Australia, in species other than the commercial *L. leucocephala*, and has been validated in these trials. A new *Leucaena* Seed Production project was started in 1988 to increase foundation seeds of resistant cultivars throughout Asia.

b. Other species. Informal collaboration continues, prompted in part by the psyllid problem, to evaluate N_2-fixing tree species for fodder use in this region. A large provenance collection has been assembled of *Gliricidia sepium* and tested in the region. Other promising woody species under evaluation in network trials with support from NFTA and the Australian Council for International Agricultural Research (ACIAR) include *Acacia*, *Calliandra*, *Desmodium*, and *Sesbania*. Primary constraints on their use include poor digestibility and yield. Persistence and pest resistance are normally not problems with the woody tropical legumes.

7. Breeding of forage legumes, including genetic engineering

Title: Interspecific hybridization of white clover (*T. repens*) with kura clover (*T. ambiguum*).

Countries: New Zealand and USA

Personnel and location:

W. Rumball
Grasslands Division
DSIR
Palmerston North
New Zealand

N. Taylor
Dep. of Agronomy
Univ. of Kentucky
Lexington, KY 40546 USA

Activity:

A study has been underway that involves improving the persistence and potential of white clover by hybridizing with kura clover. A limited level of support continues to be placed on the project by personnel in both countries.

8. Nutritional value

a. Title: Plant structural characteristics influencing breakdown of forage to small particles through the process of chewing and digestion in the rumen.

Countries: Australia and USA

Personnel and location:

D.J. Minson J.R. Wilson M.N. McLeod CSIRO, Div. of Tropical Crops & Pastures 306 Carmody Road St. Lucia Brisbane 4067 Australia	D.E. Akin USDA-ARS Richard B. Russell Agric. Res. Center Athens, GA 30613 USA

Activity:

Cooperative research between Australia and USA was completed on the physical and microbial breakdown of the leaves of temperate and tropical grasses. Reduction in particle length was almost entirely a physical process (chewing), microbial action being limited to a reduction in width by removal of the intervascular matrix (Wilson et al., 1987). Two additional papers have been submitted to Grass and Forage Science. The anatomical structure of forages has been shown to be a major factor influencing resistance to breakdown to the 1-mm critical size that can leave the rumen. This can have an important bearing on the understanding and development of future selection methods for the breeding of forages of superior intake.

b. Title: Comparison of rumen microbial activity.

Countries: Australia and USA

Personnel and location:

G. Gordon CSIRO Prospect Australia	D.E. Akin USDA-ARS Richard B. Russell Agric. Res. Center Athens, GA 30613 USA

Activity:

Collaborative work involved a comparison of rumen bacteria, protozoa, and fungi for their potential activity against grass cell walls. The work was established so that rumen fluid was spiked with cycloheximide to select for bacteria or with streptomycin and penicillin to select for fungi and protozoa. By using gravimetric analysis and microscopy, microbial groups in Australian (Prospect) and U.S. (Athens) cattle were compared. Pangolagrass (*Digitaria eriantha*) from Australia and bermudagrass ((*Cynodon dactylon* (L.)) from the USA were used as in vitro substrates, and rumen fluid was collected from cows within each country that were fed alfalfa. Results showed that protozoa and fungi from the Australian cows were more active in degrading fiber than those in the USA; bacteria were similar in activity but tended to be more active in cows from the U.S. site.

c. Title: Near infrared reflectance spectroscopy (NIRS) calibration of Australian forage samples.

Countries: Australia and USA

Personnel and location:

P.C. Flinn
Victoria Dep. of Agric. & Rural Affairs
Pastoral Research Institute
P.O. Box 180
Hamilton, Victoria 3300
Australia

J.S. Shenk
R.D. #1
109 Sellers Lane
Port Matilda, PA 16870 USA

Activity:

Exchange visits of scientists and the establishment of NIRS instrumentation at the Pastoral Research Institute was initiated in 1986 to 1987. Local calibrations of Australian forages were developed for nutritive and mineral values using U.S.-derived NIRS spectral values. Research is underway on the use of NIRS to predict the amount of legume in mixed pasture samples.

REFERENCES

Barnes, R.F. 1981. Areas of future United States-Australian cooperation in forage evaluation. p. 573-582. In J.L. Wheeler and R.D. Mochrie (ed.) Forage evaluation: Concepts and techniques. CSIRO, Canberra, Australia.

Barnes, R.F, P.R. Ball, R.W. Brougham, G.C. Marten, and D.J. Minson (ed.). 1985. Forage legumes for energy-efficient animal production. Proc. Trilateral Workshop, Palmerston North, New Zealand. 30 Apr.-4 May 1984. USDA-ARS, Washington, DC.

Hill, R.R., Jr., R.O. Clements, A.A. Hower, Jr., T.A. Jordan, and K.E. Zeiders (ed.). 1986. Proceedings international symposium on establishment of forage crops by conservation tillage: Pest management, State College, PA. 15-19 June. U.S. Regional Pasture Res. Lab., University Park, PA, USA.

Horn, F.P., J. Hodgson, J.J. Mott, and R.W. Brougham (ed.). 1987. Grazing-lands research at the plant-animal interface. Proc. Special Session, XV Int. Grassl. Congr., Kyoto, Japan. 30 Aug. Winrock Int., Morrilton, AR, USA.

Kretschmer, A.E., Jr., R. Reid, J. Gonzales R., and G.H. Snyder. 1986. Tropical forage legume collection trip in southern Mexico. Soil Crop Sci. Soc. Fla. Proc. 46:80-83.

Kretschmer, A.E., Jr., R.M. Sonoda, R.C. Bullock, G.H. Snyder, T.C. Wilson, R. Reid, and J.B. Brolmann. 1985. Diversity of Macroptilium atropurpureum (DC.) Urb. Proc. XV Int. Grassl. Congr., Kyoto, Japan.

Lawson, K.A., D.A. Johnson, and M.D. Rumbaugh. 1986. Improving symbiotic nitrogen fixation. Utah Sci. 47:114-119.

Lawson, K.A., D.A. Johnson, and M.D. Rumbaugh. 1987. A method to evaluate Trifolium germplasm response to mixed rhizobial inoculation. p. 79. In Abstracts, 11th North American Rhizobium Conf. 9-15 Aug., University Laval, Quebec, Canada.

Lowther, W.L., D.A. Johnson, and M.D. Rumbaugh. 1987a. Distribution and symbiotic effectiveness of Rhizobium meliloti in rangeland soils of the intermountain west. J. Range Manage. 40:264-267.

Lowther, W.L., M.D. Rumbaugh, and D.A. Johnson. 1987b. Populations of Rhizobium meliloti in areas with rangeland alfalfa. J. Range Manage. 40:268-271.

Rumbaugh, M.D., K.L. Lawson, and D.A. Johnson. 1987. General and specific effects of paired strain combinations of Rhizobium leguminosarum biovar trifolii on the growth of subterraneum clover. p. 100. In Agronomy abstracts. Am. Soc. Agron., Madison, WI, USA.

Smith, F.W., and W.A. Jackson. 1987a. Nitrogen enhancement of phosphate transport in roots of Zea mays L. I. Effects of ammonium and nitrate pretreatment. Plant Physiol. 84(4):1314-1318.

Smith, F.W., and W.A. Jackson. 1987b. Nitrogen enhancement of phosphate transport in roots of Zea mays L. II. Kinetic and inhibitor studies. Plant Physiol. 84(4):1319-1324.

Vance, C.P., P.E. Reibach, and C.E. Pankhurst. 1987. Symbiotic properties of Lotus pedunculatus root nodules induced by Rhizobium loti and Bradyrhizobium sp. (Lotus). Physiol. Plant. 69:435-442.

Wheeler, J.L., and R.D. Mochrie (ed.). 1981. Forage evaluation: Concepts and techniques. CSIRO, Canberra, Australia.

Wilson, J.R., M.N. McLeod, D.J. Minson, and D.E. Akin. 1987. Forage particle breakdown in cattle. p. 67. In M. Rose (ed.) Herbivore nutrition research. Aust. Soc. Anim. Prod., Brisbane, Australia.

SUMMARY OF THE TRILATERAL WORKSHOP ON PERSISTENCE OF FORAGE LEGUMES

G.C. Marten

The purpose of the workshop was to discuss the problem of poor persistence of forage legumes and to define research priorities for its solution. Six specific objectives listed in the workshop proposal to the National Science Foundation of the USA were reviewed at the close of the workshop to determine whether they had been met. These objectives are given in the Preface of these proceedings.

The organizers, delegates, and other participants in the workshop agreed that all of the objectives either had been met by the close of the workshop or had an excellent chance of being met shortly thereafter. Consensus was often easily reached regarding the generalized causes of poor forage legume persistance and the information needed to allow its modelling in the three countries.

DEFINITION OF PERSISTENCE

Perhaps the most difficult item to gain consensus upon during the workshop was the definition of legume persistence itself. It soon became apparent that one's concept of "persistence" of forage legumes was a function of: (1) location (country or region within country); (2) nutritional need of animals involved; (3) economic situation; (4) traditional production system that prevailed; (5) ecological situation and pest problems; (6) climatic situation; (7) edaphic situation; and (8) presence or absence of high-feeding-quality forage alternatives.

The USA delegation frequently differed from the New Zealand delegation in their concept of legume persistence, while the Australian delegation often had an intermediate view. For example, in the majority of USA instances (localized areas excepted), grazed perennial legumes were considered to not live or produce significant new progeny beyond about 5 or 6 yr, and some "adapted" legumes were listed as not having life expectancies beyond 2 or 3 yr in the best circumstances. In contrast, improved white clover (Trifolium repens L.) in New Zealand was considered to be perennially present, with the balance of it in mixture with ryegrass (Lolium spp.) or other species being the only real persistence question from year-to-year or season-to-season.

Thus, production was considered an integral part of the New Zealand legume persistence scene, whereas longevity was featured in the USA. A.G. Matches' (Texas) definition of persistence was "standability over time," and R.R. Smith's (Wisconsin) definition was "survival of plant material against specific stresses unique to the existing environment." On the other hand, G.W. Sheath (New Zealand) suggested that "the persistent plant is that which is most productive, yet stable, for a given environment (plus or minus modification)," while R.W. Brougham (New Zealand) suggested that "a persistent legume is one that has weed potential." After more discussion, Sheath offered "where legume populations are at a density that achieves the expectations of the specific ecosystem (e.g., economic productivity and environment/cultural stability)" as his final definition of persistence. A compromise view, offered by R.J. Clements (Australia), was that "persistence can include concepts of productivity, but the maintenance of adequate plant numbers is the essential criterion."

Furthermore, the nutritional need from the forage legume was often listed as requiring high-digestible dry matter intake for high-producing animals in intensive systems (such as for dairy cattle) in the USA. Less emphasis occurred on this need in the other countries. For the beef cow-calf enterprise, the ewe flock, and growing of weaned calves or lambs on pasture, all three nations had similar needs for forage legumes often mixed with grasses. However, for beef cattle or sheep finishing in the USA, the situation was similar to the dairy enterprise in that economics supported a traditional production system that often relied on conserved pure legumes as hay or silage grown in rotation with maize or other crops for silage or grain. Grazing, if used at all in these systems, was often of legume regrowth late in summer or early in autumn. This led to a perceived need for pure stands of legumes over years. In Australia and New Zealand, on the other hand, mixtures of legumes with grasses for grazing were desired for nearly all ruminant animal production. In Australia, however, a major function of subclover (T. subterraneum L.) and medic pastures was to provide N for succeeding cereal crops; ruminant production was often secondary.

Another situation that differed among the three countries was the frequently more rigorous and variable climate in the USA. Hence, often more legume species could survive for long periods in Australia or New Zealand. Then too, the feasible non-legume alternatives were rarely ryegrass or any high-quality grass in the USA (i.e., the feeding value of the ryegrass alternative was apprently often closer to that of the legume in New Zealand and in some non-tropical parts of Australia).

CAUSES OF POOR LEGUME PERSISTENCE

Consensus was reached on numerous generalized causes of poor legume persistence in Australia, New Zealand, and the USA. Among these (nonprioritized) were the following:

1. Poor seedling vigor or premature germination.
2. Lack of adequate percentages of hard seed in some cases.
3. Inadequate seed production of self-reseeding legumes.
4. Poor soil fertility, excess soil acidification; or other edaphic problems.

5. Poor competitive ability with associated cool- and warm-season grasses or invaders.
6. Poor shade tolerance.
7. Poor resistance to primary diseases, nematodes, or insects or poor resistance to cumulative pest stress loads.
8. Poor drought tolerance.
9. Poor heat or cold tolerance.
10. Poor tolerance to anaerobic conditions.
11. Poor tolerance to frequent defoliation or to heavy grazing pressures.
12. Lack of rhizomes, stolons, or protected growing points.
13. Inadequate management strategies.
14. Inadequate inoculation with rhizobia or low N_2 fixation.
15. Allelopathy.
16. Lack of "farmer-resistant" cultivars.
17. Inadequate root systems.
18. Lack of overall "plasticity" of current germplasm.
19. Poor extension of information to producers.
20. Lack of concentrated research effort.

Evidence was also presented which revealed localized causes of poor forage legume persistence. Among these (non-prioritized) were the following:

1. Photoperiod sensitivity.
2. High-endophyte infection in associated grasses.
3. Socio-economic restrictions.
4. Lack of definition of stresses in specific environments needed to aid plant breeding.
5. Excessive palatability.
6. Inadequate establishment machinery (for seeding or vegetative propagation).
7. Ploidy level; e.g., diploid red clover (T. pratense L.) is less persistent than tetraploid.
8. Lack of techniques needed to effectively identify stress-resistant germplasm.
9. Inadequate government policies (e.g., removal of subsidies that stimulate needed fertilizer addition by producers).

COLLABORATIVE PROJECTS AND EXCHANGES

To promote future research collaboration and exchange of information, scientists from each nation contributed to the recommendations for collaborative projects or exchanges among the three participating nations (see the lists that follow the Executive Summary at the front of these proceedings).

The Australian delegation defined several of their special strengths and weaknesses that potential collaborators should bear in mind as future plans for cooperative research and information or personnel exchanged are arranged. The Australian strengths included expertise in (i) annual legumes (including plant nutrition, genetic improvement, pest management, germplasm collection, cultural management and the legume-crop interface); (ii) tropical legumes (including plant nutrition, germplasm collection, and cultural management); and (iii) the soil acidification process and lime requirements of soils. The Australian weaknesses included need for assistance to solve problems in: (i) white clover (including diseases, cultural management in non-dairy situations, and germplasm collection, especially from the Mediterranean area); and (ii) quantification of research problems and priority setting (a weakness shared by most nations).

The New Zealand organizers concluded that the ultimate success of the workshop would depend on the number and quality of the initiatives for collaborative endeavors that result. This will require assumption of future responsibility by the workshop participants and by readers of the proceedings. A general broadening of scientific views and of local research programs on forage legumes was also an envisioned benefit of the workshop.